Advances in Physicochemical Properties of Biopolymers

(Part 2)

Edited by:

Martin Masuelli & Denis Renard

Instituto de Física Aplicada-CONICET, Universidad Nacional de San Luis, Chacabuco 917, CP 5700, San Luis, Argentina

General:

1. Any dispute or claim arising out of or in connection with this License Agreement or the Work (including non-contractual disputes or claims) will be governed by and construed in accordance with the laws of the U.A.E. as applied in the Emirate of Dubai. Each party agrees that the courts of the Emirate of Dubai shall have exclusive jurisdiction to settle any dispute or claim arising out of or in connection with this License Agreement or the Work (including non-contractual disputes or claims).
2. Your rights under this License Agreement will automatically terminate without notice and without the need for a court order if at any point you breach any terms of this License Agreement. In no event will any delay or failure by Bentham Science Publishers in enforcing your compliance with this License Agreement constitute a waiver of any of its rights.
3. You acknowledge that you have read this License Agreement, and agree to be bound by its terms and conditions. To the extent that any other terms and conditions presented on any website of Bentham Science Publishers conflict with, or are inconsistent with, the terms and conditions set out in this License Agreement, you acknowledge that the terms and conditions set out in this License Agreement shall prevail.

Bentham Science Publishers Ltd.
Executive Suite Y - 2
PO Box 7917, Saif Zone
Sharjah, U.A.E.
Email: subscriptions@benthamscience.org

BENTHAM SCIENCE

CONTENTS

PREFACE

The objective of this ebook is to provide to the readers the most recent state-of-the-art on physicochemical properties of biopolymers and their related end-uses applications. Biopolymers are usually described as polymers produced in a natural way by living species. Their molecular backbones are composed of repeating units of saccharide, nucleic acids, or amino acids and sometimes various additional side chains contributing also to their functionalities.

If the largest part of biopolymers is extracted from biomass, such as polysaccharides from cellulose and proteins from collagen, milk or wheat, biopolymers can also be produced from monomers using conventional chemical processes as polylactic acid, or directly in microorganisms or genetically modified organisms, as polyhydroxyalkanoates. The genetic manipulation of microorganisms brings a tremendous potentiality for the biotechnological production of biopolymers with tailored properties quite suitable for high-value medical applications such as tissue engineering and drug delivery.

Biopolymers from renewable sources, on the contrary display structural complexity and natural variability that need to be deeply studied and characterized before probing into the structure-function relationships for further applications. Research on natural polymers has focused on developing more environmentally friendly applications to reduce pollution caused by non-biodegradable material. Historically, biopolymers were mainly used by mankind as food, or for making clothing and furniture. Since the industrial time, fossil fuels such as oil are the greatest source in the development and manufacture of almost every commercial product, such as the plastic, which is currently used at a very large scale. But these fuels are not unlimited resources, and environmental concerns over all aspects of using fossil fuels for production and energy must be taken into account. We must act in a sustainable manner, which means that the resources must be consumed at a rate such that they can be restored by natural cycles of our planet [1].

Therefore, in recent years, the renewable nature of biopolymers leads them to a renaissance and a new considerable interest by industry due to the unique properties, including biodegradability, biocompatibility and nontoxicity, of biopolymers. To fulfil all these different functions, biopolymers must exhibit rather diverse properties. They must very specifically interact with a large variety of different substances, components and materials, and often they must have extraordinarily high affinities to them. Finally, they must have a high strength. Some of these properties are utilized directly or indirectly for various applications. This and the possibility to produce them from renewable resources, as living matter mostly does, make biopolymers interesting candidates to industry [2]. As a consequence of their properties, these biopolymers derived from natural products have found a place of choice in areas as diverse as effluent treatment, papermaking, chemical, food, cosmetic, pharmaceutical, petroleum and textile industries, as well as in analytical chemistry (biosensors) and molecular biology. However, biopolymers have to compete with polymers derived from fossil fuel not only because of their functional properties but also in terms of cost. In this respect, biopolymers are competitive when the price of oil is high and the price of feedstocks, such as starch from corn, is low.1 The continuing development of new and existing biopolymers will enable these materials to help supplement the increasing global demand for biopolymers-based products and to develop new markets with their niche applications.

The most common biopolymers used for industrial applications and thoroughly considered in this ebook are polysaccharides from plant, algal, microbial and animal origins such as starch, cellulose, lignin, arabinoxylans, sulfated polysaccharides from seaweeds, galactomannans and xyloglucans from Brazilian seeds, chitin and its derivative chitosan. Natural gums such as mesquite, tara and arabic gums are also widely used in food and non-food industry and are discussed in this ebook. Animal and plant proteins such as collagen, gelatin, albumin, dairy proteins and wheat, corn and soy proteins are also considered as sources of proteins for biomedical, microencapsulation and plastic foams applications. Nucleic acids such as DNA and RNA and their related applications in genetic engineering for instance are not considered in this ebook.

This ebook presents a comprehensive review and compile information on biopolymers in 27 chapters covering from isolation and production, properties and applications, modification, and relevant analytical methods to reveal the structure and properties of some biopolymers.

Authors write this ebook from Argentina, France, Mexico, Spain, Iran, Brazil, Egypt, Turkey, Venezuela, India, Russia, Portugal, New Zealand and Malaysia. This ebook has tried to arrange the ebook chapters in a subject order to make it easier for the readers to find what they need. However, the reader can still find information on the same subject in more than one Section.

Section A (Part 1), which includes one chapter, is mainly an introduction to biopolymers. It includes concepts and molecular weight determination.

Section B (Part 1), which includes twelve chapters, refers to some physical chemistry determinations of biopolymers.

Section C (Part 1), which consists of two chapters, deals with studies on hydrodynamic properties of biopolymers.

Section D (Part 1), which consists of one chapter, refers to theoretical models for biopolymers.

Section A (Part 2), which includes four chapters, refers to special cases of polysaccharides separation and purification.

Section B (Part 2), which includes seven chapters, deals with applications of biopolymers/hydrogels in drug delivery systems, biomaterials, biothermoplastics, bio(nano)composites, bionanostructures, biocapsules, bioadsorbents, bioelectrospinning and biopackaging. This section deserves a special attention because it forms a fascinating interdisciplinary area that brings together biology, chemistry, materials science and (nano)-technology.

This ebook is expected to be of help to many graduate and post-graduate students, professors, scientists, pharmacists, engineers and other experts in a variety of disciplines, both academic and industrial, dedicated to the determination of polymers and biopolymers properties. This ebook may not only support research and development, but also be suitable for teaching. The audience will benefit with an excellent review offering advanced knowledge about technical determinations and physicochemical properties of macromolecules, a thorough knowledge of hydrodynamics and different methods of characterization. Readers will find in this ebook a triple deal, including educational, scientific and industrial applications.

The first main objective of this e-book is therefore to highlight the progress in different techniques of molecular weight determinations and physicochemical properties of biopolymers. The last two decades have seen a number of significant advances in the methodology for evaluating the molecular weight distributions of polydisperse macromolecular systems in solution at the molecular level. These advances have centered on the coupling of chromatographic or membrane based fractionation procedures with multiple detectors on line such as multi-angle laser light scattering, refractive index, UV-Vis absorbance and intrinsic viscosity detection systems. Recent advances in SEC-MALLS (size exclusion chromatography coupled to multi-angle laser light scattering) and FFF-MALLS (field flow fractionation coupled on line to MALLS) applied to complex polymers from renewable resources are therefore presented in this e-book. Beyond molecular charcaterization using HPSEC-A4F-MALLS technique, tremendous efforts were made these last years to elucidate the structural variability and complexity of polysaccharides using matrix-assisted laser-desorption ionization (MALDI) and electrospray ionization (ESI) mass spectrometry coupled or not to nuclear magnetic resonance (NMR) spectroscopy. One chapter of this ebook in section B considers the sequence, interresidue linkage position and substitution pattern of sulfated polysaccharides after enzymatic hydrolyses.

The most widely used method for the dynamic characterization of macromolecules in solution is the capillary viscometry, as it is a simple and economic method. Although in literature there is much information on hydrodynamic measurements from intrinsic viscosity determinations, very few of them evaluate the conformation of different biopolymers. The importance of this type of study lies in the analysis of the polysaccharides or proteins behaviour in industrial processes and product quality control after extraction and purification. These physicochemical studies help to elucidate the chemical structure, macromolecular conformation and the ability biopolymers have to form gels, films, agglomerates, *etc.* A particular attention is paid in this ebook on the intrinsic viscosity determination of proteins and strong synthetic polyelectrolytes for which theoretical models always need to be implemented in order to get reliable dynamic structural informations.

The ebook also focuses on the structural analyses at the mesoscopic scale using mechanical analyses, microscopy, small angle scattering and free volume measurements and different applications related to biopolymers such as biomaterials, microcapsules, biothermoplastics, nanostructured biocomposites, super-absorbents, bioelectrospinning, biopolymers-based dermal and transdermal drug delivery systems, and biopackaging. All these applications using biopolymers aim to provide a means to reduce dependence on fossil fuels, and decrease the environmental impact of non-biodegradable materials. The main challenge to overcome with biopolymers-based materials is the control of biopolymer-biopolymer interactions, a challenge always present and discussed throughout the ebook by authors.

To conclude, the content of this ebook will bring its readers a basic understanding of the physical chemistry of biopolymers, but also the latest findings about new macromolecules recently discovered and published. Theoretical aspects of computational structural description of biopolymers are also thoroughly described. Therefore, this ebook will appeal to different readers as a great source of knowledge about the science of biopolymers.

Biopolymers Audience

- Separation, purification, characterization of biopolymers
- Hydrodynamic, molecular weight, size, shape, conformation
- Macromolecular assembly
- Molecular design and bio-nanotechnology
- Biopolymer processing and degradation

- Experimental and theoretical studies of biopolymer structures
- Three-dimensional structures of biopolymers determined by X-ray, neutrons, NMR
- Interactions and thermodynamics
- Food biocolloids
- Structure and function
- Preparation and characterization of novel biomaterials
- Capsules and microcapsules
- Biocatalysis
- Biopolymers for bioremediation
- Thin films, membranes & packaging

REFERENCES

[1] Thomas, Sabu; Durand, Dominique; Jyotishkumar, P..; Chassenieux, Christophe, editors. Handbook of Biopolymer-Based Materials: From Blends and Composites to Gels and Complex Networks. 1st. Wiley-VCH Verlag Gmbh and Co. KGaA; 2013.

[2] Steinbüchel, Alexander, editor. Biopolymers. 10. Wiley-VCH, Weinheim; 2003.

Denis Renard
Unité de Recherches Biopolymères,
Interactions, Assemblages,
INRA, Rue de la Géraudière,
France
Email: drenard@nantes.inra.fr

ACKNOWLEDEGEMENT

We first would like to thanks all the authors contributing to the realization of this ebook.

We would like to thank the Agronomic Research National Institute, France and the Área de Química Física - Departamento de Química - Facultad de Química Bioquímica y Farmacia - Universidad Nacional de San Luis – Instituto de Física Aplicada – CONICET, Argentina for giving us the time and opportunity for the writing and edition work of this ebook.

We are grateful to the editing team at Bentham eBooks for their help, especially Humaira Hashmi and Salma Sarfaraz.

Finally, we wish to express our gratitude to our spouses, Maria Gabriela and Marie-Christine, for their patience and support during the long hours of writing, reviewing, commenting, editing of this ebook.

List of Contributors

Alejandra L. Valdez	Laboratorio de Micodiversidad y Micoprospección, PROIMI-CONICET, Tucumán, Argentina Cátedra de Microbiología, Facultad de Bioquímica, Química y Farmacia (UNT), Tucumán, Argentina
Andreia Almeida	i3S – Instituto de Investigação e Inovação em Saúde, Porto, Portugal INEB – Instituto de Engenharia Biomédica, Porto, Portugal IIFACTS – Instituto de Investigação e Formação Avançada em Ciências e Tecnologias da Saúde, Gandra, Portugal
Anuradha Walallavita	School of Engineering, Faculty of Science and Engineering, University of Waikato, Hamilton, New Zealand
Athira K. Sunil	Indian Institute of Science, Bangalore – 560012, India
Bruno Sarmento	i3S – Instituto de Investigação e Inovação em Saúde, Porto, Portugal INEB – Instituto de Engenharia Biomédica, Porto, Portugal IIFACTS – Instituto de Investigação e Formação Avançada em Ciências e Tecnologias da Saúde, Gandra, Portugal
Carmen L.O. Petkowicz	Biochemistry and Molecular Biology Department, Federal University of Paraná, Curitiba-Paraná, Brazil
Casparus J.R. Verbeek	School of Engineering, Faculty of Science and Engineering, University of Waikato, Hamilton, New Zealand
Chanelle Gavin	School of Engineering, Faculty of Science and Engineering, University of Waikato, Hamilton, New Zealand
Clayton Fernandes de Souza	Chemistry Undergraduate Program, School of Education and Humanities, Pontifical Catholic University of Paraná, Curitiba-PR, Brazil
Entsar I. Rabea	Department of Plant Protection, Faculty of Agriculture, Damanhour University, 22516-Damanhour, Egypt
Jesica Gassmann	Instituto de Física Aplicada-CONICET, Universidad Nacional de San Luis, Chacabuco 917, CP 5700, San Luis, Argentina Policlínico Regional San Luis, San Luis, Argentina
Joana Léa Meira Silveira	Biochemistry and Molecular Biology Department, Federal University of Paraná, Curitiba-Paraná, Brazil
Julia I. Fariña	Laboratorio de Micodiversidad y Micoprospección, PROIMI-CONICET, Tucumán, Argentina Cátedra de Química Biológica, Facultad de Ciencias Exactas y Naturales (UNCa), Catamarca, Argentina
Juan Manuel Galdopórpora	IQUIMEFA-CONICET. Universidad de Buenos Aires, Facultad de Farmacia y Bioquímica, Cátedra de Química Analítica Instrumental. (1113) Junin 956 Piso 3. Ciudad de Buenos Aires, Argentina
João Pedro Martins	i3S – Instituto de Investigação e Inovação em Saúde, Porto, Portugal INEB – Instituto de Engenharia Biomédica, Porto, Portugal IIFACTS – Instituto de Investigação e Formação Avançada em Ciências e Tecnologias da Saúde, Gandra, Portugal

Kelvia Álvarez	Department of Analytical Chemistry, Faculty of Pharmacy, Central University of Venezuela (UCV), PO Box 40109, Caracas 1040-A, Venezuela Institute of Food Science and Technology, Faculty of Sciences, Central University of Venezuela (UCV), PO Box 47097, Caracas 1041-A, Venezuela
Lautaro M. Buffa	National Institute of Industrial Technology, Mar del Plata, Buenos Aires, Argentina
M. Fernanda Horst	South Institute of Chemistry (INQUISUR), National Council of Scientific and Technical Research (CONICET), National University of South (UNS), Bahía Blanca, Buenos Aires, Argentina
Mark C. Lay	School of Engineering, Faculty of Science and Engineering, University of Waikato, Hamilton, New Zealand
Martin Alberto Masuelli	Instituto de Física Aplicada-CONICET, Universidad Nacional de San Luis, Chacabuco 917, CP 5700, San Luis, Argentina
María Victoria Tuttolomondo	IQUIMEFA-CONICET. Universidad de Buenos Aires, Facultad de Farmacia y Bioquímica, Cátedra de Química Analítica Instrumental. (1113) Junin 956 Piso 3. Ciudad de Buenos Aires, Argentina
Maria Lucia Foglia	IQUIMEFA-CONICET. Universidad de Buenos Aires, Facultad de Farmacia y Bioquímica, Cátedra de Química Analítica Instrumental. (1113) Junin 956 Piso 3. Ciudad de Buenos Aires, Argentina
Martin Federico Desimone	IQUIMEFA-CONICET. Universidad de Buenos Aires, Facultad de Farmacia y Bioquímica, Cátedra de Química Analítica Instrumental. (1113) Junin 956 Piso 3. Ciudad de Buenos Aires, Argentina
Mariela Agotegaray	South Institute of Chemistry (INQUISUR), National Council of Scientific and Technical Research (CONICET), National University of South (UNS), Bahía Blanca, Buenos Aires, Argentina
María B. Bonecco	National Institute of Industrial Technology, Mar del Plata, Buenos Aires, Argentina
María Guadalupe García	Instituto de Física Aplicada-CONICET, Universidad Nacional de San Luis, Chacabuco 917, CP 5700, San Luis, Argentina
María G. Martínez Sáenz	National Institute of Industrial Technology, Mar del Plata, Buenos Aires, Argentina
Mohamed E. I. Badawy	Department of Pesticide Chemistry and Technology, Faculty of Agriculture, 21545-El-Shatby, Alexandria University, Alexandria, Egypt
Natalia A. Castillo	Laboratorio de Micodiversidad y Micoprospección, PROIMI-CONICET, Tucumán, Argentina Cátedra de Micología, Facultad de Bioquímica, Química y Farmacia (UNT), Tucumán, Argentina
Neoli Lucyszyn	Chemistry Undergraduate Program, School of Education and Humanities, Pontifical Catholic University of Paraná, Curitiba-PR, Brazil

Tomy J. Gutiérrez Department of Analytical Chemistry, Faculty of Pharmacy, Central University of Venezuela (UCV), PO Box 40109, Caracas 1040-A, Venezuela
Institute of Food Science and Technology, Faculty of Sciences, Central University of Venezuela (UCV), PO Box 47097, Caracas 1041-A, Venezuela
Composite Materials Group (CoMP), Research Institute of Materials Science and Technology (INTEMA), Faculty of Engineering, University of Mar del Plata and National (UNdMP) and National Council of Scientific and Technical Research (CONICET), Colón 10850, Mar del Plata 7600, Buenos Aires, Argentina

CHAPTER 1

Chitosan and its Modifications as Biologically Active Compounds in Different Applications

Mohamed E. I. Badawy[1,*] and **Entsar I. Rabea**[2]

[1] *Department of Pesticide Chemistry and Technology, Faculty of Agriculture, 21545-El-Shatby, Alexandria University, Alexandria, Egypt.*

[2] *Department of Plant Protection, Faculty of Agriculture, Damanhour University, 22516-Damanhour, Egypt.*

Abstract: A chitosan biopolymer is a reactive functional polymer, which gives possibilities for chemical modifications to generate new properties and functions. Biocompatibility, biodegradability, non-toxicity to mammals, and potential biological activities make this compound with its derivatives advantageous for many applications in different fields, such as agriculture, food and nutrition, biomedicine, pharmaceutics, and biotechnology. In this chapter, we provide collaborative studies of the biological activity of chitosan and its major derivatives in different applications. In addition, the chapter provides the latest technological applications and prospects of products based on chitosan molecule.

Keywords: Biological activity, Chemical modification, Chitosan, Physico-chemical properties, Technological applications.

1. INTRODUCTION

Chitosan is a linear biopolymer which consists of higher than 70% of β-(1-4)- 2-deoxy-β-D-glucopyranose (GlcN) and lower than 30% of β-(1-4)-2-acetam-do-2-deoxy-β-D-glucose (GlcNAc) units linked by β-1,4-glucosidic bonds. It can be obtained through a deacetylation process of purified chitin, a naturally abundant polysaccharide and the supporting material of crustaceans, insects, yeast, and fungi [1 - 4]. The main difference between chitin and chitosan is the degree of *N*-acetylation (DA) [5, 6]. Chitosan is of commercial interest due to its high percentage of nitrogen (~ 6.0-7.0%) compared to synthetically substituted

* **Corresponding author Mohamed E. I. Badawy:** Department of Pesticide Chemistry and Technology, Faculty of Agriculture, 21545-El-Shatby, Alexandria University, Alexandria, Egypt; Tel: +2039575269; Fax: +2035972780; E-mail: m_eltaher@yahoo.com

Martin Masuelli & Denis Renard (Eds.)

cellulose (1.25%) that makes it a good chelating agent for metal ions [7]. It has unique characteristics such as biocompatibility, biodegradability, low toxicity to mammals, and possesses reactive functional groups that make it useful in different areas of application related to (i) agriculture (as a biopesticide against plant diseases and pests, seed-coating, postharvest, and controlled release agrochemicals) [8 - 15], (ii) food industry and nutrition (anticholesterolemic dietary products and antimicrobial coatings for fruit and vegetables) [3, 16, 17], (iii) biotechnology (spinning, a dye-binder for textiles, strengthening additive in paper, enzyme and cell immobilization, protein separation, chromatography column matrix, and cosmetic functional ingredients) [18 - 20]; (iv) biomedicine (drug and gene delivery, blood coagulation, wound healing, bone regeneration, immunoadjuvant activity, pharmaceutics, and ophthalmology) [21, 22], and (v) combinations of chitosan with other natural or synthetic polymers (grafting, polyelectrolyte complexation, blends, and coatings) [23]. However, chitosan exhibits a limitation in its solubility and reactivity therefore, many studies have paid attention to modify its chemical structure [12, 24 - 26]. Although several works have been reported to obtain functional derivatives of chitosan by chemical modification, very few attained solubility in general organic solvents [27, 28] and some binary solvent systems [29, 30]. Other several studies modified the structure of chitosan by chemical or enzymatic methods to obtain high biologically active compounds and improve solubility in general organic solvents and aqueous solutions at wide range of the pH [25, 31 - 34]. This chapter covers the literatures dealing with the physicochemical properties of chitosan, the biological activities of chitosan and its derivatives of major importance and applications in different areas.

2. CHITOSAN HISTORY AND ORIGIN

The term "chitosan" (pronounced *Kite-O-San*) was given to deacetylated chitin by Hoppe-Seiler [35]. The name chitin is derived from Greek, meaning "tunic" or "envelope" and it was first discovered in mushrooms in 1811 by Henri Braconnot [36]. In 1823, Odier found the same material in insects and plants and named it chitin [37]. The concept of chitin was further known by Lassaigne, who demonstrated the presence of nitrogen in its structure when purified the coleopteran elytra and *Bombyx mori* exuviae, and then treated the residues with potassium at warm temperature thus obtaining potassium cyanide that confirmed the presence of nitrogen in chitin [38]. In 1859, Rouget treated chitin with strong alkali, which resulted in a substance that could, unlike chitin itself, be dissolved in acidic aqueous solutions, named as chitosan [39].

Therefore, up to date, the chitin and chitosan are mainly produced commercially from the waste products of crustaceous exoskeletons of aquatic organisms, fungi and insects. Chitin is found in the outer skeleton of insects (20-38%), crabs, and shrimps and lobsters (69-70%). It was also found that chitin is a major structural component of fungal cell wall (5-20%), which is being considered as an alternative source of chitosan and which varies between different species of fungi from 2 to 60% of dry mycelia [40 - 42]. Zygomycetes, especially *Mucor rouxii, M. mucedo, M. circinelloides, Rhizomucor miehei, Rhizopus oryzae, Phycomyces blakesleeanus*, and *Cunninghamella elegans* have high content of chitin compared to other fungi [43 - 45]. Moreover, Ascomycetes and Basidiomycetes contain significant quantities of chitin and acidic polysaccharides (26-65%) as cell wall components [46]. However, mushrooms including *Lentinus edodes, Lycophyllum shimeji, Pleurotus sajor-caju*, and *Volvariella volvacea*, contain minor content of chitinous components [47] and fungi of *Oomycetes* and the fission yeasts considered to lack chitin [48]. In contrast, chitosan occurs as a major component in walls of Zygomycetes, but probably is a minor component in many other fungi [49].

The isolation of chitin from shells waste consists of three chemical steps: deproteinization (DP), demineralization (DM), and decolorization (DC); whereby the order of the first two steps is generally considered irrelevant if protein or pigment recovery is not an objective [50]. Purified chitin is then converted to chitosan by hydrolysis of acetamide groups of chitin as a deacetylation process. This process is carried out by two main procedures: the Broussignac process [51], in which chitin is treated with a mixture of solid potassium hydroxide (50% w/w) in 96% ethanol and monoethylene glycol (1:1), and the Kurita process [52], according to which chitin is treated with hot aqueous sodium hydroxide solution (50% w/v). However, the latter is preferred for industrial purposes. In some cases, the deacetylation reaction is carried out in the presence of thiophenol [53, 54] as a scavenger of oxygen or under N_2 atmosphere to prevent chain degradation that invariably occurs due to peeling reaction under strong alkaline conditions. Here, DA took place using a calculated four times excess of NaOH for the total *N*-deacetylation of all amino groups in chitin with reaction time of 1 h at 100°C. These chemical methods are widely used in industrial scale however, the products obtained may suffer some inconsistencies such as protein contamination, inconsistent levels of DA, and hydrolysis of the polymer. In addition, there are some additional problems, such as environmental issues due to the large amount of waste concentrated alkaline solution [43, 55]. Therefore, alternatives to the chemical processes for production of chitin and chitosan are the fermentation and enzymatic techniques which are widely used on an industrial scale [56 - 61]. While chitin remained an unused natural resource for a long time, interest in chitosan polymer and its derivatives has increased in recent years due to their distinctive properties.

3. PHYSICOCHEMICAL CHARACTERIZATIONS OF CHITOSAN

3.1. Chemical Structure

Chitosan is composed of GlcN and remaining of GlcNAc units which indicate high content of free amino groups compared to chitin (Fig. **1**) [6, 62]. The polymer chains appear as linear in structure, which undergoes one full twist, every 10.1 to 10.5 Å along the polymer chain axis. A separate "left" and "right" direction can be occurred to each polymer chain as the presence of each chiral glycosidic unit in the polymer chain, and all units are connected by an oxygen atom that links C-1 of one glycosidic unit to C-4 of a neighboring unit. After heating, chitosan decomposes prior to melting, thus it has no melt point and its crystalline structure is mainly characterized by spectroscopic techniques such as X-ray, UV, FT-IR, and NMR [63 - 69]. The IR spectrum showed main absorption band at 3432 cm^{-1} due to the partially overlapped of the amine and hydroxyl (N-H, O-H, and NH_2) stretching vibrations. Weak absorption bands at 2868 and 2919 cm^{-1} represent a group -CH- and -OH stretching vibration, respectively. The absorption bands at a wave number of 1656 and 1380 cm^{-1} correspond to the C=O stretching of CO and amide group, respectively. The absorption band at wave number of 1597 cm^{-1} was due to the deformation NH of amino groups. The absorption band at 1325 cm^{-1} corresponds to amid III band. While the absorption band at a wave number of 1151, 1092 and 1024 cm^{-1} corresponds to the symmetric stretching of the C-O-C and symmetrical skeletal vibration involved in CO and CN stretching, respectively [70, 71]. In addition, the ^{1}H-NMR spectrum of chitosan is shown in Fig. (**2A**). The peak at δ 2.06-2.13 ppm is assigned to the proton of residual CH_3 in acetyl group. The peak at δ 3.10-3.28 ppm is attributed to H-2 of GlcN residue. The broad multiplet peak at δ 3.53-4.12 ppm is assigned for H-3,4,5,6 of GlcN unit and H-2,3,4,5,6 of GlcNAc unit. The intense band at 5.03-5.17 ppm is related to OH groups and D_2O (solvent). In this region, as observed more clearly from an extended spectrum, some different anomeric protons (H-1 of GlcN and GlcNAc units) are appeared at 4.80-4.91 ppm [72, 73].

Further evidence for confirmation: the chemical structure was obtained from ^{13}C-NMR spectroscopy. Strong and intense peaks for carbon atoms are found at δ 25.95 (NH(CO)$\underline{C}H_3$), 56.37 (C-2), 60.99 (C-6), 70.74 (C-3), 75.44 (C-5), 77.80 (C-4), 98.21 (C-1), and 178.24 ($\underline{C}$(O)CH_3).

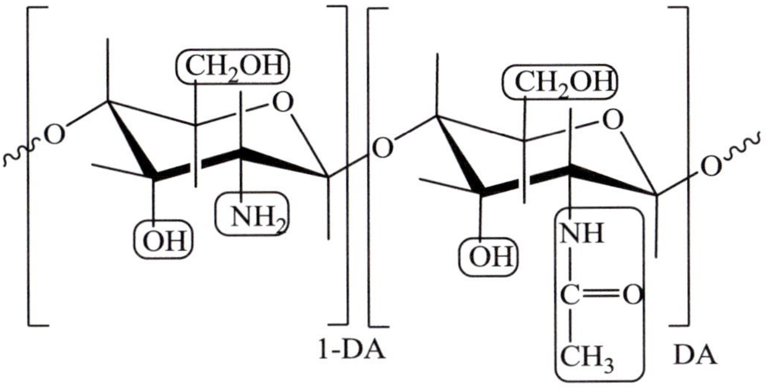

Fig. (1). Chemical structure and illustration of the possible reaction sites in chitosan (GlcN and remaining of GlcNAc units). DA is a degree of acetylation.

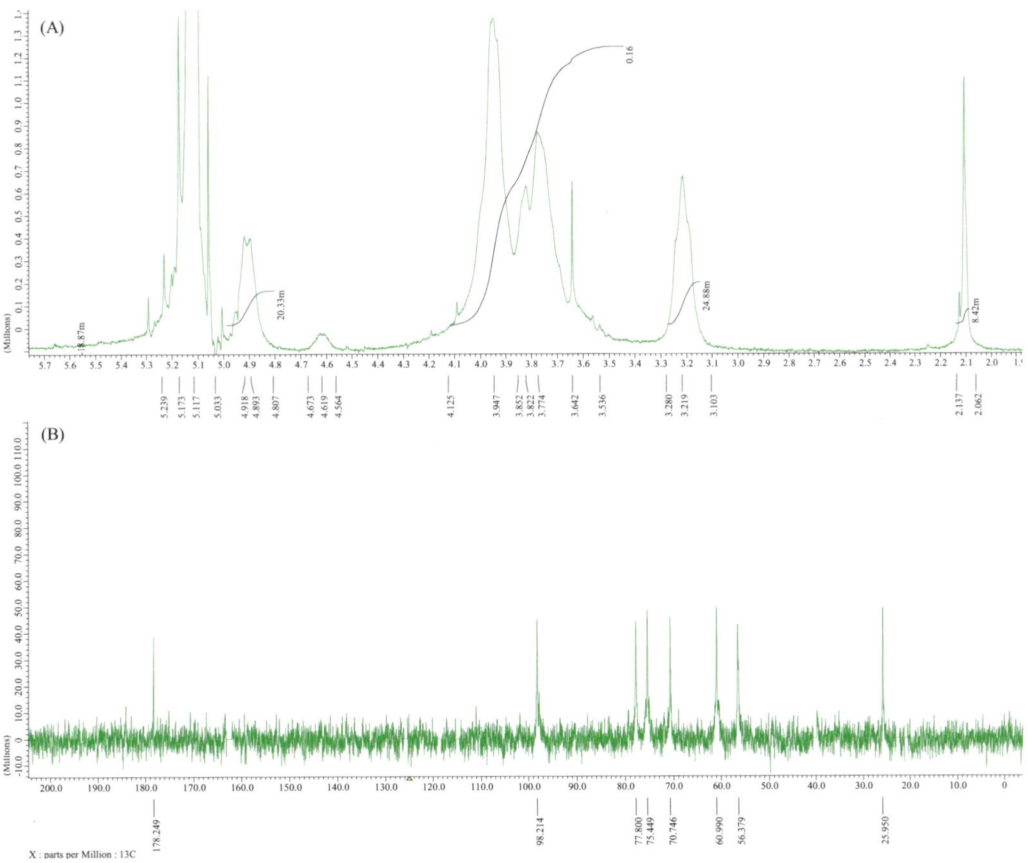

Fig. (2). ^{1}H-NMR (A) and ^{13}C-NMR (B) spectra of chitosan dissolved in 1% CD_3COOD/D_2O at 25°C.

3.2. Degree of N-acetylation (DA)

Chitosan is the fully or partially *N*-deacetylated derivative of chitin with a DA of less than 30%. This ratio has an important effect on the physicochemical and biological properties of chitosan. Determination of the DA of chitosan products has been achieved by several techniques including IR and UV spectroscopy, conductometric and potentiometric titration, elemental analysis, gel permeation chromatography (GPC), ^{1}H-, ^{13}C- and ^{15}N-NMR [63 - 66, 74 - 77]. The IR method is limited and not accurate because the result interpretation is correlated to the complexity in adopting a suitable base line [71]. However, the method of ^{1}H-NMR seems to be the most convenient technique to get the correct DA and degree of deacetylation (DDA) of the soluble samples [78]. The DA is calculated from the integral ratio between proton of acetyl group of GlcNAc at 2.06-2.13 ppm and the protons of GlcN at δ 3.10-4.12 ppm. Additionally, ^{13}C-NMR is also well-situated for DA determination in the case of pure chitin up to fully deacetylated chitosan with a good conformity between measurements in liquid phase and solid state. The most efficient UV method for determination of the DA is based on the use of dual standards of GlcN and GlcNAc [79]. In this method, several solutions with different concentrations of GlcN hydrochloride and GlcNAc are prepared and their UV absorbencies at λ_{201} nm are measured by UV-Spectrophotometer. Reading the absorption of the solutions at 201 nm, the DA is calculated by the following equation:

$$\text{DA}(\%) = \frac{161.1 \times A \times V - 0.0218 \times m}{3.3615 \times m - 42.1 \times A \times V} \times 100$$

Where **m** is the weight of chitosan in mg; **A** is the absorbance, and **V** is the volume of solution in liters.

3.3. Molecular Weight (MW)

The MW influences on the viscosity of aqueous solution of chitosan and has a significant role in the biological and industrial applications in many fields. In order to determine the MW of chitosan, different techniques methods have been widely used such as viscometry [80], gel permeation chromatographic (GPC) [81], and light scattering (LS) [82, 83] techniques. However, the LS method gives absolute values of the MW, but the technique is more difficult and sometimes the data are not easy to interpret [82, 83]. The viscometry is the most relevant technique in determination of the average MW of chitosan products which depends on the Mark-Houwink equation ($[\eta] = KM^{a}$), where the parameters of this equation are known under conditions used [84, 85]. [η] is the intrinsic viscosity, *K* and *a* are viscometric parameters and depend on the solvent type. For a chitosan dissolved in 0.5 M CH_3COOH/0.2 M CH_3COONa buffer, *K* and *a* are found to be 3.5×10^{-4} and 0.76, respectively [86, 87]. The viscosity of chitosan solutions could be measured by using Ubbelohde Viscometer immersed in a constant temperature bath at 25 °C. The running times of the solutions and solvent through the viscometer were used to calculate the relative viscosity ($\eta_{rel} = t_{ch}/t_{sol}$), specific viscosity ($\eta_{sp} = \eta_{rel} - 1$), and reduced viscosity ($\eta_{red} = \eta_{sp}/c$). Where t_{ch} is the running time of the chitosan solution, t_{sol} is the running time of the solvent, and *c* is the chitosan concentration in g/dl. The [η] value is obtained graphically by extrapolating the η_{red} *versus* concentration (g/dL) data to zero concentration ($[\eta] = C(\eta_{red})_{c=0}$) and the intercept on the ordinate is the intrinsic viscosity [84, 85, 88, 89]. Finally, the average MW is calculated from the Mark-Houwink equation.

3.4. Solubility and Solution Properties

Chitosan obtained by deacetylation process of chitin becomes soluble in all aqueous acidic media when the average DA is lower than 0.5. Formic and acetic acids are the most commonly used for solubilization of chitosan [90, 91]. The water-soluble ionic complex of chitosan and acetic acid, chitosonium acetate, can be handled like any other soluble linear polymer. The solubilization occurs by protonation of the amino group on the C-2 position of the GlcN unit, whereby the polysaccharide is converted to a polyelectrolyte in acidic media. After dissolution of chitosan, it can be cast into films, spun into fibers, precipitated into particles from suitable solvents, which revert the ionized ammonium groups back into amine functionality, and crosslink to produce chitosan fibers [27]. However, the poor solubility of unmodified chitosan in both water and organic solvents make its utilization limited in the biological and industrial applications. Being chitosan soluble in aqueous solutions, it is largely used in different applications as solutions, gels, or films and fibers. The physical properties of chitosan in a solution depend on the ionic concentration, the pH, the nature of the acid used for protonation, the distribution of acetyl groups along the polymer chain, the conditions of the production of the polymer, and the MW [88, 92]. Fully deacetylated chitosan can be reacetylated in homogeneous phase in order to get samples soluble in acidic conditions up to DA~0.6 in relation with a random distribution of the acetyl groups [93]. Under these conditions, the presence of free amino groups at C-2 position of D-glucosamine units allows performing specific reactions in homogeneous conditions [2, 91, 94, 95].

It was also reported that the solubility of chitosan in neutral water can be obtained by chemical modification such as quaternization that forms quaternary ammonium chitosan salts with permanent cationic charges on the polysaccharide backbone. These derivatives have attracted much attention and are most widely used in different areas of applications. Quaternization of chitosan can be obtained using two different ways: by addition of a substituent which contains a quaternary ammonium group [96] or by quaternization of amino groups of chitosan by means of a suitable alkylating agent [97]. The first possibility uses quaternary ammonium epoxides like glycidyltrimethylammonium chloride. The second one, quaternization of chitosan itself, utilizes alkylchlorides or dialkylsulfates such as dimethyl- or diethylsulfate.

3.5. Chemical Reactivity

Chitosan has three types of reactive functional groups including two hydroxyl groups on C-3 and C-6 in each repeating unit, and the amino group on C-2 in each deacetylated unit and the possible reaction sites are shown in Fig. (**1**). The three functional groups allow the conjugation of many substituents resulting in new modified derivatives with good physicochemical and biological properties. The presence of nucleophilic amino groups allows several reactions such as acetylation, quaternization, Schiff's bases, reductive amination, acylation, phosphorylation and chelation of metals [34, 98 - 107]. Reductive amination reaction is the most reliable procedure for preparation of Schiff bases of chitosan and *N*-alkyl or -aryl chitosan derivatives, wherein the amino group is converted to an imine (Schiff base) with an aldehyde or ketone, and subsequently was reduced by sodium borohydride or sodium cyanoborohydride to form *N*-alkyl or -aryl derivatives [32, 108 - 110]. However, the hydroxyl functional groups give various reactions such as *O*-acetylation, H-bonding with polar atoms, selective etherification and esterification, cross-linking, and graft copolymerization [24, 111]. Selective *O*-substitution can be performed by protecting the amino groups during the reaction [78, 101].

Several reactions occurring on the amino or hydroxyl groups or both of them have been studied and produced many chitosan derivatives with high solubility in water and wide range of the pH such as carboxymethyl chitosans [34, 112 - 114], carboxyalkyl and aryl chitosans [73, 115 - 117], acyl and carboxyacyl chitosans [31, 73, 78, 101, 118 - 120], hydroxyalkyl chitosans [121, 122], quaternary ammonium chitosan salts [32, 96, 109, 123 - 125], chitosan sulfate [126 - 128], phosphorylated chitosan and *N*-methylene phosphonic chitosan products [129 - 136], succinyl chitosan [137], thiolated chitosan products [138 - 141] and sugar-modified chitosan products [142 - 145]. Chitosan can also be cleaved by hydrolyzing agents due to the presence of rather unstable glycosidic bonds similar to all polysaccharides [146]. Cleavage of glycosidic bonds leads to production of low MW chitosan oligomers such as β (1→4) linked homo- or heterooligomers of GlcNAc and/or GlcN named as chitoologosaccharides (COS) with variation in the degree of polymerization (DP) as well as number and sequence of GlcN and GlcNAc [147, 148].

4. BIOLOGICAL ACTIVITIES OF CHITOSAN AND ITS DERIVATIVES IN DIFFERENT AREAS

4.1. Chitosan and Its Derivatives in Agricultural Applications

4.1.1. Control of Plant Bacteria

Chitosan is widely-used as a biocide either alone or blended with other natural products against plant pathogenic bacteria that induce decay and harmful effects of agricultural crops during the growing season and postharvest phase [11, 13, 149 - 153]. The minimum inhibitory concentration (MIC) ranged from 10 to 5000 mg/L, depending on the MW, the pH, chemical modification, DA, degree of substitution (DS), and the type of microorganism [11, 123, 151, 154]. Most of the antimicrobial activities have been investigated against bacteria of *Agrobacterium tumefaciens, Corynebacterium fascians, Erwinia amylovolora, E. carotovora, Pseudomonas solanacearum*, *Pseudomonas syringae*, *Sarcina lutea,* and *Xanthomonas axonopodis* however, nearly all of these studies were conducted *in vitro* (Table **1**) [31, 155, 156].

Chitosan, at 200 mg/L, showed strong *in vitro* antibacterial activity against nine bacterial strains of *Xanthomonas* that cause leaf spot disease of *E. pulcherrima* [156]. This obtained result was consistent with the results of Wang and co-authors, who reported that the same type of chitosan, at 100 mg/L, had strong activity against twelve bacteria of *Xanthomonas* strains up to 61.88% after 24 h of incubation compared to the control [155]. Two kinds of chitosan with different MW and DDA possess a strong antibacterial activity against rice bacterial leaf blight and leaf streak *X. oryzae* and significantly reduced the incidence and severity of disease compared to control under greenhouse conditions [173].

In addition, chitosan at 100, 200, and 400 mg/L inhibited the *in vitro* growth of *Acidovorax avenae* subsp. *avenae* that causes rice bacterial brown stripe disease [157]. Coqueiro and Di Piero added that the chitosan of different MWs, at 1000 and 3000 mg/L, reduced tomato leaf spot severity caused by *X. gardneri* under greenhouse conditions by 70%, when applied up to 3 days before inoculation [172].

Table 1. List of scientific publications reporting the antimicrobial activity of chitosan and its derivatives against bacteria attack plants.

Bacteria	Test product	Bioassay test
Acidovorax avenae	Chitosan	*in vitro* assay [157, 158].
A. citrulli	Chitosan	*in vitro* and *in vivo* assays [159].
Agrobacterium tumefaciens	Chitosan of different MW, *N*-(4-carboxybutyroyl) chitosan derivatives, quaternary *N*-alkyl chitosan derivatives, quaternary *N*-(benzyl) chitosan derivatives, *N*-(benzyl)chitosan derivatives, *N*-(cinnamyl) chitosan derivatives, *O*-(benzoyl) chitosan derivatives.	*in vitro* assay [31, 110, 160 - 165].
Corynebacterium fascians	Chitosan of different MW	*in vitro* assay [151, 163].
Erwinia amylovora	Chitosan of different MW	*in vitro* assay [151, 163].
E. carotovora	Chitosan of different MW, *N*-(4-carboxybutyroyl) chitosan derivatives, quaternary *N*-alkyl chitosan derivatives, quaternary *N*-(benzyl) chitosan derivatives, *N*-(benzyl)chitosan derivatives, *N*-(cinnamyl) chitosan derivatives, *O*-(benzoyl) chitosan derivatives.	*in vitro* assay [31, 110, 160 - 165].
Pseudomonas solanacearum	Chitosan of different MW	*in vitro* and *in vivo* assay [163, 166, 167].
P. syringae	Chitosan	*in vitro* and *in vivo* assay [168, 169].
Sarcina lutea	Chitosan of different MW	*in vitro* assay [163, 170].
Xanthomonas axonopodis	Chitosan	*in vitro* assay [155, 156].
X. campestris	Chitosan film	*in vivo* assay [171]
X.gardneri	Chitosan of different MW	*in vitro* and *in vivo* assay [172].
X. oryzae	Chitosan	*in vivo* assay [173].
X. vesicatoria	Chitosan	*in vivo* assay [174].

Chitosan with different MWs (360, 611, and 953 kDa) had a direct effect on bacterial cells of *A. tumefaciens, C. fascians, E. amylovolora, E. carotovora, P. solanacearum*, and *S. lutea* and the activity increased with the increase of the MW up to MIC 500 mg/L for *C. fascians* [175]. In an other study, chitosan of 50, 37, and 57 kDa indicated that the lowest MW (50 kDa) exhibited the highest activity against *A. tumefaciens* (MIC = 2600 mg/L), while chitosan 37 kDa was the most active against *E. carotovora* (MIC = 950 mg/L) [163].

It was reported that the antibacterial activity was increased as the MW decreased

[148, 176]. However, it is difficult to find a clear correlation between the MW and the antibacterial activity [151].

Grafting the alkyl chains or aryl moieties into chitosan backbone significantly increased the activity against plant pathogenic bacteria [161, 162, 165, 177]. However, this activity is limited to acidic conditions due to its poor solubility above pH 6.5, where chitosan start to lose its cationic nature. Therefore, many studies have focused on the preparation of water-soluble chitosan derivatives such as quaternary ammonium chitosans as antibacterial agents against plant pathogens [31, 164]. A series of *N*-(benzyl) and quaternary *N*-(benzyl) chitosan derivatives was synthesized and their activities were evaluated *in vitro* against *A. tumefaciens* and *E. carotovora* [160]. MICs of *N*-(benzyl) chitosan derivatives were ranged from 800 to 1780 mg/L against *A. tumefaciens* and 700-1550 mg/L against *E. carotovora*. However, the MICs of quaternary *N*-(benzyl) chitosan derivatives were ranged from 500 to 875 mg/L against *A. tumefaciens* and from 600 to 975 mg/L against *E. carotovora*. These compounds also inhibited some of exocellular enzymes including polygalacturonase and pectin-lyase at 1000 mg/L. *N*- (4-Carboxybutyroyl) chitosans at different DS (0.10-0.53) were also prepared and exhibited antibacterial activity with MIC ranging from 800 to 1200 mg/L against *A. tumefaciens* and 725 to 1100 mg/L against *E. carotovora* and the inhibitory effects increased with the DS increase [31]. These data confirmed that grafting of benzyl moiety or quaternization of the derivatives onto chitosan molecule was successful to inhibit bacterial growth. Moreover, increasing the water solubility of compounds by quaternization significantly increased the activity.

4.1.2. Control of Plant Fungi

The antifungal activity of chitosan and its derivatives against various species of plant pathogenic fungi has been reviewed and approved [11, 13, 149 - 152, 178]. The list of scientific publications reporting the antimicrobial activity of chitosan and its derivatives against fungi attack plants is shown in Table **2**. The MICs varied between 10 and 5000 mg/L [31, 162 - 165, 175, 177, 179, 180]. The inhibition activity was observed on different stages of fungal growth such as mycelia, sporulation, spore viability and germination, and production of fungal virulence factors. It has generally been recognized that the antifungal activity depends on the concentration applied, MW, DA, the pH of the chitosan solution, the viscosity and the target organism. In general, chitosan applied at 1000 mg/L is capable of reducing the growth of several fungi, except that there contain a chitosan molecule in their cell walls such as Zygomycetes [181]. For example, El Ghaouth and others reported that chitosan at concentrations ranging from 750 to 6000 mg/L, was high potent against spore germination and germ tube elongation of *B. cinerea* and *R. stolonifer* [179]. Chitosan caused 25-90.5% mycelial growth inhibition in *Penicillium digitatum*, *P. italicum*, *Botrydiplodia lecanidion* and *B. cinerea* depending on the type and the concentration [182]. Benhamou and co-authors found that concentrations of 500 and 1000 mg/L of chitosan were completely effective in reducing the disease incidence caused by *F. oxysporum* in tomato plants [183]. At the same time, El Ghaouth and co-authors revealed that the mycelial growth of *P. aphanidermatum* was inhibited by 400 mg/L of chitosan [184]. Mycelial growth of *A. alternata*, *B. cinerea*, *C. gloeosporioides* and *R. stolonifer* was significantly reduced at the range from 750-6000 mg/L [175]. Other studies showed a linear decrease of *R. solani* growth as the concentration of chitosan increased gradually from 500 to 6000 mg/L [185]. However, complete inhibition was recorded against *F. oxysporum*, *R. stolonifer*, *P. digitatum* and *C. gloeosporioides* at 30000 mg/L [186, 187].

The MW of chitosan has also a significant effect on the antifungal activity and several studies reported that the activity was increased as the MW decreased however, it is difficult to find a clear correlation between the MW and the antifungal activity [148, 163, 175, 180, 188]. COS are effective in inhibition of spore germination, germ tube elongation and mycelial growth of fungal phytopathogens such as *Alternaria kikuchiana, A. solani*, *B. cinerea*, *F. graminearum*, *F. oxysporum*, *R. stolonifer*, *Penicillium*, *Physalospora piricola*, *P. capsici*, *Pyricularia oryzae*, and *V.dahliae* [180, 189 - 191]. Moreover, chemical modification of chitosan with different substituents effectively enhanced the antifungal activity. The derivatives include *N*-(benzyl)chitosan derivatives against *B. cinerea*, *Botryodiplodia theobromae*, *F. oxysporum*, *P. oryzae*, and *P. infestans* [106, 110, 165, 192, 193], *N,O*-acyl chitosans against *B. cinerea* [78], heterocyclic chitosan derivatives against *P. debaryanum* and *F. oxysporum* [177], *N*-(cinnamyl) chitosan derivatives [162], ammonium dithiocarbamate chitosan and triethylenediamine dithiocarbamate chitosan [194], Schiff bases of phenyl and hydroxyphenyl chitosans against *B. cinerea* [195], (1,2,3-Triazol-4-yl) methyl nicotinate chitosan against *R. solani, Stemphylium solani*, and *A. porri* [196], diethyl dithiocarbamate chitosan against *A. porri*, *Gloeosporium theae-sinensis*, and *S. solani* [197], α-aminophosphonate chitosan derivatives against *P. asparagi* and *F. oxysporum* [198], acyl chitosan derivatives against *R. solani* and *S. rolfsii* [199], thiadiazole chitosan derivatives against *C. lagenarium, P. asparagi* and *Monilinia fructicola* [200], thiosemicarbazone chitosan derivatives against *S. solani, R. solani, A. solani,* and *P. asparagi* [201], and quaternary ammonium chitosan derivatives against *B. cinerea*, *B. theobromae*, *Cladosporium cucumerinum*, *C. lagenarium*, *F. oxysporum*, *M. fructicola, P. debaryanum*, and *Ph. infestans* [106, 160, 164, 202, 203].

Table 2. List of scientific publications reporting the antimicrobial activity of chitosan and its derivatives against fungi attack plants.

Fungi	Test product	Bioassay test
Alternaria alternata	Chitosan with different MW, *N*-(benzyl), *O*-(benzoyl), *N*-(cinnamyl), and *O*-(phenoxyacetic) chitosan derivatives	*in vitro* and *in vivo* assays [161 - 163, 165, 204].
A. mali	Hydroxypropyl chitosan derivatives	*in vitro* assay [205].
A. porri	Ammonium dithiocarbamate and triethylene diamine dithiocarbamate chitosan derivatives and (1,2,3-triazol-4-yl)methyl nicotinate chitosan	*in vitro* assay [194, 196].
A. solani	Schiff bases of carboxymethyl chitosan, acyl thiourea chitosan, sulfanilamide chitosan, and thiosemicarbazone-chitosan derivatives, and Cu-chitosan nanoparticles	*in vitro* and *in vivo* assays [206 - 211].
Botrytis cinerea	Chitosan with different MW, *N*-(4-carboxybutyroyl) chitosan, quaternary *N*-alkyl chitosan, quaternary *N*-(benzyl) chitosan, Schiff bases of chitosan, *N*-(benzyl)chitosan, *N*-(cinnamyl) chitosan, and *N*-benzylphosphoryl chitosan derivatives	*in vitro* and *in vivo* assays [31, 106, 110, 160, 162, 164, 195, 212 - 215].
B. fabae	Chitosan with different MW	[163].
Botrydiplodia lecanidion	Chitosan with different MW	*in vitro* assay [182].
B. theobromae	Chitosan of different MW, quaternary *N*-(benzyl) chitosan, *N*-(benzyl)chitosan, *N*-(cinnamyl) chitosan, and *O*-(phenoxyacetic) chitosan derivatives	*in vitro* assay [110, 160, 162, 182, 204].
Colletotrichum gloeosporioides	Chitosan *of different MW*, acyl thiourea chitosan, and thiosemicarbazone-chitosan derivatives and chitosan-silver nanoparticles	*in vitro* assay [186, 207, 211, 216 - 219].
C. lagenarium	Chitosan of different MW, quaternized chitosan derivatives, Schiff bases of chitosan, *N*-substituted chitosan.	*in vitro* assay [195, 214].
Coniella diplodiella	Hydroxypropyl chitosan derivatives	*in vitro* assay [205].
Fusarium oxysporum	Chitosan of different MW, quaternary *N*-alkyl and *N*-(benzyl) chitosan, *N*-(benzyl)chitosan, *O*-(benzoyl) chitosan, *N*-(cinnamyl) chitosan, heterocyclic chitosan, *O*-(phenoxyacetic) chitosan, Schiff bases of carboxymethyl chitosan, ammonium dithiocarbamate and triethylene diamine dithiocarbamate chitosan, acyl thiourea chitosan, Cu-chitosan, and thiosemicarbazone-chitosan derivatives, and chitosan nanoparticles	*in vitro* assay [110, 160 - 165, 177, 194, 204, 206, 207, 210, 211].
F. solani	Chitosan, *N*-(cinnamyl) chitosan, *O*-(phenoxyacetic) chitosan derivatives	*in vitro* and *in vivo* assays [162, 204, 220].
F. sulphureum	Chitosan	*in vitro* and *in vivo* assays [221 - 223]

(Table 2) contd.....

Fungi	Test product	Bioassay test
Gloeosporium theae-sinensis	*N*, *O*-(acyl)-*N*-(trimethyl) chitosan derivatives	*in vitro* assay [224]
Penicillium digitatum	Chitosan of different MW, *chitosan film*	*in vitro* and *in vivo* assays [182, 225, 226]
P. italicum	Chitosan of different MW, *chitosan film*	*in vitro* assay [182, 225, 226]
Phomopsis asparagi	Sulfanilamide chitosan, thiadiazole-functionalized chitosan, and *N*, *O*-(acyl)-*N*-(trimethyl) chitosan derivatives.	*in vitro* assay [200, 208, 224]
Pyricularia oryzae	Chitosan, hydrolyzed chitosan, *N*-benzylphosphoryl chitosan, and heterocyclic chitosan derivatives.	*in vitro* assay [177, 213, 227]
Phyllosticta zingiberi	Acyl thiourea chitosan and thiosemicarbazone-chitosan derivatives.	*in vitro* assay [207, 211]
Phytophthora infestans	*N*-(4-carboxybutyroyl) chitosan, quaternary *N*-(benzyl) chitosan, *N*-(Benzyl)chitosan, *N*-(cinnamyl) chitosan, *O*-(phenoxyacetic) chitosan derivatives.	*in vitro* assay [31, 110, 160, 162, 204]
Ph. palmivora	Chitosan	*in vitro* and *in vivo* assays [228]
Ph. botryosa	Chitosan	*in vitro* assay [229]
Ph. capsici	Oligochitosan compounds	*in vitro* assay [189]
Pythium debaryanum	Quaternary *N*-alkyl chitosan, *N*-(benzyl) chitosan, *N*-(cinnamyl) chitosan, heterocyclic chitosan, and *O*-(phenoxyacetic) chitosan derivatives.	*in vitro* assay [162, 164, 165, 177, 204]
Rhizoctonia fragariae	*Chitosan*	*in vitro* assay [185]
R. solani	*N*-(4-Carboxybutyroyl) chitosan and (1,2,3-triazol-4-yl) methyl nicotinate chitosan derivatives.	*in vitro* assay [31, 196]
Rhizopus stolonifer	Chitosan of different MW	and *in vivo* assay [180, 230 - 234]
Sclerotinia sclerotiorum	Chitosan and *O*-(benzoyl) chitosan derivatives	*in vitro* and *in vivo* assays [161, 235, 236]
Stemphylium solani	Chitosan and (1,2,3-triazol-4-yl)methyl nicotinate chitosan	*in vitro* assay [196, 237]

4.1.3. Control of Plant Viruses

Chitosan has inhibitory effect on viruses and viroids of plants. The ability of suppressing viral infections does not depend on the type of virus; because chitosan affect on the plant itself by induction of a broad spectrum of protective reactions, limiting systemic spread of viruses and viroids over the plant and leads to the development of systemic acquired resistance [238 - 241]. Pospiezny and co-authors found that the local infections produced by alfalfa mosaic virus (ALMV) were completely controlled by 1000 mg/L of chitosan either sprayed or added to the inoculum [240]. Similar inhibition was also obtained on tomato leaves treated with chitosan at the same concentration and inoculated with potato spindle tuber viroid [242]. Chitosan with different MW, at 1000 mg/L, induced resistance to viral infection in potato plants and the highest antiviral activity was obtained by chitosan of 120 kDa [238]. The treatment with 1000 mg/L chitosan enhanced tobacco inducible defenses against tobacco necrosis necrovirus (TNV) and significantly reduced the size of lesions [243]. The network of micro-hypersensitive response elicited by chitosan would then activate plant defense by mimicking a localized virus infection, thus accelerating the subsequent hypersensitive response induced by the challenge inoculation with TNV [244]. Ultimately, this would impair virus replication and translocation, reducing the number and size of virus lesions. Similarly, chitosan, depended on a lower DA, inhibited local lesion necroses induced by tobacco mosaic virus (TMV) [245].

4.1.4. Seed-Coating and Plant Growth Promotion

Chitosan has been widely used as seed coating agent and has a positive effect on germination rate, seedling growth parameters, and the yield of different cultivars including soybean sprouts [150, 246 - 249], ornamental plants [250], maize [251, 252], wheat [253, 254], lentil [255], rice [15, 256], and peanut [257]. In addition, chitosan has been used as seed coating to protect plant from pests include fungi, bacteria and insects. For example, the cotton seed coating chitosan protect it from *A. gossypii* and *H. armigera* and enhance the germination, plant growth and lint yield [258]. Previously, Benhamou and co-authors reported that chitosan at concentrations ranged from 500 to 1000 mg/L, reduced disease incidence by *F. solani* in tomato plants, when used as seed coating or admixing with soil [183]. Treatment of wheat seeds by chitosan to control a straw breaker disease (*Pseudocercosporella herpotrichoides*) had a positive effect and kept the wheat upright through to harvest [259]. Chitosan coatings of soybean seeds showed significant antifeedant effect (>80%) against *Agrotis ipsilon*, soybean pod borer (*Maruca vitrata)*, and aphids at 50000 mg/L [253]. Zeng and Shi, (2009) performed a new type of chitosan with some additives as seed coating agent for rice. By using this formulation, the antifungal efficiency was increased with increasing the dose; a 1:20 concentration was the best for inhibiting growth of *R. solani* and *F. Moniliforme* [260].

4.1.5. Control of Agrochemicals Release

Chitosan is interesting material in controlled-release of agrochemicals including fertilizers and pesticides. To improve the utilization of fertilizer and water resource, a new type of chitosan-coated nitrogen, phosphorus and potassium formulation was prepared and showed good controlled-release, water-retention capacity, degradable in soil and environment-friendly [261]. Teixeira and co-authors found a controlled-release effect by coating urea and the herbicide atrazine pellets with films of *N*-acetyl or *N*-propionyl chitosan [262]. The results showed that the release period extended by a factor of 50 for atrazine beads and a factor of 180 for urea beads compared with uncoated beads. Chitosan beads have been extensively used as devices for controlling release of many substances such as pesticides in pest control. Active compounds that have been encapsulated include, for example, methyl parathion, trifluralin, chlorpyrifos, thiocarbamates, and *B. thuringiensis israelensis* (Bti) toxins [263, 264]. Blending of chitosan with other biopolymers such as alginate, gum and starch is a convenient method to improve its properties for slow release of pesticides in agricultural applications [265 - 268]. A formulation based on chitosan and cashew gum beads containing insecticide dichlorvos was prepared and tested as a larvicide for controlling of *Aedes aegypti* larvae [265]. *in vivo* release showed that the chitosan/cashew gum beads reached a plateau after 30 h and resulted high larval mortality (60%), whereas chitosan beads caused only 42% mortality by 30 h and seemed not to have released all larvicide content up to 72 h. The persistence studies revealed that the dichlorvos-loaded beads were effective up to 336 h. The herbicide imazaquin was also encapsulated chitosan/starch beads reinforced with alginate [269]. Chitosan nanoparticles incorporated insecticidal protein beauvericin (CSNp-BV) was prepared by ionic gelation method to improve insecticidal activity against *S. litura* [270]. Loading efficiency and entrapment efficiency was found to be 82 and 85%. The *in vitro* drug release profile showed cumulative release reached almost 91% after 24 h. Pesticidal activity revealed all the life stages were susceptible to the CSNp-BV formulation and maximum mortality was recorded in early larval instars. CSNp-BV treatment reduced pupal and adult emergence.

Nanoparticles composed of chitosan and sodium tripolyphosphate (TPP) were prepared to produce an efficient herbicidal formulation for safe controlling of weeds in cultivations of maize and mustard [271]. The efficiency of nanoparticles in association with paraquat was 62.6%. Encapsulation of the herbicide controlled its diffusion, release, and sorption in soil. The nanoencapsulated herbicide was less toxic by cytotoxicity and genotoxicity assays than the pure compound, indicating its potential herbicidal activity while at the same time reducing environmental hazards. Carboxymethyl chitosan carrying ricinoleic as an emulsifier was used to encapsulate the insecticide azadirachtin in nanoparticles [272]. Nanoparticles of 200-500 nm were obtained with loading azadirachtin higher than 50% and tested for their release in outdoor condition. After five days of sun exposure, all content of control samples were lost, while the nanoparticles had a constant residual concentration during 12 days of the experiment. The insecticide etofenprox was encapsulated using a nano-sized chitosan carrier in three types according to a difference in release patterns by adjusting the molecular weight and concentration of chitosan [273]. Release properties of etofenprox and its biological activity against *S. litura* suggested that such controlled release formulation is considered to be used as a method of preventing loss etofenprox and increasing its activity against the target pest.

The volatile essential oils have promising activity against agricultural pests in different area such as insect repelling and fumigation in storred products [274, 275]. However, the release speeds of these oils are usually high and affected by the environment conditions [266]. Therefore, utilizing encapsulation technology to achieve the goal of constant release is one of the most effective methods [276]. Chitosan has been examined in the development of essential oils delivery systems [277, 278]. Chitosan possesses heat shrinking property, thus the change in pore space between chitosan wall membrane molecules resulting control of the essential oil release within the capsule or bead formulations [279, 280].

4.2. Chitosan and Its Derivatives in Food and Nutrition

4.2.1. Nutritional and Hypocholesterolemic Effect

Chitosan has attracted notable interest in food and nutrition due its superior film-forming property for fresh and industrial food products, edible, antimicrobial activity against a wide range of foodborne microorganisms, excellent compatibility with other components, hypocholesterolemic activities, and lower toxicity toward mammalian cells [281 - 285]. It has been approved by the United State Food and Drug Administration (USFDA) as a "Generally Recognized As Safe" (GRAS) food additive, as dietary fibre, and functional ingredients for the consumer based on the scientific procedures for use in foods including meat and poultry [286]. Chitosan has been approved as a food additive in Korea and Japan since 1995 and 1983, respectively [287, 288]. Biosafety of chitosan has been demonstrated in feeding trials with domestic animals [289].

Chitosan has non-digestibility in the upper gastrointestinal tract, high viscosity, and high water binding properties, which indicate its property in relation to other dietary plant fibers as highly effective hypocholesterolemic potential [282, 290 - 293]. An increase in the serum cholesterol, triacylglycerol, and free fatty acid values of these domestic animals fed cholesterol-additive diets was suppressed by feeding 1.2-1.4 g of chitosan/kg of body weight per day for hens and broilers and 0.7-0.8 g of chitosan/kg of body weight per day for rabbits [289]. Razdan and Pettersson also observed an increase of high density lipoprotein (HDL) concentrations after feeding chitosan containing diet to broiler chicken [294, 295].

Due to its ability of forming ionic bonds at acidic pH, it can bind *in vitro* to different types of anions such as bile acids or free fatty acids leading to excrete them from body [296, 297]. Chitosan forms micelles in the digestive tract with cholesterol, both endogenous and from dietary sources, in the alkaline fluids in the upper part of the intestine, resulting in the depression of absorption of dietary cholesterol and circulation of cholic acid to the liver. Because of the formation of cholic acid from blood cholesterol in the liver it tends to decrease blood cholesterol concentration [298]. The authors found that the high levels of cholesterol and triacylgiycerol in serum of rabbits were affected neither by feeding with a 20 g/kg of chitosan supplemented food, or by intravenous injection with chitosan or low MW chitosan oligosaccharide daily. Moreover, Deuchi and others found that the chitosan markedly increased the fecal lipid excretion and reduced the apparent fat digestibility in treated rats to about a half compared to the control. They proposed that chitosan is solubilized in the stomach forming an emulsion with intragastric oil droplets and begins to precipitate in the small intestine at pH 6 [299, 300]. The same research group examined the effect of the DDA of chitosan products on the fecal fat excreted from rats fed on a high-fat diet [301]. As the viscosity or DDA of a chitosan increased, the more its effect on the apparent fat digestibility by rats became noticeable that could be due to as the electrostatic attraction between chitosan and anionic substances such as fatty acids and bile acids [301, 302]. An ascorbic acid supplementation for each chitosan diets resulted in a significant reduction in the digestion and absorption of fat in the lumen. The contribution of chitosan caused a higher level of fat to be excreted in the feces of rats received corn oil than those receiving bacon, although the effect was stronger with the two diet groups [301]. The hypocholesterolemic effects of chitosans with different DDAs were further studied *in vivo*, and the effects were increased as the DDA increased [303]. Rats fed diets containing higher DDA chitosan was significantly lowered plasma triglyceride, total cholesterol and low density lipoprotein-cholesterol (LDL) and high levels of HDL. Also, the food efficiency ratio was also decreased with the increase of the DDA of the chitosan products [303, 304]. These authors also reported that chitosans of high MW limit the body weight gain in adult rats, significantly reduced the dietary level efficiency and reduced plasma lipids. The potentials of plasma cholesterol-lowering and glucose-lowering effects of high mw chitosan were investigated in diabetic rats [305]. Compared with the glucose-lowering effects of previous long-term studies [306], Yao and co-authors reported that the long-term feeding (11-20 weeks) may be required for low mw chitosan to reach a hypoglycemic effect in diabetic animals. These changes may have originated at gastrointestinal level *via* the effects that different viscosities or MWs have on the glucose and lipid metabolism [305]. Chitosan derivatives such as diethylaminoethyl chitosan and its quaternized product, and *N*-[(2-Hydroxy-3-*N*,*N*-dimethylhexadecyl ammonium) propyl] chitosan chloride having also high-capacity binding with bile acid and hypocholesterolemic potential and the LDL, TC, leptin and FFA levels were decreased after supplementation in the high-fat diet [291, 307].

It was also found that the hypocholesterolemic activity is related to the size of the chitosan particles, wherein the fine particle of a chitosan effectively reduces plasma lipid levels and liver in rats [282]. Moreover, the chitosan powder exhibited a higher rate of adsorption of oil than the flake type, while chitosan in the flake form bound low cholesterol level [308]. The microspheres of chitosan and water-soluble chitosan were prepared and tested on plasma lipids in male Sprague-Dawley rats fed with high-fat emulsions. The microspheres were effective in lowering lipid levels and plasma viscosity and increased the superoxide dismutase levels in serum [309].

4.2.2. Food Packaging and Antimicrobial Activity

The most commonly component materials of edible films are biopolymers like polysaccharides and proteins. Chitosan has been considered as an appropriate alternative as source of food preservative or coating material of natural origin for replacing the non-biodegradable and non-renewable polymers and also reducing the extensive application of harmful pesticides in food protection [17, 310 - 317]. The products can be formed into fibers, films, gels, sponges, beads or even nanoparticles. However, the chitosan films have shown potential to be used as a packaging or coating material. Chitosan coating, dipping or spraying of fresh, frozen and fabricated foods, can improve the storability of perishable foods by forming semi-permeable on the surface of the products and subsequently delaying the rate of respiration, decreasing the loss of the moisture and weight loss, and maintaining the overall quality [152, 318, 319]. The chitosan films have the advantage, over other bio-based food packaging materials, of being able to incorporate functional substances such as minerals or vitamins and as carriers releasing of antimicrobial agents [320 - 322].

Different techniques are commonly used in chitosan films preparation and coatings for food packaging applications. Among them, the solution casting method is one of the most popular techniques. Since the main function of food packaging and coating is often to avoid or decrease moisture transfer between the food and the surrounding atmosphere, or between two components of a heterogeneous food product, the water vapor permeability (WVP) should be as low as possible [323]. Many studies reported that the WVP of gelatin films were reduced when chitosan was added by ~12%. This behavior could be explained considering that chitosan might enhance the cross-linking of gelatin and decrease the free volume of the polymeric matrix, thus decreasing the diffusion rate of water molecules through the films, resulting in the lower WVP of the chitosan-gelatin films compared to chitosan or gelatin alone [324 - 326]. Similar results were also reported by Pereda and co-authors in films based on bovine hide gelatin and chitosan, but there was no statistical difference in WVP between chitosan-gelatin film and gelatin control one [318]. Low values of WVP of the films prepared from chitosan, starch, and sorbitol are may be also due to the cross-linking caused between amino groups of chitosan and the hydroxyl groups of starch molecules by thermal mixing transformed the hydrophilic-hydrophilic blend into a more hydrophobic one that exhibited relatively low water barrier characteristics [327]. Several studies by Arvanitoyannis and co-authors reported an increase of the WVP with the addition of plasticizers such as glycerol [324, 325, 328, 329]. However, Srinivasa and co-authors reported that the addition of glycerol to chitosan film reduced the WVP [330]. Chitosan has attracted notable antimicrobial activity against a wide range of foodborne pathogens and spoilage microorganisms including Gram-positive and Gram-negative bacteria, yeast, and fungi [14, 149, 317, 331 - 334]. Most microorganisms attack foodstuffs and scientific references related to the antimicrobial activity of chitosan and its derivatives are presented in Tables **3**, **4**, and **5**. Most foods are a mixture of different compounds, such as carbohydrates, proteins, fats, minerals, vitamins, salts and others, and many of them can interact with the chitosan resulting in loss or enhancement of antimicrobial activity. For example, Devlieghere and co-authors extensively studied the influence of different food components include starch, protein, oil, and NaCl on the antimicrobial effect of chitosan [313]. They reported that starch, whey protein, and NaCl had a negative effect on the antimicrobial activity of chitosan however, oil had no influence.

It was reported that the chitosan showed a stronger antibacterial activity rather than antifungal [11, 14, 149, 332, 335]. The MIC of chitosan varies among species and depending on the MW, the pH, chemical modification, DA, DS, and the type of microorganism [11, 150, 151, 154, 176, 178, 214, 332]. In general, chitosan applied at 1000 mg/L is capable of reducing the growth of several fungi, except those containing a chitosan molecule in their cell walls such as Zygomycetes [181]. It is also generally recognized that yeasts and moulds are the most sensitive group to chitosan, followed by Gram-positive and Gram-negative bacteria. Chitosan coatings designed to slow antimicrobial release from the surface, thus, smaller amount of antimicrobials would be needed. Therefore, many studies confirmed that the chitosan coatings can be used as vehicle for incorporating functional ingredients, such as antimicrobials or nutraceutical compounds that could enhance the effects chitosan coatings in application of postharvest of fresh fruit and vegetables [336 - 339].

Table 3. List of scientific publications reporting the antimicrobial activity of chitosan and its derivatives against bacteria attack food products.

Microorganism	Test product	Bioassay test
Acetobacter spp	Chitosan of different DDA	*in vitro* assay [348]
Aeromonas hydrophila	Chitosan of different DDA, chitosan of different MW	*in vitro* assay [349, 350] *in vivo* assay on meat sausage [351, 352] and fish such as *Aeromonas salmonicida* and carp (*Cyprinus carpio*) and fish fillets (*Oncorhynchus nereka*) [335, 353 - 355].
Bacillus cereus	Chitosan of different DDA, chitosan of different MW, *N*-sulfonated and *N*-sulfobenzoyl chitosan, chitosan film	*in vivo* assay on fruit and vegetables [313], meat products [356] and *in vitro* assay [317, 332, 335, 341, 357 - 360]
B. licheniformis	Chitosan	*in vitro* assay [361]
B. subtilis	COS, LMW chitosan, and *N*-/2(3)-(dode--2-enyl) succinoyl/-derivatives	*in vitro* assay [361 - 363]
Bifidobacterium adolescentis	LMW chitosan, COS	*in vitro* assay [364]
B. bifidum	COS, chitosan of different DDA	*in vitro* and *in vivo* assays [361, 364, 365]
B. breve	COS, LMW chitosan	*in vitro* assay [364]
B. catenulatum	COS, LMW chitosan	*in vitro* assay [364]
B. infantis	COS, LMW chitosan	*in vitro* assay [364]
B. longum	COS, LMW chitosan	*in vitro* assay [364]
Brochothrix thermosphacta	Chitosan	*in vitro* and *in vivo* assays [313, 366]
Clostridium historyticum	Chitosan	*in vitro* and *in vivo* assay [367, 368]
C. perfringens	Chitosan of different DDA	*in vitro* and *in vivo* assays [365, 367]
Cronobacter sakazakii	Chitosan film	*in vitro* assay [360]
Enterobacter aeromonas	Chitosan	*in vitro* assay [313]
Enterococcus faecalis	Chitosan	*in vivo* assay [369]

(Table 3) contd.....

Microorganism	Test product	Bioassay test
Escherichia coli	COS, COS/kojic acid, chitosan of different DDA, chitosan of different MW, chitosan film, low MW chitosan, *N*-/2(3)-(dode--2-enyl)succinoyl chitosan, and *N*-alkylated disaccharide chitosan derivatives	*in vitro* assay [317, 332, 335, 340, 348, 350, 357, 360, 362, 370 - 375]
E. agglomerans	LMW chitosan and *N*-/2(3)-(dode--2-enyl)succinoyl chitosan derivatives	*in vitro* assay [362]
Lactobacillus curvatus	Chitosan	*in vivo* and *in vitro* assays [313]
Lactobacillus fructivorans	Chitosan	*in vivo* assay [376]
Lactobacillus plantarum	COS, LMW chitosan	[313] *in vivo* assay [377, 378]
Lactobacillus sakei	Chitosan	*in vivo* and *in vitro* assays [313]
Lactobacillus viridescens	Chitosan	*in vivo* assay [367, 379]
Lactobacillus viridescens	Chitosan	*in vivo* and *in vitro* assays [379]
Leuconostoc mesenteroides	COS	*in vivo* assay [380]
Listeria innocua	Chitosan	*in vivo* and *in vitro* assay assays [351, 352, 379, 381]
Listeria monocytogenes	Chitosan of different DDA, chitosan of different MW, chitosan films	*in vitro* assay [335, 350, 360, 372, 373] and *in vivo* assay [351, 352, 381]
Micrococcus varians	Chitosan	*in vivo* assay [382]
Pediococcus acidilactici	Chitosan	*in vivo* and *in vitro* assays [313]
Pediococcus pentosaceus	Chitosan	*in vivo* assay [381, 382]
Photobacterium phosphoreum	Chitosan	*in vivo* and *in vitro* assays [313]
Pseudomonas aureofaciens	LMW chitosan and *N*-/2(3)-(dode--2-enyl)succinoyl chitosan derivatives	*in vitro* assay [335, 352, 362, 381]
Pseudomonas aeruginosa	Chitosan of different DDA, Chitosan of different MW	*in vitro* assay [335, 350]
Pseudomonas fluorescens	Chitosan film	*in vitro* assay [313, 359]
Pseudomonas fragi	Chitosan	*in vivo* assay [381, 382]
Salmonella enteritidis	Chitosan	*in vivo* and *in vitro* assay assays [379]
S. enterica	Chitosan, chitosan film	*in vitro* assay [383] [359]

(Table 3) contd.....

Microorganism	Test product	Bioassay test
S. typhimurium	Chitosan of different MW, chitosan films, *N*-sulfonated and *N*-sulfobenzoyl chitosan derivatives, chitosan nanoparticles	*in vitro* assay [341, 350, 360, 372, 373, 384] and *in vivo* assay [335, 351, 352, 367, 381]
Serratia liquefaciens	Chitosan	*in vivo* assay [381]
Shigella dysenteriae	Chitosan of different DDA, chitosan of different MW	*in vitro* assay [335, 350]
Staphylococcus aureus	COS, COS/kojic acid, chitosan of different DDA, chitosan of different MW, chitosan film, *N*-alkylated disaccharide chitosan derivatives, chitosan nanoparticles	*in vitro* and *in vivo* assays [335, 340, 350, 359, 360, 363, 370, 372, 375, 382, 384, 385]
S. epidermidis	COS	*in vitro* assay [363]
Yersinia enterocolitica	Chitosan	*in vitro* assay [372]
Vibrio cholerae	Chitosan of different DDA, chitosan of different MW	*in vitro* assay [335, 350] *in vivo* assay [335]
V. parahaemolyticus	Chitosan of different DDA, chitosan of different MW	*in vitro* assay [335, 350] *in vivo* assay [335]

The chemical modification of chitosan is continuously studied since it has the potential to provide an antimicrobial activity against food microorganisms. Therefore, several research groups have been modified its structure by grafting biologically active molecules or by increasing its solubility in neutral pH. Grafting the alkyl chains or aryl moieties into chitosan significantly increased its antimicrobial activity [106, 110, 162, 177, 340 - 343]. However, this activity is limited to acidic conditions due to its poor solubility above pH 6.5, where chitosan starts to lose its cationic nature. Therefore, many studies have focused on the preparation of water soluble products as antimicrobials against food pathogens [31, 164, 332, 340, 344]. Among water-soluble chitosan derivatives, the quaternary ammonium chitosan, prepared by introducing a quaternary ammonium group into a dissociative hydroxyl group or amino group of the chitosan, exhibited stronger antibacterial activity than chitosan on a range of pH values.

Table 4. List of scientific publications reporting the antimicrobial activity of chitosan and its derivatives against fungi attack food products.

Microorganism	Test product	Bioassay test
Aspergillus fumigatus	Schiff bases of chitosan, PVA/carboxymethyl chitosan nanocomposites, chitosan film	*in vitro* assay [335, 386, 387]
A. niger	COS, COS/kojic acid, Chitosan of different DDA, chitosan of different MW, chitosan film.	*in vitro* and *in vivo* assays [348, 370, 385, 388 - 391]

(Table 4) contd.....

Microorganism	Test product	Bioassay test
A. parasiticus	Chitosan of different MW, chitosan film	*in vitro* and *in vivo* assays [335, 392, 393]
A. flavus	Chitosan coating, Chitosan of different MW, Carboxymethyl chitosan, Diethylaminoethyl chitosan	*in vitro* and *in vivo* assays [342, 369, 393, 394]
Botrydiplodia lecanidion	Chitosan of different MW	*in vitro* and *in vivo* assays [182, 395]
Botrytis cinerea	Chitosan of different MW, carboxyacyl chitosans, quaternary ammonium chitosans, *N,O*-(acyl) chitosans, *N*-alkyl chitosans, *N*-(benzyl) chitosans, *O*-(acyl) chitosans, Schiff bases of chitosans, sulfate, sulfanilamide and sulfonailide chitosans.	*in vitro* and *in vivo* assays [31, 73, 106, 110, 162, 164, 182, 192, 193, 195, 208, 388, 396 - 398]
Cladosporium sp.	Chitosan of different MW	*in vivo* assay [310, 388, 399, 400]
Fusarium oxysporum	Chitosan of different DDA, chitosan of different MW, quaternary ammonium chitosans, Schiff bases of chitosans, *N*-/2(3)-(dodec-2-enyl)succinoyl chitosan derivatives, acyl thiourea chitosans, dithiocarbamate chitosans, hydroxypropyl chitosans, *N*-alkyl chitosans, *N*-benzyl chitosans, *N*-succinoyl chitosans, *O*-acyl chitosans	*in vitro* assay [31, 106, 110, 162, 164, 192, 193, 195, 198, 199, 205, 335, 362, 388]
Penicillium citrinum	Chitosan of different MW	*in vivo* assay [388]
Penicillium chrysogenum	Carboxymethyl Chitosan	*in vivo* assay [369]
Penicillium digitatum	Chitosan of different MW	*in vitro* assay [182, 395]
Penicillium expansum	Chitosan of different MW	*in vivo* assay [401 - 403]
Penicillium italicum	Chitosan of different MW	*in vitro* assay [182, 395]
Penicillium notatum	Carboxymethyl Chitosan	*in vivo* assay [369]
Rhizopus nigricans	Chitosan of different MW, carboxymethyl chitosan	*in vivo* assay [369, 388]
Rhizopus stolonifer	COS, Chitosan of different MW	*in vitro* and *in vivo* assays [180, 230, 231, 310, 389, 398, 404]

The most common mechanism of action is that the positively charged chitosan gives this one great physiological and biological polymer property. The cationic nature of chitosan leads to bind with sialic acid in phospholipids which consequently modifies the cell permeability and causes materials prevented from entering the cell and/or material being leaked from the cell [11, 14, 150, 153, 214,

331, 344 - 346]. In addition, chitosan caused a decrease in the H^+-ATPase activity on plasma membrane of *R. stolonifer*; this effect could provoke the accumulation of protons inside the cell, which would result in the inhibition of the chemiosmotic driven transport that allows the H^+/K^+ exchange [231]. Another important mechanism involves penetration of the chitosan oligomer into the cells of microorganisms and inhibits the growth of cells by preventing the transformation of DNA into RNA [11, 149, 331, 347].

Table 5. List of scientific publications reporting the antimicrobial activity of chitosan and its derivatives against yeasts attack food products.

Microorganism	Test product	Bioassay test
Candida albicans	COS, chitosan of different DDA, chitosan of different MW, carboxymethyl chitosan, *N*-(carboxybutyl) chitosan, *N*-acetyl-D-glucosamine, chitosan-essential oil	*in vitro* assay [335, 346, 405 - 409]
C. glabrata	COS, chitosan of different MW, carboxymethyl chitosan, *N*-(carboxybutyl) chitosan, and *N*-acetyl- D-glucosamine	*in vitro* assay [346, 406 - 408]
C. krusei	COS, chitosan of different MW, *N*-/2(3)-(dodec-2-enyl)succinoyl chitosan derivatives, carboxymethyl chitosan, *N*-(carboxybutyl) chitosan, and *N*-acetyl-D-glucosamine	*in vitro* assay [362, 406 - 408]
C. lambica	Chitosan, *N*-(carboxybutyl) chitosan	*in vitro* assay [313, 408]
Cryptococcus albidus	Chitosan hydrochloride	*in vitro* assay [358]
C. humiculus	Chitosan, chitosan coating	*in vitro* assay [313, 315, 410, 411]
Saccharomyces cerevisiae	Chitosan, carboxymethyl chitosan	*in vitro* and *in vivo* assays [346, 361, 369, 412 - 414]
S. exiguus	Chitosan	*in vitro* and *in vivo* assays [333, 413]
Saccharomycodes ludwigii	Chitosan	*in vitro* and *in vivo* assays [358, 379, 413]
Schizosaccharomyces pombe	Chitosan	*in vitro* and *in vivo* assays [413]
Torulaspora delbrueckii	Chitosan	*in vivo* and *in vitro* assay assays [379]
Zygosaccharomyces bailii	Chitosan, chitosan hydrochloride, chitosan film	*in vitro* and *in vivo* assays [346, 358, 413, 415, 416]

4.2.3. Antioxidant Activity

The antioxidant activity of chitosan and its derivatives has attracted the most attention because they exert strong antioxidant activity and their effects are similar to those of phenolic antioxidants [307, 417 - 419]. Among various reactive oxygen species, the chemical activity of hydroxyl radical •OH is the strongest, which can easily react with biomolecules such as amino acids, proteins, and DNA [420]. Two types of fungal chitosan have been prepared by alkaline *N*-deacetylation of crude chitin for different durations of 60, 90, and 120 min and their antioxidant properties were investigated [421]. The results showed that the chitosan has antioxidant activities of 61.6-82.4% at 1 mg/mL and showed reducing powers of 0.42-0.57 for 1,1-diphenyl-2-picrylhydrazyl (DPPH) radicals at 10000 mg/L. Also, no significant difference in the antioxidant properties between both types of chitosan has been observed and the EC_{50} values of scavenging abilities on hydroxyl radicals were below 100 mg/L whereas those of chelating abilities on ferrous ions were 580-690 mg/L. In another study by the same authors [418], who found that the chitosan prepared from crab chitin and deacetylated for 60, 90 and 120 min exhibited antioxidant activities of 58.3-70.2% at 1000 mg/L and showed reducing powers of 0.32-0.44 at 10000 mg/L. At this concentration, the scavenging ability of chitosan C60 on DPPH radicals was 28.4% whereas those of other chitosans were 46.4-52.3%. All EC_{50} values of antioxidant activity were below 1500 mg/L.

Prabu and Natarajan reported that the hydroxyl radicals scavenging activity of chitosan isolated from *Podophthalmus vigil* was found to be higher than the ascorbic acid and the range was from 23 to 93%. Superoxide anion radical was ranging from 5.21 to 65% for the concentration between 125 - 2000 mg/L [422]. In another study by Kim and Thomas, who found that three different MWs chitosans (30, 90, and 120 kDa) exhibited antioxidative activities in salmon (*Salmo salar*). Incorporation of 0.2, 0.5, and 1% chitosan, employing 2-thiobarbituric acid-reactive substances and 2,2-diphenyl-1picrylhydrazyl scavenging assays, caused reduction of lipid oxidation in salmon for seven days of storage. The free radical-scavenging activity of the 0.2 mM DPPH solution was saturated by 30 kDa chitosan at 7000 mg/L, resulting in a strong antioxidant activity of approximately 85% [423].

Hydroxypropyl chitosan and CMC were reacted with maleic acid sodium to form water-soluble derivatives and their scavenging activities against hydroxyl radical •OH were investigated by chemiluminescence technique [424]. They exhibited IC_{50} values ranging from 246 to 498 mg/L, depend on the type of substituent and the content of hydroxyl and amino groups [425 - 427]. Chitosan was irradiated in acetic acid solution (1%) with different doses (2-20 kGy) of Co-60 γ rays to investigate the enhancement of the antioxidant activity [426]. Radical mediated inhibition of lipid peroxidation, reducing power, superoxide anion and hydroxyl radical quenching assays have been used for the evaluation of the antioxidant activity of irradiated chitosan. The result indicated that the 2.1 kDa chitosan exhibited high reducing capacity and expressed good inhibition of linoleic acid peroxidation. At a concentration of 100 mg/L could trap 74.2% of the superoxide radical. *N*-CMC was studied in the prevention of warmed-over flavour (WOF) [428] and the results reported that this product was highly effective in controlling of the WOF at different temperatures tested.

It was also reported that the *N*-CMC at 5000 mg/L caused 93% inhibition of the thiobarbituric acid (TBA) in the ground beef and 99% reduction in the hexanal content in the products. These results were further confirmed by Li and co-authors, who found that the addition of 3000 mg/L of *N*-CMC to cooked pork was sufficient to prevent the oxidative rancidity of the product [429]. However, the *N,O*-CMC and its lactate, acetate and pyrrolidine carboxylate salts showed high effect in controlling the oxidation and flavour deterioration of cooked meat over a nine day storage of cold storage with inhibition percentages of 46.7, 69.9, 43.4, and 66.3% for the four products, respectively as reflected in their TBA values [430].

Several studies reported that the antioxidant activity of chitosan was enhanced by incorporation of other compounds such as sugars, essential oils and gelatin [431]. Gelatin contains amino acids or biologically active peptides which may acts as electron donors that indicate antioxidant activity of gelatin composite edible films [432 - 437]. Edible films can improve the preservation of food, principally because of their ability to act as barriers to water, prevent dehydration and to oxygen and light, reducing the oxidation of lipids [317, 438 - 440]. The packaging release antioxidant is a type of food preservation system, wherein an antioxidant or mixtures are incorporated in the package, which contributes to extending the shelf life of food.

Some studies have indicated that the antioxidant mechanism of chitosan may be a chelating action of metal ions such as copper and iron found in enzyme active sites thus to inactivate the oxidizing enzymes and/or binding with lipid such enzymes [212, 441]. In addition, chitosan films have been reported to inhibit both primary and secondary lipid oxidation in herring and Atlantic cod [442]. In all studies, it can be concluded that chitosan has a good antioxidant activity, hydroxyl radical scavenging ability, and chelating ability on ferrous ions and can be used as a source of antioxidants, as a possible supplement or ingredient in the pharmaceutical industry.

4.3. Chitosan and Its Derivatives in Biomedical Applications

Chitosan has the potential to be used in biomedical applications such as controlled drug release systems, pharmaceutical industries, gene delivery, tissue engineering, wound healing, ophthalmology, supports for immobilized enzymes, and many other applications [21, 41, 443 - 446]. Up to date, there are many chitosan formulations have been prepared and tested in biomedical applications as shown in Table **6**. These safely applications could be refer as the chitosan has unique characteristics and advantages such as nontoxicity to mammals, biocompatibility, and biodegradability without harmful of the environmental components as it breaks down slowly to amino sugars as safely products that are absorbed completely in the body.

Table 6. Selected biomedical applications of biopolymers based on chitosan.

Chitosan formulation	**Biomedical applications**
Chitosan-gelatin hydrogels	Controlled release of levamisole, chloramphenicol or cimetidine [447]. As implants blood substitutes or wound dressing material. Material supporting nerve repair [448]. Matrix containing hydroxyapatite as bone-replacing composite [449] and nerve repair [450].
Chitosan-collagen scaffold	Scaffold for culture of human epidermoid carcinoma cells and test of anticancer drugs [451].
Chitosan with silylating agent	Hydrogels for enzyme immobilisation or copper adsorption [452].
Chitosan-glutaraldehyde (GLA) membranes, scaffold, fibers, microspheres	Controlled release of of riboflavin or propanolol HCl [453]. Fibres for 5-fluorouracil release [454]. Microspheres for release of goserelin [455]. Scaffold for the immobilisation of enzymes producing dextranase [456] or of tyrosinase [457]. Material supporting nerve repair [450].
Chitosan-GLA with polyether hydrogel	Anticoagulant hydrogel containing heparin [450, 458, 459].
Chitosan-GLA with polyvinyl pyrrolidone (PVP) hydrogel	Controlled release of amoxicillin [458] Hydrogel supporting growth of epithelial cells and inhibiting fibroblast growth for wound healing [460].
Chitosan-GLA with poly(vinyl alcohol) (PVA) membrane	Membrane for the sustained release of vitamin B-12 [461]. Membrane for the pH-controlled release of insulin [462].
Chitosan-PEG dialdehyde diethyl acetal	Biocompatible and biodegradable wound dressing material [463].
Chitosan-genipin, microspheres	Biocompatible and biodegradable materials for use as implants, blood substitutes or wound dressing material [464].
Chitosan-glyoxal, hydrogels	For site-specific antibiotics delivery in the stomach [465].

(Table 6) contd.....

Chitosan formulation	**Biomedical applications**
Chitosan-glyoxal with polyethylene oxide, hydrogels	for site-specific antibiotics delivery in the stomach [465].
Carboxymethyl chitosan crosslinking with Fe(III), microspheres	Controlled release of 6-mercaptopurine [466].
Chitosan-β-glycerophosphate, hydrogels	Controlled release of chlorpheniramine or albumin [467]
Chitosan-tripolyphosphate, hydrogel beads	Controlled release of piroxicam [468] or 6- mercaptopurine [469].
IPN of chitosan-tripolyphosphate sponges	Controlled release of plateletderived growth factor-BB used in bone regeneration [470].
IPN of chitosan-sulfate-gelatin, films	For the improvement of drug sustained release performances [266].
IPN of chitosan-citrate, gel beads	For specific drug delivery in the stomach [471].
Chitosan-poly(lactic acid) (PLA) granules, films and freeze-dried scaffolds	Controlled release of ibuprofen [472].

4.3.1. Drug Delivery Systems and Pharmaceutical Industries

Controlled drug release systems are consisted of therapeutic substance enclosed in a biodegradable matrix that release required amounts of the drug in a prolonged time period. In the last few decades, considerable studies have been published on the potential use of chitosan and its derivatives in drug delivery systems in view of its biodegradability, nontoxicity, and biocompatibility. The cationic nature of chitosan is responsible for its unique properties such as controlled release of the drugs, mucoadhesion, *in situ* gellation, transfection, permeation enhancement, and efflux pump inhibitory properties [473 - 479]. In addition, improvement of these characteristics can be controlled by structural modifications of chitosan molecule. Therefore, chitosan can be used as a suitable matrix with availability in different forms such as beads, films, capsules, hydrogels, spheres, fibers, tablets, *etc.*, for controlling the release of many drugs [475, 480 - 486]. As the presence of cationic charge on chitosan molecule, it is designed as drug carrier matrix in sustained release for the anionic drug such as naproxen [487]. Moreover, using polyanionic drugs, the interactions between chitosan and the therapeutic agent are more pronounced, and based on the ionic crosslinking between them; even stable complexes are formed from which the drug can be released even over a long period. Sun and co-authors prepared enoxaparin/chitosan nanoparticles providing very stable complexes with improvement of the drug uptake [488].

Chitosan is an excellent hydrophilic biopolymer to be used for oral drug delivery that has muco/bioadhesive properties and may acts as an absorption enhancer [16,

489 - 491]. The orally administered formulations are initially exposed to the acid environment of the stomach, especially if they are administered in fasted subjects. The mucoadhesion capacity could result in formulations containing chitosan being retained gastric pH in the stomach. In addition, it has a potential in controlling of the drugs release into many biological systems. Chitosan has shown promise in the development of nonparenteral delivery systems for challenging drugs. For example, chitosan salts have been shown to increase the transport of drugs across the nasal epithelial surface. Directly compressible ketoprofen bioadhesive tablets containing chitosan and sodium alginate showed sustained release 3 h after intraoral (sublingual site in rabbits) drug administration [492]. Chitosan has a high potential in the development of a drug delivery system gastroretentive successful because it combines both floating and bioadhesion capabilities, especially for drugs that are poorly soluble in intestinal medium and easily soluble in acid. It could be ideal formulation to release drugs slowly in the stomach, since the formation of cationic chitosan gel which is manifested in the acid pH levels results in marked retarding effect on the drug release. The release of a drug depends on different parameters include the initial drug loading, thickness of the membrane, pH of media, and the type of action mechanisms [493]. For example, to localize antibiotic delivery in the stomach, a pH-sensitive swelling system was prepared by cross-linking chitosan and polyethylene glycol poly(ethylene oxide); (PEO) loaded with amoxicillin and metronidazole. The study reported that more than 65% of amoxicillin and 59% of metronidazole were released in enzyme-free simulated gastric fluid and simulated intestinal fluid [465].

Chitosan is a mucoadhesive polymer that is capable of opening the tight junctions and allowing paracellular transport of molecules across mucosal epithelium, and therefore is suitable for mucosal delivery of vaccines [494]. Mucoadhesion can be achieved as the ionic interactions between the cationic amino groups of chitosan and the anionic surface of the mucus as the presence of sialic acid and sulfonic acid substructures. In the case of unmodified chitosan, the cohesive properties, as the adhesion and Mucoadhesion, are comparatively weak therefore; chemical modifications of chitosan have been achieved such as preparation of thiolated chitosan derivatives. These products are hydrophilic macromolecules having free thiol groups and represent a new promise in the field of mucoadhesive polymers and alternative in the field of non-invasive delivery of peptide. Their chemical structures are able to form disulfide bonds with mucus glycoproteins of the mucus gel layer and also the inter- and intrachain disulfide bonds are formed within chitosan itself, resulting increase significantly of the mucoadhesion up to 130 fold [138, 140, 495 - 500]. Chitosan-cysteine, chitosan-thioglycolic acid, chitosan-4-thiobutylamidine, and chitosan-thioethylamidine conjugates are the most four types of thiolated chitosan derivatives that have been widely prepared and tested [139, 495, 499, 501 - 503]. It was reported that a degree of modification of 25-250 mmol thiol groups per one gram of chitosan showed the highest enhancement in the mucoadhesive and enhancing of the permeation properties. Thiolated chitosans have been also used as scaffold materials in tissue engineering and as coating materials for stents [139]. Quaternization of chitosan such as trimethylation provides more cationic character of the polymer lead to significantly enhance the mucoadhesive properties [504]. Carboxyalkyl chitosan derivatives, as water-soluble, were also prepared and tested as drug carriers [115, 116, 505]. *N*-(2-Carboxybenzyl) chitosan was prepared by reductive amination sequence with 2-carboxy benzaldehyde and cross-linked with glutaraldehyde to develop pH-sensitive hydrogel for colon specific drug delivery of 5-flurouracil [506, 507]. Succinyl-chitosan derivatives are also favorable as drug carriers as their biocompatibility, low toxicity and long-term retention in the body due to the presence of several carboxyl groups that exhibit solubility in different aqueous media [476, 508 - 511]. *N*-Succinyl-chitosan is obtained by introduction of succinyl moiety into chitosan *N*-terminal of the GlcN units however; the succinyl group can be introduced to *O*-terminal after protection of the free amino groups and then removal of the protecting moiety [105, 137, 508, 512 - 514]. It was reported that succinyl-chitosan showed long retained in the systemic circulation after intravenous administration and the plasma half-lives were ca. 100 h in normal mice and 43 h in Sarcoma 180-bearing mice for succinyl-chitosan (MW = 3400 kDa and degree of succinylation = 81%). The biodistribution of succinyl-chitosan into other tissues was trace apart from the prostate and lymph nodes and the maximum tolerable dose for the intraperitoneal injection was greater than 2 g/kg [105]. In previous studies, Kato and co-authors found that the biodistribution of succinyl-chitosan into the liver, kidney and spleen was below 5% of dose/g tissue [515] and was higher than 10% of that was transferred to the prostate and lymph nodes at 48 h post-administration and were less than 10% of dose/g tissue into the testes, preputial gland, intestine, femoral muscle, backbone and peritoneum [516].

Chitosan composite with poly(lactic acid) (PLA) granules, films and freeze-dried scaffolds also blending with hydroxypropyl cellulose (HPC), as well as dibutyryl chitin films were prepared and tested as controlled drug delivery systems, including transdermal and oral dosage forms. Excellent adhesion of biopolymer matrices to PLA microspheres or hydroxyapatite (HAp) particles was confirmed and the drug release kinetics of ibuprofen drug from obtained films was found to be first order [472]. Chitosan is shown to be a promising candidate vaccine supply mucosal vaccine proteins such as *B. anthracis* [517], diphtheria [518], and influenza [519, 520]. The results showed that the nasal chitosan in the influenza vaccine was both effective and protective time and the mechanism of action is considered to be due to the promoting effect of the absorption of the material rather than an effect of chitosan on the immune system [521]. Other formulations such as chitosan microspheres and microcapsules have been well established in drug delivery systems. These microparticulate carriers are considered as good potential oral delivery systems to provide a constant therapeutic level of the drug and to avoid frequent administration of the solid oral dosage form. Many active substances have been entrapped in chitosan microspheres such as diclofenac sodium, furosemide, indomethacin, methotrexate, nifedipine, prednisolone sodium phosphate, theophylline, aspirin, and griseofulvin and the results reported that the pharmacokinetic properties of microspheres were significantly affected by chitosan type, drug concentration, and the viscosity of the lipophilic phase [522 - 526].

Chitosan microcapsules have been prepared by several authors using a mild chitosan calcium alginate microencapsulation technique for drug delivery systems [484, 527 - 530]. This technique is more commonly used to manufacture oral microcapsules than other processes such as precipitation, interfacial polymerization, and spray drying, which involve organic solvents. The results of these studies suggested these microcapsules may act as a superior gastrointestinal sustained delivery system as the release rate of the drug.

4.3.2. Gene Delivery

Chitosan is the most prominent among natural polymer-based gene delivery vectors, due to its biodegradability and biocompatibility [531, 532]. However, this system is significantly limited because its low transfection efficiency and cell specificity [533]. Its role in gene delivery is supported by its ability to protonate in acidic media forming a complex with DNA through electrostatic interactions [534]. The chitosan-DNA complexes are very easy to prepare and are more effective compared to the commonly used polygalactosamine-DNA complexes [535]. The transfection efficiency of chitosan depends on the MW, DDA, pH of the transfecting media, cell type, and the charge ratio between the luciferase plasmid to chitosan [536 - 538]. It was found that the increase of the chitosan MW and concentration yielded more stable complexes, indicating that varying the chitosan chain length may provide a tool for controlling the ability of the polyplex to deliver therapeutic gene vectors to cells [539, 540]. However, other studies reported that the low MW chitosans were the best effective and enhanced adenovirus infectivity to mammalian cells in gene therapy concentration and low MW chitosans were the best effective [277, 541].

Chitosan/DNA nanoparticles may be readily formed by coacervation between the positively charged amine groups on chitosan and negatively phosphate groups on the DNA [542 - 545]. Dai *et al.* studied the gene delivery by the nanoparticles through retrograde intrabiliary infusion (RII) and tested the efficacy of liver specific targeting [545]. The transfection efficiency of chitosan-DNA nanoparticles, compared with poly(ethylenimine) (PEI)-DNA nanoparticles, was evaluated in rats by infusing into the common bile duct, portal vein, or the tail vein and the result showed that the luciferase expression was not detected when chitosan-DNA nanoparticles were administrated through the portal vein or tail vein.

Galactosylated chitosans have been reported as hepatocyte-targeting gene carriers due to the specific ligand-receptor interactions between galactose and asialoglycoprotein receptors [546, 547]. These derivatives showed low cytotoxicity and exhibited *in vitro* and *in vivo* enhanced gene transfer efficiency and specificity [548, 549]. It was also found to be coupled with spermine facilitating the transfer of the preferential gene to the liver cells [550]. Galactosylated chitosan and galactosylated chitosan-graft-dextran (GCD) were prepared and characterized for Hepatocyte-targeting and complexed with DNA [546, 551]. Galactosylated chitosan-graft-dextran/DNA complexes were only transfected with cells having asialoglycoprotein receptor and could regulate cell growth using YB-1 transcription factor decoy oligonucleotides [546].

Depolymerised chitosan with the intermediate range of MW seemed a better parent backbone for chemical conjugation that executed transfection and intracellular uptake for transgene expression. Particle size, zeta potential, buffering capacity and DNA binding ability gave encouraging result of enhanced solubility, stability, and improved transfection [550]. Katas and Alpar developed chitosan nanoparticles for siRNA delivery by simple complexation and ionic gelatin using tripolyphosphate (TPP) as a crosslinking agent. *In-vitro* studies using two types of cells lines, CHO K1 and HEK 293, revealed that the association of siRNA to chitosan plays an important role in the silencing effect [552]. Liu and co-authors reported that the physicochemical properties and *in vitro* gene silencing of chitosan/siRNA nanoparticles are significantly dependent on the chitosan MW and DDA where that the high MW chitosan with great DDA resulted in the formation of discrete stable nanoparticles of 200 nm in size [553]. Low MW of chitosan/siRNA showed almost no endogenous knockdown reinforced green fluorescent protein (EGFP) in H1299 carcinoma cells of the human lung, while those prepared from high MW and DDA showed higher gene silencing range between 45% and 65% [552, 553]. From these results, it can be concluded that there is still scope for enhancement and optimization of gene silencing using chitosan/siRNA nanoparticles and certain improvements in the polymeric characteristics would make significant differences.

4.3.3. Wound Healing and Dressing

In last few decades, many research groups are interesting in production of new and improved wound healing and dressing by modifying biocompatible products [554, 555]. The strategies are focused on the acceleration of the wound repair by systematically designed biologically derived materials such as chitosan and its derivatives that are able of accelerating the healing processes at molecular, cellular, and systemic levels [19, 446, 555, 556]. Due to the similarity between chitosan and hyaluronic acid, it was early suggested that chitosan can be used as wound healing agent [557]. It is beneficial for healing of wide open wounds in animals. The mechanisms of acceleration of wound healing have been studied *in vitro* and *in vivo* and proved that the chitosan could stimulate cell proliferation and tissue organization. Moreover, it helps in blood clotting, as it is a hemostat, and blocks nerve endings reducing the pain [558].

Chitosan is also able to activate the host defenses to prevent infection, providing an alternative antibiotic agent. In addition, chitosan depolymerizes into GlcNAc units, that initiate fibroblast proliferation, helps in ordered collagen deposition and significantly increased the level of hyaluronic acid synthesis at the wound site [446]. Chitosan improves the functions of inflammatory cells such as polymorphonuclear leukocytes, macrophages, the transforming growth factor h1, and the growth factor platelet-derived. It is also induce fibroblast to release interleukin-8, which is involved in the migration and proliferation of fibroblasts and vascular endothelial cells [559]. Chitosan hydrogel formulations have been demonstrated to effectively interact with and protect the wound, and moist healing environment. The hydrogels are based on the covalently crosslinked of chitosan and are divided into three types with respect to their structures include chitosan crosslinked with itself, hybrid polymer networks (HPN), crosslinking between chitosan chain and another type polymeric chain, and semi- or full-interpenetrating polymer networks (IPN) that contain a non-reacting polymer added to the chitosan solution before crosslinking reaction [560]. Chitosan-glutaraldehyde, chitosan-glyoxal, chitosan-epichlorohydrin, and chitosan-ethylene glycol diglycidyl ether, respectively are extensively studied and their reaction with chitosan is well-documented [449, 561, 562]. Applying gel to an open wound induced significant wound contraction and accelerated wound closure and healing. Furthermore, in cell culture studies it has been shown that chitosan hydrogel immobilized with fetal bovine serum had an influence on dermal fibroblasts and stimulated cell growth [560, 562]. In addition to the requirements for wound healing, it is important to control any infection of a wound under dressing and chitosan has been shown to provide inhibition of bacterial and fungal proliferation in the treatment of infected wounds [14, 563 - 566].

Various forms of wound dressing materials based on chitin and chitosan products are commercially available such as HemCon® Bandage (HemCon Medical Technologies Inc., Portland, OR), Syvek-Patch® (Marine Polymer Technologies), Chitopack C® and Chitopack S® (Eisai Co. Japan.), Beschitin® (Unitika Co. Japan), Chitodine® (IMS), and Trauma DEX® (Medafor) [555, 567]. These wound dressings manufactured as non-wovens, nanofibrils, composites, films, scaffolds and sponges [558, 568, 569]. The techniques of phase separation, self-assembly, and electrospinning have been developed to fabricate chitosan nanofibrous scaffolds. However, recently the electrospinning technology, by applying a high voltage to electrically charged liquid, has become the most popular one [570 - 574]. It was also reported that a combination between chitosan and collagen was used in wound dressing system with high liquid absorption, biocompatibility, and antibacterial properties [575]. In this study, various weight ratios of collagen to chitosan were immobilized on the polypropylene nonwoven fabric, which has previously been grafted with acrylic acid (AA) or *N*-isopropyl acrylamide (NIPAAm) to construct a durable sandwich membrane for dressing. The results indicated that grafted NIPAAm grafted and collagen/chitosan-immobilized polypropylene nonwoven fabric showed a better healing effect than AA-grafted and collagen/chitosan immobilized polypropylene nonwoven fabric [575].

4.3.4. Tissue Engineering

The development of biodegradable polymers such as collagen and chitosan to perform the role of a temporary matrix that mimics the native extracellular matrix of the target tissue as much as possible, is an important factor in the success of cell transplantation in tissue engineering [569, 576 - 580]. It was reported that the chitosan possesses characteristics comparable to various glycosaminoglycans (GAGs), such as chondroitin sulfate and keratin sulfate that found in the extracellular matrix of articular cartilage. Moreover, the primary monomer of chitosan has a framework similar to the predominant GAGs in cartilage. Because of this similarity, chitosan has been tested as a scaffold for cartilage repair [576, 581 - 584]. One of chitosan's most promising features is its exceptional ability to be processed into porous structures for use in cell transplantation and tissue regeneration. Porous chitosan structures can be formed by freezing and lyophilizing aqueous acetic acid chitosan solution generating a porous material with pore size that can be controlled by variation in the rate of freezing [578].

Chitosan-based hyaluronan hybrid polymer fibers indicated excellent potential as a desirable biomaterial for cartilaginous tissue scaffolds due to its excellent characters of cell adhesion, proliferation and matrix secretion [585]. Further, collagen/chitosan scaffolds were found to be as potential candidates for the dermal equivalent, with enhanced biostability and good biocompatibility [577]. Arpornmaeklong and others also reported that chitosan based sponges tended to promote growth of osteoblast cells to a greater lever than collagen-only sponges [586]. The characteristics of scaffolds composed of different ratios of type I collagen and chitosan with hyaluronic acid added was investigated to obtain the optimum conditions for the manufacture of composite porous scaffolds collagen-hyaluronan-chitosan suitable for tissue engineering applications [587]. Chitosan/nanobioactive glass ceramic (nBGC) composite scaffolds were developed by lyophilization technique and scaffolds showed adequate porosity, adequate swelling and degradation properties aside of its ability to become bioactive. Cytocompatability of chitosan/nBGC scaffolds was assessed using MT assay, direct contact test and cell attachment studies and the results indicated no signs of toxicity and cells were found to be attached to the pore walls offered by the scaffolds [588, 589].

4.3.5. Anticoagulant Activity

The phenomenon of blood clotting or coagulation involves the sequential activation of a series of serine proteinases, which results in the generation of thrombin and the subsequent thrombin catalyzed conversion of fibrinogen to insoluble fibrin. The ability of certain sulfated chitosan and glycosaminoglycans to interfere with blood clotting process is a subject of extensive medical applications [590, 591]. Compared to heparin (an efficient antithrombotic agent) [592], the sulfated chitosan has been shown to possess high anticoagulant potency. Unlike heparin, sulfated chitosan does not show anti-platelet activity, which causes excessive bleeding in patients [590]. A higher degree of sulfation has been shown to be beneficial to the anticoagulant activity, relative to the thrombin time. The disposition of sulfate groups and the chemical modification of chitosan sulfated significantly affect the process of anticoagulation [590, 591]. The anticoagulant activity of sulfated polysaccharides generally results from the interaction between the negatively charged sulfated groups and positively charged peptide sequences. *N*-acetyl groups are also shown to improve the anticoagulant activity.

4.3.6. Ophthalmology

Chitosan has been used as an alternative material to synthetic polymers in opthalmological applications not only because of its inherent biological activity, which may also have an impact in ocular therapeutics [593 - 597]. Chitosan possesses all the characteristics required for making an ideal contact lens: optical clarity, mechanical stability, and optical correction, oxygen permeability, wettability and immunological compatibility [595, 598 - 602]. Different forms of chitosan have been investigated in ophthalmology such as contact lenses, solution, coated nanocapsules, coated liposomes, microspheres, and nanoparticles [593, 598, 600, 603 - 609]. Contact lenses are made from chitosan by the technique of spin casting and the lenses are clear, hard and have other desired properties such as modulus, tensile strength, tear strength, elongation, water content, oxygen permeability, antimicrobial, and wound healing [596, 603].

Mucoadhesive chitosan was also evaluated as a potential component in ophthalmic gels to allow increased drug precorneal residence time [595]. This vehicle has been provided with cationic charge and expected to slow elimination of the drug by the lacrymal flow both by increasing the viscosity of the solution and by interacting with the negative charges of the mucus. The authors studied the effect of the MW and concentration of four types of chitosan products as parameters that may affect the eye tolerance and precorneal residence time of the formulations containing tobramycin as a therapeutic agent. They clearly demonstrate the excellent tolerance of chitosan after topical administration to the surface of the cornea with an ocular irritation test, using confocal laser scanning ophthalmoscopy combined with fluorescein staining of the cornea. At least a 3-fold increase of the corneal residence time was achieved in the presence of chitosan compared with commercial collyrium (Tobrex®), regardless of the concentration and MW [595]. A cationic chitosan is proposed for use in artificial tear formulations. Gamma scintigraphic studies have shown that chitosan formulations remain on the precorneal area as artificial tears commonly used commercial (Protagent® collyrium and Protagent-SE® unit-dose) having a 5-fold higher viscosity [600]. It has good wetting properties as well as an antibacterial effect which is desirable in the case of dry eye, which is often complicated by secondary infections. Thiolated chitosan (chitosan-*N*-acetylcysteine) product was tested in a new formulation of eye drops in two different preclinical models of dry eye mice [597]. Ocular residence time depends on the dosage, the biodistribution, a long-term irritation, and type hypersensitivity tests delayed with chitosan-*N*-acetylcysteine, eye drops formulation was studied in rabbits. It is postulated that the interaction between the thiol groups of the topical application of the chitosan-*N*-acetylcysteine and cysteine-rich mucin increased the polymer residence time (detection up to 22h) on the ocular surface and tear film stability. The results of a long-term eye irritation study in rabbits showed that the new formulation is well tolerated and non-irritating to eyes. Wowra and co-authors have developed the chitosan membrane for the preparation of scaffold providing optimal conditions for culturing cells of the corneal epithelium which could be applied in ophthalmology [602]. These materials have shown support for the adhesion, proliferation, expression of cells. *in vitro* cell culture experiments with cells of the corneal epithelium indicated that chitosan membrane prepared from mixtures provided that the laminate steady growth of epithelial cells, a good surface coverage and increased number of cell layers.

Mechanisms of action potential of chitosan as gels, coated colloidal systems, and nanoparticles to enhance the retention and biodistribution of drugs applied topically on the eye have been studied [593, 598, 599, 601, 607 - 610]. The results have shown the potential of chitosan gels to improve and prolong the retention of the drugs on the eye surface. Moreover, colloidal systems based on chitosan have been found to work as carriers transmucosal drugs, or by facilitating the transport of the drugs to the inner eye (chitosan-coated colloidal systems containing indomethacin) or their accumulation in the epithelium of the corneal/conjunctival epithelia (chitosan nanoparticles containing ciclosporin) [593]. Recently Asasutjarit and co-authors have developed new diclofenac sodium (DC) nanoparticles loaded *N*-trimethyl chitosan (DC-TMCNs) prepared by using a technique of ionic gelation, in order to enhance ocular drug biavailabiltiy and as an alternative to conventional sodium eye drops for the treatment of ophthalmic inflammation [609]. Eye irritation tests have shown that DC-TMCNs were safe for ophthalmic use. The study of the *in vivo* ophthalmic drug absorption in rabbits indicated that DC-TMCNs could improve ophthalmic bioavailability DC.

4.4. Chitosan and Its Derivatives in Biotechnological Applications

4.4.1. Enzyme Immobilization and Biosensors

Enzyme immobilization is a considerable practical utility technique, particularly to improve the catalytic potential of enzyme to specific substrate, resistance to the pH and temperature conditions, and with better stability over a considerable period of time [611 - 613]. Up to date, there are many techniques are widely used for enzyme immobilization include physical and chemical methods [613 - 618]. The physical immobilization methods depend on the adsorption [615], affinity [619, 620], and entrapment or encapsulation [614, 616, 621]. However, the chemical immobilization methods depend on the crosslinking or covalent binding [616, 622, 623]. Several other methods that are combinations of those shown or original and specific for a given carrier or an enzyme have been achieved. However, no single technique is the best for all the enzymes and their applications.

It is well known that the chitosan is an admirable material for immobilization of several enzymes, as it exhibits increased thermostability compared to the free enzyme, hence it has been extensively used in electrochemistry and biosensors in recent years [22, 624 - 628]. Chitosan proved a number of desirable characteristics of chitosan in enzyme immobilization include: good affinity to enzymes, accessibility of functional groups to rapid react with enzymes, excellent film-forming ability, high permeability, mechanical stability and rigidity, nontoxic, and biocompatible. In addition, it is also easy for the preparation in many geometric configurations which provide the permeability system and suitable for selected the surface biotransformation [622, 629 - 632]. It is widely used in different forms for enzymes immobilization technology such gels of different geometrical configurations as beads, membranes, coatings, capsules, fibers, spheres, and sponges [2, 17, 22, 572, 633, 634]. An overview of enzymes immobilized on chitosan-based materials, reported in the literatures is presented in Table 7. For example, Urease has been immobilized covalently onto chitosan-glutaraldehyde crosslinked membrane, to afford resistance to the influence of some inhibitors including thioglycolic acid, boric acid, and sodium fluoride [635]. Previous study found that the immobilized enzyme retained 94% of its original activity. The immobilized enzyme had good storage and operational stability and good reusability, properties that offer potential for practical application [636]. Likewise, resistance to mechanical stirring of D-amino acid oxidase has been obtained by enzyme immobilization on crosslinked chitosan matrix [637].

Table 7. Enzymes immobilized on chitosan-based materials.

Enzyme (EC number)	Chitosan form or its formulation	Immobilization technique	Industrial application
Acid phosphatase (3.1.3.2)	Beads, precipitate	Covalent binding	Separation, purification and recovery of enzymes, application in agriculture and medical diagnostics [634, 638 - 642].
Acetylcholinesterase (AChE)	Membrane	Adsorption followed by cross-linking	Biosensor for rapid detection of organophosphate and carbamate pesticides [643 - 646]
Alanine dehydrogenase (1.4.1.1)	Beads	Covalent binding	Determination of l-alanine (medicine) [647, 648].
Alkaline phosphatase (3.1.3.1)	Beads	Covalent binding	Hydrophobic interaction chromatography [639, 647, 649, 650].
Alkaline protease (3.4.21.62)	Beads, Powder, Microsphere	Covalent binding, adsorption	Production of laundry detergents [651 - 653].
Alcohol oxidase (1.1.3.13)	Beads	Covalent binding	Determination of ethanol [654].

(Table 7) contd.....

Enzyme (EC number)	Chitosan form or its formulation	Immobilization technique	Industrial application
Aminoacylase (3.5.1.14)	Chitosan-coated alginate beads, Electrospun non-woven cellulose-chitosan composite fibers	Gel inclusion	Production of l-phenylalanine, matrix for biological catalysts in textile industry [655, 656].
α-Amylase (3.2.1.1) and β-Amylase (3.2.1.2)	Beads	Adsorption, cross-linking	Hydrolysis of starch for glucose syrup and Production of high maltose syrup from starch [657 - 664].
Catalase (1.11.1.6)	Powder, film, membrane, particles, beads	Adsorption, covalent binding, adsorption followed by cross-linking, Gel inclusion	Removal of H_2O_2 from different food products [452, 632, 665 - 669].
Cellulase (3.2.1.4)	Beads, solutions, nanoparticles	Adsorption	Decrease in viscosity of fruit/vegetable juices and affinity chromatography [670 - 673].
α-Chymotrypsin (3.4.21.1)	Beads, nanoparticles	Adsorption	Ester and peptide synthesis [674, 675].
Galactose oxidase (1.1.3.9)	Membrane	Covalent binding	Galactose biosensor [629, 676].
β-Galactosidase (3.2.1.23)	Powder, beads, and precipitate	Adsorption, adsorption followed by cross-linking, gel inclusion	Hydrolysis of lactose (lactose-free dairy products) [676 - 683].
Glucose oxidase (1.1.3.4)	Film, membrane, particles, beads	Adsorption, covalent binding, gel inclusion	Glucose biosensors [684 - 694].
α-Glucosidase (3.2.1.20) and β-Glucosidase (3.2.1.21)	Beads, powder	Adsorption followed by cross-linking, covalent binding	Hydrolysis of maltose (food/feed additives [695 - 699].
Glutamate dehydrogenase (1.4.1.2)	Membranes, films	Covalent binding	Glutamate determination (food industry and medicine [700] and determination of ammonium in water samples [701].
Glutamate oxidase (1.4.3.11)	Membrane	Covalent binding	Glutamate biosensor [629, 702].
Invertase (3.2.1.26)	Powder, microbeads, magnetic beads, particles	Adsorption, gel inclusion	Hydrolysis of sucrose (production of invert sugar) [452, 660, 703 - 705].

(Table 7) contd.....

Enzyme (EC number)	Chitosan form or its formulation	Immobilization technique	Industrial application
Laccase (1.10.3.2)	Precipitate, microcapsules, microspheres	Adsorption followed by cross-linking, gel inclusion	Pulp and paper industry and in removal of phenols from effluents [630, 706 - 708].
Lipase (3.1.1.3)	Beads, flaks, membrane	Adsorption, covalent binding, gel inclusion	Esterifications and transesterifications and Hydrolysis of olive oil [709 - 720].
Lysozyme (3.2.1.17)	Powder, membrane	Adsorption	Affinity membrane chromatography [721 - 723].
Proteinase (3.4.24.28)	Precipitate	Adsorption followed by cross-linking	Hydrolysis of soybean protein [724, 725].
Pectinase (3.2.1.15)	Beads	Adsorption	Production of pectate oligosaccharides (inducers of flowering and antibacterial agents [726 - 728].
Pepsin (E.C. 3.4.23.1)	Beads, nanoparticles	Covalent binding	Degradation of protein [729 - 731].
Phospholipase (3.1.1.4)	Beads	Covalent binding	Lowering plasma cholesterol level [732, 733].
Proteases (3.4.groups)	Beads, membranes	Gel inclusion	Casein hydrolysate debittering [710, 734, 735].
Tannase (3.1.1.20)	Beads, precipitate	Covalent binding, gel inclusion	Hydrolysis of tea tannins [736, 737].
Transglutaminase (2.3.2.13)	Beads	Covalent binding	Deamidation of food proteins [738].
Trypsin (3.4.21.4)	Flaks, fibers, particles, membranes, microspheres	Adsorption, adsorption followed by cross-linking	Affinity purification and hydrolysis of proteins [739 - 745].
Urease (3.5.1.5)	Beads, capsules, membrane, particles	Adsorption, adsorption followed by cross-linking, covalent binding, gel inclusion	Artificial kidney, Urea biosensor, Removal of urea from beverages, food fertilizer residues [631, 635 - 637, 746 - 751].

Chitosan has also been extensively used in enzyme biosensor preparation and applications [624, 626, 627, 752]. Enzyme biosensors incorporate an immobilized enzyme on a physicochemical transducer surface which is capable of translating specific interaction of the immobilized enzyme with its corresponding binding analyte in measurable electrical signals with a concentration-dependent. Among the various forms of chitosan, chitosan films or membranes have proven to be useful transducers surface modifiers compatible with the properties of advanced sensors, at least in certain circumstances. Most of the proposed potential of chitosan in enzyme biosensor designs are complex multicomponent systems that combine plain or modified chitosan products with an specific enzyme and additional functional materials such as graphene sheets, carbon nanotubes, metal/metal oxide nanoparticles, ionic liquids, and clays [625, 692, 753 - 756]. Enzymes include oxidases, peroxidases, acetylcholine esterases, laccases, dehydrogenases, tyrosinases, reductases, hydrolases, and phosphatases have all been immobilized on chitosan formulations and used in different applications (Table 7) [22, 362, 624, 626, 643, 644, 757, 758]. In all studies, it was reported that the chitosan is as a tightly adhering and biocompatible porous polymer that provides secure fixation of the enzyme to the surface of the biosensor while allowing substrate and cofactor mobility for their continuous access to the enzyme. In addition, the most proficient technique of immobilizing enzymes on bio-based chitosan materials with avoiding leakage is the physical entrapment of the enzyme in the pores and channels of the chitosan matrix following by electrostatic and covalent enzyme bonding to the polycationic chitosan backbone [627].

Nanofibrous chitosan/PVA membrane was prepared by an electrospinning process for enzyme immobilization and was used as carrier for the immobilization of the lipase with the advantages of high enzyme loading (63.6 mg/g) and a retention of activity 49.8% [759]. Stabilities of the immobilized lipase to pH, temperature, and storage have been strengthened and the system can be used for applications of the biosensor. Amperometric biosensor for determination of glucose concentration was prepared based on a biocomposite of chitosan hydrogel, glucose oxidase, and gold nanoparticles by a direct and facile electrochemical deposition method [625]. The resulting biocomposite provided a high protection of the enzyme at experimental conditions, and this biosensor exhibited a rapid response (within 7 s) and a linear calibration range from 5.0 μM to 2.4 mM with 2.7 μM detection limit of glucose. Catalase was also immobilized into chitosan beads prepared in crosslinking solution and the result reported that the optimum pH and temperature of free and immobilized catalase were found to be 7.0 and 35°C, respectively with increase of the enzyme stability at different conditions after immobilization [632]. The K_m value of immobilized catalase (77.5 mM) was higher than that of free enzyme (35 mM). Immobilization decreased in V_{max} value from 32,000 to 122 μmol (min mg protein)$^{-1}$. Altun and Cetinus prepared chitosan beads using a crosslinking agent and the resulting beads were used in the immobilization of the pepsin. They reported that the optimum temperature range for free and immobilized pepsin was 30 to 40 and 40 to 50° C, respectively. Apparent K_m = 12.0 g/L of hemoglobin (1.56 mM tyrosine) and V_{max} = 5220 μmol (mg protein min)$^{-1}$ were obtained for the free pepsin, while apparent K_m = 20.0 g/L haemoglobin (2.16 mM tyrosine) and V_{max} = 2780 μmol (mg protein min)$^{-1}$ values were obtained for immobilized pepsin [729]. Ionic liquids have been prepared from the reaction of 1-ethyl-3-methylimidazolium hydroxide and carboxymethylated chitosan [760]. The data showed that the ionic liquids with low MW have high ionic conductivity and thermal stability. In addition, clay based biocompatible nanocomposites, chitosan, and gold nanoparticles have been prepared and used for the assembly of the horseradish peroxidase and the manufacturing of the biosensor [761]. A bioactive paper-based sensor was recently developed by Badawy and El-Aswad for detection of organophosphate and carbamate pesticides as specific inhibitors of acetylcholinesterase (AChE) [643]. Based on the colorimetric Ellman assay, the test strip comprises a paper support on which a biopolymer chitosan-gel immobilized in crosslinking with glutaraldehyde, AChE, and 5,5'-dithio bis (2-nitrobenzoic acid) (DTNB) on the paper surface and used acetylthiocholine iodide (Atchi) as a reagent outside. The test procedure consists of introducing the sample to the detection zone by immersion in a solution containing pesticides. After an incubation period, and then the paper is placed in the solution of ATChI to initiate the catalytic activity of the enzyme for the substrate, which causes a change of yellow color. The biosensor can detect organophosphate and carbamate pesticides with good limits of detection (methomyl = 6.16×10^{-4} mM and profenofos = 0.27 mM) [643].

4.4.2. Cell Immobilization

Cell immobilization has been extensively studied during the last 20 years due to the increasing potential in a wide range of fields, including medicine food, agrochemistry, environment, and cell storage [18, 611, 762 - 767]. Chitosan has been reported to use as a suitable matrix for entrapment and immobilization of living cells as it possesses good biocompatibility, nontoxicity, significant affinity to proteins, and admirable gel-forming capability [768 - 776]. There are three major methods of surface immobilization of cells: (a) physical adsorption; (b) physical 'entrapment', *i.e.* barrier systems or hydrogels; and (c) covalent attachment, *i.e.* using solid surface chemistry, hydrogels or soluble polymer conjugates [611, 763, 777 - 779]. The chitosan forming the spherical and chitosan-alginate biocatalysts and their physical characteristics such as the swelling behaviour and porous networks were studied. The applicability of different immobilization techniques for multiplying and viable cells was demonstrated [780]. A line derived pheochromocytoma cells, PC12, and two cell lines of fibroblasts, R208F and R208N.8 were encapsulated in alginate microcapsules, crosslinked chitosan microcapsules, and acrylonitrile-vinyl chloride copolymer (PAN/PVC) macrocapsules containing precipitates of chitosan as an internal matrix [772]. All three cell lines grew poorly in crosslinked chitosan microcapsules yet they have excellent viability when macrocapsules sequestered in PAN/PVC containing an internal matrix of chitosan. Chitosan nanocomposite gel and encapsulated of gold was prepared and designed for immobilization and electrochemical study of leukemic K562 cells with monitoring adhesion, proliferation and apoptosis of cells immobilized on electrodes [768]. The gel nanocomposites demonstrated a greater capacity for immobilization of cells and good biocompatibility for the preservation of the activity of immobilized living cells.

Ethanol-producing yeast cells were immobilized with cross-linked chitosan beads reached 1.2×10^{10} cells/ml-beads [773]. Although this result was almost the same as the calcium alginate, the concentration of ethanol produced with chitosan beads is about 1.5 times that of the calcium alginate. The beads chemically modified with the trimethylammonium group have a high ability to immobilize the yeast cells, but the concentration of ethanol produced was almost the same as that with the unmodified chitosan beads. Hybridoma cells were also immobilized into highly cross-linked, porous beads of chitosan-stabilized calcium alginate with viable population densities in the order of $5 \times I0^7$ cells/mL [770]. Dried baker's yeast cells were immobilized on a chitosan film crosslinked with glutaraldehyde [781]. They found that the highest amount of yeast immobilization was obtained 990 µg protein cm^{-2} film) with 5 h of incubation at 25°C and pH 6 as the optimum conditions. The growth rate of chitosan capsules immobilized with the marine cyanobacterium *Synechococcus elongatus* cells was investigated with high chlorophyll *a* and protein content in immobilized cells. They reported that this product has potential application in the removal of nitrogen and phosphorus from wastewater samples [774].

Chitosan- carboxymethylcellulose hydrogels with a fibrillar structure and characteristic pore sizes between 0.1 and 1 µm were used to immobilize yeast cells (*Saccharomyces cerevisiae*) with the cell density 1×10^9 cells/mL, about 100 times higher than that in the free culture cells. Only hydrogels with additional ionotropic gelation were sufficiently stable for fermentation continues over 10 days [771]. *Gluconobacter oxydans* was also immobilized on porous chitosan sponge with 92% activity recovery of the immobilized cells [782]. The authors reported that the morphology characterization of the immobilized cells revealed that the cells were attached to the surface of the pores (100-400 µm) which were well distributed in the chitosan sponge. Furthermore, the immobilized cells displayed enhanced pH and thermal stability compared to free cells and retained 74% of its origin activity after 12 repeated reaction cycles separated by filtration. Recently, a new immobilization process for a cell suspension of *Enterococcus faecalis* into chitosan solution mixed and cross-linked with glutaraldehyde was industrial for use in the production of the L-citrulline production [775]. A fermentation broth of *E. faecalis* cells containing the arginine deiminase has been used for the immobilization without using a step of separating cells and then compared to the immobilized cells. Cells immobilized with the fermentation broth showed an increase of the operational stability and have been used in in scale-up procedures. On an industrial scale, 210 kg of immobilized cells from 2000 L of media was used for a reaction of L-2000. After five batches, 1000 kg of l-arginine was converted to 974.6 kg of L-citrulline, yielding 733.5 kg of the final product. Consequently, *Bacillus* sp. GY19 was also immobilized on chitosan to increase cell density and facilitate lipopeptide production in the presence of palm oil as a precursor for the enzymes [783]. The immobilized cells can be reused in a fermenter stirred tank for at least 6 cycles with an optimal condition of 80 g/L of immobilized cells, 10% glycerol and 1.25% of palm oil. The highest concentration of lipopeptide observed during steady-state was 7.12 g/L. Amounts lipopeptides to foamate produced from palm oil and oil added to the media were 10.9 and 9.8 g/L, respectively.

4.4.3. Protein Purification

Several research groups have been studied the use of chitosan membranes and other forms in protein separation [784 - 786]. Different methods have been used to modify it either physically or chemically in order to improve its pore size, mechanical strength, chemical stability, hydrophilicity, and also biocompatibility [787 - 789]. A wet phase-inversion (immersion precipitation induced phase-inversion) method is a suitable technique to prepare macroporous gels with the desired morphology and pore size for protein separation [784, 787, 788, 790 - 792]. Yao and Li utilized chitosan as a run buffer additive and capillary modifier, with excellent results in capillary electrophoresis (CE) for protein separation. Their method involved permanently coating the capillary walls for CE with chitosan solution before each run to achieve the cationic analog of a regular fused silica capillary [793]. The cationic character at acidic chitosan formed a reverse electroosmotic flow (EOF) in the capillary. This capillary with an inverted EOF can be used to reduce the problem of solute-wall interaction to the positively charged proteins and the separation of the enantiomers of basic drugs [794]. The amine groups on chitosan backbone can serve as anion-exchangers and can be used for protein purification in the ion-exchange mode at pH lower than 7 which the surface is positively charged, and they can effectively adsorb proteins. Zeng and Ruckenstein prepared macroporous chitosan membranes crosslinked by ethylene glycol diglycidyl ether with controlled pore sizes and excellent mechanical properties to increase their chemical stability for protein separation [788]. Five proteins, namely ovalbumin (OVA), human serum albumin (HSA), soybean trypsin inhibitor (STI), lysozyme (LYZ) and cytochrome C were selected as model proteins to investigate their adsorption on the chitosan membranes. They found high dynamic adsorption capacities (11.6, 19 and 20.8 mg/mL membrane for HSA, OVA, and STI, respectively) and these proteins could be efficiently recovered (91-98%) from the membranes using a 1 N NaCl in 0.02 N sodium phosphate solution (pH 6) as eluent. Protein separations were also performed from binary mixtures (OVA-LYZ, HAS-cytochrome C, and STI-cytochrome C), and high purity products (~99%) were obtained [788].

Positively charged organic-inorganic hybrid membranes for ultrafiltration and separation of proteins were synthesized by mixing PVA and *N*-methyl (4-*N*, *N*-dimethylaminobenzyl) chitosan (TMBC) in combination with tetraethyl orthosilicate (TEOS) precursor by the sol-gel method [20]. The flux of water and the flux recovery ratio (*e.g.*, resistance to fouling protein) were strongly dependent membrane fraction of the TMBC. The transmission of the protein in relation to the pH and the fraction of the TMBC was investigated for OVA and LYZ and found to be controlled by size exclusion of the membrane and the charge density. The highest transmission of proteins at their isoelectric points was obtained for the membrane with 40% of TMBC and the best separation of LYZ from OVA was obtained at pH = 11 and 2 bar applied transmembrane pressure [20]. Saxena and co-authors reported, at the first time, preparation of *N*-methylene phosphonic chitosan and quaternized silica composite charged ultrafilter membranes by acid-catalyzed sol-gel methods in an aqueous medium and gelated in methanol and was used for protein separation from their mixture under pressure and electrical gradient [785]. The results showed that the introduction of highly acidic $-PO_3H_2$ or basic $-N^+(CH_3)_3$ groups caused cross-linking or covalent bonding with silica and PVA which enhanced the filtrate flux and selectivity in protein separation by several folds. 2-Hydroxypropyltrimethylammonium chloride chitosan (HACC) as quaternized product has been used as a film-forming material for a nanofiltration membrane and also has potential use for the separation of mono/divalent salts [795]. In another study, polyamide (nylon) microporous membranes were activated by bisoxiranne then linked with chitosan to enhance the hydrophilicity and increasing reactive sites [796]. In addition, polylysine (PLL) as a ligand was immobilized on nylon membranes coated with chitosan 1,1`-imidazol-carbonyldiim (CDI) activation, that has a high reactivity with various functional groups (amino, carboxyl, and hydroxyl). These membranes are useful for affinity adsorb bilirubin from bilirubin-phosphate buffer solutions and bilirubin–albumin solutions [796].Chitosan was blended with its derivative carboxymethyl chitosan (CMC) to prepare macroporous amphoteric membrane for investigation of the adsorption of OVA and LYZ with very different p*I* values [797]. The results showed that both proteins could be effectively adsorbed on the membrane, but the adsorption capacities were influenced by the pH, the protein concentration, and the CMC content in the membrane. Both OVA and LYZ were successfully separated from binary mixtures by adjusting only the pH of the feed and the desorption solutions [797]. *Consequently,* *N*-carboxymethyl chitosan (*N*-CMC) and *O*-carboxymethyl chitosan (*O*-CMC)-based amphoteric or charged membranes were prepared for protein purification [798]. The two charged membranes exhibited different natures and capacities acid/alkaline ion exchange with MW cut-off 16-19 kDa, corresponding to 0.53 and 0.48 nm pore radius at pH 7.0 for *O*-CMC and *N*-CMC, respectively. *O*-CMC membrane showed pH-tunable loaded nature and appropriate separation performance for β-casein and LYZ in alkaline media, as a representative case. Furthermore, no fouling of the membrane was detected after 50 days of operation in a protein environment.

N-Phthaloyl chitosan (NPHC) was recently prepared for the modification of poly (ether imide) (PEI) ultrafiltration membrane with higher water flux (112.2 l m^{-2} h^{-1}), higher water content (63.4%), and lower hydraulic resistance (3 kPa/l m^{-2} h^{-1}) [791]. Application studies were carried out for separation BSA, egg albumin (EA), pepsin and trypsin. The result showed that the flux and separation performances were dependent upon the NPHCs content. The significant effect of hydrophilicity of NPHCs on the fouling of PEI/NPHCs blend membranes by BSA was also obtained [791]. Sodium alginate and chitosan have a high content of chiral active sites and very good hydrophilicity therefore, they have been used for the enantioselective membranes preparation which is critical for the separation of optical isomers [799]. These membranes are applied for optical resolution of α-amino acids, especially tryptophan and tyrosine. With an increase in the degree of cross-linking, the membranes achieved better enantioselectivity. Consequently, an ion-exchange membrane matrix of poly(methacrylic acid) onto epichlorohydrin cross-linked chitosan membranes was synthesized [800]. This membrane showed high significant adsorption of trypsin (92.86 mg cm at pH 7). However, the trypsin adsorption capacity of the membrane was decreased with increasing the ionic strength. A facile and novel strategy was also developed to fabricate magnetite/graphene oxide/chitosan composite as a magnetic adsorbent for cytochrome c protein [801].

4.4.4. Biodegradable Membranes

Chitosan and its derivatives have the ability to form thin membranous films because of its high hydrophilicity and good film-forming characters (homogeneous matrix, stable structure, good permeable, good water barrier, and mechanical properties) [19, 327, 438, 802 - 804]. These films have been widely used in packaging of food products and drug delivery systems and could be replacing of the conventional synthetic plastics [481, 805 - 807]. Many techniques have been used for preparation of chitosan membranes and to improve the separation performance such as surface modification, blending, polymerization, and grafting of a selective species onto an inert film. The chitosan composite films with other polysaccharides (such as pectin, methylcellulose, gelatin and starch) or proteins have been extensively studied for improving the properties of the producing films [316, 318, 329, 808 - 811]. Gelatin and starch are very attractive biopolymers due to their abundance, biodegradability, low-cost, and functional and excellent film-forming properties [318, 330, 808, 812, 813]. However, they have low mechanical strength and dissolves relatively quickly in aqueous solution, thus limiting their practical applications. Blending of chitosan with gelatin or gelatin have been shown to be homogeneous due to good miscibility between these biopolymers leading to improve properties of the films as compared to those obtained from single polymer [326, 437, 814]. According to Xu and co-authors, the highest mechanical strength of the films occurs at the starch to chitosan ratio of 1:1 [327]. In addition, blending the PLA with chitosan improved the water barrier properties and decreased in water sensitivity of the chitosan film, however, the tensile strength and modulus of elasticity of chitosan were decreased [815, 816]. In chitosan film forming, plasticizing agents such as glycerol, sorbitol, and fatty acids are essential to overcome the weakness of the films. This weakness of the films is refers to the weak intermolecular forces of the polymer structure. Plasticizers work on the reduction of the intermolecular forces that soften the rigidity of the film structure and increase the mobility of the biopolymeric chains, thus improving the mechanical properties [323, 328, 330, 817, 818].

Chitosan membranes have been studied in different areas of applications such as dehydration of some organic solvents, separation of organic mixtures, pervaporation, ultrafiltration of biomaterials, protein adsorption/separation, and in food packaging [22, 23, 439, 784, 819]. In food applications, semi-permeable chitosan coatings decrease the loss of the moisture, respiration rate of fruit, aroma and solutes, thus increases the shelf life of agricultural products [315 - 319, 820, 821]. Chitosan coatings designed to slow antimicrobial release from the surface, thus, smaller amount of antimicrobials would be needed. They have been used on fruit and vegetables with high antimicrobial activity against several pathogens [212, 313, 315, 804, 819, 822]. In addition, these films can also use as carriers releasing of essential oils that are affected by vapor contact with pathogens [320 - 322]. Other many films based on chitosan have been prepared and used in technological applications. For example, chitosan-PVA and *N*-methylene phosphonic chitosan-PVA membranes for biomaterials separation and medical applications [574, 823, 824], a chitosan/polyacrylonitrile composite hollow fiber membrane to study the pervaporation performance of aqueous alcohol solution [825], microporous polyamide (nylon) membranes activated by bisoxirane and then bonded with chitosan for proteins separation [796], chitosan-PU/poly(*N*-isopropylacrylamide) thermosensitive membranes for wound dressing [826], composite hybrid membranes of chitosan-silica for pervaporation of methanol/dimethyl carbonate [827], quaternized chitosan-polyacrylonitrile membrane for nanofiltration [795, 828], membranes of chitosan/hybrid zeolite for use in direct methanol fuel cells [829, 830], and chitosan-trimesoyl chloride membranes for CO_2/N_2 separation, pervaporation/dehydration of isopropanol and for heavy metals removal [831, 832].

5. POSSIBLE SIDE EFFECTS OF CHITOSAN

Chitosan is not only very useful, it is also very safe and it has been used extensively in numerous [284, 833]. To determine the relative safety of various samples or foods, scientists run experiments to determine the toxic level or LD_{50}. Chitosan has been found to have an LD_{50} of over 16 g/day/kg body weight in mice [284, 834 - 836]. Chitosan is a fiber which expands to form a gel in the stomach acid environment. Problems with extremely high doses of chitosan were caused by dehydration and impaction of the stomach due to the expansion of the fiber. To put this in context, the authors compared the chitosan to the common sugars that appears indicating that chitosan is less toxic than such substances [834]. Mice are not men, for data security reasons gathered in mice is divided by 12 to get the human equivalent [837, 838]. The LD_{50} compared to man would be 1.33 g/day/kg. Given an average person weighing 150 pounds or 70 kg, which means that the toxic amount for one person would be higher than 90 g/day. Conservative, you could feel very confident below the level of 10%, or 9 g/day. Clinical studies have used amounts in the range 3-6 g/day, without adverse effects. As with any fiber, a person is well advised to drink plenty of water. Changing our diet affects our colon function. Constipation or diarrhea may occur in some people according to their individual constitutions and how the supplement chitosan was originally formulated. Although chitosan is not digestible by our enzymes, it can and is broken down by soil and many of microorganisms which makes it as environmentally friendly. This was also recognized by the US Environmental Protection Agency (EPA) when it exempted chitosan exempt from tolerance level tests [839]. Any biodegradation of chitosan by human colon microflora liberate GlcN which is itself a wonderfully beneficial nutrient for osteoarthritis patients [840]. Sometimes there is significant evidence that long-term; chitosan supplements-high doses can cause low absorption of certain essential vitamins and minerals such as calcium, magnesium, selenium and vitamins A, D, E and K. In turn, this seems to lead to a risk of osteoporosis in adults and growth retardation in children. For this reason, adults taking high doses of chitosan should also take supplements of essential fatty acids, fat soluble vitamins, and mineral supplements separate from chitosan. Taking chitosan with D- or L-ascorbic acid helps to increase the amount of fat and reduce related mineral loss

[300]. If chitosan can really prevent fat absorption, it can also prevent the absorption of other fat-soluble substances (*e.g.* soluble vitamins and essential fatty acids) and the negatively charged substance (chitosan is positively charged). However, there are still no long-term studies of the side effects of chitosan on human health that most studies have been conducted on animals. The experts also discovered other minor side effects of chitosan include breathing problems or tightness in the chest or throat, as well as chest pain and skin hives, rash or itching or swollen skin. Other possible side effects of chitosan are headaches and swollen heels and wrists. With all these possible side effects mentioned, it is no wonder that experts strongly suggest stopping taking this medicine immediately and tell your doctor if you have any of the mentioned side effects. This is because your medicine may be causing these symptoms which mean that you are allergic to products based on chitosan. So may be the best that you can protect your health before taking chitosan products in your current medication. Note that an over dose of chitosan-based products can lead to the possible side effects of chitosan and the effects depend on the amount of chitosan and the reason for taking it.

CONCLUDING REMARKS

Chitosan has unique properties, including biocompatibility, biodegradability and no toxicity to mammals. It is widely used in different fields such as biotechnology, pharmaceuticals, medicine, cosmetics and agriculture. However, poor solubility in neutral water and organic solvents with other undesirable characteristics such as low surface area and porosity are the major limiting factors in its utilization. Therefore, many scientists have been modified chitosan structure and properties producing new modified derivatives with good physicochemical and biological properties for application in different areas (agriculture, food and nutrition, biomedicine, pharmaceutics, and biotechnology). Some derivatives are suitable for agricultural applications include antimicrobial activity, seed-coating and plant growth promotion, and control of agrochemicals release. In addition, such derivatives are widely used in food and nutrition (nutritional and hypocholesterolemic effect, food packaging and antimicrobial and antioxidant activity). Other products are suitable for biomedical applications such as drug and gene delivery systems, pharmaceutical industries, wound healing and dressing, tissue engineering, anticoagulant, cholesterol-lowering, and in Ophthalmology. Recently, chitosan and its derivatives have been extensively applied in biotechnology such as enzyme and cell immobilization, biosensors, protein purification, and as biodegradable membranes.

CONFLICT OF INTEREST

The author (editor) declares no conflict of interest, financial or otherwise.

ACKNOWLEDGEMENT

The authors acknowledge and cordially thank Alexandria University Research Fund (ALEX-REP, 2010) and Misr El Kheir Foundation, Egypt: Science, Technology and Innovation (STI) Program (project code LGA05130114) to support some of this work.

REFERENCES

[1] H.K. No, and S.P. Meyers, "Preparation and characterization of chitin and chitosan-a review", *J. Aquat. Food Prod. Technol.*, vol. 4, pp. 27-52, 1995. [http://dx.doi.org/10.1300/J030v04n02_03]

[2] C. Pillai, W. Paul, and C.P. Sharma, "Chitin and chitosan polymers: Chemistry, solubility and fiber formation", *Prog. Polym. Sci.*, vol. 34, pp. 641-678, 2009. [http://dx.doi.org/10.1016/j.progpolymsci.2009.04.001]

[3] P.K. Dutta, J. Dutta, and V. Tripathi, "Chitin and chitosan: Chemistry, properties and applications", *J. Sci. Ind. Res.*, vol. 63, pp. 20-31, 2004.

[4] R.A. Muzzarelli, *Chitin.*Elsevier, 2013.

[5] J. Zikakis, W. Saylor, P. Austin, S. Hirano, and S. Takura, *Chitin and Chitosan.* The Japanese Society of Chitin and Chitosan: Tottori, 1982, p. 233.

[6] R.A. Muzzarelli, *Chitin Oxford.*Pergamon, 1977.

[7] R.A. Muzzarelli, "Natural chelating polymers; alginic acid, chitin and chitosan", In: *Natural Chelating Polymers; Alginic Acid, Chitin and Chitosan,* Pergamon Press, 1973.

[8] S. Chandrkrachang, "The application of chitin and chitosan in agriculture in Thailand", *Advances in Chitin Science,* vol. 5, pp. 458-462, 2002.

[9] L. Zhejiang, W. Jingxin, L. Yuzhu, W. Liping, and L. Wei, "Applications and prospect of chitosan in pestcide", *Pesticide Science and Administration,* vol. 6, p. 11, 2005.

[10] H. Pospieszny, H. Struszczyk, S. Chirkov, and J. Atabekov, "New applications of chitosan in agriculture", In: *Chitin world.*Wirtschaftsverlag NW: Bremerhaven, Germany, 1994, pp. 246-254.

[11] E.I. Rabea, M.E. Badawy, C.V. Stevens, G. Smagghe, and W. Steurbaut, "Chitosan as antimicrobial agent: applications and mode of action", *Biomacromolecules,* vol. 4, no. 6, pp. 1457-1465, 2003. [http://dx.doi.org/10.1021/bm034130m] [PMID: 14606868]

[12] M.E.I. Badawy, and E.I. Rabea, "A biopolymer chitosan and its derivatives as promising antimicrobial agents against plant pathogens and their applications in crop protection", *International Journal of Carbohydrate Chemistry,* vol. 2011, 2011. [http://dx.doi.org/10.1155/2011/460381]

[13] A. El Hadrami, L.R. Adam, I. El Hadrami, and F. Daayf, "Chitosan in plant protection", *Mar. Drugs,* vol. 8, no. 4, pp. 968-987, 2010. [http://dx.doi.org/10.3390/md8040968] [PMID: 20479963]

[14] M. Kong, X.G. Chen, K. Xing, and H.J. Park, "Antimicrobial properties of chitosan and mode of action: a state of the art review", *Int. J. Food Microbiol.,* vol. 144, no. 1, pp. 51-63, 2010. [http://dx.doi.org/10.1016/j.ijfoodmicro.2010.09.012] [PMID: 20951455]

[15] S. Boonlertnirun, C. Boonraung, and R. Suvanasara, "Application of chitosan in rice production", *Journal of Metals, Materials and Minerals,* vol. 18, pp. 47-52, 2008.

[16] K. Aiedeh, and M.O. Taha, "Synthesis of iron-crosslinked chitosan succinate and iron-crosslinked hydroxamated chitosan succinate and their in vitro evaluation as potential matrix materials for oral

theophylline sustained-release beads", *Eur. J. Pharm. Sci.,* vol. 13, no. 2, pp. 159-168, 2001. [http://dx.doi.org/10.1016/S0928-0987(00)00217-7] [PMID: 11297900]

[17] F. Shahidi, J.K. Arachchi, and Y-J. Jeon, "Food applications of chitin and chitosans", *Trends Food Sci. Technol.,* vol. 10, pp. 37-51, 1999. [http://dx.doi.org/10.1016/S0924-2244(99)00017-5]

[18] Q. Liu, C. Wu, H. Cai, N. Hu, J. Zhou, and P. Wang, "Cell-based biosensors and their application in biomedicine", *Chem. Rev.,* vol. 114, no. 12, pp. 6423-6461, 2014. [http://dx.doi.org/10.1021/cr2003129] [PMID: 24905074]

[19] F-L. Mi, S-S. Shyu, Y-B. Wu, S-T. Lee, J-Y. Shyong, and R-N. Huang, "Fabrication and characterization of a sponge-like asymmetric chitosan membrane as a wound dressing", *Biomaterials,* vol. 22, no. 2, pp. 165-173, 2001. [http://dx.doi.org/10.1016/S0142-9612(00)00167-8] [PMID: 11101160]

[20] M. Kumar, D. McGlade, and J. Lawler, "Functionalized chitosan derived novel positively charged organic–inorganic hybrid ultrafiltration membranes for protein separation", *RSC Advances,* vol. 4, pp. 21699-21711, 2014. [http://dx.doi.org/10.1039/C4RA02576H]

[21] V. Dodane, and V.D. Vilivalam, "Pharmaceutical applications of chitosan", *Pharm. Sci. Technol. Today,* vol. 1, pp. 246-253, 1998. [http://dx.doi.org/10.1016/S1461-5347(98)00059-5]

[22] B. Krajewska, "Application of chitin-and chitosan-based materials for enzyme immobilizations: a review", *Enzyme Microb. Technol.,* vol. 35, pp. 126-139, 2004. [http://dx.doi.org/10.1016/j.enzmictec.2003.12.013]

[23] H. Honarkar, and M. Barikani, "Applications of biopolymers I: chitosan", *Monatshefte für Chemie-Chemical Monthly,* vol. 140, pp. 1403-1420, 2009. [http://dx.doi.org/10.1007/s00706-009-0197-4]

[24] S. Nishimura, O. Kohgo, K. Kurita, and H. Kuzuhara, "Chemospecific manipulations of a rigid polysaccharide: syntheses of novel chitosan derivatives with excellent solubility in common organic solvents by regioselective chemical modifications", *Macromolecules,* vol. 24, pp. 4745-4748, 1991. [http://dx.doi.org/10.1021/ma00017a003]

[25] N.M. Alves, and J.F. Mano, *Chitosan derivatives obtained by chemical modifications for biomedical and environmental applications,* vol. 43, pp. 401-414, 2008. *International journal of biological macromolecules,* vol. 43, pp. 401-414, 2008. [http://dx.doi.org/10.1016/j.ijbiomac.2008.09.007]

[26] R.A. Venditti, J.J. Pawlak, A. Salam, and K.F. El-Tahlawy, *Modified carbohydrate-chitosan compounds, methods of making the same and methods of using the same.* Google Patents, 2015.

[27] A. Toffey, G. Samaranayake, C.E. Frazier, and W.G. Glasser, "Chitin derivatives. I. Kinetics of the heat-induced conversion of chitosan to chitin", *J. Appl. Polym. Sci.,* vol. 60, pp. 75-85, 1996. [http://dx.doi.org/10.1002/(SICI)1097-4628(19960404)60:1<75::AID-APP9>3.0.CO;2-S]

[28] H. Sashiwa, N. Kawasaki, A. Nakayama, E. Muraki, N. Yamamoto, I. Arvanitoyannis, H. Zhu, and S.-i. Aiba, *Chemical modification of chitosan 121: Synthesis of organo-soluble chitosan derivatives toward palladium absorbent for chemical plating,* pp. 598-599, 2002.*Chem. Lett.,* pp. 598-599, 2002. [http://dx.doi.org/10.1246/cl.2002.598]

[29] M. Sakamoto, H. Tseng, and K-i. Furuhata, "Regioselective chlorination of chitin with N-chlorosuccinimide-triphenylphosphine under homogeneous conditions in lithium chloride-N, N-dimethylacetamide", *Carbohydr. Res.,* vol. 265, pp. 271-280, 1994. [http://dx.doi.org/10.1016/0008-6215(94)00230-4]

[30] H. Tseng, K-i. Furuhata, and M. Sakamoto, "Bromination of regenerated chitin with N-

bromosuccinimide and triphenylphosphine under homogeneous conditions in lithium bromide-N, N-dimethylacetamide", *Carbohydr. Res.,* vol. 270, pp. 149-161, 1995. [http://dx.doi.org/10.1016/0008-6215(95)00004-D]

[31] M.E. Badawy, and E.I. Rabea, "Characterization and antimicrobial activity of water-soluble N-(--carboxybutyroyl) chitosans against some plant pathogenic bacteria and fungi", *Carbohydr. Polym.,* vol. 87, pp. 250-256, 2012. [http://dx.doi.org/10.1016/j.carbpol.2011.07.054]

[32] M.E. Badawy, and E.I. Rabea, "Synthesis and antifungal property of N-(aryl) and quaternary N-(aryl) chitosan derivatives against Botrytis cinerea", *Cellulose,* vol. 21, pp. 3121-3137, 2014. [http://dx.doi.org/10.1007/s10570-014-0333-0]

[33] K. Kurita, "Chemical modifications of chitin and chitosan", Chitin in Nature and Technology [http://dx.doi.org/10.1007/978-1-4613-2167-5_37]

[34] R.A. Muzzarelli, "Carboxymethylated chitins and chitosans", *Carbohydr. Polym.,* vol. 8, pp. 1-21, 1988. [http://dx.doi.org/10.1016/0144-8617(88)90032-X]

[35] F. Hoppe-Seyler, *Ueber chitin und cellulose.*vol. 27. Berichte der Deutschen Chemischen Gesellschaft, 1894, pp. 3329-3331.

[36] H. Braconnot, "Sur la nature des champignons", *Ann. Chim. Phys.,* vol. 79, pp. 265-304, 1811.

[37] A. Odier, *Mémoire sur la composition chimique des parties cornées des insectes,* 1823.

[38] J.L. Lassaigne, "C. r. acad. sci., Paris 16, 387 (1843)", *Ann. Chem,* vol. 48, 1843.

[39] C. Rouget, *Des substances amylacees dans le tissue des animux, specialement les Articules (Chitine).*vol. 48. Compt Rend, 1859.

[40] R.V.S. Amorim, E.S. Melo, M.G. Carneiro-da-Cunha, W.M. Ledingham, and G.M. Campos-Takaki, "Chitosan from *Syncephalastrum racemosum* used as a film support for lipase immobilization", *Bioresour. Technol.,* vol. 89, pp. 35-39, 2003. [http://dx.doi.org/10.1016/S0960-8524(03)00035-X]

[41] R.A.A. Muzzarelli, C. Jeuniaux, and G.W. Gooday, *Chitin in nature and technology,* 1986. [http://dx.doi.org/10.1007/978-1-4613-2167-5]

[42] M. Rinaudo, "Main properties and current applications of some polysaccharides as biomaterials", *Polym. Int.,* vol. 57, pp. 397-430, 2008. [http://dx.doi.org/10.1002/pi.2378]

[43] V.S. Andrade, B.B. Neto, W. Souza, and G.M. Campos-Takaki, "A factorial design analysis of chitin production by Cunninghamella elegans", *Can. J. Microbiol.,* vol. 46, no. 11, pp. 1042-1045, 2000. [http://dx.doi.org/10.1139/cjm-46-11-1042] [PMID: 11109493]

[44] G.M. Campos-Takaki, "“The fungal versatility on the copolymers chitin and chitosan production,” Chitin and Chitosan Opportunities and Challenges, India", *SSM: International Publication,* vol. 2005, pp. 69-94, 2005.

[45] W.L. Teng, E. Khor, T.K. Tan, L.Y. Lim, and S.C. Tan, "Concurrent production of chitin from shrimp shells and fungi", *Carbohydr. Res.,* vol. 332, no. 3, pp. 305-316, 2001. [http://dx.doi.org/10.1016/S0008-6215(01)00084-2] [PMID: 11376610]

[46] J. Synowiecki, and N.A. Al-Khateeb, "The recovery of protein hydrolysate during enzymatic isolation of chitin from shrimp Crangon crangon processing discards", *Food Chem.,* vol. 68, pp. 147-152, 2000. [http://dx.doi.org/10.1016/S0308-8146(99)00165-X]

[47] P.C.K. Cheung, "Dietary fiber content and composition of some cultivated edible mushroom fruiting bodies and mycelia", *J. Agric. Food Chem.,* vol. 44, pp. 468-471, 1996. [http://dx.doi.org/10.1021/jf950455l]

[48] S.M. Dietrich, "Carbohydrates from the hyphal walls of some Oomycetes", *Biochimica et Biophysica Acta (BBA)-General Subjects,* vol. 313, pp. 95-98, 1973.

[49] J. Synowiecki, and N.A.A.Q. Al-Khateeb, "Mycelia of *Mucor rouxii* as a source of chitin and chitosan", *Food Chem.,* vol. 60, pp. 605-610, 1997. [http://dx.doi.org/10.1016/S0308-8146(97)00039-3]

[50] F. Shahidi, and J. Synowiecki, "Isolation and characterization of nutrients and value-added products from snow crab (Chionoecetes opilio) and shrimp (Pandalus borealis) processing discards", *J. Agric. Food Chem.,* vol. 39, pp. 1527-1532, 1991. [http://dx.doi.org/10.1021/jf00008a032]

[51] P. Broussignac, "Chitosan: a natural polymer not well known by the industry", *Chim. Ind. Genie Chim,* vol. 99, pp. 1241-1247, 1968.

[52] K. Kurita, H. Yoshino, S-I. Nishimura, and S. Ishii, "Preparation and biodegradability of chitin derivatives having mercapto groups", *Carbohydr. Polym.,* vol. 20, pp. 239-245, 1993. [http://dx.doi.org/10.1016/0144-8617(93)90095-L]

[53] A. Domard, and M. Rinaudo, "Preparation and characterization of fully deacetylated chitosan", *International journal of biological macromolecules,* vol. 5, pp. 49-52, 1983. [http://dx.doi.org/10.1016/0141-8130(83)90078-8]

[54] P. Le Dung, M. Milas, M. Rinaudo, and J. Desbrières, "Water soluble derivatives obtained by controlled chemical modifications of chitosan", *Carbohydr. Polym.,* vol. 24, pp. 209-214, 1994. [http://dx.doi.org/10.1016/0144-8617(94)90132-5]

[55] S. Benjakul, and P. Wisitwuttikul, "Improvement of deacetylation of chitin from black tiger shrimp (Penaeus monodon) carapace and shell", *ASEAN Food J.,* vol. 9, pp. 136-136, 1994.

[56] E.S. Abdou, K.S. Nagy, and M.Z. Elsabee, "Extraction and characterization of chitin and chitosan from local sources", *Bioresour. Technol.,* vol. 99, no. 5, pp. 1359-1367, 2008. [http://dx.doi.org/10.1016/j.biortech.2007.01.051] [PMID: 17383869]

[57] H.K. No, and S.P. Meyers, *Preparation of chitin and chitosan.*vol. 1. Chitin Handbook, 1997, pp. 475-489.

[58] W.J. Jung, G.H. Jo, J.H. Kuk, Y.J. Kim, K.T. Oh, and R.D. Park, "Production of chitin from red crab shell waste by successive fermentation with Lactobacillus paracasei KCTC-3074 and Serratia marcescens FS-3", *Carbohydr. Polym.,* vol. 68, pp. 746-750, 2007. [http://dx.doi.org/10.1016/j.carbpol.2006.08.011]

[59] G-H. Jo, R-D. Park, W-J. Jung, and K. Se-Kwon, and their Derivatives, "Enzymatic production of chitin from crustacean shell waste", In: *Biological Activities and Applications,* Chitin Chitosan, R-D. Park, Oligosaccharides, Eds., vol. 2010. CRC Press Taylor & Francis Group: Boca Raton, London, New York, 2010, pp. 37-45.

[60] C. Harkin, W.M. Brück, and C. Lynch, "Isolation and identification of bacteria for the treatment of brown crab (*Cancer pagurus*) waste to produce chitinous material", *J. Appl. Microbiol.,* 2015. [http://dx.doi.org/10.1111/jam.12768]

[61] A. Sila, N. Mlaik, N. Sayari, R. Balti, and A. Bougatef, "Chitin and chitosan extracted from shrimp waste using fish proteases aided process: Efficiency of chitosan in the treatment of unhairing effluents", *J. Polym. Environ.,* vol. 22, pp. 78-87, 2014. [http://dx.doi.org/10.1007/s10924-013-0598-7]

[62] G. Mayer, and M. Sarikaya, "Rigid biological composite materials: structural examples for biomimetic design", *Experimental Mechanics,* vol. 42, pp. 395-403, 2002. [http://dx.doi.org/10.1007/BF02412144]

[63] L. Heux, J. Brugnerotto, J. Desbrières, M.F. Versali, and M. Rinaudo, "Solid state NMR for determination of degree of acetylation of chitin and chitosan", *Biomacromolecules,* vol. 1, no. 4, pp.

746-751, 2000.
[http://dx.doi.org/10.1021/bm000070y] [PMID: 11710206]

[64] A. Hirai, H. Odani, and A. Nakajima, "Determination of degree of deacetylation of chitosan by 1H NMR spectroscopy", *Polym. Bull.,* vol. 26, pp. 87-94, 1991.
[http://dx.doi.org/10.1007/BF00299352]

[65] M.R. Kasaai, "A review of several reported procedures to determine the degree of N-acetylation for chitin and chitosan using infrared spectroscopy", *Carbohydr. Polym.,* vol. 71, pp. 497-508, 2008.
[http://dx.doi.org/10.1016/j.carbpol.2007.07.009]

[66] M.R. Kasaai, "Determination of the degree of N-acetylation for chitin and chitosan by various NMR spectroscopy techniques: A review", *Carbohydr. Polym.,* vol. 79, pp. 801-810, 2010.
[http://dx.doi.org/10.1016/j.carbpol.2009.10.051]

[67] M. Lavertu, Z. Xia, A.N. Serreqi, M. Berrada, A. Rodrigues, D. Wang, M.D. Buschmann, and A. Gupta, "A validated 1H NMR method for the determination of the degree of deacetylation of chitosan", *J. Pharm. Biomed. Anal.,* vol. 32, no. 6, pp. 1149-1158, 2003.
[http://dx.doi.org/10.1016/S0731-7085(03)00155-9] [PMID: 12907258]

[68] L. Raymond, F.G. Morin, and R.H. Marchessault, "Degree of deacetylation of chitosan using conductometric titration and solid-state NMR", *Carbohydrate research,* vol. 246, pp. 331-336, 1993.
[http://dx.doi.org/10.1016/0008-6215(93)84044-7]

[69] K. Van de Velde, and P. Kiekens, "Structure analysis and degree of substitution of chitin, chitosan and dibutyrylchitin by FT-IR spectroscopy and solid state 13C NMR", *Carbohydr. Polym.,* vol. 58, pp. 409-416, 2004.
[http://dx.doi.org/10.1016/j.carbpol.2004.08.004]

[70] M.E. Badawy, and E.I. Rabea, "Synthesis and structure-activity relationship of N-(cinnamyl) chitosan analogs as antimicrobial agents", *Int. J. Biol. Macromol.,* vol. 57, pp. 185-192, 2013.
[http://dx.doi.org/10.1016/j.ijbiomac.2013.03.028] [PMID: 23511055]

[71] J. Brugnerotto, J. Lizardi, F.M. Goycoolea, W. Argüelles-Monal, J. Desbrieres, and M. Rinaudo, "An infrared investigation in relation with chitin and chitosan characterization", *Polymer,* vol. 42, pp. 3569-3580, 2001.
[http://dx.doi.org/10.1016/S0032-3861(00)00713-8]

[72] H. Sashiwa, and Y. Shigemasa, "Chemical modification of chitin and chitosan 2: preparation and water soluble property of N-acylated or N-alkylated partially deacetylated chitins", *Carbohydr. Polym.,* vol. 39, pp. 127-138, 1999.
[http://dx.doi.org/10.1016/S0144-8617(98)00167-2]

[73] M.E. Badawy, E.I. Rabea, T.M. Rogge, C.V. Stevens, G. Smagghe, W. Steurbaut, and M. Höfte, "Synthesis and fungicidal activity of new N,O-acyl chitosan derivatives", *Biomacromolecules,* vol. 5, no. 2, pp. 589-595, 2004.
[http://dx.doi.org/10.1021/bm0344295] [PMID: 15003025]

[74] A. Baxter, M. Dillon, K.D.A. Taylor, and G.A.F. Roberts, "Improved method for ir determination of the degree of N-acetylation of chitosan", *International journal of biological macromolecules,* vol. 14, pp. 166-169, 1992.

[75] R.A. Muzzarelli, F. Tanfani, G. Scarpini, and G. Laterza, "The degree of acetylation of chitins by gas chromatography and infrared spectroscopy", *J. Biochem. Biophys. Methods,* vol. 2, no. 5, pp. 299-306, 1980.
[http://dx.doi.org/10.1016/0165-022X(80)90054-8] [PMID: 7052947]

[76] Y. Shigemasa, H. Matsuura, H. Sashiwa, and H. Saimato, "An improved IR spectroscopic determination of degree of deacetylation of chitin", *Advances in Chitin Science,* vol. 1, pp. 204-209, 1996.

[77] S.C. Tan, E. Khor, T.K. Tan, and S.M. Wong, "The degree of deacetylation of chitosan: advocating the

first derivative UV-spectrophotometry method of determination", *Talanta,* vol. 45, no. 4, pp. 713-719, 1998.
[http://dx.doi.org/10.1016/S0039-9140(97)00288-9] [PMID: 18967053]

[78] M.E. Badawy, E.I. Rabea, T.M. Rogge, C.V. Stevens, G. Smagghe, W. Steurbaut, and M. Höfte, "Synthesis and fungicidal activity of new N,O-acyl chitosan derivatives", *Biomacromolecules,* vol. 5, no. 2, pp. 589-595, 2004.
[http://dx.doi.org/10.1021/bm0344295] [PMID: 15003025]

[79] D. Liu, Y. Wei, P. Yao, and L. Jiang, "Determination of the degree of acetylation of chitosan by UV spectrophotometry using dual standards", *Carbohydrate research,* vol. 314, pp. 782-785, 2006.
[http://dx.doi.org/10.1016/j.carres.2006.01.008]

[80] M. Bohdanecky, and J. Kovar, *Viscosity of polymer solutions.* Elsevier Scientific Pub. Co., 1982.

[81] W.W. Yau, J.J. Kirkland, and D.D. Bly, *Modern size-exclusion liquid chromatography,* vol. 1979, 1979.

[82] R.A.A. Muzzarelli, C. Lough, and M. Emanuelli, "The molecular weight of chitosans studied by laser light-scattering", *Carbohydrate research,* vol. 164, pp. 433-442, 1987.
[http://dx.doi.org/10.1016/0008-6215(87)80146-5]

[83] C. Schatz, C. Viton, T. Delair, C. Pichot, and A. Domard, "Typical physicochemical behaviors of chitosan in aqueous solution", *Biomacromolecules,* vol. 4, no. 3, pp. 641-648, 2003.
[http://dx.doi.org/10.1021/bm025724c] [PMID: 12741780]

[84] G.G. Allan, and M. Peyron, "Molecular weight manipulation of chitosan II: prediction and control of extent of depolymerization by nitrous acid", *Carbohydrate research,* vol. 277, pp. 273-282, 1995.
[http://dx.doi.org/10.1016/0008-6215(95)00208-B]

[85] G.G. Allan, and M. Peyron, "Molecular weight manipulation of chitosan. I: Kinetics of depolymerization by nitrous acid", *Carbohydr. Res.,* vol. 277, no. 2, pp. 257-272, 1995.
[http://dx.doi.org/10.1016/0008-6215(95)00207-A] [PMID: 8556735]

[86] P.J. Flory, *Principles of polymer chemistry.*Cornell University Press, 1953.

[87] C. Tanford, *Physical chemistry of macromolecules,* 1961.

[88] M. Rinaudo, and A. Domard, "Solution properties of chitosan", *Chitin and Chitosan,* vol. 1989, pp. 71-86, 1989.

[89] W. Wang, S.Q. Bo, S.Q. Li, and W. Qin, "Determination of the Mark-Houwink equation for chitosans with different degrees of deacetylation", *Int. J. Biol. Macromol.,* vol. 13, no. 5, pp. 281-285, 1991.
[http://dx.doi.org/10.1016/0141-8130(91)90027-R] [PMID: 1801901]

[90] R.N. Tharanathan, and F.S. Kittur, *Chitin-the undisputed biomolecule of great potential,* 2003.
[http://dx.doi.org/10.1080/10408690390826455]

[91] M. Rinaudo, "Chitin and chitosan: properties and applications", *Prog. Polym. Sci.,* vol. 31, pp. 603-632, 2006.
[http://dx.doi.org/10.1016/j.progpolymsci.2006.06.001]

[92] M. Rinaudo, G. Pavlov, and J. Desbrieres, "Solubilization of chitosan in strong acid medium", *Int. J. Polym. Anal. Charact.,* vol. 5, pp. 267-276, 1999.

[93] G.G. Maghami, and G.A. Roberts, "Evaluation of the viscometric constants for chitosan", *Makromol. Chem.,* vol. 189, pp. 195-200, 1988.
[http://dx.doi.org/10.1002/macp.1988.021890118]

[94] I. Younes, and M. Rinaudo, "Chitin and chitosan preparation from marine sources. Structure, properties and applications", *Mar. Drugs,* vol. 13, no. 3, pp. 1133-1174, 2015.
[http://dx.doi.org/10.3390/md13031133] [PMID: 25738328]

[95] J. Desbrières, C. Martinez, and M. Rinaudo, "Hydrophobic derivatives of chitosan: characterization

and rheological behaviour", *Int. J. Biol. Macromol.,* vol. 19, no. 1, pp. 21-28, 1996. [http://dx.doi.org/10.1016/0141-8130(96)01095-1] [PMID: 8782715]

[96] V.A. Spinelli, M. Laranjeira, and V.T. Fávere, "Preparation and characterization of quaternary chitosan salt: adsorption equilibrium of chromium (VI) ion", *React. Funct. Polym.,* vol. 61, pp. 347-352, 2004. [http://dx.doi.org/10.1016/j.reactfunctpolym.2004.06.010]

[97] R.A. Muzzarelli, and F. Tanfani, "N-(o-carboxybenzyl) chitosan, N-carboxymethyl chitosan and dithiocarbamate chitosan: new chelating derivatives of chitosan", *Pure Appl. Chem.,* vol. 54, pp. 2141-2150, 1982. [http://dx.doi.org/10.1351/pac198254112141]

[98] S. Aiba, "Studies on chitosan: 2. Solution stability and reactivity of partially N-acetylated chitosan derivatives in aqueous media", *Int. J. Biol. Macromol.,* vol. 11, no. 4, pp. 249-252, 1989. [http://dx.doi.org/10.1016/0141-8130(89)90077-9] [PMID: 2489089]

[99] E. Loubaki, M. Ourevitch, and S. Sicsic, "Chemical modification of chitosan by glycidyl trimethylammonium chloride. characterization of modified chitosan by 13C-and 1H-NMR spectroscopy", *Eur. Polym. J.,* vol. 27, pp. 311-317, 1991. [http://dx.doi.org/10.1016/0014-3057(91)90111-Z]

[100] H. Sashiwa, N. Kawasaki, A. Nakayama, E. Muraki, N. Yamamoto, H. Zhu, H. Nagano, Y. Omura, H. Saimoto, Y. Shigemasa, and S. Aiba, "Chemical modification of chitosan. 13.(1) Synthesis of organosoluble, palladium adsorbable, and biodegradable chitosan derivatives toward the chemical plating on plastics", *Biomacromolecules,* vol. 3, no. 5, pp. 1120-1125, 2002. [http://dx.doi.org/10.1021/bm0200478] [PMID: 12217062]

[101] H. Sashiwa, N. Kawasaki, A. Nakayama, E. Muraki, N. Yamamoto, and S. Aiba, "Chemical modification of chitosan. 14:(1) Synthesis of water-soluble chitosan derivatives by simple acetylation", *Biomacromolecules,* vol. 3, no. 5, pp. 1126-1128, 2002. [http://dx.doi.org/10.1021/bm0200480] [PMID: 12217063]

[102] H. Sashiwa, N. Kawasaki, A. Nakayama, E. Muraki, H. Yajima, N. Yamamori, Y. Ichinose, J. Sunamoto, and S.-i. Aiba, "Chemical modification of chitosan. Part 15: Synthesis of novel chitosan derivatives by substitution of hydrophilic amine using N-carboxyethylchitosan ethyl ester as an intermediate", *Carbohydrate research,* vol. 338, no. 5, pp. 557-561, 2003. [http://dx.doi.org/10.1016/S0008-6215(02)00492-5]

[103] E-R. Kenawy, F.I. Abdel-Hay, M.M. Eldin, T. Tamer, and E.M. Abo-Elghit, "Novel aminated chitosan-aromatic aldehydes schiff bases: Synthesis, characterization and bio-evaluation", *Int. J.,* vol. 3, pp. 563-572, 2015.

[104] S. Hirano, Y. Yamaguchi, and M. Kamiya, "Novel N-saturated-fatty-acyl derivatives of chitosan soluble in water and in aqueous acid and alkaline solutions", *Carbohydr. Polym.,* vol. 48, pp. 203-207, 2002. [http://dx.doi.org/10.1016/S0144-8617(01)00243-0]

[105] Y. Kato, H. Onishi, and Y. Machida, "N-succinyl-chitosan as a drug carrier: water-insoluble and water-soluble conjugates", *Biomaterials,* vol. 25, no. 5, pp. 907-915, 2004. [http://dx.doi.org/10.1016/S0142-9612(03)00598-2] [PMID: 14609679]

[106] M.E. Badawy, and E.I. Rabea, "Synthesis and antifungal property of N-(aryl) and quaternary N-(aryl) chitosan derivatives against Botrytis cinerea", *Cellulose,* vol. 21, pp. 3121-3137, 2014. [http://dx.doi.org/10.1007/s10570-014-0333-0]

[107] W. Sajomsang, S. Tantayanon, V. Tangpasuthadol, M. Thatte, and W.H. Daly, "Synthesis and characterization of N-aryl chitosan derivatives", *Int. J. Biol. Macromol.,* vol. 43, no. 2, pp. 79-87, 2008. [http://dx.doi.org/10.1016/j.ijbiomac.2008.03.010] [PMID: 18514305]

[108] M.E.I. Badawy, "Chemical modification of chitosan: Synthesis and biological activity of new

heterocyclic chitosan derivatives", *Polym. Int.,* vol. 57, pp. 254-261, 2008. [http://dx.doi.org/10.1002/pi.2333]

[109] R.A. Muzzarelli, and F. Tanfani, "The< i> N</i>-permethylation of chitosan and the preparation of N-trimethyl chitosan iodide", *Carbohydr. Polym.,* vol. 5, pp. 297-307, 1985. [http://dx.doi.org/10.1016/0144-8617(85)90037-2]

[110] E.I. Rabea, M.E. Badawy, W. Steurbaut, and C.V. Stevens, "in vitro assessment of N-(benzyl) chitosan derivatives against some plant pathogenic bacteria and fungi", *Eur. Polym. J.,* vol. 45, pp. 237-245, 2009. [http://dx.doi.org/10.1016/j.eurpolymj.2008.10.021]

[111] M. Rinaudo, and J. Reguant, *Polysaccharide derivatives.* Natural Polymers and Agrofibers Composites, 2000, pp. 15-39.

[112] H.-C. Ge, and D.-K. Luo, "Preparation of carboxymethyl chitosan in aqueous solution under microwave irradiation", *Carbohydrate research,* vol. 34, pp. 13-39, 2005. [http://dx.doi.org/10.1016/j.carres.2005.02.025]

[113] S. Allen, C. Carolan, H. Blair, and G. Mckay, *N, O-Carboxymethyl chitosan, a water soluble derivative.*Chem. E. Res. Conf: Cambridge, 1991.

[114] S-C. Chen, Y-C. Wu, F-L. Mi, Y-H. Lin, L-C. Yu, and H-W. Sung, "A novel pH-sensitive hydrogel composed of N,O-carboxymethyl chitosan and alginate cross-linked by genipin for protein drug delivery", *J. Control. Release,* vol. 96, no. 2, pp. 285-300, 2004. [http://dx.doi.org/10.1016/j.jconrel.2004.02.002] [PMID: 15081219]

[115] C.H. Kim, and K.S. Choi, "Synthesis and properties of carboxyalkyl chitosan derivatives", *J. Ind. Eng. Chem.,* vol. 4, pp. 19-25, 1998.

[116] A. Yoshida, K. Kuroda, Y. Kurauchi, T. Inoue, and K. Ohga, "N-(1-carboxyalkyl) chitosan as a chiral selector", *Kichin, Kitosan Kenkyu,* vol. 5, pp. 188-189, 1999.

[117] Y. Shigemasa, A. Ishida, H. Sashiwa, H. Saimoto, Y. Okamoto, and S. Minami, "Synthesis of a new chitin derivative,(1-carboxyethyl) chitosan", *Chem. Lett.,* vol. 1995, pp. 623-624, 1995. [http://dx.doi.org/10.1246/cl.1995.623]

[118] T. Seo, Y. Ikeda, K. Torada, Y. Nakata, and Y. Shimomura, "Synthesis of N, O-acylated chitosan and its sorptivity", *Chitin and Chitosan Research,* vol. 7, pp. 212-213, 2001.

[119] H. Sashiwa, Y. Shigemasa, and R. Roy, "Homogeneous N, O-acylation of chitosan in dimethyl sulfoxide with cyclic acid anhydrides", *Chem. Lett.,* vol. •••, pp. 1186-1187, 2000. [http://dx.doi.org/10.1246/cl.2000.1186]

[120] S. Hirano, and T. Moriyasu, "Some novel N-(carboxyacyl) chitosan filaments", *Carbohydr. Polym.,* vol. 55, pp. 245-248, 2004. [http://dx.doi.org/10.1016/j.carbpol.2003.07.003]

[121] G. Lang, G. Maresch, H. Wendel, E. Konrad, H-R. Lenz, and J. Titze, *Cosmetic compositions based upon n-hydroxypropyl-chitosans, new n-hydroxypropyl-chitosans, as well as processes for the production thereof.*1989. Google Patents

[122] P.K. Swain, M. Dash, and P. Nayak, "Synthesis and characterization of chitosan-based novel superabsorbent hydrogel", *Middle-East Journal of Scientific Research,* vol. 23, pp. 259-276, 2015.

[123] H. Tan, R. Ma, C. Lin, Z. Liu, and T. Tang, "Quaternized chitosan as an antimicrobial agent: antimicrobial activity, mechanism of action and biomedical applications in orthopedics", *Int. J. Mol. Sci.,* vol. 14, no. 1, pp. 1854-1869, 2013. [http://dx.doi.org/10.3390/ijms14011854] [PMID: 23325051]

[124] H. Feng, W. Xia, C. Shan, T. Zhou, W. Cai, and W. Zhang, "Quaternized chitosan oligomers as novel elicitors inducing protection against B. cinerea in Arabidopsis", *Int. J. Biol. Macromol.,* vol. 72, pp. 364-369, 2015.

[http://dx.doi.org/10.1016/j.ijbiomac.2014.06.060] [PMID: 25034756]

[125] D. de Britto, R. Celi Goy, S.P. Campana Filho, and O.B. Assis, "Quaternary salts of chitosan: history, antimicrobial features, and prospects", *International Journal of Carbohydrate Chemistry,* vol. 2011, 2011.
[http://dx.doi.org/10.1155/2011/312539]

[126] A. Gamzazade, A. Sklyar, S. Nasibov, I. Sushkov, A. Shashkov, and Y. Knirel, "Structural features of sulfated chitosans", *Carbohydr. Polym.,* vol. 34, pp. 113-116, 1997.
[http://dx.doi.org/10.1016/S0144-8617(97)00067-2]

[127] K.R. Holme, and A.S. Perlin, "Chitosan N-sulfate. A water-soluble polyelectrolyte", *Carbohydrate research,* vol. 302, pp. 7-12, 1997.
[http://dx.doi.org/10.1016/S0008-6215(97)00117-1]

[128] J.-Y. Je, P.-J. Park, and S.-K. Kim, "Prolyl endopeptidase inhibitory activity of chitosan sulfates with different degree of deacetylation", *Carbohydr. Polym.,* vol. 60, pp. 553-556, 2005.
[http://dx.doi.org/10.1016/j.carbpol.2005.03.007]

[129] V. Ramos, N. Rodrıguez, M. Dıaz, M. Rodrıguez, A. Heras, and E. Agullo, "N-methylene phosphonic chitosan. Effect of preparation methods on its properties", *Carbohydr. Polym.,* vol. 52, pp. 39-46, 2003.
[http://dx.doi.org/10.1016/S0144-8617(02)00264-3]

[130] A. Heras, N. Rodriguez, V. Ramos, and E. Agullo, "N-methylene phosphonic chitosan: a novel soluble derivative", *Carbohydr. Polym.,* vol. 44, pp. 1-8, 2001.
[http://dx.doi.org/10.1016/S0144-8617(00)00195-8]

[131] N. Nishi, Muzzarelli et al., "Preparation and characterization of phosphorylated chitin and chitosan", In: *Chitin in Nature and Technology* Plenum Press: New York, 1986.

[132] X. Wang, J. Ma, Y. Wang, and B. He, "Structural characterization of phosphorylated chitosan and their applications as effective additives of calcium phosphate cements", *Biomaterials,* vol. 22, no. 16, pp. 2247-2255, 2001.
[http://dx.doi.org/10.1016/S0142-9612(00)00413-0] [PMID: 11456064]

[133] I.F. Amaral, P.L. Granja, and M.A. Barbosa, "Chemical modification of chitosan by phosphorylation: an XPS, FT-IR and SEM study", *J. Biomater. Sci. Polym. Ed.,* vol. 16, no. 12, pp. 1575-1593, 2005.
[http://dx.doi.org/10.1163/156856205774576736] [PMID: 16366338]

[134] R. Jayakumar, N. Selvamurugan, S.V. Nair, S. Tokura, and H. Tamura, "Preparative methods of phosphorylated chitin and chitosanan overview", *Int. J. Biol. Macromol.,* vol. 43, no. 3, pp. 221-225, 2008.
[http://dx.doi.org/10.1016/j.ijbiomac.2008.07.004] [PMID: 18657571]

[135] K. Wang, and Q. Liu, "Chemical structure analyses of phosphorylated chitosan", *Carbohydr. Res.,* vol. 386, pp. 48-56, 2014.
[http://dx.doi.org/10.1016/j.carres.2013.12.021] [PMID: 24486421]

[136] P. Dadhich, B. Das, and S. Dhara, "Microwave assisted rapid synthesis of N-methylene phosphonic chitosan via Mannich-type reaction", *Carbohydr. Polym.,* vol. 133, pp. 345-352, 2015.
[http://dx.doi.org/10.1016/j.carbpol.2015.06.105] [PMID: 26344290]

[137] C. Zhang, Q. Ping, H. Zhang, and J. Shen, "Synthesis and characterization of water-soluble O-succinyl-chitosan", *Eur. Polym. J.,* vol. 39, pp. 1629-1634, 2003.
[http://dx.doi.org/10.1016/S0014-3057(03)00068-5]

[138] C.E. Kast, and A. Bernkop-Schnürch, "Thiolated polymers-thiomers: development and *in vitro* evaluation of chitosan–thioglycolic acid conjugates", *Biomaterials,* vol. 22, pp. 2345-2352, 2001.
[http://dx.doi.org/10.1016/S0142-9612(00)00421-X]

[139] A. Bernkop-Schnürch, M. Hornof, and D. Guggi, "Thiolated chitosans", *Eur. J. Pharm. Biopharm.,* vol. 57, no. 1, pp. 9-17, 2004.

[http://dx.doi.org/10.1016/S0939-6411(03)00147-4] [PMID: 14729077]

[140] A. Bernkop-Schnürch, M. Hornof, and T. Zoidl, "Thiolated polymersthiomers: synthesis and in vitro evaluation of chitosan-2-iminothiolane conjugates", *Int. J. Pharm.,* vol. 260, no. 2, pp. 229-237, 2003. [http://dx.doi.org/10.1016/S0378-5173(03)00271-0] [PMID: 12842342]

[141] A. Bernkop-Schnürch, A.E. Clausen, and M. Hnatyszyn, "Thiolated polymers: synthesis and in vitro evaluation of polymer-cysteamine conjugates", *Int. J. Pharm.,* vol. 226, no. 1-2, pp. 185-194, 2001. [http://dx.doi.org/10.1016/S0378-5173(01)00807-9] [PMID: 11532581]

[142] J.H. Park, Y.W. Cho, H. Chung, I.C. Kwon, and S.Y. Jeong, "Synthesis and characterization of sugar-bearing chitosan derivatives: aqueous solubility and biodegradability", *Biomacromolecules,* vol. 4, pp. 1087-1091, 2003. [http://dx.doi.org/10.1021/bm034094r]

[143] L.D. Hall, and M. Yalpani, "Formation of branched-chain, soluble polysaccharides from chitosan", *J. Chem. Soc. Chem. Commun.,* pp. 1153-1154, 1980. [http://dx.doi.org/10.1039/c39800001153]

[144] M. Yalpani, and L.D. Hall, "Some chemical and analytical aspects of polysaccharide modifications. III. Formation of branched-chain, soluble chitosan derivatives", *Macromolecules,* vol. 17, pp. 272-281, 1984. [http://dx.doi.org/10.1021/ma00133a003]

[145] K.R. Holme, and L. Hall, "Chitosan derivatives bearing C10-alkyl glycoside branches: a temperature-induced gelling polysaccharide", *Macromolecules,* vol. 24, pp. 3828-3833, 1991. [http://dx.doi.org/10.1021/ma00013a014]

[146] K. Kurita, "Chitin and chitosan: functional biopolymers from marine crustaceans", *Mar. Biotechnol. (NY),* vol. 8, no. 3, pp. 203-226, 2006. [http://dx.doi.org/10.1007/s10126-005-0097-5] [PMID: 16532368]

[147] A. Domard, and N. Cartier, "Glucosamine oligomers: 1. Preparation and characterization", *Int. J. Biol. Macromol.,* vol. 11, no. 5, pp. 297-302, 1989. [http://dx.doi.org/10.1016/0141-8130(89)90023-8] [PMID: 2489094]

[148] S-K. Kim, and N. Rajapakse, "Enzymatic production and biological activities of chitosan oligosaccharides (COS): A review", *Carbohydr. Polym.,* vol. 62, pp. 357-368, 2005. [http://dx.doi.org/10.1016/j.carbpol.2005.08.012]

[149] R.C. Goy, D.d. Britto, and O.B. Assis, "A review of the antimicrobial activity of chitosan", *Polímeros,* vol. 19, pp. 241-247, 2009. [http://dx.doi.org/10.1590/S0104-14282009000300013]

[150] K. Xing, X. Zhu, X. Peng, and S. Qin, "Chitosan antimicrobial and eliciting properties for pest control in agriculture: a review", *Agron. Sustain. Dev.,* vol. 35, pp. 569-588, 2015. [http://dx.doi.org/10.1007/s13593-014-0252-3]

[151] M.E. Badawy, and E.I. Rabea, "A biopolymer chitosan and its derivatives as promising antimicrobial agents against plant pathogens and their applications in crop protection", *International Journal of Carbohydrate Chemistry,* 2011. [http://dx.doi.org/10.1155/2011/460381]

[152] S. Bautista-Baños, A. Hernández-Lauzardo, M. Velázquez-del Valle, M. Hernández-López, E.A. Barka, E. Bosquez-Molina, and C. Wilson, "Chitosan as a potential natural compound to control pre and postharvest diseases of horticultural commodities", *Crop Prot.,* vol. 25, pp. 108-118, 2006. [http://dx.doi.org/10.1016/j.cropro.2005.03.010]

[153] I.M. Helander, E-L. Nurmiaho-Lassila, R. Ahvenainen, J. Rhoades, and S. Roller, "Chitosan disrupts the barrier properties of the outer membrane of gram-negative bacteria", *Int. J. Food Microbiol.,* vol. 71, no. 2-3, pp. 235-244, 2001. [http://dx.doi.org/10.1016/S0168-1605(01)00609-2] [PMID: 11789941]

[154] I. Younes, S. Sellimi, M. Rinaudo, K. Jellouli, and M. Nasri, "Influence of acetylation degree and molecular weight of homogeneous chitosans on antibacterial and antifungal activities", *Int. J. Food Microbiol.,* vol. 185, pp. 57-63, 2014. [http://dx.doi.org/10.1016/j.ijfoodmicro.2014.04.029] [PMID: 24929684]

[155] Y. Wang, L. Li, B. Li, G. Wu, Q. Tang, M. Ibrahim, H. Li, G. Xie, and G. Sun, "Action of chitosan against Xanthomonas pathogenic bacteria isolated from Euphorbia pulcherrima", *Molecules,* vol. 17, no. 6, pp. 7028-7041, 2012. [http://dx.doi.org/10.3390/molecules17067028] [PMID: 22678416]

[156] B. Li, X. Wang, R. Chen, W. Huangfu, and G. Xie, "Antibacterial activity of chitosan solution against Xanthomonas pathogenic bacteria isolated from Euphorbia pulcherrima", *Carbohydr. Polym.,* vol. 72, pp. 287-292, 2008. [http://dx.doi.org/10.1016/j.carbpol.2007.08.012]

[157] C. Yang, B. Li, M. Ge, K. Zhou, Y. Wang, J. Luo, M. Ibrahim, G. Xie, and G. Sun, "Inhibitory effect and mode of action of chitosan solution against rice bacterial brown stripe pathogen Acidovorax avenae subsp. avenae RS-1", *Carbohydr. Res.,* vol. 391, pp. 48-54, 2014. [http://dx.doi.org/10.1016/j.carres.2014.02.025] [PMID: 24785387]

[158] K.U. Kakar, Z. Nawaz, Z. Cui, A.A. Almoneafy, B. Zhu, and G-L. Xie, "Characterizing the mode of action of Brevibacillus laterosporus B4 for control of bacterial brown strip of rice caused by A. avenae subsp. avenae RS-1", *World J. Microbiol. Biotechnol.,* vol. 30, no. 2, pp. 469-478, 2014. [http://dx.doi.org/10.1007/s11274-013-1469-z] [PMID: 23990042]

[159] B. Li, Y. Shi, C. Shan, Q. Zhou, M. Ibrahim, Y. Wang, G. Wu, H. Li, G. Xie, and G. Sun, "Effect of chitosan solution on the inhibition of Acidovorax citrulli causing bacterial fruit blotch of watermelon", *J. Sci. Food Agric.,* vol. 93, no. 5, pp. 1010-1015, 2013. [http://dx.doi.org/10.1002/jsfa.5812] [PMID: 23400827]

[160] M.E. Badawy, E.I. Rabea, and N.E. Taktak, "Antimicrobial and inhibitory enzyme activity of N-(benzyl) and quaternary N-(benzyl) chitosan derivatives on plant pathogens", *Carbohydr. Polym.,* vol. 111, pp. 670-682, 2014. [http://dx.doi.org/10.1016/j.carbpol.2014.04.098] [PMID: 25037402]

[161] M.E. Badawy, and E.I. Rabea, "Preparation and antimicrobial activity of O-(benzoyl) chitosan derivatives against some plant pathogens", *Afr. J. Microbiol. Res.,* vol. 7, pp. 2259-2268, 2013. [http://dx.doi.org/10.5897/AJMR12.1185]

[162] M.E. Badawy, and E.I. Rabea, "Synthesis and structure-activity relationship of N-(cinnamyl) chitosan analogs as antimicrobial agents", *Int. J. Biol. Macromol.,* vol. 57, pp. 185-192, 2013. [http://dx.doi.org/10.1016/j.ijbiomac.2013.03.028] [PMID: 23511055]

[163] M. Badawy, "Effect of depolymerization degree of the natural biopolymer chitosan on some plant pathogenic bacteria and fungi", *Journal of Pest Control and Environmental Sciences,* vol. 15, pp. 69-85, 2007.

[164] M.E. Badawy, "Structure and antimicrobial activity relationship of quaternary N-alkyl chitosan derivatives against some plant pathogens", *J. Appl. Polym. Sci.,* vol. 117, pp. 960-969, 2010. [http://dx.doi.org/10.1002/app.31492]

[165] E.I. Rabea, and W. Steurbaut, "Chemically modified chitosans as antimicrobial agents against some plant pathogenic bacteria and fungi", *Plant Prot. Sci.,* vol. 46, pp. 149-158, 2010.

[166] S. Algam, G. Xie, B. Li, S. Yu, T. Su, and J. Larsen, "Effects of Paenibacillus strains and chitosan on plant growth promotion and control of Ralstonia wilt in tomato", *J. Plant Pathol.,* pp. 593-600, 2010.

[167] L.M. Kiirika, F. Stahl, and K. Wydra, "Phenotypic and molecular characterization of resistance induction by single and combined application of chitosan and silicon in tomato against Ralstonia solanacearum", *Physiol. Mol. Plant Pathol.,* vol. 81, pp. 1-12, 2013. [http://dx.doi.org/10.1016/j.pmpp.2012.11.002]

[168] C.M. Ryu, J.F. Murphy, M.S. Reddy, and J.W. Kloepper, "A two-strain mixture of rhizobacteria elicits induction of systemic resistance against *Pseudomonas syringae* and Cucumber mosaic virus coupled to promotion of plant growth on *Arabidopsis thaliana* ", *J. Microbiol. Biotechnol.*, vol. 17, no. 2, pp. 280-286, 2007.
[PMID: 18051759]

[169] A.Y. Mansilla, L. Albertengo, M.S. Rodríguez, A. Debbaudt, A. Zúñiga, and C.A. Casalongué, "Evidence on antimicrobial properties and mode of action of a chitosan obtained from crustacean exoskeletons on Pseudomonas syringae pv. tomato DC3000", *Appl. Microbiol. Biotechnol.*, vol. 97, no. 15, pp. 6957-6966, 2013.
[http://dx.doi.org/10.1007/s00253-013-4993-8] [PMID: 23703326]

[170] A.A. Tayel, S. Moussa, K. Opwis, D. Knittel, E. Schollmeyer, and A. Nickisch-Hartfiel, "Inhibition of microbial pathogens by fungal chitosan", *Int. J. Biol. Macromol.*, vol. 47, no. 1, pp. 10-14, 2010.
[http://dx.doi.org/10.1016/j.ijbiomac.2010.04.005] [PMID: 20416334]

[171] C. Mohamed, K.A. Clementine, M. Didier, L. Gérard, and D-C. Noëlle, "Antimicrobial and physical properties of edible chitosan films enhanced by lactoperoxidase system", *Food Hydrocoll.*, vol. 30, pp. 576-580, 2013.
[http://dx.doi.org/10.1016/j.foodhyd.2012.07.018]

[172] D. Coqueiro, and R. Di Piero, "Antibiotic activity against Xanthomonas gardneri and protection of tomato plants by chitosan", *J. Plant Pathol.*, pp. 337-344, 2011.

[173] B. Li, B. Liu, C. Shan, M. Ibrahim, Y. Lou, Y. Wang, G. Xie, H.Y. Li, and G. Sun, "Antibacterial activity of two chitosan solutions and their effect on rice bacterial leaf blight and leaf streak", *Pest Manag. Sci.*, vol. 69, no. 2, pp. 312-320, 2013.
[http://dx.doi.org/10.1002/ps.3399] [PMID: 23129534]

[174] F. Cavalcanti, M. Resende, C. Carvalho, J. Silveira, and J. Oliveira, "An aqueous suspension of *Crinipellis perniciosa* mycelium activates tomato defence responses against *Xanthomonas vesicatoria*", *Crop Prot.*, vol. 26, pp. 729-738, 2007.
[http://dx.doi.org/10.1016/j.cropro.2006.06.012]

[175] M. Badawy, S. Ahmed, and E. Rabea, "Bactericidal and fungicidal activities of different molecular weight chitosan samples", *Journal of Pest Control and Environmental Sciences,* vol. 14, pp. 19-34, 2006.

[176] Y. Uchida, "Preparation of chitosan oligomers with purified chitosanase and its application", *Proc. 4th Int. Conf. Chitin/Chitosan,* 1989

[177] M.E. Badawy, "Chemical modification of chitosan: Synthesis and biological activity of new heterocyclic chitosan derivatives", *Polym. Int.,* vol. 57, pp. 254-261, 2008.
[http://dx.doi.org/10.1002/pi.2333]

[178] A.N. Hernández-Lauzardo, M.G. Velázquez-del Valle, and M.G. Guerra-Sánchez, "Current status of action mode and effect of chitosan against phytopathogens fungi", *Afr. J. Microbiol. Res.*, vol. 5, pp. 4243-4247, 2011.

[179] A. El Ghaouth, J. Arul, A. Asselin, and N. Benhamou, "Antifungal activity of chitosan on post-harvest pathogens: induction of morphological and cytological alterations in Rhizopus stolonifer", *Mycol. Res.*, vol. 96, pp. 769-779, 1992.
[http://dx.doi.org/10.1016/S0953-7562(09)80447-4]

[180] A.N. Hernández-Lauzardo, S. Bautista-Baños, M.G. Velázquez-Del Valle, M.G. Méndez-Montealvo, M.M. Sánchez-Rivera, and L.A. Bello-Pérez, "Antifungal effects of chitosan with different molecular weights on in vitro development of *Rhizopus stolonifer* (Ehrenb.:Fr.) Vuill", *Carbohydr. Polym.*, vol. 73, no. 4, pp. 541-547, 2008.
[http://dx.doi.org/10.1016/j.carbpol.2007.12.020] [PMID: 26048219]

[181] C.R. Allan, and L.A. Hadwiger, "The fungicidal effect of chitosan on fungi of varying cell wall

composition", *Exp. Mycol.,* vol. 3, pp. 285-287, 1979.
[http://dx.doi.org/10.1016/S0147-5975(79)80054-7]

[182] P.J. Chien, and C.C. Chou, "Antifungal activity of chitosan and its application to control post-harvest quality and fungal rotting of Tankan citrus fruit (Citrus tankan Hayata)", *J. Sci. Food Agric.,* vol. 86, pp. 1964-1969, 2006.
[http://dx.doi.org/10.1002/jsfa.2570]

[183] N. Benhamou, P. Lafontaine, and M. Nicole, "Induction of systemic resistance to Fusarium crown and root rot in tomato plants by seed treatment with chitosan", *Phytopathology,* vol. 84, pp. 1432-1444, 1994.
[http://dx.doi.org/10.1094/Phyto-84-1432]

[184] A. El Ghaouth, J. Arul, C. Wilson, and N. Benhamou, "Ultrastructural and cytochemical aspects of the effect of chitosan on decay of bell pepper fruit", *Physiol. Mol. Plant Pathol.,* vol. 44, pp. 417-432, 1994.
[http://dx.doi.org/10.1016/S0885-5765(05)80098-0]

[185] W. Elmer, and J. Lamondia, *Chitosan inhibits Rhizoctonia fragariae but not strawberry black root rot.*Advances in Strawberry Research: USA, 1994.

[186] S. Bautista-Baños, M. Hernández-López, E. Bosquez-Molina, and C. Wilson, "Effects of chitosan and plant extracts on growth of Colletotrichum gloeosporioides, anthracnose levels and quality of papaya fruit", *Crop Prot.,* vol. 22, pp. 1087-1092, 2003.
[http://dx.doi.org/10.1016/S0261-2194(03)00117-0]

[187] S. Bautista-Baños, M. Hernández-López, and E. Bosquez-Molina, "Growth inhibition of selected fungi by chitosan and plant extracts", *Mexican Journal of Phytopathology,* vol. 22, pp. 178-186, 2004.

[188] E.N. Oliveira Jr, N.E. El Gueddari, B.M. Moerschbacher, M.G. Peter, and T.T. Franco, "Growth of phytopathogenic fungi in the presence of partially acetylated chitooligosaccharides", *Mycopathologia,* vol. 166, no. 3, pp. 163-174, 2008.
[http://dx.doi.org/10.1007/s11046-008-9125-0] [PMID: 18523865]

[189] J. Xu, X. Zhao, X. Han, and Y. Du, "Antifungal activity of oligochitosan against Phytophthora capsici and other plant pathogenic fungi in vitro", *Pestic. Biochem. Physiol.,* vol. 87, pp. 220-228, 2007.
[http://dx.doi.org/10.1016/j.pestbp.2006.07.013]

[190] X. Meng, L. Yang, J.F. Kennedy, and S. Tian, "Effects of chitosan and oligochitosan on growth of two fungal pathogens and physiological properties in pear fruit", *Carbohydr. Polym.,* vol. 81, pp. 70-75, 2010.
[http://dx.doi.org/10.1016/j.carbpol.2010.01.057]

[191] "L. YANG, and X. MENG, “Study on the inhibition of chitosan and oligochitosan against the growth and development of Penicillium", *Food Science,* p. S1, 2011.

[192] E.I. Rabea, M.E. Badawy, T.M. Rogge, C.V. Stevens, M. Höfte, W. Steurbaut, and G. Smagghe, "Insecticidal and fungicidal activity of new synthesized chitosan derivatives", *Pest Manag. Sci.,* vol. 61, no. 10, pp. 951-960, 2005.
[http://dx.doi.org/10.1002/ps.1085] [PMID: 15999339]

[193] E.I. Rabea, M. El Badawy, T.M. Rogge, C.V. Stevens, W. Steurbaut, M. Höfte, and G. Smagghe, "Enhancement of fungicidal and insecticidal activity by reductive alkylation of chitosan", *Pest Manag. Sci.,* vol. 62, no. 9, pp. 890-897, 2006.
[http://dx.doi.org/10.1002/ps.1263] [PMID: 16847817]

[194] Y. Qin, S. Liu, R. Xing, H. Yu, K. Li, X. Meng, R. Li, and P. Li, "Synthesis and characterization of dithiocarbamate chitosan derivatives with enhanced antifungal activity", *Carbohydr. Polym.,* vol. 89, no. 2, pp. 388-393, 2012.
[http://dx.doi.org/10.1016/j.carbpol.2012.03.018] [PMID: 24750734]

[195] Z. Guo, R. Xing, S. Liu, Z. Zhong, X. Ji, L. Wang, and P. Li, "Antifungal properties of Schiff bases of

chitosan, N-substituted chitosan and quaternized chitosan", *Carbohydr. Res.,* vol. 342, no. 10, pp. 1329-1332, 2007.
[http://dx.doi.org/10.1016/j.carres.2007.04.006] [PMID: 17485075]

[196] Y. Qin, S. Liu, R. Xing, K. Li, H. Yu, and P. Li, "Synthesis and antifungal evaluation of (1,2,3-triazo--4-yl)methyl nicotinate chitosan", *Int. J. Biol. Macromol.,* vol. 61, pp. 58-62, 2013.
[http://dx.doi.org/10.1016/j.ijbiomac.2013.05.023] [PMID: 23732332]

[197] Y. Qin, R. Xing, S. Liu, K. Li, L. Hu, H. Yu, X. Chen, and P. Li, "Synthesis of chitosan derivative with diethyldithiocarbamate and its antifungal activity", *Int. J. Biol. Macromol.,* vol. 65, pp. 369-374, 2014.
[http://dx.doi.org/10.1016/j.ijbiomac.2014.01.072] [PMID: 24530333]

[198] Y. Qin, R. Xing, S. Liu, H. Yu, K. Li, L. Hu, and P. Li, "Synthesis and antifungal properties of (4-tolyloxy)-pyrimidyl-α-aminophosphonates chitosan derivatives", *Int. J. Biol. Macromol.,* vol. 63, pp. 83-91, 2014.
[http://dx.doi.org/10.1016/j.ijbiomac.2013.10.023] [PMID: 24183805]

[199] S.S. Elkholy, H.A. Salem, M. Eweis, and M.Z. Elsabee, "Synthesis and characterization of some acyl thiourea derivatives of chitosan and their biocidal activities", *Int. J. Biol. Macromol.,* vol. 70, pp. 199-207, 2014.
[http://dx.doi.org/10.1016/j.ijbiomac.2014.05.068] [PMID: 25002014]

[200] Q. Li, J. Ren, F. Dong, Y. Feng, G. Gu, and Z. Guo, "Synthesis and antifungal activity of thiadiazole-functionalized chitosan derivatives", *Carbohydr. Res.,* vol. 373, pp. 103-107, 2013.
[http://dx.doi.org/10.1016/j.carres.2013.03.001] [PMID: 23624516]

[201] Y. Qin, R. Xing, S. Liu, K. Li, X. Meng, R. Li, J. Cui, B. Li, and P. Li, "Novel thiosemicarbazone chitosan derivatives: preparation, characterization, and antifungal activity", *Carbohydr. Polym.,* vol. 87, pp. 2664-2670, 2012.
[http://dx.doi.org/10.1016/j.carbpol.2011.11.048]

[202] Z. Guo, R. Xing, S. Liu, Z. Zhong, X. Ji, L. Wang, and P. Li, "The influence of the cationic of quaternized chitosan on antifungal activity", *Int. J. Food Microbiol.,* vol. 118, no. 2, pp. 214-217, 2007.
[http://dx.doi.org/10.1016/j.ijfoodmicro.2007.07.003] [PMID: 17692420]

[203] R. Li, Z. Guo, and P. Jiang, "Synthesis, characterization, and antifungal activity of novel quaternary chitosan derivatives", *Carbohydr. Res.,* vol. 345, no. 13, pp. 1896-1900, 2010.
[http://dx.doi.org/10.1016/j.carres.2010.05.029] [PMID: 20615498]

[204] E.I. Rabea, "in vitro assessment of antimicrobial property of O-(phenoxyacetic) chitosan compounds on plant pathogens", *Journal of Chitin and Chitosan Science,* vol. 2, pp. 293-298, 2014.
[http://dx.doi.org/10.1166/jcc.2014.1081]

[205] Y. Peng, B. Han, W. Liu, and X. Xu, "Preparation and antimicrobial activity of hydroxypropyl chitosan", *Carbohydr. Res.,* vol. 340, no. 11, pp. 1846-1851, 2005.
[http://dx.doi.org/10.1016/j.carres.2005.05.009] [PMID: 15979054]

[206] Z. Guo, R. Chen, R. Xing, S. Liu, H. Yu, P. Wang, C. Li, and P. Li, "Novel derivatives of chitosan and their antifungal activities in vitro", *Carbohydr. Res.,* vol. 341, no. 3, pp. 351-354, 2006.
[http://dx.doi.org/10.1016/j.carres.2005.11.002] [PMID: 16343460]

[207] Z. Zhong, R. Xing, S. Liu, L. Wang, S. Cai, and P. Li, "Synthesis of acyl thiourea derivatives of chitosan and their antimicrobial activities in vitro", *Carbohydr. Res.,* vol. 343, no. 3, pp. 566-570, 2008.
[http://dx.doi.org/10.1016/j.carres.2007.11.024] [PMID: 18083151]

[208] Z. Zhong, R. Chen, R. Xing, X. Chen, S. Liu, Z. Guo, X. Ji, L. Wang, and P. Li, "Synthesis and antifungal properties of sulfanilamide derivatives of chitosan", *Carbohydr. Res.,* vol. 342, no. 16, pp. 2390-2395, 2007.
[http://dx.doi.org/10.1016/j.carres.2007.07.015] [PMID: 17765881]

[209] M. Sathiyabama, G. Akila, and R.E. Charles, "Chitosan-induced defence responses in tomato plants against early blight disease caused by Alternaria solani (Ellis and Martin) Sorauer", *Arch. Phytopathol. Plant Protect.,* vol. 47, pp. 1963-1973, 2014. [http://dx.doi.org/10.1080/03235408.2013.863497]

[210] V. Saharan, G. Sharma, M. Yadav, M.K. Choudhary, S.S. Sharma, A. Pal, R. Raliya, and P. Biswas, "Synthesis and in vitro antifungal efficacy of Cu-chitosan nanoparticles against pathogenic fungi of tomato", *Int. J. Biol. Macromol.,* vol. 75, pp. 346-353, 2015. [http://dx.doi.org/10.1016/j.ijbiomac.2015.01.027] [PMID: 25617841]

[211] Z. Zhong, B. Aotegen, H. Xu, and S. Zhao, "The influence of chemical structure on the antimicrobial activities of thiosemicarbazone-chitosan", *Cellulose,* vol. 21, pp. 105-114, 2014. [http://dx.doi.org/10.1007/s10570-013-0154-6]

[212] M.E. Badawy, and E.I. Rabea, "Potential of the biopolymer chitosan with different molecular weights to control postharvest gray mold of tomato fruit", *Postharvest Biol. Technol.,* vol. 51, pp. 110-117, 2009. [http://dx.doi.org/10.1016/j.postharvbio.2008.05.018]

[213] M. Badawy, and E.I. Rabea, "Synthesis of some N-benzylphosphoryl chitosan derivatives and their fungicidal and insecticidal activity", *Journal of Pest Control and Environmental Sciences,* vol. 13, pp. 43-56, 2005.

[214] Z. Guo, R. Xing, S. Liu, Z. Zhong, X. Ji, L. Wang, and P. Li, "The influence of molecular weight of quaternized chitosan on antifungal activity", *Carbohydr. Polym.,* vol. 71, pp. 694-697, 2008. [http://dx.doi.org/10.1016/j.carbpol.2007.06.027]

[215] G. Romanazzi, F. Nigro, A. Ippolito, D. Di Venere, and M. Salerno, "Effects of pre-and postharvest chitosan treatments to control storage grey mold of table grapes", *Journal of Food Science-Chicago,* vol. 67, pp. 1862-1867, 2002. [http://dx.doi.org/10.1111/j.1365-2621.2002.tb08737.x]

[216] S. Bautista-Baños, M. Hernández-López, A.N. Hernández-Lauzardo, J.L. Trejo-Espino, M. Bautista-Cerón, and G. Melo-Giorgana, "Effect of chitosan on in vitro development and morphology of two isolates of Colletotrichum gloeosporioides (Penz.) Penz. and Sacc", *Mexican Journal of Phytopathology,* vol. 23, pp. 62-67, 2005.

[217] A. Ali, M.T. Muhammad, K. Sijam, and Y. Siddiqui, "Potential of chitosan coating in delaying the postharvest anthracnose (Colletotrichum gloeosporioides Penz.) of Eksotika II papaya", *Int. J. Food Sci. Technol.,* vol. 45, pp. 2134-2140, 2010. [http://dx.doi.org/10.1111/j.1365-2621.2010.02389.x]

[218] P. Chowdappa, S. Gowda, C. Chethana, and S. Madhura, "Antifungal activity of chitosan-silver nanoparticle composite against Colletotrichum gloeosporioides associated with mango anthracnose", *Afr. J. Microbiol. Res.,* vol. 8, pp. 1803-1812, 2014. [http://dx.doi.org/10.5897/AJMR2013.6584]

[219] A. Ali, and T. Mahmud, "The potential use of locally prepared chitosan to control *in vitro* growth of *Colletotrichum gloeosporioides* isolated from papaya fruits", *Europe-Asia Symposium on Quality Management in Postharvest Systems-Eurasia 2007 804,* 2007pp. 177-182

[220] B. Prapagdee, K. Kotchadat, A. Kumsopa, and N. Visarathanonth, "The role of chitosan in protection of soybean from sudden death syndrome caused by Fusarium solani f. sp. glycines", *Bioresour. Technol.,* vol. 98, no. 7, pp. 1353-1358, 2007. [http://dx.doi.org/10.1016/j.biortech.2006.05.029] [PMID: 16828285]

[221] Y-C. Li, X-J. Sun, B. Yang, Y-H. Ge, and W. Yi, "Antifungal activity of chitosan on Fusarium sulphureum in relation to dry rot of potato tuber", *Agric. Sci. China,* vol. 8, pp. 597-604, 2009. [http://dx.doi.org/10.1016/S1671-2927(08)60251-5]

[222] Y-J. Sun, B. Yang, Y.-C. Li, R.-F. Han, and Y.-H. Ge, "Postharvest chitosan treatment induces

resistance in potato against Fusarium sulphureum", *Agric. Sci. China,* vol. 7, pp. 615-621, 2008. [http://dx.doi.org/10.1016/S1671-2927(08)60060-7]

[223] S. Xiao-juan, B. Yang, L. Yong-cai, H. Rui-feng, and G. Yonghong, "Postharvest chitosan treatment induces resistance in potato against Fusarium sulphureum", *Sci. China,* vol. 7, pp. 615-621, 2008.

[224] R. Li, and Z. Guo, "Synthesis, characterization and antifungal properties of N, O-(acyl)-N-(trimethyl) chitosan chloride", In: *e-Polymers*vol. 10. 2010, pp. 1273-1278.

[225] R.S. El-Mohamedy, N.G. El-Gamal, and A.R. Bakeer, "Application of Chitosan and essential oils as alternatives fungicides to control green and blue moulds of citrus fruits", *Int. J. Curr. Microbiol. App. Sci,* vol. 4, pp. 629-643, 2015.

[226] L. Sánchez-González, M. Cháfer, A. Chiralt, and C. González-Martínez, "Physical properties of edible chitosan films containing bergamot essential oil and their inhibitory action on Penicillium italicum", *Carbohydr. Polym.,* vol. 82, pp. 277-283, 2010. [http://dx.doi.org/10.1016/j.carbpol.2010.04.047]

[227] A. Rodríguez, M. Ramírez, R. Cárdenas, A. Hernández, M. Velázquez, and S. Bautista, "Induction of defense response of Oryza sativa L. against Pyricularia grisea (Cooke) Sacc. by treating seeds with chitosan and hydrolyzed chitosan", *Pestic. Biochem. Physiol.,* vol. 89, pp. 206-215, 2007. [http://dx.doi.org/10.1016/j.pestbp.2007.06.007]

[228] A. Sunpapao, and C. Pornsuriya, "Effects of chitosan treatments on para rubber leaf fall disease caused by Phytophthora palmivora Butler-a laboratory study", *Songklanakarin Journal of Science & Technology,* vol. 36, , 2014.

[229] A. Sunpapao, and C. Pornsuriya, "Chitosan inhibits the growth of Phytophthora botryosa: The causal agent of Para rubber leaf fall disease", *Plant Pathol. J.,* vol. 12, pp. 92-97, 2013. [http://dx.doi.org/10.3923/ppj.2013.92.97]

[230] M. Guerra-Sánchez, J. Vega-Pérez, M. Velázquez-del Valle, and A. Hernández-Lauzardo, "Antifungal activity and release of compounds on Rhizopus stolonifer (Ehrenb.: Fr.) Vuill. by effect of chitosan with different molecular weights", *Pestic. Biochem. Physiol.,* vol. 93, pp. 18-22, 2009. [http://dx.doi.org/10.1016/j.pestbp.2008.09.001]

[231] J. García-Rincón, J. Vega-Pérez, M. Guerra-Sánchez, A. Hernández-Lauzardo, A. Peña-Díaz, and M. Velázquez-Del Valle, "Effect of chitosan on growth and plasma membrane properties of Rhizopus stolonifer (Ehrenb.: Fr.) Vuill", *Pestic. Biochem. Physiol.,* vol. 97, pp. 275-278, 2010. [http://dx.doi.org/10.1016/j.pestbp.2010.03.008]

[232] A.N. Hernández-Lauzardo, M.G. Velázquez-del Valle, L. Veranza-Castelán, G.E. Melo-Giorgana, and M.G. Guerra-Sánchez, "Effect of chitosan on three isolates of Rhizopus stolonifer obtained from peach, papaya and tomato", *Fruits,* vol. 65, pp. 245-253, 2010. [http://dx.doi.org/10.1051/fruits/2010020]

[233] M.G. Velázquez-del Valle, A.N. Hernández-Lauzardo, M.G. Guerra-Sánchez, and G.I. Mariaca-Gaspar, "Chitosan as an alternative to control phytopathogenic fungi on fruits and vegetables in Mexico", *Afr. J. Microbiol. Res.,* vol. 6, pp. 6606-6611, 2012. [http://dx.doi.org/10.5897/AJMR12.1164]

[234] A.N. Hernández-Lauzardo, M.G. Guerra-Sánchez, A. Hernández-Rodríguez, M. Heydrich-Pérez, J. Vega-Pérez, and M.G. Velázquez-del Valle, "Assessment of the effect of chitosan of different molecular weights in controlling Rhizopus rots in tomato fruits", *Arch. Phytopathol. Plant Protect.,* vol. 45, pp. 33-41, 2012. [http://dx.doi.org/10.1080/03235400903345190]

[235] L. Cheah, B. Page, and R. Shepherd, *Chitosan coating for inhibition of sclerotinia rot of carrots,* 1997. [http://dx.doi.org/10.1080/01140671.1997.9513992]

[236] C. Molloy, L-H. Cheah, and J.P. Koolaard, "Induced resistance against Sclerotinia sclerotiorum in carrots treated with enzymatically hydrolysed chitosan", *Postharvest Biol. Technol.,* vol. 33, pp. 61-65,

2004.
[http://dx.doi.org/10.1016/j.postharvbio.2004.01.009]

[237] K. Li, R. Xing, S. Liu, Y. Qin, X. Meng, and P. Li, "Microwave-assisted degradation of chitosan for a possible use in inhibiting crop pathogenic fungi", *Int. J. Biol. Macromol.*, vol. 51, no. 5, pp. 767-773, 2012.
[http://dx.doi.org/10.1016/j.ijbiomac.2012.07.021] [PMID: 22829054]

[238] S. Chirkov, A. Il'ina, N. Surgucheva, E. Letunova, Y.A. Varitsev, N.Y. Tatarinova, and V. Varlamov, "Effect of chitosan on systemic viral infection and some defense responses in potato plants", *Russ. J. Plant Physiol.*, vol. 48, pp. 774-779, 2001.
[http://dx.doi.org/10.1023/A:1012508625017]

[239] S.N. Chirkov, "[The antiviral activity of chitosan (review)]", *Prikl. Biokhim. Mikrobiol.*, vol. 38, no. 1, pp. 5-13, 2002.
[PMID: 11852567]

[240] H. Pospieszny, S. Chirkov, and J. Atabekov, "Induction of antiviral resistance in plants by chitosan", *Plant Sci.*, vol. 79, pp. 63-68, 1991.
[http://dx.doi.org/10.1016/0168-9452(91)90070-O]

[241] F. Faoro, "Induced systemic resistance against systemic viruses: a feasible approach", *IOBC WPRS Bull.*, vol. 89, pp. 199-203, 2013.

[242] H. Pospieszny, "Antiviroid activity of chitosan", *Crop Prot.*, vol. 16, pp. 105-106, 1997.
[http://dx.doi.org/10.1016/S0261-2194(96)00077-4]

[243] M. Iriti, M. Sironi, S. Gomarasca, A.P. Casazza, C. Soave, and F. Faoro, "Cell death-mediated antiviral effect of chitosan in tobacco", *Plant Physiol. Biochem.*, vol. 44, no. 11-12, pp. 893-900, 2006.
[http://dx.doi.org/10.1016/j.plaphy.2006.10.009] [PMID: 17097297]

[244] A. Levine, R. Tenhaken, R. Dixon, and C. Lamb, "H2O2 from the oxidative burst orchestrates the plant hypersensitive disease resistance response", *Cell,* vol. 79, no. 4, pp. 583-593, 1994.
[http://dx.doi.org/10.1016/0092-8674(94)90544-4] [PMID: 7954825]

[245] V.N. Davydova, V.P. Nagorskaia, V.I. Gorbach, A.A. Kalitnik, A.V. Reunov, T.F. Soloveva, and I.M. Ermak, "Chitosan antiviral activity: dependence on structure and depolymerization method", *Prikl. Biokhim. Mikrobiol.*, vol. 47, no. 1, pp. 113-118, 2011.
[PMID: 21442928]

[246] D. Zeng, X. Luo, and R. Tu, "Application of bioactive coatings based on chitosan for soybean seed protection", *International Journal of Carbohydrate Chemistry,* 2012.
[http://dx.doi.org/10.1155/2012/104565]

[247] H. No, K. Lee, I. Kim, M. Park, S. Kim, and S. Meyers, "Chitosan treatment affects yield, ascorbic acid content, and hardness of soybean sprouts", *J. Food Sci.*, vol. 68, pp. 680-685, 2003.
[http://dx.doi.org/10.1111/j.1365-2621.2003.tb05731.x]

[248] Y-S. Lee, Y-H. Kim, and S-B. Kim, "Changes in the respiration, growth, and vitamin C content of soybean sprouts in response to chitosan of different molecular weights", *HortScience,* vol. 40, pp. 1333-1335, 2005.

[249] V. Mancini, and G. Romanazzi, "Seed treatments to control seedborne fungal pathogens of vegetable crops", *Pest Manag. Sci.*, vol. 70, no. 6, pp. 860-868, 2014.
[http://dx.doi.org/10.1002/ps.3693] [PMID: 24293285]

[250] P. Wanichpongpan, K. Suriyachan, and S. Chandrkrachang, "Effect of chitosan on the growth of Gerbera flower plant (*Gerbera jamesonii*)", In: *Chitin and Chitosan: Chitin and Chitosan in Life Science, Yamaguchi, Japan*2001.

[251] Y.J. Guan, J. Hu, X.J. Wang, and C.X. Shao, "Seed priming with chitosan improves maize germination and seedling growth in relation to physiological changes under low temperature stress", *J. Zhejiang Univ. Sci. B,* vol. 10, no. 6, pp. 427-433, 2009.

[http://dx.doi.org/10.1631/jzus.B0820373] [PMID: 19489108]

[252] E-G. Lizárraga-Paulín, S-P. Miranda-Castro, E. Moreno-Martínez, A-V. Lara-Sagahón, and I. Torres-Pacheco, "Maize seed coatings and seedling sprayings with chitosan and hydrogen peroxide: their influence on some phenological and biochemical behaviors", *J. Zhejiang Univ. Sci. B,* vol. 14, no. 2, pp. 87-96, 2013.
[http://dx.doi.org/10.1631/jzus.B1200270] [PMID: 23365007]

[253] D. Zeng, and X. Luo, "Physiological effects of chitosan coating on wheat growth and activities of protective enzyme with drought tolerance", *Open Journal of Soil Science,* vol. 2, p. 282, 2012.
[http://dx.doi.org/10.4236/ojss.2012.23034]

[254] L. Orzali, C. Forni, and L. Riccioni, "Effect of chitosan seed treatment as elicitor of resistance to Fusarium graminearum in wheat", *Seed Science and Technology,* vol. 42, pp. 132-149, 2014.
[http://dx.doi.org/10.15258/sst.2014.42.2.03]

[255] M. Janmohammadi, H. Mostafavi, H. Kazemi, G-R. Mahdavinia, and N. Sabaghnia, "Effect of chitosan application on the performance of lentil genotypes under rainfed conditions", *Acta Technologica Agriculturae,* vol. 17, pp. 86-90, 2014.
[http://dx.doi.org/10.2478/ata-2014-0020]

[256] R.S.-L.X. Qing-Zhong, "Effects of chitosan coating on seed germination and salt-tolerance of seedlings in hybrid rice (*Oryza sativa L.*)", *Acta Agronomica Sinica,* vol. 6, p. 14, 2002.

[257] Y. Zhou, Y. Yang, Y. Qi, Z. Zhang, X. Wang, and X. Hu, "Effects of chitosan on some physiological activity in germinating seed of peanut", *J. Peanut Sci.,* vol. 31, pp. 22-25, 2002.

[258] D. Zeng, and X. Mei, "Application of natural aminopolysaccharide in seed film-coating for pest control and cotton growth", *Russ. Agric. Sci.,* vol. 37, pp. 20-24, 2011.
[http://dx.doi.org/10.3103/S106836741101023X]

[259] L. Hadwiger, B. Fristensky, and R. Riggleman, "Chitosan, a natural regulator in plant-fungal pathogen interactions, increases crop yields", *Chitin, Chitosan and Related Enzymes,* vol. 1, pp. 291-302, 1984.
[http://dx.doi.org/10.1016/B978-0-12-780950-2.50024-1]

[260] D. Zeng, and Y. Shi, "Preparation and application of a novel environmentally friendly organic seed coating for rice", *J. Sci. Food Agric.,* vol. 89, pp. 2181-2185, 2009.
[http://dx.doi.org/10.1002/jsfa.3700]

[261] L. Wu, and M. Liu, "Preparation and properties of chitosan-coated NPK compound fertilizer with controlled-release and water-retention", *Carbohydr. Polym.,* vol. 72, pp. 240-247, 2008.
[http://dx.doi.org/10.1016/j.carbpol.2007.08.020]

[262] M.A. Teixeira, W.J. Paterson, E.J. Dunn, Q. Li, B.K. Hunter, and M.F. Goosen, "Assessment of chitosan gels for the controlled release of agrochemicals", *Ind. Eng. Chem. Res.,* vol. 29, pp. 1205-1209, 1990.
[http://dx.doi.org/10.1021/ie00103a019]

[263] A.R. Kulkarni, K.S. Soppimath, T.M. Aminabhavi, A.M. Dave, and M.H. Mehta, "Urea–formaldehyde crosslinked starch and guar gum matrices for encapsulation of natural liquid pesticide [Azadirachta Indica A. Juss.(neem) seed oil]: swelling and release kinetics", *J. Appl. Polym. Sci.,* vol. 73, pp. 2437-2446, 1999.
[http://dx.doi.org/10.1002/(SICI)1097-4628(19990919)73:12<2437::AID-APP12>3.0.CO;2-7]

[264] P. Chamberlain, and K. Symes, "Encapsulation and controlled release", *Karsa, DR,* p. 131, .

[265] H.C. Paula, R. de Paula, and S.K. Bezerra, "Swelling and release kinetics of larvicide-containing chitosan/cashew gum beads", *J. Appl. Polym. Sci.,* vol. 102, pp. 395-400, 2006.
[http://dx.doi.org/10.1002/app.24009]

[266] X.Z. Shu, and K.J. Zhu, "A novel approach to prepare tripolyphosphate/chitosan complex beads for controlled release drug delivery", *Int. J. Pharm.,* vol. 201, no. 1, pp. 51-58, 2000.
[http://dx.doi.org/10.1016/S0378-5173(00)00403-8] [PMID: 10867264]

[267] L. Vollner, *Agrochemical formulations for improved efficacy and reduced environmental impacts: slow release formulations with natural polymers.*Study and Prediction of Pesticides Behaviour in Soils, Plants and Aquatic Aystems. GSF: Neuherberg, Germany, 1990, pp. 136-145.

[268] T.W. Wong, L.W. Chan, S.B. Kho, and P.W. Sia Heng, "Design of controlled-release solid dosage forms of alginate and chitosan using microwave", *J. Control. Release,* vol. 84, no. 3, pp. 99-114, 2002. [http://dx.doi.org/10.1016/S0168-3659(02)00237-7] [PMID: 12468214]

[269] L.A. Nnamonu, R. Sha'Ato, and I. Onyido, "Alginate reinforced chitosan and starch beads in slow release formulation of imazaquin herbicide-preparation and characterization",

[270] R.A. Bharani, S.K.R. Namasivayam, and S.S. Shankar, "Biocompatible chitosan nanoparticles incorporated pesticidal protein beauvericin (Csnp-Bv) preparation for the improved pesticidal activity against major groundnut defoliator *Spodoptera Litura* (Fab.)(Lepidoptera; Noctuidae)", *International Journal of ChemTech Research,* vol. 6, pp. 5007-5012, 2014.

[271] R. Grillo, A.E. Pereira, C.S. Nishisaka, R. de Lima, K. Oehlke, R. Greiner, and L.F. Fraceto, "Chitosan/tripolyphosphate nanoparticles loaded with paraquat herbicide: an environmentally safer alternative for weed control", *J. Hazard. Mater.,* vol. 278, pp. 163-171, 2014. [http://dx.doi.org/10.1016/j.jhazmat.2014.05.079] [PMID: 24968252]

[272] B-H. Feng, and L-F. Peng, "Synthesis and characterization of carboxymethyl chitosan carrying ricinoleic functions as an emulsifier for azadirachtin", *Carbohydr. Polym.,* vol. 88, pp. 576-582, 2012. [http://dx.doi.org/10.1016/j.carbpol.2012.01.002]

[273] I.C. Hwang, T.H. Kim, S.H. Bang, K.S. Kim, H.R. Kwon, M.J. Seo, Y.N. Youn, H.J. Park, C. Yasunaga-Aoki, and Y.M. Yu, "Insecticidal effect of controlled release formulations of etofenprox based on nano-bio technique", *J. Fac. Agric. Kyushu Univ.,* vol. 56, pp. 33-40, 2011.

[274] S. Karaman, M. Digrak, U. Ravid, and A. Ilcim, "Antibacterial and antifungal activity of the essential oils of Thymus revolutus Celak from Turkey", *J. Ethnopharmacol.,* vol. 76, no. 2, pp. 183-186, 2001. [http://dx.doi.org/10.1016/S0378-8741(01)00238-0] [PMID: 11390134]

[275] A.O. Oyedeji, O. Ekundayo, O.N. Olawore, B.A. Adeniyi, and W.A. Koenig, "Antimicrobial activity of the essential oils of five Eucalyptus species growing in Nigeria", *Fitoterapia,* vol. 70, pp. 526-528, 1999.
[http://dx.doi.org/10.1016/S0367-326X(99)00083-0]

[276] S. Freitas, H.P. Merkle, and B. Gander, "Microencapsulation by solvent extraction/evaporation: reviewing the state of the art of microsphere preparation process technology", *J. Control. Release,* vol. 102, no. 2, pp. 313-332, 2005.
[http://dx.doi.org/10.1016/j.jconrel.2004.10.015] [PMID: 15653154]

[277] E. Khor, and L.Y. Lim, "Implantable applications of chitin and chitosan", *Biomaterials,* vol. 24, no. 13, pp. 2339-2349, 2003.
[http://dx.doi.org/10.1016/S0142-9612(03)00026-7] [PMID: 12699672]

[278] J.H. Park, M. Ye, and K. Park, "Biodegradable polymers for microencapsulation of drugs", *Molecules,* vol. 10, no. 1, pp. 146-161, 2005.
[http://dx.doi.org/10.3390/10010146] [PMID: 18007283]

[279] T.M. Don, S.C. Hsu, and W.Y. Chiu, "Structures and thermal properties of chitosan-modified poly (methyl methacrylate)", *J. Polym. Sci. A Polym. Chem.,* vol. 39, pp. 1646-1655, 2001.
[http://dx.doi.org/10.1002/pola.1142]

[280] W-C. Hsieh, C-P. Chang, and Y-L. Gao, "Controlled release properties of Chitosan encapsulated volatile Citronella Oil microcapsules by thermal treatments", *Colloids Surf. B Biointerfaces,* vol. 53, no. 2, pp. 209-214, 2006.
[http://dx.doi.org/10.1016/j.colsurfb.2006.09.008] [PMID: 17049821]

[281] M. Sugano, K. Yoshida, M. Hashimoto, K. Enomoto, and S. Hirano, "Hypocholesterolemic activity of partially hydrolyzed chitosan in rats", *Advances in Chitin and Chitosan,* vol. 2, pp. 472-478, 1992.

[http://dx.doi.org/10.1007/978-94-011-5942-5_53]

[282] M. Sugano, T. Fujikawa, Y. Hiratsuji, K. Nakashima, N. Fukuda, and Y. Hasegawa, "A novel use of chitosan as a hypocholesterolemic agent in rats", *Am. J. Clin. Nutr.,* vol. 33, no. 4, pp. 787-793, 1980. [PMID: 7361697]

[283] M. Sugano, S. Watanabe, A. Kishi, M. Izume, and A. Ohtakara, "Hypocholesterolemic action of chitosans with different viscosity in rats", *Lipids,* vol. 23, no. 3, pp. 187-191, 1988. [http://dx.doi.org/10.1007/BF02535456] [PMID: 2836688]

[284] P. Baldrick, "The safety of chitosan as a pharmaceutical excipient", *Regul. Toxicol. Pharmacol.,* vol. 56, no. 3, pp. 290-299, 2010. [http://dx.doi.org/10.1016/j.yrtph.2009.09.015] [PMID: 19788905]

[285] A. Lagarto, N. Merino, O. Valdes, J. Dominguez, E. Spencer, N. de la Paz, and G. Aparicio, "Safety evaluation of chitosan and chitosan acid salts from Panurilus argus lobster", *Int. J. Biol. Macromol.,* vol. 72, pp. 1343-1350, 2015. [http://dx.doi.org/10.1016/j.ijbiomac.2014.10.030] [PMID: 25450835]

[286] U. Food, and D. Administration, "Generally recognized as safe (GRAS) notice inventory",

[287] KFDA, *Food additives code.* Korea Food Drug Admin, 1995, p. 449.

[288] M. Weiner, *An overview of the regulatory status and of the safety of chitin and chitosan as food and pharmaceutical ingredients.*Advances in Chitin and Chitosan, 1992, pp. 663-670. [http://dx.doi.org/10.1007/978-94-011-5942-5_76]

[289] S. Hirano, C. Itakura, H. Seino, Y. Akiyama, I. Nonaka, N. Kanbara, and T. Kawakami, "Chitosan as an ingredient for domestic animal feeds", *J. Agric. Food Chem.,* vol. 38, pp. 1214-1217, 1990. [http://dx.doi.org/10.1021/jf00095a012]

[290] M. Sugano, T. Fujikawa, Y. Hiratsuji, and Y. Hasegawa, "Hypocholesterolemic effects of chitosan in cholesterol-fed rats", *Nutr. Rep. Int.,* vol. 18, pp. 531-537, 1978.

[291] J.K. Lee, S.U. Kim, and J.H. Kim, "Modification of chitosan to improve its hypocholesterolemic capacity", *Biosci. Biotechnol. Biochem.,* vol. 63, no. 5, pp. 833-839, 1999. [http://dx.doi.org/10.1271/bbb.63.833] [PMID: 10380626]

[292] F. Karadeniz, and S-K. Kim, "Antidiabetic activities of chitosan and its derivatives: a mini review", *Adv. Food Nutr. Res.,* vol. 73, pp. 33-44, 2014. [http://dx.doi.org/10.1016/B978-0-12-800268-1.00003-2] [PMID: 25300541]

[293] G. Kerch, "The potential of chitosan and its derivatives in prevention and treatment of age-related diseases", *Mar. Drugs,* vol. 13, no. 4, pp. 2158-2182, 2015. [http://dx.doi.org/10.3390/md13042158] [PMID: 25871293]

[294] A. Razdan, and D. Pettersson, "Hypolipidaemic, gastrointestinal and related responses of broiler chickens to chitosans of different viscosity", *Br. J. Nutr.,* vol. 76, no. 3, pp. 387-397, 1996. [http://dx.doi.org/10.1079/BJN19960044] [PMID: 8881711]

[295] A. Razdan, D. Pettersson, and J. Pettersson, "Broiler chicken body weights, feed intakes, plasma lipid and small-intestinal bile acid concentrations in response to feeding of chitosan and pectin", *Br. J. Nutr.,* vol. 78, no. 2, pp. 283-291, 1997. [http://dx.doi.org/10.1079/BJN19970146] [PMID: 9301417]

[296] R. Muzzarelli, and M. De Vincenzi, "Chitosans as dietary food additives", In: *Applications of Chitin and Chitosan.*Technomic Publishing: Lancaster, Pa, 1997, pp. 115-127.

[297] A.C. Zacour, M.E. Silva, P.R. Cecon, E.A. Bambirra, and E.C. Vieira, "Effect of dietary chitin on cholesterol absorption and metabolism in rats", *J. Nutr. Sci. Vitaminol. (Tokyo),* vol. 38, no. 6, pp. 609-613, 1992. [http://dx.doi.org/10.3177/jnsv.38.609] [PMID: 1304604]

[298] S. Hirano, and Y. Akiyama, "Absence of a hypocholesterolaemic action of chitosan in high-seru-

-cholesterol rabbits", *J. Sci. Food Agric.,* vol. 69, pp. 91-94, 1995.
[http://dx.doi.org/10.1002/jsfa.2740690114]

[299] O. Kanauchi, K. Deuchi, Y. Imasato, and E. Kobayashi, "Increasing effect of a chitosan and ascorbic acid mixture on fecal dietary fat excretion", *Biosci. Biotechnol. Biochem.,* vol. 58, pp. 1617-1620, 1994.
[http://dx.doi.org/10.1271/bbb.58.1617]

[300] O. Kanauchi, K. Deuchi, Y. Imasato, M. Shizukuishi, and E. Kobayashi, "Mechanism for the inhibition of fat digestion by chitosan and for the synergistic effect of ascorbate", *Biosci. Biotechnol. Biochem.,* vol. 59, no. 5, pp. 786-790, 1995.
[http://dx.doi.org/10.1271/bbb.59.786] [PMID: 7787293]

[301] K. Deuchi, O. Kanauchi, Y. Imasato, and E. Kobayashi, "Effect of the viscosity or deacetylation degree of chitosan on fecal fat excreted from rats fed on a high-fat diet", *Biosci. Biotechnol. Biochem.,* vol. 59, no. 5, pp. 781-785, 1995.
[http://dx.doi.org/10.1271/bbb.59.781] [PMID: 7787292]

[302] G.V. Vahouny, S. Satchithanandam, M.M. Cassidy, F.B. Lightfoot, and I. Furda, "Comparative effects of chitosan and cholestyramine on lymphatic absorption of lipids in the rat", *Am. J. Clin. Nutr.,* vol. 38, no. 2, pp. 278-284, 1983.
[PMID: 6881083]

[303] J. Liu, J. Zhang, and W. Xia, "Hypocholesterolaemic effects of different chitosan samples in vitro and in vivo", *Food Chem.,* vol. 107, pp. 419-425, 2008.
[http://dx.doi.org/10.1016/j.foodchem.2007.08.044]

[304] L. Li, J. Zhang, and J. Liu, "Effects of chitosan on serum lipids and fatty liver", *Chinese Journal of Marine Drugs,* vol. 26, p. 7, 2007.

[305] H-T. Yao, S-Y. Huang, and M-T. Chiang, "A comparative study on hypoglycemic and hypocholesterolemic effects of high and low molecular weight chitosan in streptozotocin-induced diabetic rats", *Food Chem. Toxicol.,* vol. 46, no. 5, pp. 1525-1534, 2008.
[http://dx.doi.org/10.1016/j.fct.2007.12.012] [PMID: 18255211]

[306] K. Hayashi, and M. Ito, "Antidiabetic action of low molecular weight chitosan in genetically obese diabetic KK-Ay mice", *Biol. Pharm. Bull.,* vol. 25, no. 2, pp. 188-192, 2002.
[http://dx.doi.org/10.1248/bpb.25.188] [PMID: 11853163]

[307] X. Liu, A. Zeng, T. Song, L. Li, F. Yang, Q. Wang, B. Wu, Y. Liu, and X. Zhi, "Hypocholesterolemic effects of N-[(2-Hydroxy-3-N, N-Dimethylhexadecyl Ammonium) propyl] chitosan chloride in high-fat-diet-induced rats", *J. Biomater. Sci. Polym. Ed.,* vol. 23, no. 8, pp. 1107-1114, 2012.
[http://dx.doi.org/10.1163/092050611X575540] [PMID: 21639973]

[308] A.L. Ahmad, S. Sumathi, and B.H. Hameed, "Adsorption of residue oil from palm oil mill effluent using powder and flake chitosan: equilibrium and kinetic studies", *Water Res.,* vol. 39, no. 12, pp. 2483-2494, 2005.
[http://dx.doi.org/10.1016/j.watres.2005.03.035] [PMID: 15985277]

[309] Y. Tao, H-L. Zhang, Y-M. Hu, S. Wan, and Z-Q. Su, "Preparation of chitosan and water-soluble chitosan microspheres via spray-drying method to lower blood lipids in rats fed with high-fat diets", *Int. J. Mol. Sci.,* vol. 14, no. 2, pp. 4174-4184, 2013.
[http://dx.doi.org/10.3390/ijms14024174] [PMID: 23429200]

[310] S.I. Park, S.D. Stan, M.A. Daeschel, and Y. Zhao, "Antifungal coatings on fresh strawberries (Fragaria× ananassa) to control mold growth during cold storage", *J. Food Sci.,* vol. 70, pp. M202-M207, 2005.
[http://dx.doi.org/10.1111/j.1365-2621.2005.tb07189.x]

[311] S.Y. Wang, and H. Gao, "Effect of chitosan-based edible coating on antioxidants, antioxidant enzyme system, and postharvest fruit quality of strawberries (Fragaria x aranassa Duch.)", *LWT-Food Science and Technology,* vol. 52, pp. 71-79, 2013.

[http://dx.doi.org/10.1016/j.lwt.2012.05.003]

[312] N.B. Gol, P.R. Patel, and T.R. Rao, "Improvement of quality and shelf-life of strawberries with edible coatings enriched with chitosan", *Postharvest Biol. Technol.,* vol. 85, pp. 185-195, 2013. [http://dx.doi.org/10.1016/j.postharvbio.2013.06.008]

[313] F. Devlieghere, A. Vermeulen, and J. Debevere, "Chitosan: antimicrobial activity, interactions with food components and applicability as a coating on fruit and vegetables", *Food Microbiol.,* vol. 21, pp. 703-714, 2004. [http://dx.doi.org/10.1016/j.fm.2004.02.008]

[314] E.A. Baldwin, "Edible coatings for fresh fruits and vegetables: past, present, and future", *Edible Coatings and Films to Improve Food Quality,* vol. 1, p. 25, 1994.

[315] H.K. No, S.P. Meyers, W. Prinyawiwatkul, and Z. Xu, "Applications of chitosan for improvement of quality and shelf life of foods: a review", *J. Food Sci.,* vol. 72, no. 5, pp. R87-R100, 2007. [http://dx.doi.org/10.1111/j.1750-3841.2007.00383.x] [PMID: 17995743]

[316] M.Z. Elsabee, and E.S. Abdou, "Chitosan based edible films and coatings: a review", *Mater Sci Eng C Mater Biol Appl,* vol. 33, no. 4, pp. 1819-1841, 2013. [http://dx.doi.org/10.1016/j.msec.2013.01.010] [PMID: 23498203]

[317] P. Dutta, S. Tripathi, G. Mehrotra, and J. Dutta, "Perspectives for chitosan based antimicrobial films in food applications", *Food Chem.,* vol. 114, pp. 1173-1182, 2009. [http://dx.doi.org/10.1016/j.foodchem.2008.11.047]

[318] M. Pereda, A. Ponce, N. Marcovich, R. Ruseckaite, and J. Martucci, "Chitosan-gelatin composites and bi-layer films with potential antimicrobial activity", *Food Hydrocoll.,* vol. 25, pp. 1372-1381, 2011. [http://dx.doi.org/10.1016/j.foodhyd.2011.01.001]

[319] H. Zhang, M.-l. Wang, and Y.-t. Mao, "Advances in the application of chitosan coating in fresh-keeping of fruits and vegetables", *Guizhou Agricultural Sciences,* vol. 10, p. 40, 2011.

[320] R. Avila-Sosa, E. Palou, M.T. Jiménez Munguía, G.V. Nevárez-Moorillón, A.R. Navarro Cruz, and A. López-Malo, "Antifungal activity by vapor contact of essential oils added to amaranth, chitosan, or starch edible films", *Int. J. Food Microbiol.,* vol. 153, no. 1-2, pp. 66-72, 2012. [http://dx.doi.org/10.1016/j.ijfoodmicro.2011.10.017] [PMID: 22100789]

[321] S. Zivanovic, S. Chi, and A.F. Draughon, "Antimicrobial activity of chitosan films enriched with essential oils", *J. Food Sci.,* vol. 70, pp. M45-M51, 2005. [http://dx.doi.org/10.1111/j.1365-2621.2005.tb09045.x]

[322] A. Acevedo-Fani, L. Salvia-Trujillo, M.A. Rojas-Graü, and O. Martín-Belloso, "Edible films from essential-oil-loaded nanoemulsions: Physicochemical characterization and antimicrobial properties", *Food Hydrocoll.,* vol. 47, pp. 168-177, 2015. [http://dx.doi.org/10.1016/j.foodhyd.2015.01.032]

[323] N. Gontard, S. Guilbert, and J.L. Cuq, "Water and glycerol as plasticizers affect mechanical and water vapor barrier properties of an edible wheat gluten film", *J. Food Sci.,* vol. 58, pp. 206-211, 1993. [http://dx.doi.org/10.1111/j.1365-2621.1993.tb03246.x]

[324] I. Arvanitoyannis, A. Nakayama, and S-i. Aiba, "Edible films made from hydroxypropyl starch and gelatin and plasticized by polyols and water", *Carbohydr. Polym.,* vol. 36, pp. 105-119, 1998. [http://dx.doi.org/10.1016/S0144-8617(98)00017-4]

[325] I. Arvanitoyannis, E. Psomiadou, A. Nakayama, S. Aiba, and N. Yamamoto, "Edible films made from gelatin, soluble starch and polyols, Part 3", *Food Chem.,* vol. 60, pp. 593-604, 1997. [http://dx.doi.org/10.1016/S0308-8146(97)00038-1]

[326] D. Kowalczyk, M. Kordowska-Wiater, J. Nowak, and B. Baraniak, "Characterization of films based on chitosan lactate and its blends with oxidized starch and gelatin", *Int. J. Biol. Macromol.,* vol. 77, pp. 350-359, 2015. [http://dx.doi.org/10.1016/j.ijbiomac.2015.03.032] [PMID: 25841370]

[327] Y. Xu, K.M. Kim, M.A. Hanna, and D. Nag, "Chitosan–starch composite film: preparation and characterization", *Ind. Crops Prod.,* vol. 21, pp. 185-192, 2005.
[http://dx.doi.org/10.1016/j.indcrop.2004.03.002]

[328] I. Arvanitoyannis, I. Kolokuris, A. Nakayama, N. Yamamoto, and S-i. Aiba, "Physico-chemical studies of chitosan-poly (vinyl alcohol) blends plasticized with sorbitol and sucrose", *Carbohydr. Polym.,* vol. 34, pp. 9-19, 1997.
[http://dx.doi.org/10.1016/S0144-8617(97)00089-1]

[329] I.S. Arvanitoyannis, A. Nakayama, and S-i. Aiba, "Chitosan and gelatin based edible films: state diagrams, mechanical and permeation properties", *Carbohydr. Polym.,* vol. 37, pp. 371-382, 1998.
[http://dx.doi.org/10.1016/S0144-8617(98)00083-6]

[330] P. Srinivasa, M. Ramesh, and R. Tharanathan, "Effect of plasticizers and fatty acids on mechanical and permeability characteristics of chitosan films", *Food Hydrocoll.,* vol. 21, pp. 1113-1122, 2007.
[http://dx.doi.org/10.1016/j.foodhyd.2006.08.005]

[331] N. Sudarshan, D. Hoover, and D. Knorr, "Antibacterial action of chitosan", *Food Biotechnol.,* vol. 6, pp. 257-272, 1992.
[http://dx.doi.org/10.1080/08905439209549838]

[332] H.K. No, N.Y. Park, S.H. Lee, and S.P. Meyers, "Antibacterial activity of chitosans and chitosan oligomers with different molecular weights", *Int. J. Food Microbiol.,* vol. 74, no. 1-2, pp. 65-72, 2002.
[http://dx.doi.org/10.1016/S0168-1605(01)00717-6] [PMID: 11929171]

[333] S.K. Sagoo, R. Board, and S. Roller, "Chitosan potentiates the antimicrobial action of sodium benzoate on spoilage yeasts", *Lett. Appl. Microbiol.,* vol. 34, no. 3, pp. 168-172, 2002.
[http://dx.doi.org/10.1046/j.1472-765x.2002.01067.x] [PMID: 11874536]

[334] J. Gómez-Estaca, A. López de Lacey, M.E. López-Caballero, M.C. Gómez-Guillén, and P. Montero, "Biodegradable gelatin-chitosan films incorporated with essential oils as antimicrobial agents for fish preservation", *Food Microbiol.,* vol. 27, no. 7, pp. 889-896, 2010.
[http://dx.doi.org/10.1016/j.fm.2010.05.012] [PMID: 20688230]

[335] G.J. Tsai, W.H. Su, H.C. Chen, and C.L. Pan, "Antimicrobial activity of shrimp chitin and chitosan from different treatments and applications of fish preservation", *Fish. Sci.,* vol. 68, pp. 170-177, 2002.
[http://dx.doi.org/10.1046/j.1444-2906.2002.00404.x]

[336] K. Vu, R. Hollingsworth, E. Leroux, S. Salmieri, and M. Lacroix, "Development of edible bioactive coating based on modified chitosan for increasing the shelf life of strawberries", *Food Res. Int.,* vol. 44, pp. 198-203, 2011.
[http://dx.doi.org/10.1016/j.foodres.2010.10.037]

[337] A. Perdones, L. Sánchez-González, A. Chiralt, and M. Vargas, "Effect of chitosan–lemon essential oil coatings on storage-keeping quality of strawberry", *Postharvest Biol. Technol.,* vol. 70, pp. 32-41, 2012.
[http://dx.doi.org/10.1016/j.postharvbio.2012.04.002]

[338] S.M. Ojagh, M. Rezaei, S.H. Razavi, and S.M. Hosseini, "Development and evaluation of a novel biodegradable film made from chitosan and cinnamon essential oil with low affinity toward water", *Food Chem.,* vol. 122, pp. 161-166, 2010.
[http://dx.doi.org/10.1016/j.foodchem.2010.02.033]

[339] Z. Shen, and D.P. Kamdem, "Development and characterization of biodegradable chitosan films containing two essential oils", *Int. J. Biol. Macromol.,* vol. 74, pp. 289-296, 2015.
[http://dx.doi.org/10.1016/j.ijbiomac.2014.11.046] [PMID: 25542174]

[340] T-C. Yang, C-C. Chou, and C-F. Li, "Antibacterial activity of N-alkylated disaccharide chitosan derivatives", *Int. J. Food Microbiol.,* vol. 97, no. 3, pp. 237-245, 2005.
[http://dx.doi.org/10.1016/S0168-1605(03)00083-7] [PMID: 15582734]

[341] C-S. Chen, W-Y. Liau, and G-J. Tsai, "Antibacterial effects of N-sulfonated and N-sulfobenzoyl

chitosan and application to oyster preservation", *J. Food Prot.,* vol. 61, no. 9, pp. 1124-1128, 1998. [http://dx.doi.org/10.4315/0362-028X-61.9.1124] [PMID: 9766062]

[342] Jdos.S. Gabriel, M.J. Tiera, and V.A. Tiera, "Synthesis, characterization, and antifungal activities of amphiphilic derivatives of diethylaminoethyl chitosan against Aspergillus flavus", *J. Agric. Food Chem.,* vol. 63, no. 24, pp. 5725-5731, 2015. [http://dx.doi.org/10.1021/acs.jafc.5b00278] [PMID: 26039463]

[343] D-H. Ngo, T-S. Vo, D-N. Ngo, K-H. Kang, J-Y. Je, H.N Pham, H-G. Byun, and S-K. Kim, "Biological effects of chitosan and its derivatives", *Food Hydrocoll.,* 2015. [http://dx.doi.org/10.1016/j.foodhyd.2015.05.023]

[344] X. Fei Liu, Y. Lin Guan, D. Zhi Yang, Z. Li, and K. De Yao, "Antibacterial action of chitosan and carboxymethylated chitosan", *J. Appl. Polym. Sci.,* vol. 79, pp. 1324-1335, 2001. [http://dx.doi.org/10.1002/1097-4628(20010214)79:7<1324::AID-APP210>3.0.CO;2-L]

[345] J-Y. Je, and S-K. Kim, "Chitosan derivatives killed bacteria by disrupting the outer and inner membrane", *J. Agric. Food Chem.,* vol. 54, no. 18, pp. 6629-6633, 2006. [http://dx.doi.org/10.1021/jf061310p] [PMID: 16939319]

[346] A. Zakrzewska, A. Boorsma, S. Brul, K.J. Hellingwerf, and F.M. Klis, "Transcriptional response of Saccharomyces cerevisiae to the plasma membrane-perturbing compound chitosan", *Eukaryot. Cell,* vol. 4, no. 4, pp. 703-715, 2005. [http://dx.doi.org/10.1128/EC.4.4.703-715.2005] [PMID: 15821130]

[347] L. Hadwiger, D. Kendra, B. Fristensky, and W. Wagoner, "Chitosan both activates genes in plants and inhibits RNA synthesis in fungi", In: *Chitin in Nature and Technology*Springer, 1986, pp. 209-214. [http://dx.doi.org/10.1007/978-1-4613-2167-5_28]

[348] N. Wanni Arachchi, *A Study on the extraction and applications of chitosan in the food industry,* 2015.

[349] S. Taha, and H.M. Swailam, "Antibacterial activity of chitosan against Aeromonas hydrophila", *Food/Nahrung,* vol. 46, pp. 337-340, 2002.

[350] G-J. Tsai, Z-Y. Wu, and W-H. Su, "Antibacterial activity of a chitooligosaccharide mixture prepared by cellulase digestion of shrimp chitosan and its application to milk preservation", *J. Food Prot.,* vol. 63, no. 6, pp. 747-752, 2000. [http://dx.doi.org/10.4315/0362-028X-63.6.747] [PMID: 10852568]

[351] S-K. Youn, S-M. Park, and D-H. Ahn, "Studies on the improvement of storage property in meat sausage using chitosan-II. Difference of storage property by molecular weight of chitosan", *Journal-Korean Society of Food Science and Nutrition,* vol. 29, pp. 849-853, 2000.

[352] S. Park, S. Youn, H. Kim, and D. Ahn, "Studies on the improvement of storage property in meat sausage using chitosan-IJ Korean Soc", *Food Sci. Nutr.,* vol. 28, pp. 167-171, 1999.

[353] D.P. Anderson, and A.K. Siwicki, "Duration of protection against Aeromonas salmonicida in brook trout immunostimulated with glucan or chitosan by injection or immersion", *Prog. Fish-Cult.,* vol. 56, pp. 258-261, 1994. [http://dx.doi.org/10.1577/1548-8640(1994)056<0258:DOPAAS>2.3.CO;2]

[354] A. Gopalakannan, and V. Arul, "Immunomodulatory effects of dietary intake of chitin, chitosan and levamisole on the immune system of Cyprinus carpio and control of Aeromonas hydrophila infection in ponds", *Aquaculture,* vol. 255, pp. 179-187, 2006. [http://dx.doi.org/10.1016/j.aquaculture.2006.01.012]

[355] G. Yin, L. Ardó, K.D. Thompson, A. Adams, Z. Jeney, and G. Jeney, "Chinese herbs (Astragalus radix and Ganoderma lucidum) enhance immune response of carp, Cyprinus carpio, and protection against Aeromonas hydrophila", *Fish Shellfish Immunol.,* vol. 26, no. 1, pp. 140-145, 2009. [http://dx.doi.org/10.1016/j.fsi.2008.08.015] [PMID: 18817878]

[356] M.S. Rao, R. Chander, and A. Sharma, "Development of shelf-stable intermediate-moisture meat products using active edible chitosancoating and irradiation", *J. Food Sci.,* vol. 70, pp. m325-m331,

2005.
[http://dx.doi.org/10.1111/j.1365-2621.2005.tb11475.x]

[357] A.B. Vishu Kumar, M.C. Varadaraj, L.R. Gowda, and R.N. Tharanathan, "Characterization of chito-oligosaccharides prepared by chitosanolysis with the aid of papain and Pronase, and their bactericidal action against Bacillus cereus and Escherichia coli", *Biochem. J.,* vol. 391, no. Pt 2, pp. 167-175, 2005. [http://dx.doi.org/10.1042/BJ20050093] [PMID: 15932346]

[358] J. Rhoades, and S. Roller, "Antimicrobial actions of degraded and native chitosan against spoilage organisms in laboratory media and foods", *Appl. Environ. Microbiol.,* vol. 66, no. 1, pp. 80-86, 2000. [http://dx.doi.org/10.1128/AEM.66.1.80-86.2000] [PMID: 10618206]

[359] B. Orgaz, M.M. Lobete, C.H. Puga, and C. San Jose, "Effectiveness of chitosan against mature biofilms formed by food related bacteria", *Int. J. Mol. Sci.,* vol. 12, no. 1, pp. 817-828, 2011. [http://dx.doi.org/10.3390/ijms12010817] [PMID: 21340015]

[360] E. Torlak, and D. Sert, "Antibacterial effectiveness of chitosan-propolis coated polypropylene films against foodborne pathogens", *Int. J. Biol. Macromol.,* vol. 60, pp. 52-55, 2013. [http://dx.doi.org/10.1016/j.ijbiomac.2013.05.013] [PMID: 23707735]

[361] J-S. Lee, H-Y. Lee, S-M. Park, and D-H. Ahn, "Effect of chitosan on drained solution and quality of soybean curd", *Journal of Chitin and Chitosan,* vol. 7, pp. 201-207, 2002.

[362] V.E. Tikhonov, E.A. Stepnova, V.G. Babak, I.A. Yamskov, J. Palma-Guerrero, H-B. Jansson, L.V. Lopez-Llorca, J. Salinas, D.V. Gerasimenko, and I.D. Avdienko, "Bactericidal and antifungal activities of a low molecular weight chitosan and its N-/2 (3)-(dodec-2-enyl) succinoyl/-derivatives", *Carbohydr. Polym.,* vol. 64, pp. 66-72, 2006. [http://dx.doi.org/10.1016/j.carbpol.2005.10.021]

[363] Y-J. Jeon, P-J. Park, and S-K. Kim, "Antimicrobial effect of chitooligosaccharides produced by bioreactor", *Carbohydr. Polym.,* vol. 44, pp. 71-76, 2001. [http://dx.doi.org/10.1016/S0144-8617(00)00200-9]

[364] J. Simůnek, I. Koppová, L. Filip, G. Tishchenko, and G. Bełzecki, "The antimicrobial action of low-molar-mass chitosan, chitosan derivatives and chitooligosaccharides on bifidobacteria", *Folia Microbiol. (Praha),* vol. 55, no. 4, pp. 379-382, 2010. [http://dx.doi.org/10.1007/s12223-010-0063-0] [PMID: 20680576]

[365] S.P. Hwang, "in vitro and in vivo antibacterial activity of shrimp chitosan against some intestinal bacteria", *Fish. Sci.,* vol. 70, pp. 675-681, 2004. [http://dx.doi.org/10.1111/j.1444-2906.2004.00856.x]

[366] N. Soultos, Z. Tzikas, A. Abrahim, D. Georgantelis, and I. Ambrosiadis, "Chitosan effects on quality properties of Greek style fresh pork sausages", *Meat Sci.,* vol. 80, no. 4, pp. 1150-1156, 2008. [http://dx.doi.org/10.1016/j.meatsci.2008.05.008] [PMID: 22063850]

[367] S-K. Youn, S-M. Park, Y-J. Kim, and D-H. Ahn, "Studies on substitution effect of chitosan against sodium nitrite in pork sausage", *Korean Journal of Food Science and Technology,* vol. 33, pp. 551-559, 2001.

[368] D. Campaniello, and M.R. Corbo, *Chitosan: a polysaccharide with antimicrobial action.*Antonio Bevilacqua, Maria Rosaria Corbo and Milena Sinigaglia, 2010, p. 92.

[369] K-H. Lee, and Y-C. Lee, "Effect of carboxymethyl chitosan on quality of fermented pan bread", *Korean Journal of Food Science and Technology,* vol. 29, pp. 96-100, 1997.

[370] X. Liu, W. Xia, Q. Jiang, Y. Xu, and P. Yu, "Synthesis, characterization, and antimicrobial activity of kojic acid grafted chitosan oligosaccharide", *J. Agric. Food Chem.,* vol. 62, no. 1, pp. 297-303, 2014. [http://dx.doi.org/10.1021/jf404026f] [PMID: 24364425]

[371] S. Lin, L. Chen, L. Huang, S. Cao, X. Luo, K. Liu, and Z. Huang, "Preparation and characterization of chitosan/cellulose blend films using ZnCl2• 3H2O as a solvent", *BioResources,* vol. 7, pp. 5488-5499, 2012.

[http://dx.doi.org/10.15376/biores.7.4.5488-5499]

[372] G-H. Wang, "Inhibition and inactivation of five species of foodborne pathogens by chitosan", *J. Food Prot.,* vol. 55, pp. 916-919, 1992. [http://dx.doi.org/10.4315/0362-028X-55.11.916]

[373] K.W. Kim, B. Min, Y-T. Kim, R.M. Kimmel, K. Cooksey, and S. Park, "Antimicrobial activity against foodborne pathogens of chitosan biopolymer films of different molecular weights", *LWT-Food Science and Technology,* vol. 44, pp. 565-569, 2011. [http://dx.doi.org/10.1016/j.lwt.2010.08.001]

[374] G-J. Tsai, and W-H. Su, "Antibacterial activity of shrimp chitosan against *Escherichia coli*", *J. Food Prot.,* vol. 62, no. 3, pp. 239-243, 1999. [http://dx.doi.org/10.4315/0362-028X-62.3.239] [PMID: 10090242]

[375] L-Y. Zheng, and J-F. Zhu, "Study on antimicrobial activity of chitosan with different molecular weights", *Carbohydr. Polym.,* vol. 54, pp. 527-530, 2003. [http://dx.doi.org/10.1016/j.carbpol.2003.07.009]

[376] S. Roller, and N. Covill, "The antimicrobial properties of chitosan in mayonnaise and mayonnaise-based shrimp salads", *J. Food Prot.,* vol. 63, no. 2, pp. 202-209, 2000. [http://dx.doi.org/10.4315/0362-028X-63.2.202] [PMID: 10678425]

[377] S-H. Lee, and O-K. Jo, "Effect of Lithospermum erythrorhizon, Glycyrrhiza uralensis and dipping of chitosan on shelf-life of kimchi", *Korean Journal of Food Science and Technology,* vol. 30, pp. 1367-1372, 1998.

[378] Y-M. Son, K-O. Kim, D-W. Jeon, and K-H. Kyung, "The effect of low molecular weight chitosan with and without other preservatives on the characteristics of kimchi during fermentation", *Korean Journal of Food Science and Technology,* vol. 28, pp. 888-896, 1996.

[379] S. Sagoo, R. Board, and S. Roller, "Chitosan inhibits growth of spoilage micro-organisms in chilled pork products", *Food Microbiol.,* vol. 19, pp. 175-182, 2002. [http://dx.doi.org/10.1006/fmic.2001.0474]

[380] E. Yoo, H. Lim, J. Kim, S. Song, and M. Choi, "The investigation of chitosanoligosaccharide for prolongating fermentation period of kimchi", *Journal of the Korean Society of Food Science and Nutrition,* 1998.

[381] H. Lee, S. Park, and D. Ahn, "Effect of storage properties of pork dipped in chitosan solution", *Journal of the Korean Society of Food Science and Nutrition,* 2003.

[382] P. Darmadji, and M. Izumimoto, "Effect of chitosan in meat preservation", *Meat Sci.,* vol. 38, no. 2, pp. 243-254, 1994. [http://dx.doi.org/10.1016/0309-1740(94)90114-7] [PMID: 22059662]

[383] A. Marques, S. Encarnação, S. Pedro, and M.L. Nunes, "*in vitro* antimicrobial activity of garlic, oregano and chitosan against *Salmonella enterica*", *World J. Microbiol. Biotechnol.,* vol. 24, pp. 2357-2360, 2008. [http://dx.doi.org/10.1007/s11274-008-9721-7]

[384] A.A. Tayel, "W.F. EL-TRAS, S. Moussa, A.F. EL-BAZ, H. Mahrous, M.F. Salem, and L. Brimer, "Antibacterial action of zinc oxide nanoparticles against foodborne pathogens", *J. Food Saf.,* vol. 31, pp. 211-218, 2011. [http://dx.doi.org/10.1111/j.1745-4565.2010.00287.x]

[385] J. Kingkaew, S. Kirdponpattara, N. Sanchavanakit, P. Pavasant, and M. Phisalaphong, "Effect of molecular weight of chitosan on antimicrobial properties and tissue compatibility of chitosan-impregnated bacterial cellulose films", *Biotechnology and Bioprocess Engineering,* vol. 19, pp. 534-544, 2014. [http://dx.doi.org/10.1007/s12257-014-0081-x]

[386] H.E. Salama, G.R. Saad, and M.W. Sabaa, "Synthesis, characterization and biological activity of

Schiff bases based on chitosan and arylpyrazole moiety", *Int. J. Biol. Macromol.,* vol. 79, pp. 996-1003, 2015.
[http://dx.doi.org/10.1016/j.ijbiomac.2015.06.009] [PMID: 26067768]

[387] M.W. Sabaa, H.M. Abdallah, N.A. Mohamed, and R.R. Mohamed, "Synthesis, characterization and application of biodegradable crosslinked carboxymethyl chitosan/poly(vinyl alcohol) clay nanocomposites", *Mater Sci Eng C Mater Biol Appl,* vol. 56, pp. 363-373, 2015.
[http://dx.doi.org/10.1016/j.msec.2015.06.043] [PMID: 26249602]

[388] S. Hirano, and N. Nagao, "Effects of chitosan, pectic acid, lysozyme, and chitinase on the growth of several phytopathogens", *Agric. Biol. Chem.,* vol. 53, pp. 3065-3066, 1989.

[389] N.S. dos Santos, A.J. Athayde Aguiar, C.E. de Oliveira, C. Veríssimo de Sales, S. de Melo E Silva, R. Sousa da Silva, T.C. Stamford, and E.L. de Souza, "Efficacy of the application of a coating composed of chitosan and Origanum vulgare L. essential oil to control Rhizopus stolonifer and Aspergillus niger in grapes (Vitis labrusca L.)", *Food Microbiol.,* vol. 32, no. 2, pp. 345-353, 2012.
[http://dx.doi.org/10.1016/j.fm.2012.07.014] [PMID: 22986200]

[390] S.W. Fang, C.F. Li, and D.Y. Shih, "Antifungal activity of chitosan and its preservative effect on low-sugar candied kumquat", *J. Food Prot.,* vol. 57, pp. 136-140, 1994.
[http://dx.doi.org/10.4315/0362-028X-57.2.136]

[391] X-F. Li, X-Q. Feng, S. Yang, T-P. Wang, and Z-X. Su, "Effects of molecular weight and concentration of chitosan on antifungal activity against Aspergillus niger", *Iranian Polymer Journal,* vol. 17, pp. 843-852, 2008.

[392] O. Cota-Arriola, M.O. Cortez-Rocha, E.C. Rosas-Burgos, A. Burgos-Hernández, Y.L. López-Franco, and M. Plascencia-Jatomea, "Antifungal effect of chitosan on the growth of Aspergillus parasiticus and production of aflatoxin B1", *Polym. Int.,* vol. 60, pp. 937-944, 2011.
[http://dx.doi.org/10.1002/pi.3054]

[393] R.H. Viegas de Souza, M. Takaki, R. de Oliveira Pedro, J. dos Santos Gabriel, M.J. Tiera, and V.A. de Oliveira Tiera, "Hydrophobic effect of amphiphilic derivatives of chitosan on the antifungal activity against Aspergillus flavus and Aspergillus parasiticus", *Molecules,* vol. 18, no. 4, pp. 4437-4450, 2013.
[http://dx.doi.org/10.3390/molecules18044437] [PMID: 23591924]

[394] E.L. de Souza, C.V. Sales, C.E. de Oliveira, L.A. Lopes, M.L. da Conceição, L.R. Berger, and T.C. Stamford, *Efficacy of a coating composed of chitosan from Mucor circinelloides and carvacrol to control Aspergillus flavus and the quality of cherry tomato fruits.*vol. 6. Frontiers in Microbiology, 2015.

[395] P-J. Chien, F. Sheu, and H-R. Lin, "Coating citrus (Murcott tangor) fruit with low molecular weight chitosan increases postharvest quality and shelf life", *Food Chem.,* vol. 100, pp. 1160-1164, 2007.
[http://dx.doi.org/10.1016/j.foodchem.2005.10.068]

[396] M.E. Badawy, "Structure and antimicrobial activity relationship of quaternary N-alkyl chitosan derivatives against some plant pathogens", *J. Appl. Polym. Sci.,* vol. 117, pp. 960-969, 2010.
[http://dx.doi.org/10.1002/app.31492]

[397] Z. Zhong, P. Li, R. Xing, and S. Liu, "Antimicrobial activity of hydroxylbenzenesulfonailides derivatives of chitosan, chitosan sulfates and carboxymethyl chitosan", *Int. J. Biol. Macromol.,* vol. 45, no. 2, pp. 163-168, 2009.
[http://dx.doi.org/10.1016/j.ijbiomac.2009.04.020] [PMID: 19414029]

[398] A. El Ghaouth, J. Arul, J. Grenier, and A. Asselin, "Antifungal activity of chitosan on two postharvest pathogens of strawberry fruits", *Phytopathology,* vol. 82, pp. 398-402, 1992.
[http://dx.doi.org/10.1094/Phyto-82-398]

[399] M. Rodriguez, V. Ramos, and E. Agullo, "Antimicrobial action of chitosan against spoilage organisms in precooked pizza", *J. Food Sci.,* vol. 68, pp. 271-274, 2003.
[http://dx.doi.org/10.1111/j.1365-2621.2003.tb14151.x]

[400] G. Romanazzi, "Chitosan treatment for the control of postharvest decay of table grapes, strawberries and sweet cherries", *Fresh Produce–Special Issues: New Trends in Postharvest Management of Fresh Produce,* vol. 4, pp. 111-115, 2010.

[401] T. Yu, H.Y. Li, and X.D. Zheng, "Synergistic effect of chitosan and Cryptococcus laurentii on inhibition of Penicillium expansum infections", *Int. J. Food Microbiol.,* vol. 114, no. 3, pp. 261-266, 2007.
[http://dx.doi.org/10.1016/j.ijfoodmicro.2006.09.008] [PMID: 17107729]

[402] J. Liu, S. Tian, X. Meng, and Y. Xu, "Effects of chitosan on control of postharvest diseases and physiological responses of tomato fruit", *Postharvest Biol. Technol.,* vol. 44, pp. 300-306, 2007.
[http://dx.doi.org/10.1016/j.postharvbio.2006.12.019]

[403] L. Wang, H. Wu, G. Qin, and X. Meng, "Chitosan disrupts Penicillium expansum and controls postharvest blue mold of jujube fruit", *Food Contr.,* vol. 41, pp. 56-62, 2014.
[http://dx.doi.org/10.1016/j.foodcont.2013.12.028]

[404] I.C. Alfaro-Gutiérrez, M.G. Guerra-Sánchez, A.N. Hernández-Lauzardo, and M.G. Velázquez-del Valle, "Morphological and Physiological Changes on Rhizopus stolonifer by Effect of Chitosan, Oligochitosan or Essential Oils", *J. Phytopathol.,* vol. 162, pp. 723-730, 2014.
[http://dx.doi.org/10.1111/jph.12252]

[405] C. Juliano, C. Demurtas, and L. Piu, "in vitro study on the anticandidal activity of Melaleuca alternifolia (tea tree) essential oil combined with chitosan", *Flavour Fragrance J.,* vol. 23, pp. 227-231, 2008.
[http://dx.doi.org/10.1002/ffj.1871]

[406] F. Seyfarth, S. Schliemann, P. Elsner, and U-C. Hipler, "Antifungal effect of high- and low-molecula--weight chitosan hydrochloride, carboxymethyl chitosan, chitosan oligosaccharide and N-acetyl-D-glucosamine against Candida albicans, Candida krusei and Candida glabrata", *Int. J. Pharm.,* vol. 353, no. 1-2, pp. 139-148, 2008.
[PMID: 18164151]

[407] A. Palmeira-de-Oliveira, M.P. Ribeiro, R. Palmeira-de-Oliveira, C. Gaspar, S. Costa-de-Oliveira, I.J. Correia, C. Pina Vaz, J. Martinez-de-Oliveira, J.A. Queiroz, and A.G. Rodrigues, "Anti-Candida activity of a chitosan hydrogel: mechanism of action and cytotoxicity profile", *Gynecol. Obstet. Invest.,* vol. 70, no. 4, pp. 322-327, 2010.
[http://dx.doi.org/10.1159/000314023] [PMID: 21109742]

[408] R. Muzzarelli, R. Tarsi, O. Filippini, E. Giovanetti, G. Biagini, and P.E. Varaldo, "Antimicrobial properties of N-carboxybutyl chitosan", *Antimicrob. Agents Chemother.,* vol. 34, no. 10, pp. 2019-2023, 1990.
[http://dx.doi.org/10.1128/AAC.34.10.2019] [PMID: 2291669]

[409] A. Balicka-Ramisz, A. Wojtasz-Pajak, B. Pilarczyk, A. Ramisz, and L. Laurans, "Antibacterial and antifungal activity of chitosan", *ISAH,* vol. 2, pp. 406-408, 2005.

[410] D. Spadaro, A. Garibaldi, and M. Gullino, "Discovery, development and technology transfer of biocontrol agents for postharvest disease control", *II International Symposium on Discovery and Development of Innovative Strategies for Postharvest Disease Management 1053,* 2013pp. 23-36

[411] S.L. Bico, M.F. de Jesus Raposo, R.M. De Morais, and A.M. De Morais, "Chemical dips and edible coatings to retard softening and browning of fresh-cut banana", *International Journal of Postharvest Technology and Innovation,* vol. 2, pp. 13-24, 2010.
[http://dx.doi.org/10.1504/IJPTI.2010.038185]

[412] T-J. Ha, and S-H. Lee, "Utilization of chitosan to improve the quality of processed milk", *Journal-Korean Society of Food Science and Nutrition,* vol. 30, pp. 630-634, 2001.

[413] S. Roller, and N. Covill, "The antifungal properties of chitosan in laboratory media and apple juice", *Int. J. Food Microbiol.,* vol. 47, no. 1-2, pp. 67-77, 1999.

[http://dx.doi.org/10.1016/S0168-1605(99)00006-9] [PMID: 10357275]

[414] L. Gómez-Rivas, B.I. Escudero-Abarca, M.G. Aguilar-Uscanga, P.M. Hayward-Jones, P. Mendoza, and M. Ramírez, "Selective antimicrobial action of chitosan against spoilage yeasts in mixed culture fermentations", *J. Ind. Microbiol. Biotechnol.*, vol. 31, no. 1, pp. 16-22, 2004. [http://dx.doi.org/10.1007/s10295-004-0112-2] [PMID: 14747932]

[415] M.B. Vásconez, S.K. Flores, C.A. Campos, J. Alvarado, and L.N. Gerschenson, "Antimicrobial activity and physical properties of chitosan–tapioca starch based edible films and coatings", *Food Res. Int.*, vol. 42, pp. 762-769, 2009. [http://dx.doi.org/10.1016/j.foodres.2009.02.026]

[416] H.I. Oh, Y.J. Kim, E.J. Chang, and J.Y. Kim, "Antimicrobial characteristics of chitosans against food spoilage microorganisms in liquid media and mayonnaise", *Biosci. Biotechnol. Biochem.*, vol. 65, no. 11, pp. 2378-2383, 2001. [http://dx.doi.org/10.1271/bbb.65.2378] [PMID: 11791708]

[417] P-J. Park, J-Y. Je, and S-K. Kim, "Free radical scavenging activities of differently deacetylated chitosans using an ESR spectrometer", *Carbohydr. Polym.*, vol. 55, pp. 17-22, 2004. [http://dx.doi.org/10.1016/j.carbpol.2003.05.002]

[418] M-T. Yen, J-H. Yang, and J-L. Mau, "Antioxidant properties of chitosan from crab shells", *Carbohydr. Polym.*, vol. 74, pp. 840-844, 2008. [http://dx.doi.org/10.1016/j.carbpol.2008.05.003]

[419] D-H. Ngo, and S-K. Kim, *Antioxidant effects of chitin, chitosan and their derivatives.*Elsevier Inc.: Oxford, UK, 2014. [http://dx.doi.org/10.1016/B978-0-12-800268-1.00002-0]

[420] M.A. Cacciuttolo, L. Trinh, J.A. Lumpkin, and G. Rao, "Hyperoxia induces DNA damage in mammalian cells", *Free Radic. Biol. Med.*, vol. 14, no. 3, pp. 267-276, 1993. [http://dx.doi.org/10.1016/0891-5849(93)90023-N] [PMID: 8458584]

[421] M-T. Yen, Y-H. Tseng, R-C. Li, and J-L. Mau, "Antioxidant properties of fungal chitosan from shiitake stipes", *LWT-Food Science and Technology,* vol. 40, pp. 255-261, 2007. [http://dx.doi.org/10.1016/j.lwt.2005.08.006]

[422] K. Prabu, and E. Natarajan, *in vitro antimicrobial and antioxidant activity of chitosan isolated from podophthalmus vigil,* 2012. [http://dx.doi.org/10.7324/JAPS.2012.2916]

[423] K.W. Kim, and R. Thomas, "Antioxidative activity of chitosans with varying molecular weights", *Food Chem.*, vol. 101, pp. 308-313, 2007. [http://dx.doi.org/10.1016/j.foodchem.2006.01.038]

[424] W. Xie, P. Xu, and Q. Liu, "Antioxidant activity of water-soluble chitosan derivatives", *Bioorg. Med. Chem. Lett.*, vol. 11, no. 13, pp. 1699-1701, 2001. [http://dx.doi.org/10.1016/S0960-894X(01)00285-2] [PMID: 11425541]

[425] M.A. Garcıa, N. de la Paz, C. Castro, J.L. Rodrıguez, M. Rapado, R. Zuluaga, P. Ganán, and A. Casariego, *Effect of molecular weight reduction by gamma irradiation on the antioxidant capacity of chitosan from lobster shells,* 2015. [http://dx.doi.org/10.1016/j.jrras.2015.01.003]

[426] T. Feng, Y. Du, J. Li, Y. Hu, and J.F. Kennedy, "Enhancement of antioxidant activity of chitosan by irradiation", *Carbohydr. Polym.*, vol. 73, pp. 126-132, 2008. [http://dx.doi.org/10.1016/j.carbpol.2007.11.003]

[427] W. Pasanphan, T. Rattanawongwiboon, S. Choofong, O. Güven, and K.K. Katti, "Irradiated chitosan nanoparticle as a water-based antioxidant and reducing agent for a green synthesis of gold nanoplatforms", *Radiat. Phys. Chem.*, vol. 106, pp. 360-370, 2015. [http://dx.doi.org/10.1016/j.radphyschem.2014.08.023]

[428] A.J. Angelo, and J.R. Vercellotti, "Inhibition of warmed-over flavor and preserving of uncured meat containing materials", In: *Google Patents*1989.

[429] K. Li, Y. Hwang, T. Tsai, and S. Chi, "Chelation of iron ion and antioxidative effect on cooked salted ground pork by N-carboxymethylchitosan (NCMC)", *Food Science Taiwan,* vol. 23, pp. 608-616, 1996.

[430] F. Shahidi, "Role of chemistry and biotechnology in value-added utilization of shellfish processing discards", *ChemInform,* vol. 27, 1996.

[431] D. Altiok, E. Altiok, and F. Tihminlioglu, "Physical, antibacterial and antioxidant properties of chitosan films incorporated with thyme oil for potential wound healing applications", *J. Mater. Sci. Mater. Med.,* vol. 21, no. 7, pp. 2227-2236, 2010. [http://dx.doi.org/10.1007/s10856-010-4065-x] [PMID: 20372985]

[432] E. Mendis, N. Rajapakse, and S-K. Kim, "Antioxidant properties of a radical-scavenging peptide purified from enzymatically prepared fish skin gelatin hydrolysate", *J. Agric. Food Chem.,* vol. 53, no. 3, pp. 581-587, 2005. [http://dx.doi.org/10.1021/jf048877v] [PMID: 15686405]

[433] J-I. Yang, H-Y. Ho, Y-J. Chu, and C-J. Chow, "Characteristic and antioxidant activity of retorted gelatin hydrolysates from cobia (Rachycentron canadum) skin", *Food Chem.,* vol. 110, no. 1, pp. 128-136, 2008. [http://dx.doi.org/10.1016/j.foodchem.2008.01.072] [PMID: 26050175]

[434] A. Alemán, B. Giménez, P. Montero, and M. Gómez-Guillén, "Antioxidant activity of several marine skin gelatins", *LWT-Food Science and Technology,* vol. 44, pp. 407-413, 2011. [http://dx.doi.org/10.1016/j.lwt.2010.09.003]

[435] J. Gómez-Estaca, L. Bravo, M. Gómez-Guillén, A. Alemán, and P. Montero, "Antioxidant properties of tuna-skin and bovine-hide gelatin films induced by the addition of oregano and rosemary extracts", *Food Chem.,* vol. 112, pp. 18-25, 2009. [http://dx.doi.org/10.1016/j.foodchem.2008.05.034]

[436] J. Martucci, L. Gende, L. Neira, and R. Ruseckaite, "Oregano and lavender essential oils as antioxidant and antimicrobial additives of biogenic gelatin films", *Ind. Crops Prod.,* vol. 71, pp. 205-213, 2015. [http://dx.doi.org/10.1016/j.indcrop.2015.03.079]

[437] M. Jridi, S. Hajji, H.B. Ayed, I. Lassoued, A. Mbarek, M. Kammoun, N. Souissi, and M. Nasri, "Physical, structural, antioxidant and antimicrobial properties of gelatin-chitosan composite edible films", *Int. J. Biol. Macromol.,* vol. 67, pp. 373-379, 2014. [http://dx.doi.org/10.1016/j.ijbiomac.2014.03.054] [PMID: 24709012]

[438] E.A. Baldwin, R. Hagenmaier, and J. Bai, *Edible coatings and films to improve food quality.*CRC Press, 2011.

[439] M. Aider, "Chitosan application for active bio-based films production and potential in the food industry: Review", *LWT-Food Science and Technology,* vol. 43, pp. 837-842, 2010. [http://dx.doi.org/10.1016/j.lwt.2010.01.021]

[440] S-Y. Cheng, B-J. Wang, and Y-M. Weng, "Antioxidant and antimicrobial edible zein/chitosan composite films fabricated by incorporation of phenolic compounds and dicarboxylic acids", *LWT-Food Science and Technology,* vol. 63, pp. 115-121, 2015. [http://dx.doi.org/10.1016/j.lwt.2015.03.030]

[441] C. Xue, G. Yu, T. Hirata, J. Terao, and H. Lin, "Antioxidative activities of several marine polysaccharides evaluated in a phosphatidylcholine-liposomal suspension and organic solvents", *Biosci. Biotechnol. Biochem.,* vol. 62, no. 2, pp. 206-209, 1998. [http://dx.doi.org/10.1271/bbb.62.206] [PMID: 9532776]

[442] Y-J. Jeon, J.Y. Kamil, and F. Shahidi, "Chitosan as an edible invisible film for quality preservation of herring and atlantic cod", *J. Agric. Food Chem.,* vol. 50, no. 18, pp. 5167-5178, 2002.

[http://dx.doi.org/10.1021/jf011693l] [PMID: 12188625]

[443] R.A. Muzzarelli, F. Tanfani, M. Emanuelli, E. Chiurazzi, and M. Piani, "Sulfated N-carboxymethyl chitosans as blood anticoagulants", In: *Chitin in Nature and Technology* Springer, 1986, pp. 469-476. [http://dx.doi.org/10.1007/978-1-4613-2167-5_55]

[444] A. Niekraszewicz, "Chitosan medical dressings", *Fibres & Textiles in Eastern Europe,* vol. 13, p. 54, 2005.

[445] T. Anthonsen, P.A. Sandford, and G. Skjåk-Braek, *Chitin and chitosan: sources, chemistry, biochemistry, physical properties, and applications.*Elsevier Applied Science, 1989.

[446] W. Paul, and C.P. Sharma, "Chitosan and alginate wound dressings: a short review", *Trends Biomater. Artif. Organs,* vol. 18, pp. 18-23, 2004.

[447] K.D. Yao, Y.J. Yin, M.X. Xu, and Y.F. Wang, "Investigation of pH-sensitive drug delivery system of chitosan/gelatin hybrid polymer network", *Polym. Int.,* vol. 38, pp. 77-82, 1995. [http://dx.doi.org/10.1002/pi.1995.210380110]

[448] B. Nies, "Cross-linking products of amino-group-containing biopolymers", *Patent WO,* vol. 97, p. 29132, 1997.

[449] Y. Ji Yin, F. Zhao, X. Feng Song, K. De Yao, W.W. Lu, and J. Chiyan Leong, "Preparation and characterization of hydroxyapatite/chitosan–gelatin network composite", *J. Appl. Polym. Sci.,* vol. 77, pp. 2929-2938, 2000. [http://dx.doi.org/10.1002/1097-4628(20000923)77:13<2929::AID-APP16>3.0.CO;2-Q]

[450] G. Haipeng, Z. Yinghui, L. Jianchun, G. Yandao, Z. Nanming, and Z. Xiufang, "Studies on nerve cell affinity of chitosan-derived materials", *J. Biomed. Mater. Res.,* vol. 52, no. 2, pp. 285-295, 2000. [http://dx.doi.org/10.1002/1097-4636(200011)52:2<285::AID-JBM7>3.0.CO;2-G] [PMID: 10951367]

[451] N. Shanmugasundaram, P. Ravichandran, P.N. Reddy, N. Ramamurty, S. Pal, and K.P. Rao, "Collagen-chitosan polymeric scaffolds for the in vitro culture of human epidermoid carcinoma cells", *Biomaterials,* vol. 22, no. 14, pp. 1943-1951, 2001. [http://dx.doi.org/10.1016/S0142-9612(00)00220-9] [PMID: 11426872]

[452] C. Airoldi, and O.A. Monteiro, "Chitosan–organosilane hybrids-syntheses, characterization, copper adsorption, and enzyme immobilization", *J. Appl. Polym. Sci.,* vol. 77, pp. 797-804, 2000. [http://dx.doi.org/10.1002/(SICI)1097-4628(20000725)77:4<797::AID-APP12>3.0.CO;2-Z]

[453] A.S. Aly, "Self-dissolving chitosan, I. Preparation, characterization and evaluation for drug delivery system", *Die Angewandte Makromolekulare Chemie,* vol. 259, pp. 13-18, 1998. [http://dx.doi.org/10.1002/(SICI)1522-9505(19981001)259:1<13::AID-APMC13>3.0.CO;2-T]

[454] E. Denkbaş, M. Seyyal, and E. Pişkin, "Implantable 5-fluorouracil loaded chitosan scaffolds prepared by wet spinning", *J. Membr. Sci.,* vol. 172, pp. 33-38, 2000. [http://dx.doi.org/10.1016/S0376-7388(00)00314-8]

[455] L. Illum, P. Watts, A. Fisher, I. Jabbal Gill, and S. Davis, "Novel chitosan-based delivery systems for the nasal administration of a LHRH-analogue", *STP Pharma Sci.,* vol. 10, pp. 89-94, 2000.

[456] M.A. Abdel-Naby, A-M. Ismail, A.M. Abdel-Fattah, and A.F. Abdel-Fattah, "Preparation and some properties of immobilized Penicillium funiculosum 258 dextranase", *Process Biochem.,* vol. 34, pp. 391-398, 1999. [http://dx.doi.org/10.1016/S0032-9592(98)00127-7]

[457] G.M. Carvalho, T.L. Alves, and D.M. Freire, "L-DOPA production by immobilized tyrosinase", *Twenty-First Symposium on Biotechnology for Fuels and Chemicals,* 2000pp. 791-800 [http://dx.doi.org/10.1007/978-1-4612-1392-5_62]

[458] M.V. Risbud, A.A. Hardikar, S.V. Bhat, and R.R. Bhonde, "pH-sensitive freeze-dried chitosan-polyvinyl pyrrolidone hydrogels as controlled release system for antibiotic delivery", *J. Control. Release,* vol. 68, no. 1, pp. 23-30, 2000.

[http://dx.doi.org/10.1016/S0168-3659(00)00208-X] [PMID: 10884576]

[459] M.S. Beena, T. Chandy, and C.P. Sharma, "Heparin immobilized chitosanpoly ethylene glycol interpenetrating network: antithrombogenicity", *Artif. Cells Blood Substit. Immobil. Biotechnol.*, vol. 23, no. 2, pp. 175-192, 1995.
[http://dx.doi.org/10.3109/10731199509117937] [PMID: 7767440]

[460] M. Risbud, A. Hardikar, and R. Bhonde, "Growth modulation of fibroblasts by chitosan-polyvinyl pyrrolidone hydrogel: implications for wound management?", *J. Biosci.*, vol. 25, no. 1, pp. 25-31, 2000.
[http://dx.doi.org/10.1007/BF02985178] [PMID: 10824195]

[461] S. Nakatsuka, and A.L. Andrady, "Permeability of vitamin B-12 in chitosan membranes. Effect of crosslinking and blending with poly (vinyl alcohol) on permeability", *J. Appl. Polym. Sci.*, vol. 44, pp. 17-28, 1992.
[http://dx.doi.org/10.1002/app.1992.070440103]

[462] J.H. Kim, J.Y. Kim, Y.M. Lee, and K.Y. Kim, "Controlled release of riboflavin and insulin through crosslinked poly (vinyl alcohol)/chitosan blend membrane", *J. Appl. Polym. Sci.*, vol. 44, pp. 1823-1828, 1992.
[http://dx.doi.org/10.1002/app.1992.070441015]

[463] A. Dal Pozzo, L. Vanini, M. Fagnoni, M. Guerrini, A. De Benedittis, and R. Muzzarelli, "Preparation and characterization of poly (ethylene glycol)-crosslinked reacetylated chitosans", *Carbohydr. Polym.*, vol. 42, pp. 201-206, 2000.
[http://dx.doi.org/10.1016/S0144-8617(99)00134-4]

[464] H.W. Sung, R.N. Huang, L.L. Huang, C.C. Tsai, and C.T. Chiu, "Feasibility study of a natural crosslinking reagent for biological tissue fixation", *J. Biomed. Mater. Res.*, vol. 42, no. 4, pp. 560-567, 1998.
[http://dx.doi.org/10.1002/(SICI)1097-4636(19981215)42:4<560::AID-JBM12>3.0.CO;2-I] [PMID: 9827680]

[465] V.R. Patel, and M.M. Amiji, "Preparation and characterization of freeze-dried chitosan-poly(ethylene oxide) hydrogels for site-specific antibiotic delivery in the stomach", *Pharm. Res.*, vol. 13, no. 4, pp. 588-593, 1996.
[http://dx.doi.org/10.1023/A:1016054306763] [PMID: 8710751]

[466] F-L. Mi, C-T. Chen, Y-C. Tseng, C-Y. Kuan, and S-S. Shyu, "Iron (III)-carboxymethylchitin microsphere for the pH-sensitive release of 6-mercaptopurine", *J. Control. Release,* vol. 44, pp. 19-32, 1997.
[http://dx.doi.org/10.1016/S0168-3659(96)01502-7]

[467] E. Ruel-Gariépy, A. Chenite, C. Chaput, S. Guirguis, and J. Leroux, "Characterization of thermosensitive chitosan gels for the sustained delivery of drugs", *Int. J. Pharm.*, vol. 203, no. 1-2, pp. 89-98, 2000.
[http://dx.doi.org/10.1016/S0378-5173(00)00428-2] [PMID: 10967431]

[468] A. Sezer, and J. Akbuğa, "Controlled release of piroxicam from chitosan beads", *Int. J. Pharm.*, vol. 121, pp. 113-116, 1995.
[http://dx.doi.org/10.1016/0378-5173(94)00413-Y]

[469] F.L. Mi, S.S. Shyu, T.B. Wong, S.F. Jang, S.T. Lee, and K.T. Lu, "Chitosan–polyelectrolyte complexation for the preparation of gel beads and controlled release of anticancer drug. II. Effect of pH-dependent ionic crosslinking or interpolymer complex using tripolyphosphate or polyphosphate as reagent", *J. Appl. Polym. Sci.*, vol. 74, pp. 1093-1107, 1999.
[http://dx.doi.org/10.1002/(SICI)1097-4628(19991031)74:5<1093::AID-APP6>3.0.CO;2-C]

[470] J-Y. Lee, S-H. Nam, S-Y. Im, Y-J. Park, Y-M. Lee, Y-J. Seol, C-P. Chung, and S-J. Lee, "Enhanced bone formation by controlled growth factor delivery from chitosan-based biomaterials", *J. Control. Release,* vol. 78, no. 1-3, pp. 187-197, 2002.

[http://dx.doi.org/10.1016/S0168-3659(01)00498-9] [PMID: 11772460]

[471] X.Z. Shu, and K.J. Zhu, "Controlled drug release properties of ionically cross-linked chitosan beads: the influence of anion structure", *Int. J. Pharm.*, vol. 233, no. 1-2, pp. 217-225, 2002. [http://dx.doi.org/10.1016/S0378-5173(01)00943-7] [PMID: 11897426]

[472] M. Mucha, J. Balcerzak, I. Michalak, and M. Tylman, "Biopolymeric matrices based on chitosan for medical applications", *e-Polymers,* vol. 11, pp. 21-28, 2011.

[473] A. Bernkop-Schnürch, and S. Dünnhaupt, "Chitosan-based drug delivery systems", *Eur. J. Pharm. Biopharm.*, vol. 81, no. 3, pp. 463-469, 2012. [http://dx.doi.org/10.1016/j.ejpb.2012.04.007] [PMID: 22561955]

[474] A. Fini, and I. Orienti, "The role of chitosan in drug delivery", *Am. J. Drug Deliv.*, vol. 1, pp. 43-59, 2003. [http://dx.doi.org/10.2165/00137696-200301010-00004]

[475] S.A. Agnihotri, N.N. Mallikarjuna, and T.M. Aminabhavi, "Recent advances on chitosan-based micro- and nanoparticles in drug delivery", *J. Control. Release,* vol. 100, no. 1, pp. 5-28, 2004. [http://dx.doi.org/10.1016/j.jconrel.2004.08.010] [PMID: 15491807]

[476] M. Prabaharan, "Review paper: chitosan derivatives as promising materials for controlled drug delivery", *J. Biomater. Appl.*, vol. 23, pp. 5-36, 2008. [http://dx.doi.org/10.1177/0885328208091562]

[477] M.N. Kumar, "A review of chitin and chitosan applications", *React. Funct. Polym.*, vol. 46, pp. 1-27, 2000. [http://dx.doi.org/10.1016/S1381-5148(00)00038-9]

[478] J.S. Boateng, K.H. Matthews, H.N. Stevens, and G.M. Eccleston, "Wound healing dressings and drug delivery systems: a review", *J. Pharm. Sci.*, vol. 97, no. 8, pp. 2892-2923, 2008. [http://dx.doi.org/10.1002/jps.21210] [PMID: 17963217]

[479] D. Duttagupta, V. Jadhav, and V. Kadam, *Chitosan: a propitious biopolymer for drug delivery.* Curr. Drug Del, 2015.

[480] D. Singh, and A.R. Ray, "Graft copolymerization of 2-hydroxyethylmethacrylate onto chitosan films and their blood compatibility", *J. Appl. Polym. Sci.*, vol. 53, pp. 1115-1121, 1994. [http://dx.doi.org/10.1002/app.1994.070530814]

[481] D.K. Singh, and A.R. Ray, "Controlled release of glucose through modified chitosan membranes", *J. Membr. Sci.*, vol. 155, pp. 107-112, 1999. [http://dx.doi.org/10.1016/S0376-7388(98)00311-1]

[482] T. Nagai, Y. Sawayanagi, and N. Nambu, *Application of chitin and chitosan to pharmaceutical preparations.*Chitin, Chitosan and Related Enzymes, 1984, pp. 21-39. [http://dx.doi.org/10.1016/B978-0-12-780950-2.50008-3]

[483] N. Bhattarai, J. Gunn, and M. Zhang, "Chitosan-based hydrogels for controlled, localized drug delivery", *Adv. Drug Deliv. Rev.*, vol. 62, no. 1, pp. 83-99, 2010. [http://dx.doi.org/10.1016/j.addr.2009.07.019] [PMID: 19799949]

[484] M.L. Lorenzo-Lamosa, C. Remuñán-López, J.L. Vila-Jato, and M.J. Alonso, "Design of microencapsulated chitosan microspheres for colonic drug delivery", *J. Control. Release,* vol. 52, no. 1-2, pp. 109-118, 1998. [http://dx.doi.org/10.1016/S0168-3659(97)00203-4] [PMID: 9685941]

[485] J.R. Franca, M.P. De Luca, T.G. Ribeiro, R.O. Castilho, A.N. Moreira, V.R. Santos, and A.A. Faraco, "Propolisbased chitosan varnish: drug delivery, controlled release and antimicrobial activity against oral pathogen bacteria", *BMC Complement. Altern. Med.,* vol. 14, p. 478, 2014. [http://dx.doi.org/10.1186/1472-6882-14-478] [PMID: 25495921]

[486] J. Xu, S. Strandman, J.X. Zhu, J. Barralet, and M. Cerruti, "Genipin-crosslinked catechol-chitosan

mucoadhesive hydrogels for buccal drug delivery", *Biomaterials,* vol. 37, pp. 395-404, 2015. [http://dx.doi.org/10.1016/j.biomaterials.2014.10.024] [PMID: 25453967]

[487] K.S. Bhise, R.S. Dhumal, A.R. Paradkar, and S.S. Kadam, "Effect of drying methods on swelling, erosion and drug release from chitosan-naproxen sodium complexes", *AAPS PharmSciTech,* vol. 9, no. 1, pp. 1-12, 2008. [http://dx.doi.org/10.1208/s12249-007-9001-0] [PMID: 18446455]

[488] W. Sun, S. Mao, Y. Wang, V.B. Junyaprasert, T. Zhang, L. Na, and J. Wang, "Bioadhesion and oral absorption of enoxaparin nanocomplexes", *Int. J. Pharm.,* vol. 386, no. 1-2, pp. 275-281, 2010. [http://dx.doi.org/10.1016/j.ijpharm.2009.11.025] [PMID: 19958824]

[489] T. Sonia, and C.P. Sharma, "Chitosan and its derivatives for drug delivery perspective", In: *Chitosan for Biomaterials I*Springer, 2011, pp. 23-53. [http://dx.doi.org/10.1007/12_2011_117]

[490] J. Shaji, V. Jain, and S. Lodha, "Chitosan: a novel pharmaceutical excipient", *Int J Pharm Appl Sci,* vol. 1, pp. 11-28, 2010.

[491] A. Mitra, and B. Dey, "Chitosan microspheres in novel drug delivery systems", *Indian J. Pharm. Sci.,* vol. 73, no. 4, pp. 355-366, 2011. [PMID: 22707817]

[492] D. Guggi, and A. Bernkop-Schnürch, "in vitro evaluation of polymeric excipients protecting calcitonin against degradation by intestinal serine proteases", *Int. J. Pharm.,* vol. 252, no. 1-2, pp. 187-196, 2003. [http://dx.doi.org/10.1016/S0378-5173(02)00631-2] [PMID: 12550794]

[493] T. Ouchi, K. Inosaka, T. Banba, Y. Ohya, C. Brine, P. Sanford, and J. Zikakis, "Advances in chitin and chitosan", In: *Brine, CJ*1992, p. 106.

[494] I.M. van der Lubben, J.C. Verhoef, G. Borchard, and H.E. Junginger, "Chitosan and its derivatives in mucosal drug and vaccine delivery", *Eur. J. Pharm. Sci.,* vol. 14, no. 3, pp. 201-207, 2001. [http://dx.doi.org/10.1016/S0928-0987(01)00172-5] [PMID: 11576824]

[495] A. Bernkop-Schnürch, U. Brandt, and A. Clausen, "Synthesis and *in vitro* evaluation of chitosan-cysteine conjugates", *Sci. Pharm.,* vol. 67, pp. 196-208, 1999.

[496] I. Bravo-Osuna, C. Vauthier, A. Farabollini, G.F. Palmieri, and G. Ponchel, "Mucoadhesion mechanism of chitosan and thiolated chitosan-poly(isobutyl cyanoacrylate) core-shell nanoparticles", *Biomaterials,* vol. 28, no. 13, pp. 2233-2243, 2007. [http://dx.doi.org/10.1016/j.biomaterials.2007.01.005] [PMID: 17261330]

[497] B. Andreas, M. Hornof, and T. Zoidl, "Thiolated polymers–thiomers: modification of chitosan with 2-iminothiolane", *Int. J. Pharm.,* vol. 260, pp. 229-237, 2003. [http://dx.doi.org/10.1016/S0378-5173(03)00271-0] [PMID: 12842342]

[498] M. Roldo, M. Hornof, P. Caliceti, and A. Bernkop-Schnürch, "Mucoadhesive thiolated chitosans as platforms for oral controlled drug delivery: synthesis and *in vitro* evaluation", *Eur. J. Pharm. Biopharm.,* vol. 57, no. 1, pp. 115-121, 2004. [http://dx.doi.org/10.1016/S0939-6411(03)00157-7] [PMID: 14729087]

[499] M. Werle, and A. Bernkop-Schnürch, "Thiolated chitosans: useful excipients for oral drug delivery", *J. Pharm. Pharmacol.,* vol. 60, no. 3, pp. 273-281, 2008. [http://dx.doi.org/10.1211/jpp.60.3.3001] [PMID: 18284806]

[500] S. Dünnhaupt, J. Barthelmes, J. Iqbal, G. Perera, C.C. Thurner, H. Friedl, and A. Bernkop-Schnürch, "*in vivo* evaluation of an oral drug delivery system for peptides based on S-protected thiolated chitosan", *J. Control. Release,* vol. 160, no. 3, pp. 477-485, 2012. [http://dx.doi.org/10.1016/j.jconrel.2012.04.020] [PMID: 22542699]

[501] S. Sreenivas, and K. Pai, "Thiolated chitosans: novel polymers for mucoadhesive drug delivery–a review", *Trop. J. Pharm. Res.,* vol. 7, pp. 1077-1088, 2008. [http://dx.doi.org/10.4314/tjpr.v7i3.14694]

[502] K. Kafedjiiski, F. Föger, M. Werle, and A. Bernkop-Schnürch, "Synthesis and in vitro evaluation of a novel chitosan-glutathione conjugate", *Pharm. Res.,* vol. 22, no. 9, pp. 1480-1488, 2005. [http://dx.doi.org/10.1007/s11095-005-6248-6] [PMID: 16132360]

[503] F. Sarti, and A. Bernkop-Schnürch, "Chitosan and thiolated chitosan", In: *Chitosan for Biomaterials* /Springer, 2011, pp. 93-110. [http://dx.doi.org/10.1007/12_2011_109]

[504] A. Jintapattanakit, V.B. Junyaprasert, and T. Kissel, "The role of mucoadhesion of trimethyl chitosan and PEGylated trimethyl chitosan nanocomplexes in insulin uptake", *J. Pharm. Sci.,* vol. 98, no. 12, pp. 4818-4830, 2009. [http://dx.doi.org/10.1002/jps.21783] [PMID: 19408295]

[505] V. Mourya, N.N. Inamdar, and A. Tiwari, "Carboxymethyl chitosan and its applications", *Advanced Materials Letters,* vol. 1, pp. 11-33, 2010. [http://dx.doi.org/10.5185/amlett.2010.3108]

[506] Y. Lin, Q. Chen, and H. Luo, "Preparation and characterization of N-(2-carboxybenzyl)chitosan as a potential pH-sensitive hydrogel for drug delivery", *Carbohydr. Res.,* vol. 342, no. 1, pp. 87-95, 2007. [http://dx.doi.org/10.1016/j.carres.2006.11.002] [PMID: 17125756]

[507] M.N. Kumar, R.A. Muzzarelli, C. Muzzarelli, H. Sashiwa, and A.J. Domb, "Chitosan chemistry and pharmaceutical perspectives", *Chem. Rev.,* vol. 104, no. 12, pp. 6017-6084, 2004. [http://dx.doi.org/10.1021/cr030441b] [PMID: 15584695]

[508] R. Yamaguchi, Y. Arai, T. Itoh, and S. Hirano, "Preparation of partially N-succinylated chitosans and their cross-linked gels", *Carbohydr. Res.,* vol. 88, pp. 172-175, 1981. [http://dx.doi.org/10.1016/S0008-6215(00)84614-5]

[509] M.R. Rekha, and C.P. Sharma, "Synthesis and evaluation of lauryl succinyl chitosan particles towards oral insulin delivery and absorption", *J. Control. Release,* vol. 135, no. 2, pp. 144-151, 2009. [http://dx.doi.org/10.1016/j.jconrel.2009.01.011] [PMID: 19331862]

[510] F.H. Nasr, and S. Khoee, "Design, characterization and in vitro evaluation of novel shell crosslinked poly(butylene adipate)-co-N-succinyl chitosan nanogels containing loteprednol etabonate: A new system for therapeutic effect enhancement via controlled drug delivery", *Eur. J. Med. Chem.,* vol. 102, pp. 132-142, 2015. [http://dx.doi.org/10.1016/j.ejmech.2015.07.045] [PMID: 26263245]

[511] P. Mukhopadhyay, K. Sarkar, S. Bhattacharya, A. Bhattacharyya, R. Mishra, and P.P. Kundu, "pH sensitive N-succinyl chitosan grafted polyacrylamide hydrogel for oral insulin delivery", *Carbohydr. Polym.,* vol. 112, pp. 627-637, 2014. [http://dx.doi.org/10.1016/j.carbpol.2014.06.045] [PMID: 25129792]

[512] Z. Aiping, C. Tian, Y. Lanhua, W. Hao, and L. Ping, "Synthesis and characterization of N-succiny--chitosan and its self-assembly of nanospheres", *Carbohydr. Polym.,* vol. 66, pp. 274-279, 2006. [http://dx.doi.org/10.1016/j.carbpol.2006.03.014]

[513] L. Li, X-y. Xu, and J-p. Zhou, "Preparation and characterization of N-octyl-N'-succinyl chitosan micelles", *CHINESE NEW DRUGS JOURNAL,* vol. 16, p. 543, 2007.

[514] T. Woraphatphadung, W. Sajomsang, P. Gonil, S. Saesoo, and P. Opanasopit, "Synthesis and characterization of pH-responsive N-naphthyl-N,O-succinyl chitosan micelles for oral meloxicam delivery", *Carbohydr. Polym.,* vol. 121, pp. 99-106, 2015. [http://dx.doi.org/10.1016/j.carbpol.2014.12.039] [PMID: 25659677]

[515] Y. Kato, H. Onishi, and Y. Machida, "Evaluation of N-succinyl-chitosan as a systemic long-circulating polymer", *Biomaterials,* vol. 21, no. 15, pp. 1579-1585, 2000. [http://dx.doi.org/10.1016/S0142-9612(00)00044-2] [PMID: 10885730]

[516] Y. Kato, H. Onishi, and Y. Machida, "Lactosaminated N-succinyl-chitosan: Preparation and biodistribution into the intestine, bone, lymph nodes and male genital organs after Iv administration",

Macromol. Res., vol. 11, pp. 382-386, 2003. [http://dx.doi.org/10.1007/BF03218380]

[517] M. Bivas-Benita, M. Laloup, S. Versteyhe, J. Dewit, J. De Braekeleer, E. Jongert, and G. Borchard, "Generation of Toxoplasma gondii GRA1 protein and DNA vaccine loaded chitosan particles: preparation, characterization, and preliminary in vivo studies", *Int. J. Pharm.*, vol. 266, no. 1-2, pp. 17-27, 2003. [http://dx.doi.org/10.1016/S0378-5173(03)00377-6] [PMID: 14559390]

[518] Y. Xie, N-J. Zhou, Y-F. Gong, X-J. Zhou, J. Chen, S-J. Hu, N-H. Lu, and X-H. Hou, "Th immune response induced by H pylori vaccine with chitosan as adjuvant and its relation to immune protection", *World J. Gastroenterol.*, vol. 13, no. 10, pp. 1547-1553, 2007. [http://dx.doi.org/10.3748/wjg.v13.i10.1547] [PMID: 17461447]

[519] H. Hasegawa, T. Ichinohe, S.-i. Tamura, and T. Kurata, *Development of a mucosal vaccine for influenza viruses: preparation for a potential influenza pandemic*, 2007. [http://dx.doi.org/10.1586/14760584.6.2.193]

[520] M. Amidi, S.G. Romeijn, J.C. Verhoef, H.E. Junginger, L. Bungener, A. Huckriede, D.J. Crommelin, and W. Jiskoot, "N-trimethyl chitosan (TMC) nanoparticles loaded with influenza subunit antigen for intranasal vaccination: biological properties and immunogenicity in a mouse model", *Vaccine*, vol. 25, no. 1, pp. 144-153, 2007. [http://dx.doi.org/10.1016/j.vaccine.2006.06.086] [PMID: 16973248]

[521] L. Illum, I. Jabbal-Gill, M. Hinchcliffe, A.N. Fisher, and S.S. Davis, "Chitosan as a novel nasal delivery system for vaccines", *Adv. Drug Deliv. Rev.*, vol. 51, no. 1-3, pp. 81-96, 2001. [http://dx.doi.org/10.1016/S0169-409X(01)00171-5] [PMID: 11516781]

[522] M.L. González-Rodríguez, M.A. Holgado, C. Sánchez-Lafuente, A.M. Rabasco, and A. Fini, "Alginate/chitosan particulate systems for sodium diclofenac release", *Int. J. Pharm.*, vol. 232, no. 1-2, pp. 225-234, 2002. [http://dx.doi.org/10.1016/S0378-5173(01)00915-2] [PMID: 11790506]

[523] F.L. Mi, H.W. Sung, and S.S. Shyu, "Release of indomethacin from a novel chitosan microsphere prepared by a naturally occurring crosslinker: Examination of crosslinking and polycation–anionic drug interaction", *J. Appl. Polym. Sci.*, vol. 81, pp. 1700-1711, 2001. [http://dx.doi.org/10.1002/app.1602]

[524] Y. Sun, F. Cui, K. Shi, J. Wang, M. Niu, and R. Ma, "The effect of chitosan molecular weight on the characteristics of spray-dried methotrexate-loaded chitosan microspheres for nasal administration", *Drug Dev. Ind. Pharm.*, vol. 35, no. 3, pp. 379-386, 2009. [http://dx.doi.org/10.1080/03639040802395185] [PMID: 18951272]

[525] A.P. Rokhade, N.B. Shelke, S.A. Patil, and T.M. Aminabhavi, "Novel interpenetrating polymer network microspheres of chitosan and methylcellulose for controlled release of theophylline", *Carbohydr. Polym.*, vol. 69, pp. 678-687, 2007. [http://dx.doi.org/10.1016/j.carbpol.2007.02.008]

[526] V.R. Sinha, A.K. Singla, S. Wadhawan, R. Kaushik, R. Kumria, K. Bansal, and S. Dhawan, "Chitosan microspheres as a potential carrier for drugs", *Int. J. Pharm.*, vol. 274, no. 1-2, pp. 1-33, 2004. [http://dx.doi.org/10.1016/j.ijpharm.2003.12.026] [PMID: 15072779]

[527] P.R. Hari, T. Chandy, and C.P. Sharma, "Chitosan/calcium alginate microcapsules for intestinal delivery of nitrofurantoin", *J. Microencapsul.*, vol. 13, no. 3, pp. 319-329, 1996. [http://dx.doi.org/10.3109/02652049609026019] [PMID: 8860687]

[528] Q. Zhao, B. Han, Z. Wang, C. Gao, C. Peng, and J. Shen, "Hollow chitosan-alginate multilayer microcapsules as drug delivery vehicle: doxorubicin loading and in vitro and in vivo studies", *Nanomedicine (Lond.)*, vol. 3, no. 1, pp. 63-74, 2007. [PMID: 17379170]

[529] P-L. Lam, K-H. Lee, Y-W. Ho, R-M. Wong, S-W. Tong, C-H. Cheng, K-H. Lam, J-O. Tang, Z-X.

Bian, and R. Gambari, "The development of chitosan based microcapsules as delivery vehicles for orally administered daunorubicin", *RSC Advances,* vol. 4, pp. 14109-14114, 2014. [http://dx.doi.org/10.1039/c4ra00195h]

[530] H. Gurruchaga, L. Saenz del Burgo, J. Ciriza, G. Orive, R.M. Hernández, and J.L. Pedraz, "Advances in cell encapsulation technology and its application in drug delivery", *Expert Opin. Drug Deliv.,* vol. 12, no. 8, pp. 1251-1267, 2015. [http://dx.doi.org/10.1517/17425247.2015.1001362] [PMID: 25563077]

[531] M. Dash, F. Chiellini, R.M. Ottenbrite, and E. Chiellini, "Chitosan-A versatile semi-synthetic polymer in biomedical applications", *Prog. Polym. Sci.,* vol. 36, pp. 981-1014, 2011. [http://dx.doi.org/10.1016/j.progpolymsci.2011.02.001]

[532] D. Ibraheem, A. Elaissari, and H. Fessi, "Gene therapy and DNA delivery systems", *Int. J. Pharm.,* vol. 459, no. 1-2, pp. 70-83, 2014. [http://dx.doi.org/10.1016/j.ijpharm.2013.11.041] [PMID: 24286924]

[533] S. Mansouri, Y. Cuie, F. Winnik, Q. Shi, P. Lavigne, M. Benderdour, E. Beaumont, and J.C. Fernandes, "Characterization of folate-chitosan-DNA nanoparticles for gene therapy", *Biomaterials,* vol. 27, no. 9, pp. 2060-2065, 2006. [http://dx.doi.org/10.1016/j.biomaterials.2005.09.020] [PMID: 16202449]

[534] N. Duceppe, and M. Tabrizian, "Advances in using chitosan-based nanoparticles for in vitro and in vivo drug and gene delivery", *Expert Opin. Drug Deliv.,* vol. 7, no. 10, pp. 1191-1207, 2010. [http://dx.doi.org/10.1517/17425247.2010.514604] [PMID: 20836623]

[535] M. Prabaharan, and J.F. Mano, "Chitosan-based particles as controlled drug delivery systems", *Drug Deliv.,* vol. 12, no. 1, pp. 41-57, 2005. [http://dx.doi.org/10.1080/10717540590889781] [PMID: 15801720]

[536] A.K. Sailaja, P. Amareshwar, and P. Chakravarty, "Chitosan nanoparticles as a drug delivery system", *Res. J. Pharm. Biol. Chem. Sci,* vol. 1, pp. 474-484, 2010.

[537] F. Bordi, L. Chronopoulou, C. Palocci, F. Bomboi, A. Di Martino, N. Cifani, B. Pompili, F. Ascenzioni, and S. Sennato, "Chitosan–DNA complexes: effect of molecular parameters on the efficiency of delivery", *Colloids Surf. Physicochem. Eng. Aspects,* vol. 460, pp. 184-190, 2014. [http://dx.doi.org/10.1016/j.colsurfa.2013.12.022]

[538] T. Sato, T. Ishii, and Y. Okahata, "in vitro gene delivery mediated by chitosan. effect of pH, serum, and molecular mass of chitosan on the transfection efficiency", *Biomaterials,* vol. 22, no. 15, pp. 2075-2080, 2001. [http://dx.doi.org/10.1016/S0142-9612(00)00385-9] [PMID: 11432586]

[539] S. Danielsen, S. Strand, C. de Lange Davies, and B.T. Stokke, "“Glycosaminoglycan destabilization of DNA–chitosan polyplexes for gene delivery depends on chitosan chain length and GAG properties,” Biochimica et Biophysica Acta (BBA)-", *General Subjects,* vol. 1721, pp. 44-54, 2005. [http://dx.doi.org/10.1016/j.bbagen.2004.10.011]

[540] S. Mansouri, P. Lavigne, K. Corsi, M. Benderdour, E. Beaumont, and J.C. Fernandes, "Chitosan-DNA nanoparticles as non-viral vectors in gene therapy: strategies to improve transfection efficacy", *Eur. J. Pharm. Biopharm.,* vol. 57, no. 1, pp. 1-8, 2004. [http://dx.doi.org/10.1016/S0939-6411(03)00155-3] [PMID: 14729076]

[541] Y. Kawamata, Y. Nagayama, K. Nakao, H. Mizuguchi, T. Hayakawa, T. Sato, and N. Ishii, "Receptor-independent augmentation of adenovirus-mediated gene transfer with chitosan *in vitro*", *Biomaterials,* vol. 23, no. 23, pp. 4573-4579, 2002. [http://dx.doi.org/10.1016/S0142-9612(02)00203-X] [PMID: 12322978]

[542] R.N. Mitra, Z. Han, M. Merwin, M. Al Taai, S.M. Conley, and M.I. Naash, "Synthesis and characterization of glycol chitosan DNA nanoparticles for retinal gene delivery", *ChemMedChem,* vol. 9, no. 1, pp. 189-196, 2014. [http://dx.doi.org/10.1002/cmdc.201300371] [PMID: 24203490]

[543] H. Lu, Y. Dai, L. Lv, and H. Zhao, *Chitosan-graft-polyethylenimine/DNA nanoparticles as novel non-viral gene delivery vectors targeting osteoarthritis,* 2014. [http://dx.doi.org/10.1371/journal.pone.0084703]

[544] H-Q. Mao, K. Roy, V.L. Troung-Le, K.A. Janes, K.Y. Lin, Y. Wang, J.T. August, and K.W. Leong, "Chitosan-DNA nanoparticles as gene carriers: synthesis, characterization and transfection efficiency", *J. Control. Release,* vol. 70, no. 3, pp. 399-421, 2001. [http://dx.doi.org/10.1016/S0168-3659(00)00361-8] [PMID: 11182210]

[545] H. Dai, X. Jiang, G.C. Tan, Y. Chen, M. Torbenson, K.W. Leong, and H-Q. Mao, "Chitosan-DNA nanoparticles delivered by intrabiliary infusion enhance liver-targeted gene delivery", *Int. J. Nanomedicine,* vol. 1, no. 4, pp. 507-522, 2006. [http://dx.doi.org/10.2147/nano.2006.1.4.507] [PMID: 17369870]

[546] I.K. Park, Y.H. Park, B.A. Shin, E.S. Choi, Y.R. Kim, T. Akaike, and C.S. Cho, "Galactosylated chitosan-graft-dextran as hepatocyte-targeting DNA carrier", *J. Control. Release,* vol. 69, no. 1, pp. 97-108, 2000. [http://dx.doi.org/10.1016/S0168-3659(00)00298-4] [PMID: 11018549]

[547] S. Gao, J. Chen, X. Xu, Z. Ding, Y-H. Yang, Z. Hua, and J. Zhang, "Galactosylated low molecular weight chitosan as DNA carrier for hepatocyte-targeting", *Int. J. Pharm.,* vol. 255, no. 1-2, pp. 57-68, 2003. [http://dx.doi.org/10.1016/S0378-5173(03)00082-6] [PMID: 12672602]

[548] H.L. Jiang, J.T. Kwon, Y.K. Kim, E.M. Kim, R. Arote, H.J. Jeong, J.W. Nah, Y.J. Choi, T. Akaike, M-H. Cho, and C.S. Cho, "Galactosylated chitosan-graft-polyethylenimine as a gene carrier for hepatocyte targeting", *Gene Ther.,* vol. 14, no. 19, pp. 1389-1398, 2007. [http://dx.doi.org/10.1038/sj.gt.3302997] [PMID: 17637795]

[549] Q. Wang, L. Zhang, W. Hu, Z-H. Hu, Y-Y. Bei, J-Y. Xu, W-J. Wang, X-N. Zhang, and Q. Zhang, "Norcantharidin-associated galactosylated chitosan nanoparticles for hepatocyte-targeted delivery", *Nanomedicine (Lond.),* vol. 6, no. 2, pp. 371-381, 2010. [PMID: 19699319]

[550] S.M. Alex, M.R. Rekha, and C.P. Sharma, "Spermine grafted galactosylated chitosan for improved nanoparticle mediated gene delivery", *Int. J. Pharm.,* vol. 410, no. 1-2, pp. 125-137, 2011. [http://dx.doi.org/10.1016/j.ijpharm.2011.02.067] [PMID: 21396993]

[551] J.H. Park, G. Saravanakumar, K. Kim, and I.C. Kwon, "Targeted delivery of low molecular drugs using chitosan and its derivatives", *Adv. Drug Deliv. Rev.,* vol. 62, no. 1, pp. 28-41, 2010. [http://dx.doi.org/10.1016/j.addr.2009.10.003] [PMID: 19874862]

[552] H. Katas, and H.O. Alpar, "Development and characterisation of chitosan nanoparticles for siRNA delivery", *J. Control. Release,* vol. 115, no. 2, pp. 216-225, 2006. [http://dx.doi.org/10.1016/j.jconrel.2006.07.021] [PMID: 16959358]

[553] X. Liu, K.A. Howard, M. Dong, M.Ø. Andersen, U.L. Rahbek, M.G. Johnsen, O.C. Hansen, F. Besenbacher, and J. Kjems, "The influence of polymeric properties on chitosan/siRNA nanoparticle formulation and gene silencing", *Biomaterials,* vol. 28, no. 6, pp. 1280-1288, 2007. [http://dx.doi.org/10.1016/j.biomaterials.2006.11.004] [PMID: 17126901]

[554] H. Shibata, N. Shioya, and Y. Kuroyanagi, "Development of new wound dressing composed of spongy collagen sheet containing dibutyryl cyclic AMP", *J. Biomater. Sci. Polym. Ed.,* vol. 8, no. 8, pp. 601-621, 1997. [http://dx.doi.org/10.1163/156856297X00209] [PMID: 9241584]

[555] R. Jayakumar, M. Prabaharan, P.T. Sudheesh Kumar, S.V. Nair, and H. Tamura, "Biomaterials based on chitin and chitosan in wound dressing applications", *Biotechnol. Adv.,* vol. 29, no. 3, pp. 322-337, 2011. [http://dx.doi.org/10.1016/j.biotechadv.2011.01.005] [PMID: 21262336]

[556] J.G. Medina, and E.G. Sherman, "Chitosan paste wound dressing", In: *Google Patents* 2014.

[557] J.F. Prudden, P. Migel, P. Hanson, L. Friedrich, and L. Balassa, "The discovery of a potent pure chemical wound-healing accelerator", *Am. J. Surg.*, vol. 119, no. 5, pp. 560-564, 1970. [http://dx.doi.org/10.1016/0002-9610(70)90175-3] [PMID: 5441746]

[558] H. Ueno, T. Mori, and T. Fujinaga, "Topical formulations and wound healing applications of chitosan", *Adv. Drug Deliv. Rev.*, vol. 52, no. 2, pp. 105-115, 2001. [http://dx.doi.org/10.1016/S0169-409X(01)00189-2] [PMID: 11718934]

[559] H. Ueno, F. Nakamura, M. Murakami, M. Okumura, T. Kadosawa, and T. Fujinag, "Evaluation effects of chitosan for the extracellular matrix production by fibroblasts and the growth factors production by macrophages", *Biomaterials,* vol. 22, no. 15, pp. 2125-2130, 2001. [http://dx.doi.org/10.1016/S0142-9612(00)00401-4] [PMID: 11432592]

[560] J. Berger, M. Reist, J.M. Mayer, O. Felt, and R. Gurny, "Structure and interactions in chitosan hydrogels formed by complexation or aggregation for biomedical applications", *Eur. J. Pharm. Biopharm.*, vol. 57, no. 1, pp. 35-52, 2004. [http://dx.doi.org/10.1016/S0939-6411(03)00160-7] [PMID: 14729079]

[561] O. Zholobko, I. Tarnavchyk, A. Voronov, O. Budishevska, A. Kohut, and S. Voronov, "Covalent bond formation in the reaction of glucosamine as monomer unit of chitosan macromolecules and poly (ethylene glycol) disuccinate at elevated temperature", *Journal of Chitin and Chitosan Science,* vol. 2, pp. 299-305, 2014. [http://dx.doi.org/10.1166/jcc.2014.1072]

[562] M. Ishihara, K. Nakanishi, K. Ono, M. Sato, M. Kikuchi, Y. Saito, H. Yura, T. Matsui, H. Hattori, M. Uenoyama, and A. Kurita, "Photocrosslinkable chitosan as a dressing for wound occlusion and accelerator in healing process", *Biomaterials,* vol. 23, no. 3, pp. 833-840, 2002. [http://dx.doi.org/10.1016/S0142-9612(01)00189-2] [PMID: 11771703]

[563] S-Y. Ong, J. Wu, S.M. Moochhala, M-H. Tan, and J. Lu, "Development of a chitosan-based wound dressing with improved hemostatic and antimicrobial properties", *Biomaterials,* vol. 29, no. 32, pp. 4323-4332, 2008. [http://dx.doi.org/10.1016/j.biomaterials.2008.07.034] [PMID: 18708251]

[564] T. Dai, M. Tanaka, Y.-Y. Huang, and M.R. Hamblin, "Chitosan preparations for wounds and burns: antimicrobial and wound-healing effects", *Expert Review of Anti-infective Therapy,* vol. 9, no. 7, pp. 857-879, 2011. [http://dx.doi.org/10.1586/eri.11.59]

[565] R. Edwards, and K.G. Harding, "Bacteria and wound healing", *Curr. Opin. Infect. Dis.,* vol. 17, no. 2, pp. 91-96, 2004. [http://dx.doi.org/10.1097/00001432-200404000-00004] [PMID: 15021046]

[566] C. Wiegand, M. Abel, P. Ruth, P. Elsner, and U-C. Hipler, "*in vitro* assessment of the antimicrobial activity of wound dressings: influence of the test method selected and impact of the pH", *J. Mater. Sci. Mater. Med.,* vol. 26, no. 1, p. 5343, 2015. [http://dx.doi.org/10.1007/s10856-014-5343-9] [PMID: 25578697]

[567] M.A. Brown, M.R. Daya, and J.A. Worley, "Experience with chitosan dressings in a civilian EMS system", *J. Emerg. Med.,* vol. 37, no. 1, pp. 1-7, 2009. [http://dx.doi.org/10.1016/j.jemermed.2007.05.043] [PMID: 18024069]

[568] H. Tamura, T. Furuike, S. Nair, and R. Jayakumar, "Biomedical applications of chitin hydrogel membranes and scaffolds", *Carbohydr. Polym.,* vol. 84, pp. 820-824, 2011. [http://dx.doi.org/10.1016/j.carbpol.2010.06.001]

[569] I.-Y. Kim, S.-J. Seo, H.-S. Moon, M.-K. Yoo, I.-Y. Park, B.-C. Kim, and C.-S. Cho, "Chitosan and its derivatives for tissue engineering applications", *Biotechnol. Adv.,* vol. 26, pp. 1-21, 2008. [http://dx.doi.org/10.1016/j.biotechadv.2007.07.009]

[570] W-E. Teo, and S. Ramakrishna, "Electrospun nanofibers as a platform for multifunctional, hierarchically organized nanocomposite", *Compos. Sci. Technol.*, vol. 69, pp. 1804-1817, 2009. [http://dx.doi.org/10.1016/j.compscitech.2009.04.015]

[571] Z-M. Huang, Y-Z. Zhang, M. Kotaki, and S. Ramakrishna, "A review on polymer nanofibers by electrospinning and their applications in nanocomposites", *Compos. Sci. Technol.*, vol. 63, pp. 2223-2253, 2003.
[http://dx.doi.org/10.1016/S0266-3538(03)00178-7]

[572] J.D. Schiffman, and C.L. Schauer, "One-step electrospinning of cross-linked chitosan fibers", *Biomacromolecules,* vol. 8, no. 9, pp. 2665-2667, 2007.
[http://dx.doi.org/10.1021/bm7006983] [PMID: 17696400]

[573] H. Homayoni, S.A. Ravandi, and M. Valizadeh, "Electrospinning of chitosan nanofibers: Processing optimization", *Carbohydr. Polym.*, vol. 77, pp. 656-661, 2009.
[http://dx.doi.org/10.1016/j.carbpol.2009.02.008]

[574] Y. Zhou, D. Yang, X. Chen, Q. Xu, F. Lu, and J. Nie, "Electrospun water-soluble carboxyethyl chitosan/poly(vinyl alcohol) nanofibrous membrane as potential wound dressing for skin regeneration", *Biomacromolecules,* vol. 9, no. 1, pp. 349-354, 2008.
[http://dx.doi.org/10.1021/bm7009015] [PMID: 18067266]

[575] C.C. Wang, C.H. Su, and C.C. Chen, "Water absorbing and antibacterial properties of N-isopropyl acrylamide grafted and collagen/chitosan immobilized polypropylene nonwoven fabric and its application on wound healing enhancement", *J. Biomed. Mater. Res. A,* vol. 84, no. 4, pp. 1006-1017, 2008.
[http://dx.doi.org/10.1002/jbm.a.31482] [PMID: 17647243]

[576] J-K. Suh, and H.W. Matthew, "Application of chitosan-based polysaccharide biomaterials in cartilage tissue engineering: a review", *Biomaterials,* vol. 21, no. 24, pp. 2589-2598, 2000.
[http://dx.doi.org/10.1016/S0142-9612(00)00126-5] [PMID: 11071608]

[577] L. Ma, C. Gao, Z. Mao, J. Zhou, J. Shen, X. Hu, and C. Han, "Collagen/chitosan porous scaffolds with improved biostability for skin tissue engineering", *Biomaterials,* vol. 24, no. 26, pp. 4833-4841, 2003.
[http://dx.doi.org/10.1016/S0142-9612(03)00374-0] [PMID: 14530080]

[578] S.V. Madihally, and H.W. Matthew, "Porous chitosan scaffolds for tissue engineering", *Biomaterials,* vol. 20, no. 12, pp. 1133-1142, 1999.
[http://dx.doi.org/10.1016/S0142-9612(99)00011-3] [PMID: 10382829]

[579] S.L. Levengood, and M. Zhang, "Chitosan-based scaffolds for bone tissue engineering", *J. Mater. Chem. B Mater. Biol. Med.,* vol. 2, no. 21, pp. 3161-3184, 2014.
[http://dx.doi.org/10.1039/c4tb00027g] [PMID: 24999429]

[580] A. Martínez, M.D. Blanco, N. Davidenko, and R.E. Cameron, "Tailoring chitosan/collagen scaffolds for tissue engineering: Effect of composition and different crosslinking agents on scaffold properties", *Carbohydr. Polym.*, vol. 132, pp. 606-619, 2015.
[http://dx.doi.org/10.1016/j.carbpol.2015.06.084] [PMID: 26256388]

[581] N. Vázquez, M. Chacón, Á. Meana, Y. Menéndez-Menéndez, A. Ferrero-Gutierrez, D. Cereijo-Martín, M. Naveiras, and J. Merayo-Lloves, "Keratin-chitosan membranes as scaffold for tissue engineering of human cornea", *Histol. Histopathol.*, vol. 30, no. 7, pp. 813-821, 2015.
[PMID: 25587895]

[582] J.E. Lee, K.E. Kim, I.C. Kwon, H.J. Ahn, S-H. Lee, H. Cho, H.J. Kim, S.C. Seong, and M.C. Lee, "Effects of the controlled-released TGF-β 1 from chitosan microspheres on chondrocytes cultured in a collagen/chitosan/glycosaminoglycan scaffold", *Biomaterials,* vol. 25, no. 18, pp. 4163-4173, 2004.
[http://dx.doi.org/10.1016/j.biomaterials.2003.10.057] [PMID: 15046906]

[583] W.W. Thein-Han, and R.D. Misra, "Biomimetic chitosan-nanohydroxyapatite composite scaffolds for bone tissue engineering", *Acta Biomater.*, vol. 5, no. 4, pp. 1182-1197, 2009.

[http://dx.doi.org/10.1016/j.actbio.2008.11.025] [PMID: 19121983]

[584] T. Chandy, and C.P. Sharma, "Chitosan-as a biomaterial", *Artif. Cells Blood Substit. Biotechnol,* vol. 18, pp. 1-24, 1990.

[585] S. Yamane, N. Iwasaki, T. Majima, T. Funakoshi, T. Masuko, K. Harada, A. Minami, K. Monde, and S. Nishimura, "Feasibility of chitosan-based hyaluronic acid hybrid biomaterial for a novel scaffold in cartilage tissue engineering", *Biomaterials,* vol. 26, no. 6, pp. 611-619, 2005. [http://dx.doi.org/10.1016/j.biomaterials.2004.03.013] [PMID: 15282139]

[586] P. Arpornmaeklong, N. Suwatwirote, P. Pripatnanont, and K. Oungbho, "Growth and differentiation of mouse osteoblasts on chitosan-collagen sponges", *Int. J. Oral Maxillofac. Surg.,* vol. 36, no. 4, pp. 328-337, 2007. [http://dx.doi.org/10.1016/j.ijom.2006.09.023] [PMID: 17223012]

[587] Y-C. Lin, F.J. Tan, K.G. Marra, S-S. Jan, and D-C. Liu, "Synthesis and characterization of collagen/hyaluronan/chitosan composite sponges for potential biomedical applications", *Acta Biomater.,* vol. 5, no. 7, pp. 2591-2600, 2009. [http://dx.doi.org/10.1016/j.actbio.2009.03.038] [PMID: 19427824]

[588] M. Peter, N. Binulal, S. Soumya, S. Nair, T. Furuike, H. Tamura, and R. Jayakumar, "Nanocomposite scaffolds of bioactive glass ceramic nanoparticles disseminated chitosan matrix for tissue engineering applications", *Carbohydr. Polym.,* vol. 79, pp. 284-289, 2010. [http://dx.doi.org/10.1016/j.carbpol.2009.08.001]

[589] M. Peter, P.T. Kumar, N.S. Binulal, S.V. Nair, H. Tamura, and R. Jayakumar, "Development of novel α-chitin/nanobioactive glass ceramic composite scaffolds for tissue engineering applications", *Carbohydr. Polym.,* vol. 78, pp. 926-931, 2009. [http://dx.doi.org/10.1016/j.carbpol.2009.07.016]

[590] R. Huang, Y. Du, J. Yang, and L. Fan, "Influence of functional groups on the *in vitro* anticoagulant activity of chitosan sulfate", *Carbohydr. Res.,* vol. 338, no. 6, pp. 483-489, 2003. [http://dx.doi.org/10.1016/S0008-6215(02)00505-0] [PMID: 12668103]

[591] G. Vikhoreva, G. Bannikova, P. Stolbushkina, A. Panov, N. Drozd, V. Makarov, V. Varlamov, and L. Gal'braikh, "Preparation and anticoagulant activity of a low-molecular-weight sulfated chitosan", *Carbohydr. Polym.,* vol. 62, pp. 327-332, 2005. [http://dx.doi.org/10.1016/j.carbpol.2005.05.022]

[592] J. Hirsh, J.E. Dalen, D. Deykin, and L. Poller, "Heparin: mechanism of action, pharmacokinetics, dosing considerations, monitoring, efficacy, and safety", *Chest,* vol. 102, no. 4, suppl. Suppl., pp. 337S-351S, 1992. [http://dx.doi.org/10.1378/chest.102.4_Supplement.337S] [PMID: 1327666]

[593] M.J. Alonso, and A. Sánchez, "The potential of chitosan in ocular drug delivery", *J. Pharm. Pharmacol.,* vol. 55, no. 11, pp. 1451-1463, 2003. [http://dx.doi.org/10.1211/0022357022476] [PMID: 14713355]

[594] J.A. Calles, J. Bermúdez, E. Vallés, D. Allemandi, and S. Palma, "Polymers in Ophthalmology", In: *Advanced Polymers in Medicine*Springer, 2015, pp. 147-176. [http://dx.doi.org/10.1007/978-3-319-12478-0_6]

[595] O. Felt, P. Furrer, J.M. Mayer, B. Plazonnet, P. Buri, and R. Gurny, "Topical use of chitosan in ophthalmology: tolerance assessment and evaluation of precorneal retention", *Int. J. Pharm.,* vol. 180, no. 2, pp. 185-193, 1999. [http://dx.doi.org/10.1016/S0378-5173(99)00003-4] [PMID: 10370189]

[596] Y. Machida, and T. Nagai, *Chitin/chitosan as pharmaceutical excipients.* Topics in Pharmaceutical Sciences. FIP: The Hague, 1989, pp. 211-220.

[597] L. Schmetterer, S. Höller, and M. Hornof, "Chitosan in the treatment of dry eye", *Acta Ophthalmol. (Copenh.),* vol. 88, 2010.

[598] A.M. De Campos, A. Sánchez, and M.J. Alonso, "Chitosan nanoparticles: a new vehicle for the improvement of the delivery of drugs to the ocular surface. Application to cyclosporin A", *Int. J. Pharm.*, vol. 224, no. 1-2, pp. 159-168, 2001. [http://dx.doi.org/10.1016/S0378-5173(01)00760-8] [PMID: 11472825]

[599] M.E. Braga, M.T. Pato, H.S. Silva, E.I. Ferreira, M.H. Gil, C.M. Duarte, and H.C. de Sousa, "Supercritical solvent impregnation of ophthalmic drugs on chitosan derivatives", *J. Supercrit. Fluids,* vol. 44, pp. 245-257, 2008. [http://dx.doi.org/10.1016/j.supflu.2007.10.002]

[600] O. Felt, A. Carrel, P. Baehni, P. Buri, and R. Gurny, "Chitosan as tear substitute: a wetting agent endowed with antimicrobial efficacy", *J. Ocul. Pharmacol. Ther.*, vol. 16, no. 3, pp. 261-270, 2000. [http://dx.doi.org/10.1089/jop.2000.16.261] [PMID: 10872923]

[601] R.C. Nagarwal, P.N. Singh, S. Kant, P. Maiti, and J.K. Pandit, "Chitosan coated PLA nanoparticles for ophthalmic delivery: characterization, *in-vitro* and *in-vivo* study in rabbit eye", *J. Biomed. Nanotechnol.,* vol. 6, no. 6, pp. 648-657, 2010. [http://dx.doi.org/10.1166/jbn.2010.1168] [PMID: 21361129]

[602] B. Wowra, M. Grolik, K. Szczubialka, M. Nowakowska, D. Dobrowolski, B. Orzechowska Wylegala, and E. Wylegala, "Chitosan-based biocompatible materials in ophthalmology", In: *Acta Ophthalmol. (Copenh.)* vol. 91. 2013.

[603] M.L. Markey, L.M. Bowman, and M. Bergamini, "Contact lenses made of chitosan", In: *Chitin and Chitosan*1989, p. 713.

[604] I. Genta, B. Conti, P. Perugini, F. Pavanetto, A. Spadaro, and G. Puglisi, "Bioadhesive microspheres for ophthalmic administration of acyclovir", *J. Pharm. Pharmacol.,* vol. 49, no. 8, pp. 737-742, 1997. [http://dx.doi.org/10.1111/j.2042-7158.1997.tb06103.x] [PMID: 9379347]

[605] G. Di Colo, Y. Zambito, S. Burgalassi, A. Serafini, and M.F. Saettone, "Effect of chitosan on *in vitro* release and ocular delivery of ofloxacin from erodible inserts based on poly(ethylene oxide)", *Int. J. Pharm.,* vol. 248, no. 1-2, pp. 115-122, 2002. [http://dx.doi.org/10.1016/S0378-5173(02)00421-0] [PMID: 12429465]

[606] G. Abdelbary, "Ocular ciprofloxacin hydrochloride mucoadhesive chitosan-coated liposomes", *Pharm. Dev. Technol.,* vol. 16, no. 1, pp. 44-56, 2011. [http://dx.doi.org/10.3109/10837450903479988] [PMID: 20025433]

[607] N. Li, C. Zhuang, M. Wang, X. Sun, S. Nie, and W. Pan, "Liposome coated with low molecular weight chitosan and its potential use in ocular drug delivery", *Int. J. Pharm.,* vol. 379, no. 1, pp. 131-138, 2009. [http://dx.doi.org/10.1016/j.ijpharm.2009.06.020] [PMID: 19559775]

[608] V. Andrés-Guerrero, M. Zong, E. Ramsay, B. Rojas, S. Sarkhel, B. Gallego, R. de Hoz, A.I. Ramírez, J.J. Salazar, A. Triviño, J.M. Ramírez, E.M. Del Amo, N. Cameron, B. de-Las-Heras, A. Urtti, G. Mihov, A. Dias, and R. Herrero-Vanrell, "Novel biodegradable polyesteramide microspheres for controlled drug delivery in Ophthalmology", *J. Control. Release,* vol. 211, pp. 105-117, 2015. [http://dx.doi.org/10.1016/j.jconrel.2015.05.279] [PMID: 26003040]

[609] R. Asasutjarit, T. Theerachayanan, P. Kewsuwan, S. Veeranodha, A. Fuongfuchat, and G.C. Ritthidej, "Development and Evaluation of Diclofenac Sodium Loaded-N-Trimethyl Chitosan Nanoparticles for Ophthalmic Use", *AAPS PharmSciTech,* vol. 16, no. 5, pp. 1013-1024, 2015. [http://dx.doi.org/10.1208/s12249-015-0290-4] [PMID: 25609376]

[610] F. Dilnawaz, and S.K. Sahoo, "Nanotechnology-based ophthalmic drug delivery system", In: *Focal Controlled Drug Delivery*Springer, 2014, pp. 225-241. [http://dx.doi.org/10.1007/978-1-4614-9434-8_10]

[611] G.F. Bickerstaff Jr, *Immobilization of enzymes and cells.*Springer, 1997.

[612] B. Mattiasson, "Immobilization methods", *Immobilized Cells and Organelles,* vol. 1, pp. 3-25, 1983.

[613] J. Cabral, and J. Kennedy, *Immobilization techniques for altering thermal stability of enzymes.*Springer-Verlag: Narosa Publishing House: New Delhi, 1993.

[614] L. Cao, "Immobilised enzymes: science or art?", *Curr. Opin. Chem. Biol.,* vol. 9, no. 2, pp. 217-226, 2005.
[http://dx.doi.org/10.1016/j.cbpa.2005.02.014] [PMID: 15811808]

[615] C. Spahn, and S.D. Minteer, "Enzyme immobilization in biotechnology", *Recent Patents on Engineering,* vol. 2, pp. 195-200, 2008.
[http://dx.doi.org/10.2174/187221208786306333]

[616] R.A. Sheldon, "Enzyme immobilization: the quest for optimum performance", *Adv. Synth. Catal.,* vol. 349, pp. 1289-1307, 2007.
[http://dx.doi.org/10.1002/adsc.200700082]

[617] J. Woodward, *Immobilised cells and enzymes.*IRL press, 1985.

[618] C. Bullock, "Immobilised enzymes", *Sci. Prog.,* vol. 78, pp. 119-134, 1995.

[619] B. Mattiasson, "Affinity immobilization", *Methods Enzymol.,* vol. 137, pp. 647-656, 1988.
[http://dx.doi.org/10.1016/0076-6879(88)37062-X] [PMID: 3374362]

[620] S. Andreescu, B. Bucur, and J-L. Marty, "Affinity immobilization of tagged enzymes", In: *Immobilization of Enzymes and Cells*Springer, 2006, pp. 97-106.
[http://dx.doi.org/10.1007/978-1-59745-053-9_9]

[621] D. Brady, and J. Jordaan, "Advances in enzyme immobilisation", *Biotechnol. Lett.,* vol. 31, no. 11, pp. 1639-1650, 2009.
[http://dx.doi.org/10.1007/s10529-009-0076-4] [PMID: 19590826]

[622] P. Wang, S. Dai, S.D. Waezsada, A.Y. Tsao, and B.H. Davison, "Enzyme stabilization by covalent binding in nanoporous sol-gel glass for nonaqueous biocatalysis", *Biotechnol. Bioeng.,* vol. 74, no. 3, pp. 249-255, 2001.
[http://dx.doi.org/10.1002/bit.1114] [PMID: 11400098]

[623] S. D'souza, "Immobilized enzymes in bioprocess", *Curr. Sci.,* vol. 77, pp. 69-79, 1999.

[624] M. Diaconu, S.C. Litescu, and G.L. Radu, "Laccase–MWCNT–chitosan biosensor-A new tool for total polyphenolic content evaluation from in vitro cultivated plants", *Sens. Actuators B Chem.,* vol. 145, pp. 800-806, 2010.
[http://dx.doi.org/10.1016/j.snb.2010.01.064]

[625] X-L. Luo, J-J. Xu, Y. Du, and H-Y. Chen, "A glucose biosensor based on chitosan-glucose oxidase-gold nanoparticles biocomposite formed by one-step electrodeposition", *Anal. Biochem.,* vol. 334, no. 2, pp. 284-289, 2004.
[http://dx.doi.org/10.1016/j.ab.2004.07.005] [PMID: 15494135]

[626] S.K. Shukla, A.K. Mishra, O.A. Arotiba, and B.B. Mamba, "Chitosan-based nanomaterials: a state-o--the-art review", *Int. J. Biol. Macromol.,* vol. 59, pp. 46-58, 2013.
[http://dx.doi.org/10.1016/j.ijbiomac.2013.04.043] [PMID: 23608103]

[627] W. Suginta, P. Khunkaewla, and A. Schulte, "Electrochemical biosensor applications of polysaccharides chitin and chitosan", *Chem. Rev.,* vol. 113, no. 7, pp. 5458-5479, 2013.
[http://dx.doi.org/10.1021/cr300325r] [PMID: 23557137]

[628] M.M. Rahman, A.J. Ahammad, J-H. Jin, S.J. Ahn, and J-J. Lee, "A comprehensive review of glucose biosensors based on nanostructured metal-oxides", *Sensors (Basel),* vol. 10, no. 5, pp. 4855-4886, 2010.
[http://dx.doi.org/10.3390/s100504855] [PMID: 22399911]

[629] Y. Wang, J. Zhu, R. Zhu, Z. Zhu, Z. Lai, and Z. Chen, "Chitosan/Prussian blue-based biosensors", *Meas. Sci. Technol.,* vol. 14, p. 831, 2003.
[http://dx.doi.org/10.1088/0957-0233/14/6/317]

[630] R. Vazquez-Duhalt, R. Tinoco, P. DAntonio, L.D. Topoleski, and G.F. Payne, "Enzyme conjugation to the polysaccharide chitosan: smart biocatalysts and biocatalytic hydrogels", *Bioconjug. Chem.*, vol. 12, no. 2, pp. 301-306, 2001.
[http://dx.doi.org/10.1021/bc000095u] [PMID: 11312692]

[631] J.M. Magalhães, and A.A. Machado, "Enzyme immobilization on chitin and chitosan for construction of enzymatic sensors", In: *Uses of Immobilized Biological Compounds*Springer, 1993, pp. 191-200.
[http://dx.doi.org/10.1007/978-94-011-1932-0_19]

[632] Ş.A. Çetinus, and H.N. Öztop, "Immobilization of catalase into chemically crosslinked chitosan beads", *Enzyme Microb. Technol.*, vol. 32, pp. 889-894, 2003.
[http://dx.doi.org/10.1016/S0141-0229(03)00065-6]

[633] P.K. Dutta, M. Ravikumar, and J. Dutta, "Chitin and chitosan for versatile applications", *J. Macromol. Sci. Part C Polym. Rev.*, vol. 42, pp. 307-354, 2002.
[http://dx.doi.org/10.1081/MC-120006451]

[634] R-S. Juang, F-C. Wu, and R-L. Tseng, "Use of chemically modified chitosan beads for sorption and enzyme immobilization", *Adv. Environ. Res.*, vol. 6, pp. 171-177, 2002.
[http://dx.doi.org/10.1016/S1093-0191(00)00078-2]

[635] W. Zaborska, "Competitive inhibitors of free and chitosan-immobilized urease", *Acta Biochim. Pol.*, vol. 42, no. 1, pp. 115-118, 1995.
[PMID: 7653153]

[636] B. Krajewska, M. Leszko, and W. Zaborska, "Urease immobilized on chitosan membrane: preparation and properties", *J. Chem. Technol. Biotechnol.*, vol. 48, no. 3, pp. 337-350, 1990.
[http://dx.doi.org/10.1002/jctb.280480309] [PMID: 1366452]

[637] A. Lemainque, J. Braun, and F. Le Goffic, "Influence of polymerization of D-amino acid oxidase on the behavior of the enzyme immobilized on chitosan by covalent fixation", *European Journal of Biochemistry/FEBS*, vol. 174, pp. 171-176, 1988.

[638] R-S. Juang, F-C. Wu, and R-L. Tseng, "Solute adsorption and enzyme immobilization on chitosan beads prepared from shrimp shell wastes", *Bioresour. Technol.*, vol. 80, no. 3, pp. 187-193, 2001.
[http://dx.doi.org/10.1016/S0960-8524(01)00090-6] [PMID: 11601542]

[639] R. Agarwal, and M. Gupta, "Evaluation of gluteraldehyde-modified chitosan as a matrix for hydrophobic interaction chromatography", *Anal. Chim. Acta*, vol. 313, pp. 253-257, 1995.
[http://dx.doi.org/10.1016/0003-2670(95)00241-Q]

[640] M-Y. Chang, and R-S. Juang, "Stability and catalytic kinetics of acid phosphatase immobilized on composite beads of chitosan and activated clay", *Process Biochem.*, vol. 39, pp. 1087-1091, 2004.
[http://dx.doi.org/10.1016/S0032-9592(03)00221-8]

[641] P.K. Srivastava, and A. Anand, "Immobilization of acid phosphatase from Vigna aconitifolia seeds on chitosan beads and its characterization", *Int. J. Biol. Macromol.*, vol. 64, pp. 150-154, 2014.
[http://dx.doi.org/10.1016/j.ijbiomac.2013.11.023] [PMID: 24309514]

[642] K. Belho, S. Nongpiur, and P. Ambasht, "Immobilization of acid phosphatase (type I) from wheat germ on glutaraldehyde activated chitosan beads: Optimization and characterization", *Journal of Proteins and Proteomics*, vol. 5, 2014.

[643] M.E. Badawy, and A.F. El-Aswad, "Bioactive paper sensor based on the acetylcholinesterase for the rapid detection of organophosphate and carbamate pesticides", *Int. J. Anal. Chem.*, 2014.
[http://dx.doi.org/10.1155/2014/536823]

[644] D. Du, X. Huang, J. Cai, and A. Zhang, "Amperometric detection of triazophos pesticide using acetylcholinesterase biosensor based on multiwall carbon nanotube–chitosan matrix", *Sens. Actuators B Chem.*, vol. 127, pp. 531-535, 2007.
[http://dx.doi.org/10.1016/j.snb.2007.05.006]

[645] N. Xia, Y. Zhang, K. Chang, X. Gai, Y. Jing, S. Li, L. Liu, and G. Qu, "Ferrocene-phenylalanine hydrogels for immobilization of acetylcholinesterase and detection of chlorpyrifos", *J. Electroanal. Chem.,* vol. 746, pp. 68-74, 2015.
[http://dx.doi.org/10.1016/j.jelechem.2015.03.030]

[646] J. Gong, T. Liu, D. Song, X. Zhang, and L. Zhang, "One-step fabrication of three-dimensional porous calcium carbonate–chitosan composite film as the immobilization matrix of acetylcholinesterase and its biosensing on pesticide", *Electrochem. Commun.,* vol. 11, pp. 1873-1876, 2009.
[http://dx.doi.org/10.1016/j.elecom.2009.08.005]

[647] N. Kiba, Y. Oyama, and M. Furusawa, "Determination of aliphatic amino acids in serum by HPLC with fluorimetric detection using co-immobilized enzyme reactor", *Talanta,* vol. 40, no. 5, pp. 657-660, 1993.
[http://dx.doi.org/10.1016/0039-9140(93)80274-U] [PMID: 18965682]

[648] B. Krajewska, "Membrane-based processes performed with use of chitin/chitosan materials", *Separ. Purif. Tech.,* vol. 41, pp. 305-312, 2005.
[http://dx.doi.org/10.1016/j.seppur.2004.03.019]

[649] A. Zubriene, S. Budriene, J. Lubiene, and G. Dienys, "Immobilized alkaline phosphatase for molecular cloning", *Biocatalysis Biotransform.,* vol. 20, pp. 423-427, 2002.
[http://dx.doi.org/10.1080/1024242021000040873]

[650] S. Hirano, and O. Miura, "Alkaline phosphatase and pepsin immobilized in gels", *Biotechnol. Bioeng.,* vol. 21, no. 4, pp. 711-714, 1979.
[http://dx.doi.org/10.1002/bit.260210415] [PMID: 105773]

[651] M.A. Abdel-Naby, A.S. Ismail, S. Ahmed, and A.F. Fattah, "Production and immobilization of alkaline protease from Bacillus mycoides", *Bioresour. Technol.,* vol. 64, pp. 205-210, 1998.
[http://dx.doi.org/10.1016/S0960-8524(97)00160-0]

[652] S. Bhandari, V.K. Gupta, and H. Singh, "Enhanced stabilization of mungbean thiol protease immobilized on glutaraldehyde-activated chitosan beads", *Biocatalysis Biotransform.,* vol. 27, pp. 71-77, 2009.
[http://dx.doi.org/10.1080/10242420802607918]

[653] S.-Y. Lin, L.-F. Song, J.-B. Liu, and Z.-P. Yu, "Study on immobilization of alkaline protease with chitosan microspheres", *Food Science,* vol. 11, p. 81, 2008.

[654] T. Taniai, A. Sukuragawa, and T. Okutani, "Fluorometric determination of ethanol in liquor samples by flow-injection analysis using an immobilized enzyme-reactor column with packing prepared by coupling alcohol oxidase and peroxidase onto chitosan beads", *J. AOAC Int.,* vol. 84, no. 5, pp. 1475-1483, 2001.
[PMID: 11601467]

[655] K.H. Lee, P.M. Lee, and Y.S. Siaw, "Studies of L-phenylalanine production by immobilised aminoacylase in stabilized calcium alginate beads", *J. Chem. Technol. Biotechnol.,* vol. 54, pp. 375-382, 1992.
[http://dx.doi.org/10.1002/jctb.280540411]

[656] K. Devarayan, M. Miyamoto, M. Hachisu, J. Araki, V. Periasamy, and K. Ohkawa, "Cationic derivative of electrospun non-woven cellulose-chitosan composite fabrics for immobilization of aminoacylase-I", *Text. Res. J.,* vol. 83, pp. 1918-1925, 2013.
[http://dx.doi.org/10.1177/0040517513478457]

[657] M.A. Abdel-Naby, A.M. Hashem, M.A. Esawy, and A.F. Abdel-Fattah, "Immobilization of Bacillus subtilis α-amylase and characterization of its enzymatic properties", *Microbiol. Res.,* vol. 153, pp. 319-325, 1999.
[http://dx.doi.org/10.1016/S0944-5013(99)80044-5]

[658] M. Reiss, A. Heibges, J. Metzger, and W. Hartmeier, "Determination of BOD-values of starch-

containing waste water by a BOD-biosensor", *Biosens. Bioelectron.*, vol. 13, no. 10, pp. 1083-1090, 1998.
[http://dx.doi.org/10.1016/S0956-5663(98)00071-2] [PMID: 9842703]

[659] T. Morita, H.J. Lim, and I. Karube, "Enzymatic hydrolysis of polysaccharides in water-immiscible organic solvent, biphasic systems", *J. Biotechnol.*, vol. 38, pp. 253-261, 1995.
[http://dx.doi.org/10.1016/0168-1656(94)00131-U]

[660] M.G. Siso, E. Lang, B. Carrenõ-Gómez, M. Becerra, F.O. Espinar, and J.B. Méndez, "Enzyme encapsulation on chitosan microbeads", *Process Biochem.*, vol. 32, pp. 211-216, 1997.
[http://dx.doi.org/10.1016/S0032-9592(96)00064-7]

[661] T. Noda, S. Furuta, and I. Suda, "Sweet potato β-amylase immobilized on chitosan beads and its application in the semi-continuous production of maltose", *Carbohydr. Polym.*, vol. 44, pp. 189-195, 2001.
[http://dx.doi.org/10.1016/S0144-8617(00)00226-5]

[662] L. Zhulai, C. Sheng, and K. Jie, "Immobilized α-Amylase in Chitosan", *Journal of Fujian Medical University*, p. 3, 1994.

[663] A. Kumari, and A.M. Kayastha, "Immobilization of soybean (Glycine max) α-amylase onto Chitosan and Amberlite MB-150 beads: optimization and characterization", *J. Mol. Catal., B Enzym.*, vol. 69, pp. 8-14, 2011.
[http://dx.doi.org/10.1016/j.molcatb.2010.12.003]

[664] G. Srivastava, S. Roy, and A.M. Kayastha, "Immobilisation of Fenugreek β-amylase on chitosan/PVP blend and chitosan coated PVC beads: a comparative study", *Food Chem.*, vol. 172, pp. 844-851, 2015.
[http://dx.doi.org/10.1016/j.foodchem.2014.09.145] [PMID: 25442629]

[665] P. Pifferi, V. Bonora, G. Spagna, and M. Tramontini, "Immobilization of catalase on macromolecular supports activated with acid dyes", *Process Biochem.*, vol. 28, pp. 29-38, 1993.
[http://dx.doi.org/10.1016/0032-9592(94)80033-2]

[666] H. Huang, N. Hu, Y. Zeng, and G. Zhou, "Electrochemistry and electrocatalysis with heme proteins in chitosan biopolymer films", *Anal. Biochem.*, vol. 308, no. 1, pp. 141-151, 2002.
[http://dx.doi.org/10.1016/S0003-2697(02)00242-7] [PMID: 12234475]

[667] H.N. Öztop, and H.N. Öztop, "Immobilization of catalase on chitosan film", *Enzyme Microb. Technol.*, vol. 26, no. 7, pp. 497-501, 2000.
[http://dx.doi.org/10.1016/S0141-0229(99)00189-1] [PMID: 10771052]

[668] V. Ramakrishnan, K.M. Lathika, S.J. DSouza, B.B. Singh, and K.G. Raghavan, "Investigation with chitosan-oxalate oxidase-catalase conjugate for degrading oxalate from hyperoxaluric rat chyme", *Indian J. Biochem. Biophys.*, vol. 34, no. 4, pp. 373-378, 1997.
[PMID: 9491647]

[669] Çetinus.S. Akkuş, and H.N. Öztop, "Immobilization of catalase on chitosan film", *Enzyme Microb. Technol.*, vol. 26, no. 7, pp. 497-501, 2000.
[http://dx.doi.org/10.1016/S0141-0229(99)00189-1] [PMID: 10771052]

[670] I. Roy, M. Sardar, and M.N. Gupta, "Exploiting unusual affinity of usual polysaccharides for separation of enzymes on fluidized beds", *Enzyme Microb. Technol.*, vol. 27, no. 1-2, pp. 53-65, 2000.
[http://dx.doi.org/10.1016/S0141-0229(00)00148-4] [PMID: 10862902]

[671] R. Darias, and R. Villalonga, "Functional stabilization of cellulase by covalent modification with chitosan", *J. Chem. Technol. Biotechnol.*, vol. 76, pp. 489-493, 2001.
[http://dx.doi.org/10.1002/jctb.386]

[672] A. Dinçer, and A. Telefoncu, "Improving the stability of cellulase by immobilization on modified polyvinyl alcohol coated chitosan beads", *J. Mol. Catal., B Enzym.*, vol. 45, pp. 10-14, 2007.
[http://dx.doi.org/10.1016/j.molcatb.2006.10.005]

[673] L. Zang, J. Qiu, X. Wu, W. Zhang, E. Sakai, and Y. Wei, "Preparation of magnetic chitosan nanoparticles as support for cellulase immobilization", *Ind. Eng. Chem. Res.,* vol. 53, pp. 3448-3454, 2014.
[http://dx.doi.org/10.1021/ie404072s]

[674] H. Kise, and A. Hayakawa, "Immobilization of proteases to porous chitosan beads and their catalysis for ester and peptide synthesis in organic solvents", *Enzyme Microb. Technol.,* vol. 13, no. 7, pp. 584-588, 1991.
[http://dx.doi.org/10.1016/0141-0229(91)90094-Q] [PMID: 1367641]

[675] H-Y. Ju, C-H. Kuo, J-R. Too, H-Y. Huang, Y-K. Twu, C-M. Chang, Y-C. Liu, and C-J. Shieh, "Optimal covalent immobilization of α-chymotrypsin on Fe 3 O 4-chitosan nanoparticles", *J. Mol. Catal., B Enzym.,* vol. 78, pp. 9-15, 2012.
[http://dx.doi.org/10.1016/j.molcatb.2012.01.015]

[676] C.R. Carrara, and A.C. Rubiolo, "Immobilization of. beta.-galactosidase on chitosan", *Biotechnol. Prog.,* vol. 10, pp. 220-224, 1994.
[http://dx.doi.org/10.1021/bp00026a012]

[677] A.R. Figueroa, H. Talavera, and A. Colominas, "Flow optimization in a class of enzymatic plug-flow reactor", *Biotechnol. Prog.,* vol. 13, pp. 109-112, 1997.
[http://dx.doi.org/10.1021/bp9600905]

[678] S. Rejikumar, and S. Devi, "Immobilization of β-galactosidase onto polymeric supports", *J. Appl. Polym. Sci.,* vol. 55, pp. 871-878, 1995.
[http://dx.doi.org/10.1002/app.1995.070550605]

[679] S. Rejikumar, and S. Devi, "Hydrolysis of lactose and milk whey using a fixed-bed reactor containing β-galactosidase covalently bound onto chitosan and cross-linked poly (vinyl alcohol)", *Int. J. Food Sci. Technol.,* vol. 36, pp. 91-98, 2001.
[http://dx.doi.org/10.1046/j.1365-2621.2001.00425.x]

[680] H-J. Shin, J-M. Park, and J-W. Yang, "Continuous production of galacto-oligosaccharides from lactose by Bullera singularis β-galactosidase immobilized in chitosan beads", *Process Biochem.,* vol. 33, pp. 787-792, 1998.
[http://dx.doi.org/10.1016/S0032-9592(98)00045-4]

[681] D-C. Sheu, S-Y. Li, K-J. Duan, and C.W. Chen, "Production of galactooligosaccharides by β-galactosidase immobilized on glutaraldehyde-treated chitosan beads", *Biotechnol. Tech.,* vol. 12, pp. 273-276, 1998.
[http://dx.doi.org/10.1023/A:1008894029377]

[682] A. Ghanem, and D. Skonberg, "Effect of preparation method on the capture and release of biologically active molecules in chitosan gel beads", *J. Appl. Polym. Sci.,* vol. 84, pp. 405-413, 2002.
[http://dx.doi.org/10.1002/app.10393]

[683] M. Portaccio, S. Stellato, S. Rossi, U. Bencivenga, M.M. Eldin, F. Gaeta, and D. Mita, "Galactose competitive inhibition of β-galactosidase (Aspergillus oryzae) immobilized on chitosan and nylon supports", *Enzyme Microb. Technol.,* vol. 23, pp. 101-106, 1998.
[http://dx.doi.org/10.1016/S0141-0229(98)00018-0]

[684] T. Taniai, A. Sakuragawa, and T. Okutani, "Fluorometric determination of glucose by flow injection analysis using immobilized enzylne-reactor columns prepared by coupling glucose oxidase and peroxidase onto chitosan beads", *Anal. Sci.,* vol. 16, pp. 517-521, 2000.
[http://dx.doi.org/10.2116/analsci.16.517]

[685] J. Zhu, Z. Zhu, Z. Lai, R. Wang, X. Guo, X. Wu, G. Zhang, Z. Zhang, Y. Wang, and Z. Chen, "Planar amperometric glucose sensor based on glucose oxidase immobilized by chitosan film on Prussian blue layer", *Sensors (Basel Switzerland),* vol. 2, pp. 127-136, 2002.
[http://dx.doi.org/10.3390/s20400127]

[686] B-C. Hsieh, T-J. Cheng, T-Y. Wang, and R.L. Chen, "Use of chitosan membrane from the carapace of the soldier crab Mictyris brevidactylus for biosensor construction", *Mar. Biotechnol. (NY),* vol. 5, no. 2, pp. 119-125, 2003.
[http://dx.doi.org/10.1007/s10126-002-0094-x] [PMID: 12876646]

[687] Y. Miao, L. Chia, N. Goh, and S. Tan, "Amperometric glucose biosensor based on immobilization of glucose oxidase in chitosan matrix cross-linked with glutaraldehyde", *Electroanalysis,* vol. 13, pp. 347-349, 2001.
[http://dx.doi.org/10.1002/1521-4109(200103)13:4<347::AID-ELAN347>3.0.CO;2-Z]

[688] L. Chen, and W. Gorski, "Bioinorganic composites for enzyme electrodes", *Anal. Chem.,* vol. 73, no. 13, pp. 2862-2868, 2001.
[http://dx.doi.org/10.1021/ac010009z] [PMID: 11467528]

[689] X. Wei, J. Cruz, and W. Gorski, "Integration of enzymes and electrodes: spectroscopic and electrochemical studies of chitosan-enzyme films", *Anal. Chem.,* vol. 74, no. 19, pp. 5039-5046, 2002.
[http://dx.doi.org/10.1021/ac020216e] [PMID: 12380828]

[690] X.H. Xu, B. Han, Y.S. Fu, J. Han, H.B. Shi, B. Wu, S. Han, and Q. Chen, "Preparation of chitosan/glucose oxidase nanolayered films for electrode modification by the technique of layer-b--layer self-assembly", *J. Mater. Sci. Lett.,* vol. 22, pp. 695-697, 2003.
[http://dx.doi.org/10.1023/A:1023627329738]

[691] X. Chen, J. Jia, and S. Dong, "Organically modified sol-gel/chitosan composite based glucose biosensor", *Electroanalysis,* vol. 15, pp. 608-612, 2003.
[http://dx.doi.org/10.1002/elan.200390076]

[692] Y. Yang, J. Wang, and R. Tan, "Immobilization of glucose oxidase on chitosan–SiO2 gel", *Enzyme Microb. Technol.,* vol. 34, pp. 126-131, 2004.
[http://dx.doi.org/10.1016/j.enzmictec.2003.09.007]

[693] S. Wang, M. Yoshimoto, K. Fukunaga, and K. Nakao, "Optimal covalent immobilization of glucose oxidase-containing liposomes for highly stable biocatalyst in bioreactor", *Biotechnol. Bioeng.,* vol. 83, no. 4, pp. 444-453, 2003.
[http://dx.doi.org/10.1002/bit.10684] [PMID: 12800138]

[694] J.R. Anusha, A.T. Fleming, H-J. Kim, B.C. Kim, K-H. Yu, and C.J. Raj, "Effective immobilization of glucose oxidase on chitosan submicron particles from gladius of Todarodes pacificus for glucose sensing", *Bioelectrochemistry,* vol. 104, pp. 44-50, 2015.
[http://dx.doi.org/10.1016/j.bioelechem.2015.02.004] [PMID: 25727854]

[695] D-C. Sheu, C-I. Huang, and K-J. Duan, "Production of isomaltooligosaccharides by a-glucosidase immobilized in chitosan beads and by polyethyleneimine-glutaraldehyde treated mycelia of Aspergillus carbonarious", *Biotechnol. Tech.,* vol. 11, pp. 287-291, 1997.
[http://dx.doi.org/10.1023/A:1018407009503]

[696] A. Martino, M. Durante, P. Pifferi, G. Spagna, and G. Bianchi, "Immobilization of β-glucosidase from a commercial preparation. Part 1. A comparative study of natural supports", *Process Biochem.,* vol. 31, pp. 281-285, 1996.
[http://dx.doi.org/10.1016/0032-9592(95)00065-8]

[697] A. Gallifuoco, L. D'Ercole, F. Alfani, M. Cantarella, G. Spagna, and P. Pifferi, "On the use of chitosan-immobilized β-glucosidase in wine-making: kinetics and enzyme inhibition", *Process Biochem.,* vol. 33, pp. 163-168, 1998.
[http://dx.doi.org/10.1016/S0032-9592(97)00041-1]

[698] Y.Z.P.Z.X. Jianxiang, and Z.M.Z. Zhixu, "Immobilization of α-Glucosidase on Chitosan", *Food and Fermentation Industries,* vol. 4, p. 8, 2001.

[699] Y. Zhenfeng, P. Zhiying, and X. Jianxiang, "Immobilization of alpha-Glucosidase on Chitosan", *Food and Fermentation Industries,* vol. 27, pp. 20-24, 2001.

[700] H.H. Petach, and J. Driscoll, "Transparent chitosan derivatives for the immobilization of glutamate dehydrogenase", *Biotechnol. Bioeng.,* vol. 44, no. 8, pp. 1018-1022, 1994. [http://dx.doi.org/10.1002/bit.260440820] [PMID: 18618922]

[701] N.E. Azmi, M. Ahmad, J. Abdullah, H. Sidek, L.Y. Heng, and N. Karuppiah, "Biosensor based on glutamate dehydrogenase immobilized in chitosan for the determination of ammonium in water samples", *Anal. Biochem.,* vol. 388, no. 1, pp. 28-32, 2009. [http://dx.doi.org/10.1016/j.ab.2009.02.005] [PMID: 19454217]

[702] M. Zhang, C. Mullens, and W. Gorski, "Chitosan-glutamate oxidase gels: Synthesis, characterization, and glutamate determination", *Electroanalysis,* vol. 17, pp. 2114-2120, 2005. [http://dx.doi.org/10.1002/elan.200503348]

[703] H-J. Hsieh, P-C. Liu, and W-J. Liao, "Immobilization of invertase via carbohydrate moiety on chitosan to enhance its thermal stability", *Biotechnol. Lett.,* vol. 22, pp. 1459-1464, 2000. [http://dx.doi.org/10.1023/A:1005602812037]

[704] L. Gómez, H.L. Ramírez, and R. Villalonga, "Stabilization of invertase by modification of sugar chains with chitosan", *Biotechnol. Lett.,* vol. 22, pp. 347-350, 2000. [http://dx.doi.org/10.1023/A:1005664432575]

[705] M. Ghosh, R. Tyagi, and M. Gupta, "Preparation of trypsin free chymotrypsin", *Biotechnol. Tech.,* vol. 9, pp. 149-152, 1995. [http://dx.doi.org/10.1007/BF00224416]

[706] A. D'Annibale, S.R. Stazi, V. Vinciguerra, E. Di Mattia, and G.G. Sermanni, "Characterization of immobilized laccase from Lentinula edodes and its use in olive-mill wastewater treatment", *Process Biochem.,* vol. 34, pp. 697-706, 1999. [http://dx.doi.org/10.1016/S0032-9592(98)00144-7]

[707] D-S. Jiang, S-Y. Long, J. Huang, H-Y. Xiao, and J-Y. Zhou, "Immobilization of Pycnoporus sanguineus laccase on magnetic chitosan microspheres", *Biochem. Eng. J.,* vol. 25, pp. 15-23, 2005. [http://dx.doi.org/10.1016/j.bej.2005.03.007]

[708] L. Lu, M. Zhao, and Y. Wang, "Immobilization of laccase by alginate–chitosan microcapsules and its use in dye decolorization", *World J. Microbiol. Biotechnol.,* vol. 23, pp. 159-166, 2007. [http://dx.doi.org/10.1007/s11274-006-9205-6]

[709] E.B. Pereira, H.F. De Castro, F.F. De Moraes, and G.M. Zanin, "Kinetic studies of lipase from Candida rugosa: a comparative study between free and immobilized enzyme onto porous chitosan beads", *Appl. Biochem. Biotechnol.,* vol. 91-93, pp. 739-752, 2001. [http://dx.doi.org/10.1385/ABAB:91-93:1-9:739] [PMID: 11963902]

[710] E. Chornet, and S. Dumitriu, "Inclusion and release of proteins from polysaccharide-based polyion complexes", *Adv. Drug Deliv. Rev.,* vol. 31, no. 3, pp. 223-246, 1998. [http://dx.doi.org/10.1016/S0169-409X(97)00120-8] [PMID: 10837627]

[711] E.B. Pereira, H.F. de Castro, F.F. De Moraes, and G.M. Zanin, "Esterification activity and stability of *Candida rugosa* lipase immobilized into chitosan", In: *Biotechnology for Fuels and Chemicals*Springer, 2002, pp. 977-986. [http://dx.doi.org/10.1007/978-1-4612-0119-9_79]

[712] K. Itoyama, S. Tokura, and T. Hayashi, "Lipoprotein lipase immobilization onto porous chitosan beads", *Biotechnol. Prog.,* vol. 10, no. 2, pp. 225-229, 1994. [http://dx.doi.org/10.1021/bp00026a013] [PMID: 7764679]

[713] T-C. Hung, R. Giridhar, S-H. Chiou, and W-T. Wu, "Binary immobilization of Candida rugosa lipase on chitosan", *J. Mol. Catal., B Enzym.,* vol. 26, pp. 69-78, 2003. [http://dx.doi.org/10.1016/S1381-1177(03)00167-X]

[714] S-H. Chiou, and W-T. Wu, "Immobilization of Candida rugosa lipase on chitosan with activation of the hydroxyl groups", *Biomaterials,* vol. 25, no. 2, pp. 197-204, 2004.

[http://dx.doi.org/10.1016/S0142-9612(03)00482-4] [PMID: 14585707]

[715] S.S. Betigeri, and S.H. Neau, "Immobilization of lipase using hydrophilic polymers in the form of hydrogel beads", *Biomaterials,* vol. 23, no. 17, pp. 3627-3636, 2002. [http://dx.doi.org/10.1016/S0142-9612(02)00095-9] [PMID: 12109688]

[716] I.A. Alsarra, S.S. Betigeri, H. Zhang, B.A. Evans, and S.H. Neau, "Molecular weight and degree of deacetylation effects on lipase-loaded chitosan bead characteristics", *Biomaterials,* vol. 23, no. 17, pp. 3637-3644, 2002. [http://dx.doi.org/10.1016/S0142-9612(02)00096-0] [PMID: 12109689]

[717] T. Tan, F. Wang, and H. Zhang, "Preparation of PVA/chitosan lipase membrane reactor and its application in synthesis of monoglyceride", *J. Mol. Catal., B Enzym.,* vol. 18, pp. 325-331, 2002. [http://dx.doi.org/10.1016/S1381-1177(02)00113-3]

[718] R.V. Amorim, E.S. Melo, M.G. Carneiro-da-Cunha, W.M. Ledingham, and G.M. Campos-Takaki, "Chitosan from Syncephalastrum racemosum used as a film support for lipase immobilization", *Bioresour. Technol.,* vol. 89, no. 1, pp. 35-39, 2003. [http://dx.doi.org/10.1016/S0960-8524(03)00035-X] [PMID: 12676498]

[719] D. Magnin, S. Dumitriu, P. Magny, and E. Chornet, "Lipase immobilization into porous chitoxan beads: activities in aqueous and organic media and lipase localization", *Biotechnol. Prog.,* vol. 17, no. 4, pp. 734-737, 2001. [http://dx.doi.org/10.1021/bp0100528] [PMID: 11485436]

[720] S. Dumitriu, E. Chornet, P. Vidal, and C. Moresoli, "Polyionic hydrogels as support for immobilization of lipase", *Biotechnol. Tech.,* vol. 9, pp. 833-836, 1995. [http://dx.doi.org/10.1007/BF00159410]

[721] A. Crapisi, A. Lante, G. Pasini, and P. Spettoli, "Enhanced microbial cell lysis by the use of lysozyme immobilized on different carriers", *Process Biochem.,* vol. 28, pp. 17-21, 1993. [http://dx.doi.org/10.1016/0032-9592(94)80031-6]

[722] G. Bayramoğlu, and M.Y. Arıca, "Procion Green H-4G immobilized on a new IPN hydrogel membrane composed of poly (2-hydroxyethylmethacrylate)/chitosan: preparation and its application to the adsorption of lysozyme", *Colloids Surf. Physicochem. Eng. Aspects,* vol. 202, pp. 41-52, 2002. [http://dx.doi.org/10.1016/S0927-7757(01)01055-X]

[723] G. Bayramoğlu, M. Yılmaz, and M.Y. Arıca, "Affinity dye–ligand poly (hydroxyethyl methacrylate)/chitosan composite membrane for adsorption lysozyme and kinetic properties", *Biochem. Eng. J.,* vol. 13, pp. 35-42, 2003. [http://dx.doi.org/10.1016/S1369-703X(02)00098-0]

[724] M-L. Guo, Y-M. Jiang, Z-L. Ma, and Y-L. Li, "Hydrolytic characteristics of chitosan-immobilized As 1.398 neutral proteinase (from B. subtilis) to soybean protein", *Food Chem.,* vol. 55, pp. 373-377, 1996. [http://dx.doi.org/10.1016/0308-8146(95)00137-9]

[725] J-L. Leuba, and F. Widmer, "Immobilization of proteinases on Chitosan", *Biotechnol. Lett.,* vol. 1, pp. 109-114, 1979. [http://dx.doi.org/10.1007/BF01386708]

[726] K. Iwasaki, M. Inoue, and Y. Matsubara, "Continuous hydrolysis of pectate by immobilized endo-polygalacturonase in a continuously stirred tank reactor", *Biosci. Biotechnol. Biochem.,* vol. 62, no. 2, pp. 262-267, 1998. [http://dx.doi.org/10.1271/bbb.62.262] [PMID: 27388517]

[727] H.L. Ramirez, L. Gómez Brizuela, J. Úbeda Iranzo, M. Arevalo-Villena, and A.I. Briones Pérez, "Pectinase immobilization on a chitosan-coated chitin support", *J. Food Process Eng.,* 2015.

[728] L. Xiujin, "Study on immobilization of pectinase on chitosan", *Food Science,* vol. 10, p. 8, 2002.

[729] G.D. Altun, and S.A. Cetinus, "Immobilization of pepsin on chitosan beads", *Food Chem.,* vol. 100,

pp. 964-971, 2007.
[http://dx.doi.org/10.1016/j.foodchem.2005.11.005]

[730] M.M. Lakouraj, A. Maashsani, R-S. Norouzian, and M. Mohadjerani, "Immobilization of pepsin on chitosan magnetic nanoparticles and its application in deacetylation of amides", *J. Carbohydr. Chem.*, vol. 34, pp. 103-119, 2015.
[http://dx.doi.org/10.1080/07328303.2015.1009534]

[731] V.S. Nithianandam, K.S. Srinivasan, K.T. Joseph, and M. Santappa, "Immobilization of pepsin on chitosan", *Indian J. Biochem. Biophys.*, vol. 16, no. 2, pp. 119-121, 1979.
[PMID: 120305]

[732] J-P. Chen, and J-Y. Chen, "Preparation and characterization of immobilized phospholipase A 2 on chitosan beads for lowering serum cholesterol concentration", *J. Mol. Catal., B Enzym.*, vol. 5, pp. 483-490, 1998.
[http://dx.doi.org/10.1016/S1381-1177(98)00077-0]

[733] J. Zhang, Y. Song, and Y. Wang, "X.-y. ZOU, and D.-y. YU, "Study on phospholipase A2 immobilized by sodium alginate-chitosan", *Sci Technol Food Industry,* vol. 33, pp. 201-205, 2012.

[734] S. Dumitriu, P. Magny, D. Montane, P. Vidal, and E. Chornet, "Polyionic hydrogels obtained by complexation between xanthan and chitosan: their properties as supports for enzyme immobilization", *J. Bioact. Compat. Polym.*, vol. 9, pp. 184-209, 1994.
[http://dx.doi.org/10.1177/088391159400900205]

[735] P-H. Elchinger, C. Delattre, S. Faure, O. Roy, S. Badel, T. Bernardi, C. Taillefumier, and P. Michaud, "Immobilization of proteases on chitosan for the development of films with anti-biofilm properties", *Int. J. Biol. Macromol.*, vol. 72, pp. 1063-1068, 2015.
[http://dx.doi.org/10.1016/j.ijbiomac.2014.09.061] [PMID: 25451753]

[736] M. Abdel-Naby, A. Sherif, A. El-Tanash, and A. Mankarios, "Immobilization of Aspergillus oryzae tannase and properties of the immobilized enzyme", *J. Appl. Microbiol.*, vol. 87, pp. 108-114, 1999.
[http://dx.doi.org/10.1046/j.1365-2672.1999.00799.x]

[737] D.K. Boadi, and R.J. Neufeld, "Encapsulation of tannase for the hydrolysis of tea tannins", *Enzyme Microb. Technol.*, vol. 28, no. 7-8, pp. 590-595, 2001.
[http://dx.doi.org/10.1016/S0141-0229(01)00295-2] [PMID: 11339939]

[738] M. Nonaka, A. Sawa, Y. Matsuura, M. Motoki, and N. Nio, "Deamidation of several food proteins using free and immobilized Ca2+-independent microbial transglutaminase", *Biosci. Biotechnol. Biochem.*, vol. 60, no. 3, pp. 532-533, 1996.
[http://dx.doi.org/10.1271/bbb.60.532] [PMID: 27299561]

[739] S-J. Ge, H. Bai, and L-X. Zhang, "Trypsin immobilization on shrimp chitin with formaldehyde and its application to continuous hydrolysis of casein", *Biotechnol. Appl. Biochem.*, vol. 24, pp. 1-6, 1996.

[740] X. An, and Z. Su, "Characterization and application of high magnetic property chitosan particles", *J. Appl. Polym. Sci.*, vol. 81, pp. 1175-1181, 2001.
[http://dx.doi.org/10.1002/app.1539]

[741] H. Bao, Q. Chen, L. Zhang, and G. Chen, "Immobilization of trypsin in the layer-by-layer coating of graphene oxide and chitosan on in-channel glass fiber for microfluidic proteolysis", *Analyst (Lond.),* vol. 136, no. 24, pp. 5190-5196, 2011.
[http://dx.doi.org/10.1039/c1an15690j] [PMID: 22013584]

[742] Z. Li, J. Kang, H. Luo, and L. Wang, "Immobilization of trypsin on chitosan", *Journal of Fujian Medical College,* vol. 31, pp. 205-208A, 1996.

[743] A. Manrich, C.M. Galvão, C.D. Jesus, R.C. Giordano, and R.L. Giordano, "Immobilization of trypsin on chitosan gels: use of different activation protocols and comparison with other supports", *Int. J. Biol. Macromol.*, vol. 43, no. 1, pp. 54-61, 2008.
[http://dx.doi.org/10.1016/j.ijbiomac.2007.11.007] [PMID: 18187189]

[744] G. Peng, C. Zhao, B. Liu, F. Ye, and H. Jiang, "Immobilized trypsin onto chitosan modified monodisperse microspheres: A different way for improving carrier's surface biocompatibility", *Appl. Surf. Sci.,* vol. 258, pp. 5543-5552, 2012. [http://dx.doi.org/10.1016/j.apsusc.2012.01.071]

[745] L. Zhulai, K. Jie, and L. Hongbin, "Immobilization of Trypsin on Chitosan", *Journal of Fujian Medical University,* p. 9, 1997.

[746] A.M. Kayastha, and P.K. Srivastava, "Pigeonpea (Cajanus cajan L.) urease immobilized on glutaraldehyde-activated chitosan beads and its analytical applications", *Appl. Biochem. Biotechnol.,* vol. 96, no. 1-3, pp. 41-53, 2001. [http://dx.doi.org/10.1385/ABAB:96:1-3:041] [PMID: 11783899]

[747] J. Chen, and S. Chiu, "Preparation and characterization of urease immobilized onto porous chitosan beads for urea hydrolysis", *Bioprocess Eng.,* vol. 21, pp. 323-330, 1999. [http://dx.doi.org/10.1007/s004490050683]

[748] J.M. Magalhães, and A.A. Machado, "Urea potentiometric biosensor based on urease immobilized on chitosan membranes", *Talanta,* vol. 47, no. 1, pp. 183-191, 1998. [http://dx.doi.org/10.1016/S0039-9140(98)00066-6] [PMID: 18967317]

[749] M.J. Miguez, B.C. Rodrigues, Mde.N. Sanchez, and M.C. Laranjeira, "Preparation and scanning electronic microscopy study of chitosan/polyvinyl (alcohol)-encapsulated crude urease extract", *J. Microencapsul.,* vol. 14, no. 5, pp. 639-646, 1997. [http://dx.doi.org/10.3109/02652049709006816] [PMID: 9292439]

[750] M. Chellapandian, and M. Krishnan, "Chitosan-poly (glycidyl methacrylate) copolymer for immobilization of urease", *Process Biochem.,* vol. 33, pp. 595-600, 1998. [http://dx.doi.org/10.1016/S0032-9592(98)80001-0]

[751] A.R. DeGroot, and R.J. Neufeld, "Encapsulation of urease in alginate beads and protection from α-chymotrypsin with chitosan membranes", *Enzyme Microb. Technol.,* vol. 29, pp. 321-327, 2001. [http://dx.doi.org/10.1016/S0141-0229(01)00393-3]

[752] M.E. Ghica, R. Pauliukaite, O. Fatibello-Filho, and C.M. Brett, "Application of functionalised carbon nanotubes immobilised into chitosan films in amperometric enzyme biosensors", *Sens. Actuators B Chem.,* vol. 142, pp. 308-315, 2009. [http://dx.doi.org/10.1016/j.snb.2009.08.012]

[753] Z. Zhao, W. Lei, X. Zhang, B. Wang, and H. Jiang, "ZnO-based amperometric enzyme biosensors", *Sensors (Basel),* vol. 10, no. 2, pp. 1216-1231, 2010. [http://dx.doi.org/10.3390/s100201216] [PMID: 22205864]

[754] A. Turner, I. Karube, and G.S. Wilson, *Biosensors: fundamentals and applications,* 1987. [http://dx.doi.org/10.1016/S0003-2670(00)85361-1]

[755] Y. Shao, J. Wang, H. Wu, J. Liu, I.A. Aksay, and Y. Lin, "Graphene based electrochemical sensors and biosensors: a review", *Electroanalysis,* vol. 22, pp. 1027-1036, 2010. [http://dx.doi.org/10.1002/elan.200900571]

[756] H. Zhao, "Methods for stabilizing and activating enzymes in ionic liquids-a review", *J. Chem. Technol. Biotechnol.,* vol. 85, pp. 891-907, 2010. [http://dx.doi.org/10.1002/jctb.2375]

[757] Z. Zhu, L. Garcia-Gancedo, A.J. Flewitt, H. Xie, F. Moussy, and W.I. Milne, "A critical review of glucose biosensors based on carbon nanomaterials: carbon nanotubes and graphene", *Sensors (Basel),* vol. 12, no. 5, pp. 5996-6022, 2012. [http://dx.doi.org/10.3390/s120505996] [PMID: 22778628]

[758] S. Wang, Y. Tan, D. Zhao, and G. Liu, "Amperometric tyrosinase biosensor based on Fe3O4 nanoparticles-chitosan nanocomposite", *Biosens. Bioelectron.,* vol. 23, no. 12, pp. 1781-1787, 2008. [http://dx.doi.org/10.1016/j.bios.2008.02.014] [PMID: 18387292]

[759] X-J. Huang, D. Ge, and Z-K. Xu, "Preparation and characterization of stable chitosan nanofibrous membrane for lipase immobilization", *Eur. Polym. J.,* vol. 43, pp. 3710-3718, 2007. [http://dx.doi.org/10.1016/j.eurpolymj.2007.06.010]

[760] L. Huang, M. Zhai, J. Peng, L. Xu, J. Li, and G. Wei, "Synthesis of room temperature ionic liquids from carboxymethylated chitosan", *Carbohydr. Polym.,* vol. 71, pp. 690-693, 2008. [http://dx.doi.org/10.1016/j.carbpol.2007.06.020]

[761] X. Zhao, Z. Mai, X. Kang, and X. Zou, "Direct electrochemistry and electrocatalysis of horseradish peroxidase based on clay-chitosan-gold nanoparticle nanocomposite", *Biosens. Bioelectron.,* vol. 23, no. 7, pp. 1032-1038, 2008. [http://dx.doi.org/10.1016/j.bios.2007.10.012] [PMID: 18054482]

[762] S. Romo, and C. Pérez-Martínez, "The use of immobilization in alginate beads for long-term storage of pseudanabaena galeata (cyanobacteria) in the laboratory1", *J. Phycol.,* vol. 33, pp. 1073-1076, 1997. [http://dx.doi.org/10.1111/j.0022-3646.1997.01073.x]

[763] A.C. Jen, M.C. Wake, and A.G. Mikos, "Review: Hydrogels for cell immobilization", *Biotechnol. Bioeng.,* vol. 50, no. 4, pp. 357-364, 1996. [http://dx.doi.org/10.1002/(SICI)1097-0290(19960520)50:4<357::AID-BIT2>3.3.CO;2-F] [PMID: 18626984]

[764] P.J. Verbelen, D.P. De Schutter, F. Delvaux, K.J. Verstrepen, and F.R. Delvaux, "Immobilized yeast cell systems for continuous fermentation applications", *Biotechnol. Lett.,* vol. 28, no. 19, pp. 1515-1525, 2006. [http://dx.doi.org/10.1007/s10529-006-9132-5] [PMID: 16937245]

[765] M. Cassidy, H. Lee, and J. Trevors, "Environmental applications of immobilized microbial cells: a review", *J. Ind. Microbiol.,* vol. 16, pp. 79-101, 1996. [http://dx.doi.org/10.1007/BF01570068]

[766] I. Chibata, and L.B. Wingard, *Immobilized microbial cells: applied biochemistry and bioengineering.*vol. 4. Elsevier, 2014.

[767] G. Orive, E. Santos, J.L. Pedraz, and R.M. Hernández, "Application of cell encapsulation for controlled delivery of biological therapeutics", *Adv. Drug Deliv. Rev.,* vol. 67-68, pp. 3-14, 2014. [http://dx.doi.org/10.1016/j.addr.2013.07.009] [PMID: 23886766]

[768] L. Ding, C. Hao, Y. Xue, and H. Ju, "A bio-inspired support of gold nanoparticles-chitosan nanocomposites gel for immobilization and electrochemical study of K562 leukemia cells", *Biomacromolecules,* vol. 8, no. 4, pp. 1341-1346, 2007. [http://dx.doi.org/10.1021/bm061224y] [PMID: 17375952]

[769] S. Fierro, Mdel.P. Sánchez-Saavedra, and C. Copalcúa, "Nitrate and phosphate removal by chitosan immobilized Scenedesmus", *Bioresour. Technol.,* vol. 99, no. 5, pp. 1274-1279, 2008. [http://dx.doi.org/10.1016/j.biortech.2007.02.043] [PMID: 17531478]

[770] S. Overgaard, J. Scharer, M. Moo-Young, and N. Bols, "Immobilization of hybridoma cells in chitosan alginate beads", *Can. J. Chem. Eng.,* vol. 69, pp. 439-443, 1991. [http://dx.doi.org/10.1002/cjce.5450690206]

[771] D.D. Long, and D.V. Luyen, "Chitosan-carboxymethylcellulose hydrogels as supports for cell immobilization", *J. Macromol. Sci. Part A Pure Appl. Chem.,* vol. 33, pp. 1875-1884, 1996. [http://dx.doi.org/10.1080/10601329608011013]

[772] B.A. Zielinski, and P. Aebischer, "Chitosan as a matrix for mammalian cell encapsulation", *Biomaterials,* vol. 15, no. 13, pp. 1049-1056, 1994. [http://dx.doi.org/10.1016/0142-9612(94)90090-6] [PMID: 7888575]

[773] M-A. Shinonaga, Y. Kawamura, and T. Yamane, "Immobilization of yeast cells with cross-linked chitosan beads", *J. Ferment. Bioeng.,* vol. 74, pp. 90-94, 1992.

[http://dx.doi.org/10.1016/0922-338X(92)80007-6]

[774] B. Aguilar-May, and M. del Pilar Sánchez-Saavedra, "Growth and removal of nitrogen and phosphorus by free-living and chitosan-immobilized cells of the marine cyanobacterium Synechococcus elongatus", *J. Appl. Phycol.,* vol. 21, pp. 353-360, 2009. [http://dx.doi.org/10.1007/s10811-008-9376-7]

[775] Z. Yang, "Immobilization of Enterococcus faecalis cells with chitosan: A new process for the industrial production of l-citrulline", *Process Biochem.,* 2015. [http://dx.doi.org/10.1016/j.procbio.2015.02.010]

[776] N.D. Leipzig, R.G. Wylie, H. Kim, and M.S. Shoichet, "Differentiation of neural stem cells in three-dimensional growth factor-immobilized chitosan hydrogel scaffolds", *Biomaterials,* vol. 32, no. 1, pp. 57-64, 2011. [http://dx.doi.org/10.1016/j.biomaterials.2010.09.031] [PMID: 20934216]

[777] T. Jesionowski, J. Zdarta, and B. Krajewska, "Enzyme immobilization by adsorption: a review", *Adsorption,* vol. 20, pp. 801-821, 2014. [http://dx.doi.org/10.1007/s10450-014-9623-y]

[778] C.J. Wilson, R.E. Clegg, D.I. Leavesley, and M.J. Pearcy, "Mediation of biomaterial-cell interactions by adsorbed proteins: a review", *Tissue Eng.,* vol. 11, no. 1-2, pp. 1-18, 2005. [http://dx.doi.org/10.1089/ten.2005.11.1] [PMID: 15738657]

[779] H.J. Kong, M.K. Smith, and D.J. Mooney, "Designing alginate hydrogels to maintain viability of immobilized cells", *Biomaterials,* vol. 24, no. 22, pp. 4023-4029, 2003. [http://dx.doi.org/10.1016/S0142-9612(03)00295-3] [PMID: 12834597]

[780] K-D. Vorlop, and J. Klein, "New developments in the field of cell immobilization-formation of biocatalysts by ionotropic gelation", In: *Enzyme Technology*Springer, 1983, pp. 219-235. [http://dx.doi.org/10.1007/978-3-642-69148-5_20]

[781] H.N. Öztop, D. Saraydin, and S. Cetinus, "pH-sensitive chitosan films for bakers yeast immobilization", *Appl. Biochem. Biotechnol.,* vol. 101, no. 3, pp. 239-249, 2002. [http://dx.doi.org/10.1385/ABAB:101:3:239] [PMID: 12109819]

[782] C. Wang, K. Ni, X. Zhou, D. Wei, and Y. Ren, "Immobilization of Gluconobacter oxydans by entrapment in porous chitosan sponge", *J. Bioprocess. Biotech.,* vol. 3, p. 2, 2013. [http://dx.doi.org/10.4172/2155-9821.1000132]

[783] N. Khondee, S. Tathong, O. Pinyakong, R. Müller, S. Soonglerdsongpha, C. Ruangchainikom, C. Tongcumpou, and E. Luepromchai, "Lipopeptide biosurfactant production by chitosan-immobilized Bacillus sp. GY19 and their recovery by foam fractionation", *Biochem. Eng. J.,* vol. 93, pp. 47-54, 2015. [http://dx.doi.org/10.1016/j.bej.2014.09.001]

[784] D. Musale, A. Kumar, and G. Pleizier, "Formation and characterization of poly (acrylonitrile)/chitosan composite ultrafiltration membranes", *J. Membr. Sci.,* vol. 154, pp. 163-173, 1999. [http://dx.doi.org/10.1016/S0376-7388(98)00265-8]

[785] A. Saxena, B.P. Tripathi, and V.K. Shahi, "An improved process for separation of proteins using modified chitosan-silica cross-linked charged ultrafilter membranes under coupled driving forces: isoelectric separation of proteins", *J. Colloid Interface Sci.,* vol. 319, no. 1, pp. 252-262, 2008. [http://dx.doi.org/10.1016/j.jcis.2007.11.035] [PMID: 18068717]

[786] E. Casal, A. Montilla, F.J. Moreno, A. Olano, and N. Corzo, "Use of chitosan for selective removal of β-lactoglobulin from whey", *J. Dairy Sci.,* vol. 89, no. 5, pp. 1384-1389, 2006. [http://dx.doi.org/10.3168/jds.S0022-0302(06)72206-8] [PMID: 16606709]

[787] X. Zeng, and E. Ruckenstein, "Control of pore sizes in macroporous chitosan and chitin membranes", *Ind. Eng. Chem. Res.,* vol. 35, pp. 4169-4175, 1996. [http://dx.doi.org/10.1021/ie960270j]

[788] X. Zeng, and E. Ruckenstein, "Cross-linked macroporous chitosan anion-exchange membranes for protein separations", *J. Membr. Sci.,* vol. 148, pp. 195-205, 1998. [http://dx.doi.org/10.1016/S0376-7388(98)00183-5]

[789] J.Z. Knaul, S.M. Hudson, and K.A. Creber, "Improved mechanical properties of chitosan fibers", *J. Appl. Polym. Sci.,* vol. 72, pp. 1721-1732, 1999. [http://dx.doi.org/10.1002/(SICI)1097-4628(19990624)72:13<1721::AID-APP8>3.0.CO;2-V]

[790] F. Xi, J. Wu, Z. Jia, and X. Lin, "Preparation and characterization of trypsin immobilized on silica gel supported macroporous chitosan bead", *Process Biochem.,* vol. 40, pp. 2833-2840, 2005. [http://dx.doi.org/10.1016/j.procbio.2004.12.013]

[791] P. Kanagaraj, A. Nagendran, D. Rana, T. Matsuura, S. Neelakandan, T. Karthikkumar, and A. Muthumeenal, "Influence of N-phthaloyl chitosan on poly (ether imide) ultrafiltration membranes and its application in biomolecules and toxic heavy metal ion separation and their antifouling properties", *Appl. Surf. Sci.,* vol. 329, pp. 165-173, 2015. [http://dx.doi.org/10.1016/j.apsusc.2014.12.082]

[792] T-Y. Guo, Y-Q. Xia, G-J. Hao, B-H. Zhang, G-Q. Fu, Z. Yuan, B-L. He, and J.F. Kennedy, "Chemically modified chitosan beads as matrices for adsorptive separation of proteins by molecularly imprinted polymer", *Carbohydr. Polym.,* vol. 62, pp. 214-221, 2005. [http://dx.doi.org/10.1016/j.carbpol.2005.03.012]

[793] Y. Yao, and S. Li, "Capillary zone electrophoresis of basic proteins with chitosan as a capillary modifier", *J. Chromatogr. A,* vol. 663, pp. 97-104, 1994. [http://dx.doi.org/10.1016/0021-9673(94)80500-8]

[794] P. Sun, A. Landman, and R.A. Hartwick, "Chitosan coated capillary with reversed electroosmotic flow in capillary electrophoresis for the separation of basic drugs and proteins", *J. Microcolumn Sep.,* vol. 6, pp. 403-407, 1994. [http://dx.doi.org/10.1002/mcs.1220060409]

[795] R. Huang, G. Chen, M. Sun, Y. Hu, and C. Gao, "Studies on nanofiltration membrane formed by diisocyanate cross-linking of quaternized chitosan on poly (acrylonitrile)(PAN) support", *J. Membr. Sci.,* vol. 286, pp. 237-244, 2006. [http://dx.doi.org/10.1016/j.memsci.2006.09.045]

[796] W. Shi, F. Zhang, and G. Zhang, "Adsorption of bilirubin with polylysine carrying chitosan-coated nylon affinity membranes", *J. Chromatogr. B Analyt. Technol. Biomed. Life Sci.,* vol. 819, no. 2, pp. 301-306, 2005. [http://dx.doi.org/10.1016/j.jchromb.2005.02.018] [PMID: 15833294]

[797] Z. Feng, Z. Shao, J. Yao, Y. Huang, and X. Chen, "Protein adsorption and separation with chitosan-based amphoteric membranes", *Polymer (Guildf.),* vol. 50, pp. 1257-1263, 2009. [http://dx.doi.org/10.1016/j.polymer.2008.12.046]

[798] T. Chakrabarty, and V.K. Shahi, "Modified chitosan-based, pH-responsive membrane for protein separation", *RSC Advances,* vol. 4, pp. 53245-53252, 2014. [http://dx.doi.org/10.1039/C4RA05314A]

[799] J.H. Kim, J.H. Kim, J. Jegal, and K-H. Lee, "Optical resolution of α-amino acids through enantioselective polymeric membranes based on polysaccharides", *J. Membr. Sci.,* vol. 213, pp. 273-283, 2003. [http://dx.doi.org/10.1016/S0376-7388(02)00534-3]

[800] G. Bayramoğlu, H. Erdogan, and M.Y. Arica, "Studies of adsorption of alkaline trypsin by poly (methacrylic acid) brushes on chitosan membranes", *J. Appl. Polym. Sci.,* vol. 108, pp. 456-465, 2008. [http://dx.doi.org/10.1002/app.27690]

[801] N. Ye, Y. Xie, P. Shi, T. Gao, and J. Ma, "Synthesis of magnetite/graphene oxide/chitosan composite and its application for protein adsorption", *Mater Sci Eng C Mater Biol Appl,* vol. 45, pp. 8-14, 2014.

[http://dx.doi.org/10.1016/j.msec.2014.08.064] [PMID: 25491795]

[802] X.Z. Shu, K.J. Zhu, and W. Song, "Novel pH-sensitive citrate cross-linked chitosan film for drug controlled release", *Int. J. Pharm.,* vol. 212, no. 1, pp. 19-28, 2001. [http://dx.doi.org/10.1016/S0378-5173(00)00582-2] [PMID: 11165817]

[803] C.R. Huei, and H-D. Hwa, "Effect of molecular weight of chitosan with the same degree of deacetylation on the thermal, mechanical, and permeability properties of the prepared membrane", *Carbohydr. Polym.,* vol. 29, pp. 353-358, 1996. [http://dx.doi.org/10.1016/S0144-8617(96)00007-0]

[804] L.A. van den Broek, R.J. Knoop, F.H. Kappen, and C.G. Boeriu, "Chitosan films and blends for packaging material", *Carbohydr. Polym.,* vol. 116, pp. 237-242, 2015. [http://dx.doi.org/10.1016/j.carbpol.2014.07.039] [PMID: 25458295]

[805] S. Puttipipatkhachorn, J. Nunthanid, K. Yamamoto, and G.E. Peck, "Drug physical state and drug-polymer interaction on drug release from chitosan matrix films", *J. Control. Release,* vol. 75, no. 1-2, pp. 143-153, 2001. [http://dx.doi.org/10.1016/S0168-3659(01)00389-3] [PMID: 11451504]

[806] M. Zhang, X.H. Li, Y.D. Gong, N.M. Zhao, and X.F. Zhang, "Properties and biocompatibility of chitosan films modified by blending with PEG", *Biomaterials,* vol. 23, no. 13, pp. 2641-2648, 2002. [http://dx.doi.org/10.1016/S0142-9612(01)00403-3] [PMID: 12059013]

[807] C. Tual, E. Espuche, M. Escoubes, and A. Domard, "Transport properties of chitosan membranes: influence of crosslinking", *J. Polym. Sci., B, Polym. Phys.,* vol. 38, pp. 1521-1529, 2000. [http://dx.doi.org/10.1002/(SICI)1099-0488(20000601)38:11<1521::AID-POLB120>3.0.CO;2-#]

[808] C. Valenzuela, L. Abugoch, and C. Tapia, "Quinoa protein–chitosan–sunflower oil edible film: Mechanical, barrier and structural properties", *LWT-Food Science and Technology,* vol. 50, pp. 531-537, 2013. [http://dx.doi.org/10.1016/j.lwt.2012.08.010]

[809] F. Debeaufort, J-A. Quezada-Gallo, and A. Voilley, "Edible films and coatings: tomorrows packagings: a review", *Crit. Rev. Food Sci. Nutr.,* vol. 38, no. 4, pp. 299-313, 1998. [http://dx.doi.org/10.1080/10408699891274219] [PMID: 9626488]

[810] A. Alishahi, and M. Aïder, "Applications of chitosan in the seafood industry and aquaculture: a review", *Food and Bioprocess Technology,* vol. 5, pp. 817-830, 2012. [http://dx.doi.org/10.1007/s11947-011-0664-x]

[811] D.S. Cha, and M.S. Chinnan, "Biopolymer-based antimicrobial packaging: a review", *Crit. Rev. Food Sci. Nutr.,* vol. 44, no. 4, pp. 223-237, 2004. [http://dx.doi.org/10.1080/10408690490464276] [PMID: 15462127]

[812] V. Coma, A. Martial-Gros, S. Garreau, A. Copinet, F. Salin, and A. Deschamps, "Edible antimicrobial films based on chitosan matrix", *J. Food Sci.,* vol. 67, pp. 1162-1169, 2002. [http://dx.doi.org/10.1111/j.1365-2621.2002.tb09470.x]

[813] C.A. Cozzolino, G. Cerri, A. Brundu, and S. Farris, "Microfibrillated cellulose (MFC): pullulan bionanocomposite films", *Cellulose,* vol. 21, pp. 4323-4335, 2014. [http://dx.doi.org/10.1007/s10570-014-0433-x]

[814] S. Fakhreddin Hosseini, M. Rezaei, M. Zandi, and F.F. Ghavi, "Preparation and functional properties of fish gelatin-chitosan blend edible films", *Food Chem.,* vol. 136, no. 3-4, pp. 1490-1495, 2013. [http://dx.doi.org/10.1016/j.foodchem.2012.09.081] [PMID: 23194553]

[815] F. Sébastien, G. Stéphane, A. Copinet, and V. Coma, "Novel biodegradable films made from chitosan and poly (lactic acid) with antifungal properties against mycotoxinogen strains", *Carbohydr. Polym.,* vol. 65, pp. 185-193, 2006. [http://dx.doi.org/10.1016/j.carbpol.2006.01.006]

[816] H. Wang, H. Liu, C. Chu, Y. She, S. Jiang, L. Zhai, S. Jiang, and X. Li, *Diffusion and antibacterial*

*properties of nisin-loaded chitosan/poly (L-lactic acid) towards development of active food packaging film.*Food and Bioprocess Technology, 2015, pp. 1-11.

[817] L.F. Boesel, "Effect of plasticizers on the barrier and mechanical properties of biomimetic composites of chitosan and clay", *Carbohydr. Polym.,* vol. 115, pp. 356-363, 2015. [http://dx.doi.org/10.1016/j.carbpol.2014.08.064] [PMID: 25439905]

[818] Q. Meng, M-C. Heuzey, and P.J. Carreau, "Hierarchical structure and physicochemical properties of plasticized chitosan", *Biomacromolecules,* vol. 15, no. 4, pp. 1216-1224, 2014. [http://dx.doi.org/10.1021/bm401792u] [PMID: 24564751]

[819] V. Coma, A. Martial-Gros, S. Garreau, A. Copinet, F. Salin, and A. Deschamps, "Edible antimicrobial films based on chitosan matrix", *Journal of Food Science-Chicago,* vol. 67, pp. 1162-1169, 2002. [http://dx.doi.org/10.1111/j.1365-2621.2002.tb09470.x]

[820] A. El Ghaouth, J. Arul, R. Ponnampalam, and M. Boulet, "Chitosan coating effect on storability and quality of fresh strawberries", *J. Food Sci.,* vol. 56, pp. 1618-1620, 1991. [http://dx.doi.org/10.1111/j.1365-2621.1991.tb08655.x]

[821] M.A. Rojas-Graü, R. Soliva-Fortuny, and O. Martín-Belloso, "Edible coatings to incorporate active ingredients to fresh-cut fruits: a review", *Trends Food Sci. Technol.,* vol. 20, pp. 438-447, 2009. [http://dx.doi.org/10.1016/j.tifs.2009.05.002]

[822] B. Ouattar, R.E. Simard, G. Piett, A. Bégin, and R.A. Holley, "Inhibition of surface spoilage bacteria in processed meats by application of antimicrobial films prepared with chitosan", *Int. J. Food Microbiol.,* vol. 62, no. 1-2, pp. 139-148, 2000. [http://dx.doi.org/10.1016/S0168-1605(00)00407-4] [PMID: 11139014]

[823] Y. Zhang, X. Huang, B. Duan, L. Wu, S. Li, and X. Yuan, "Preparation of electrospun chitosan/poly (vinyl alcohol) membranes", *Colloid Polym. Sci.,* vol. 285, pp. 855-863, 2007. [http://dx.doi.org/10.1007/s00396-006-1630-4]

[824] V. Binsu, R. Nagarale, V.K. Shahi, and P. Ghosh, "Studies on N-methylene phosphonic chitosan/poly (vinyl alcohol) composite proton-exchange membrane", *React. Funct. Polym.,* vol. 66, pp. 1619-1629, 2006. [http://dx.doi.org/10.1016/j.reactfunctpolym.2006.06.003]

[825] H-A. Tsai, W-H. Chen, C-Y. Kuo, K-R. Lee, and J-Y. Lai, "Study on the pervaporation performance and long-term stability of aqueous iso-propanol solution through chitosan/polyacrylonitrile hollow fiber membrane", *J. Membr. Sci.,* vol. 309, pp. 146-155, 2008. [http://dx.doi.org/10.1016/j.memsci.2007.10.018]

[826] J.M. Yang, S.J. Yang, H.T. Lin, T-H. Wu, and H-J. Chen, "Chitosan containing PU/Poly (NIPAAm) thermosensitive membrane for wound dressing", *Mater. Sci. Eng. C,* vol. 28, pp. 150-156, 2008. [http://dx.doi.org/10.1016/j.msec.2007.01.011]

[827] J.H. Chen, Q.L. Liu, J. Fang, A.M. Zhu, and Q.G. Zhang, "Composite hybrid membrane of chitosan-silica in pervaporation separation of MeOH/DMC mixtures", *J. Colloid Interface Sci.,* vol. 316, no. 2, pp. 580-588, 2007. [http://dx.doi.org/10.1016/j.jcis.2007.09.022] [PMID: 17915245]

[828] R. Huang, G. Chen, M. Sun, and C. Gao, "Preparation and characterization of quaterinized chitosan/poly (acrylonitrile) composite nanofiltration membrane from anhydride mixture cross-linking", *Separ. Purif. Tech.,* vol. 58, pp. 393-399, 2008. [http://dx.doi.org/10.1016/j.seppur.2007.05.013]

[829] J. Wang, X. Zheng, H. Wu, B. Zheng, Z. Jiang, X. Hao, and B. Wang, "Effect of zeolites on chitosan/zeolite hybrid membranes for direct methanol fuel cell", *J. Power Sources,* vol. 178, pp. 9-19, 2008. [http://dx.doi.org/10.1016/j.jpowsour.2007.12.063]

[830] H. Wu, B. Zheng, X. Zheng, J. Wang, W. Yuan, and Z. Jiang, "Surface-modified Y zeolite-filled

chitosan membrane for direct methanol fuel cell", *J. Power Sources,* vol. 173, pp. 842-852, 2007. [http://dx.doi.org/10.1016/j.jpowsour.2007.08.020]

[831] S. Xiao, X. Feng, and R.Y. Huang, "Trimesoyl chloride crosslinked chitosan membranes for CO2/N2 separation and pervaporation dehydration of isopropanol", *J. Membr. Sci.,* vol. 306, pp. 36-46, 2007. [http://dx.doi.org/10.1016/j.memsci.2007.08.021]

[832] S. Zhang, M.H. Peh, Z. Thong, and T-S. Chung, "Thin film interfacial cross-linking approach to fabricate a chitosan rejecting layer over poly (ether sulfone) support for heavy metal removal", *Ind. Eng. Chem. Res.,* vol. 54, pp. 472-479, 2014. [http://dx.doi.org/10.1021/ie503809c]

[833] J. Winterowd, and P. Sandford, *Food polysaccharides and their applications.*Chitin and Chitosan, 1995, pp. 441-462.

[834] K. Arai, T. Kinumaki, and T. Fujita, *"Toxicity of chitosan," Bull.* Region. Fish. Res. Lab: Tokai, 1968, pp. 89-94.

[835] D. Jones, and H. Mawhinney, "Chitosan", In: *Handbook of Pharmaceutical Excipients.*5th ed Pharmaceutical Press and American Pharmacist Association: London, 2006, pp. 159-162.

[836] S. Minami, M. Oh-Oka, Y. Okamoto, K. Miyatake, A. Matsuhashi, Y. Shigemasa, and Y. Fukumoto, "Chitosan-inducing hemorrhagic pneumonia in dogs", *Carbohydr. Polym.,* vol. 29, pp. 241-246, 1996. [http://dx.doi.org/10.1016/0144-8617(95)00157-3]

[837] B. Prajapati, "Chitosan a marine medical polymer and its lipid lowering capacity", *Int J Health,* vol. 9, 2009.

[838] E.J. Freireich, E.A. Gehan, D.P. Rall, L.H. Schmidt, and H.E. Skipper, "Quantitative comparison of toxicity of anticancer agents in mouse, rat, hamster, dog, monkey, and man", *Cancer Chemother. Rep.,* vol. 50, no. 4, pp. 219-244, 1966. [PMID: 4957125]

[839] J.G. Wlnterowd, "Chitin and chitosan", *Food Polysaccharides and Their Applications,* vol. 67, p. 441, 1995.

[840] A. Bignamini, "Therapeutic activity of oral glucosamine sulfate in osteoarthrosis: A Placebo-controlled double-blind investigation", In: *Clin. Ther*vol. 3. 1980.

CHAPTER 2

Biopolymers of Microbial Origin

Natalia A. Castillo[1,2], **Alejandra L. Valdez**[1,3] and **Julia I. Fariña**[1,4,*]

[1] *Laboratorio de Micodiversidad y Micoprospección, PROIMI-CONICET, Tucumán, Argentina*

[2] *Cátedra de Micología, Facultad de Bioquímica, Química y Farmacia (UNT), Tucumán, Argentina*

[3] *Cátedra de Microbiología, Facultad de Bioquímica, Química y Farmacia (UNT), Tucumán, Argentina*

[4] *Cátedra de Química Biológica, Facultad de Ciencias Exactas y Naturales (UNCa), Catamarca, Argentina*

Abstract: Synthetic polymers play a significant role in the technological development of humankind and are obtained from petroleum by-products. Because of their physical properties, they can be used for the manufacture of a diversity of products ranging from simple garbage bags or contact lenses up to products for construction. Based on plastic skills such as low price, low weight, resistance to abrasion, impact and corrosion, inertia and versatility, they gradually replaced traditional materials like wood, stone and metal. However, these same advantages nowadays have become their worst drawback, turning them into wastes of difficult disposal and consequently, in a serious environmental problem. Additionally, their multiple usage in daily life has caused a massive increase in consumption and the consequent pollution problems. In this context, biopolymers have emerged as an ideal alternative to the synthetic polymer industry. Biopolymers provide a solution to the origin of the problem as coming from renewable resources, practically all of them being biodegradable, which is not the case for most synthetic polymers. Eco-friendly waste disposal of biopolymers takes advantage of their property of being degraded by soil microbiota, which significantly reduces CO_2 emission as compared to conventional incineration. Therefore, the use of biodegradable biopolymers is also relevant from the point of view of global warming prevention. Based on this rationale, biopolymers based on renewable resources are generating an increasing interest, both in the overall society and particularly, in the plastic industry.

Keywords: Applications, Biopolymers, Downstream processing, Exopolysaccharides, Fermentation, Physicochemical properties, Sustainable production.

[*] **Corresponding author Julia I. Fariña:** Laboratorio de Micodiversidad y Micoprospección, PROIMI-CONICET, Tucumán, Argentina; Tel: +54 381 4344888 #23; Fax: +54 381 4344887; E-mails: jifarina@yahoo.com, jifarina@proimi.org.ar

Martin Masuelli & Denis Renard (Eds.)

INTRODUCTION

Although a great controversy exists concerning the definition and scope of the "biopolymer" term, it may be properly ascribed to all macromolecules produced by biological systems including plants, animals and microorganisms. In a step further, polymers chemically synthesized from biological building blocks (*e.g.* amino acids, sugars or lipids) can also be considered as biopolymers.

In the recent years, polysaccharides of microbial origin have gained special interest, compared to the gums from plants and algae. This can be mainly due to the following advantages:

- Specific biosynthetic processes allow obtaining polysaccharides of controlled purity and molecular weight.
- Production is not subjected to geographical and/or climatic variations as happens with plants and algae. These factors can profoundly affect the quality and quantity of the product obtained and the availability of the natural source.
- The possibility to use regional agro-industrial by-products as raw materials for biopolymer synthesis.
- The potential to change the characteristics of the producing organism in order to increase production, or to produce polysaccharides with a special (tailor-made) property.
- The ability to produce a wide variety of polysaccharides, some with high similarities to those of plant or animal origin, although most with unique and distinctive features.

Microorganisms can synthesize biopolymers of diverse chemical classes, *e.g.* nucleic acids, polyamides (proteins and poly-amino acids), polysaccharides, organic polyoxoesters (such as poly-hydroxyalkanoic acids), polythioesters, inorganic polyesters with polyphosphate, and others. These biopolymers fulfil quite different essential functions for the microorganisms such as conservation and expression of genetic information, catalysis of bioreactions, storage of energy and nutrients, communication with other microorganisms, resistance to hazardous environments and structural stability, among others. It has been widely described that many of these biopolymers, more frequently polysaccharides, also possess biological effects when administered by different routes to animals and/or humans. Based on this feature, these are called 'biological response modifiers' (BRMs). These biopolymers bind to pattern recognition receptors in the host, triggering the immune response. Exogenous BRMs have been reported to have anti-viral, anti-bacterial, anti-fungal, anti-parasitic, and anti-tumor activities. Another functional property of polysaccharides is their ability to alter the rheological properties of water, such as viscosity. Finally, bioplastics constitute a relevant group within biopolymers, which can be processed using the same technologies that are used with conventional thermoplastic materials such as extrusion, injection or blow. Because of all these diverse and unique properties, biopolymers are used by different industries for a wide spectrum of applications, covering areas such as medicine, pharmacy, agriculture, food, and many others.

Biotechnological production of these polymers is, at present, mostly achieved by fermentation with selected microorganisms in stirred tank bioreactors. Depending on the polymer and the producing microorganism, they can be obtained as extracellular or intracellular compounds. Alternatively, biopolymers can also be produced by enzymatic *in vitro* processes. However, some of these biopolymers still face a number of challenges, such as the optimization of production and recovery processes to reduce production costs, as well as the standardization and strict control of the process in order to obtain products with homogeneous features.

In this chapter, we will review some of these biopolymers of microbial origin that have become of commercial interest, because of their outstanding applications and economic competitiveness.

Biopolymers: Towards a Comprehensive and Appropriate Definition

Nowadays, the long-standing definition of biopolymer as "a molecule of high molecular weight formed by monomers that come from a biological system" is still the most accepted one [1]. Accordingly, bio-based biopolymers may come from plants, animals or microorganisms (Table **1**). Nevertheless, there is some discrepancy concerning this definition, and other authors also consider those polymers as biopolymers which come from fossil fuels and exhibit the condition of being biodegradable. Regarding the first definition, another common misunderstanding is to believe that all the biopolymers are biodegradable. Although this is a frequent feature among biopolymers, it is not an absolute fact (Table **1**); the non-degradability ratio among bio-based biopolymers has increased [2].

According to IUPAC recommendations, when biopolymers are man-made, they should be named synthetic biopolymers (Table **1**), to make a distinction with true biopolymers [3].With respect to the natural sources, biopolymers can be extracted from plants and animals, but their industrial production from microbial sources has currently become more relevant due to the advantages of these processes and their high commercial interest [4, 5].

Table 1. General overview of different biodegradable and non-biodegradable biopolymers and synthetic biopolymers.

Biopolymer	Biodegradable			Non-biodegradable
Source	**Plant**	**Microorganism**	**Animal**	**Bio-based sources**
Polysaccharides	Alginate	Acetan	Chitin	
	Carrageenan	Alginate	Chitosan	
	Cellulose and derivatives	Bacterial cellulose	Hyaluronan	
	Lignin	Chitin		
	Starch and derivatives	Chitosan		
		Curdlan		
		Dextran		
		Hyaluronan		
		Pullulan		
		Scleroglucan		
		Sphingan		
		Xanthan		
Proteins	Corn		Albumin	
	Pea		Casein	
	Potato		Collagen	
	Wheat		Keratin	
			Leather	
			Whey	
Polyesters	PLA (from starch or sugarcane)	PHAs		PET, PTT
Polyurethanes				PU
Polycarbonate				PC
Polyamides				PA 11
Vinyl Polymers				PE, PP, PVC
Triglycerides	Lipids			

PLA= Poly(ethylene adipate), PHAs= Polyhydroxyalkanoates, PET= Poly(ethylene terephthalate), PTT=Poly(trimethylene terephthalate), PU= Polyurethane, PC= Polycarbonate, PA= Polyamides, PE= Polyethylene, PP= Polypropylene, PVC= Poly(vinyl chloride)

On the basis of polymer backbone, true or synthetic biopolymers (coming from different sources) can be roughly classified as follows [2]:

Polyesters (Polyhydroxy Acids)

- *Poly(α-hydroxyalkanoic acid)s*
  - Polylactide (PLA)
- *Poly(β-, γ-, δ-hydroxyalkanoate)s (PHAs)*
  - Poly(3-hydroxybutyrate (P3HB or PHB)
  - Poly(4-hydroxybutyrate) (P4HB)
  - Poly(3-hydroxybutyrate-co-3-hydroxyhexanoate) (P3HB/P3HV or PHB/PHV or PHBHx)
  - Poly(3-hydroxybutyrate-co-3-hydroxyvalerate) (P3HB/P3HV or PHBHV)

Aromatic (Co)Polyesters

- Poly(ethylene terephthalate) (bio-based PET)
- Poly(trimethylene terephthalate) (bio-based PTT)

Polyamides

- Polyamide 11 (PA 11)

Proteins

- Albumin
- Casein
- Collagen
- Keratin
- Silk

Polyurethanes (Bio-Based PU)

- *Poly(ester urethane)s*
- *Poly(ether urethane)s*

Polysaccharides

- *Linear polysaccharides*
  - Alginate
  - Carrageenan
  - Cellulose and derivatives, *e.g.*: methyl cellulose, ethyl cellulose, propyl cellulose, hydroxyethyl cellulose, carboxymethyl cellulose, cellulose acetate, cellulose acetate butyrate, cellulose acetate propionate, cellulose nitrate
  - Chitin
  - Chitosan
  - Curdlan
  - Hyaluronan

    - Pullulan
- *Branched Polysaccharides*
    - Dextran
    - Gellan (slightly branched)
    - Levan
    - Lignin
    - Scleroglucan
    - Starch
    - Xanthan

Lipids (Triglycerides)

- *Vinyl Polymers*
    - Polyolefins (bio-based polyethylene, PE, LDPE,HDPE; bio-based polypropylene, PP)
    - Poly(vinyl chloride) (bio-based PVC)
- *Diene Polymers*
    - cis-1, 4-Polyisoprene

Additionally, based on the ways in which biopolymers respond to heat, they can also be classified as thermoplastics, thermosets or elastomers. Today, the volume of bio-based thermoset biopolymers exceeds the volume of bio-based thermoplastic biopolymers [2].

In this chapter, we will focus on the microbial biopolymers more recognized up to date (polysaccharides and polyhydroxyalkanoates) based on the current knowledge and taking into account their industrial applications. Particular attention will be paid to aspects such as the producing microorganisms, the monomer composition and structural organization, as well as the most significant physicochemical properties and applications.

Microbial Polysaccharides

In a general context, polysaccharides are defined as natural, non-toxic, and biodegradable polymers of biological origin consisting of carbohydrate units attached by glycosidic linkages. They are widely distributed in living organisms, such as plants, fungi, bacteria, shellfish and mammals, where they play important roles for numerous biological functions such as immune response, adhesion, infection, and signal transduction. Because of their unique physicochemical and biological properties, academics, researchers and industrialists from different fields believe that polysaccharides will be the leading global resource for the production of sustainable fuel, food, medicines and biomaterials [4].

Microbes are able to produce at least three main types of polysaccharides [6], namely:

- ***Somatic or Intracellular:*** located within the cytoplasm, in the form of glycogen for example, which the microorganism can use after depletion of external carbon sources.
- ***Structural:*** as components of cell structure (cytoplasmic membrane and cell wall).
- ***Extracellular:*** in some cases, covalently linked to the cell as a true capsule (capsular EPS), whilst in others, secreted into the surrounding medium thus increasing in the case of liquid cultures, the viscosity of the fermentation broth (free mucoid EPS). These latter ones have always provoked more interest because of their lower cost of purification and the highest possibilities for industrial production at larger scale.

On the basis of their shape, microbial polysaccharides can be divided in:

- ***Linear*** (*e.g.* alginate, curdlan, cellulose, chitin, chitosan, pullulan), or
- ***Branched*** (*e.g.* dextran, levan, scleroglucan, xanthan, and to a lesser degree, gellan).

According to the type of their monomeric units, they can be classified as:

- ***Homoglycans*** (cellulose, chitin, curdlan, dextran, levan, and pullulan)
- ***Di-heteroglycans*** (alginates and chitosan)
- ***Tri-heteroglycans*** (xanthan and gellan)

Depending on the net charge, they can behave as:

- ***Neutral*** (cellulose, curdlan, dextran, levan, pullulan, and scleroglucan)
- ***Anionic*** (alginate, gellan, xanthan)
- ***Cationic*** (chitin and chitosan)

The aforementioned characteristics and further structural features for the main microbial polysaccharides are resumed in Table (**2**).

Because of their structural differences, polysaccharides can greatly vary in their physicochemical properties, particularly rheological behavior. As widely described, a number of exopolysaccharides (EPSs) have a pseudoplastic behavior, which involves a drop in the apparent viscosity as the deformation of the fluid increases. This typical behavior is characteristic of xanthan, scleroglucan and alginate. Another important property of various microbial polysaccharides of industrial interest is their ability to form gels.

Table 2. Elemental features of most studied microbial polysaccharides.

Producing organism	Net charge	Shape	Monomers	Polysaccharide	Glycosidic linkages
Bacteria	Neutral	Linear	- Glucose	Cellulose	β-(1,4)
	Neutral	Linear	- Glucose	Curdlan	β-(1,3)
	Neutral	Branched	- Glucose	Dextran	Backbone: α-(1,6) Side chain: α-(1,2) α-(1,3) α-(1,4)
	Neutral	Branched	- Fructose	Levan	Backbone: β-(2,6) Side chain: β-(1,2)
	Anionic	Linear	- Glucuronic acid - Mannuronic acid	Alginate	β-(1,4) α-(1,4)
	Anionic	Linear	- Glucuronic acid - N-acetylglucosamine	Hyaluronan	β-(1,4) β-(1,3)
	Anionic	Slightly branched	- Glucose - Rhamnose - Glucuronic acid	Gellan	β-(1,4)
	Anionic	Branched	Backbone: - Glucose Side chain: - Raffinose - Glucose - Glucuronic acid - Mannose	Acetan	Backbone: β-(1,4) Side chain: β-(1,6) α-(1,6) β-(1,4) α-(1,2) (1,3)
	Anionic	Branched	- Glucose - Mannose - Glucuronic acid	Xanthan	Backbone: β-(1,4) Side chain: β-(1,2) α-(1,3)
Fungi	Neutral	Linear	- Glucose	Pullulan	α-(1,4), α-(1,6)
	Neutral	Branched	- Glucose	Scleroglucan	Backbone: β-(1,3) Side chain: β-(1,6)
	Cationic	Linear	- N-acetylglucosamine - Glucosamine	Chitin	β-(1,4)
	Cationic	Linear	- Glucosamine - N-acetylglucosamine	Chitosan	β-(1,4)

To achieve this behavior, however, it is known that some polymers require the presence of ions. For example, alginates form gels in the presence of divalent cations, particularly Ca^{2+}. Gel properties are normally associated to the formation of crosslinked polymer chain networks. Further biotechnological properties of these biopolymers are widely exploited by the industry, such as emulsifying, adhesive, flocculant, lubricant, stabilizing, thickening, texturizing, viscosifying, and film-forming abilities. Many of these unique properties, along with biocompatibility and low production costs, led to the application of natural gums in various areas of modern life.

Depending on the microbial polysaccharide involved, industrial large-scale production may present numerous **advantages,** such as:

- Obtainment of polysaccharides with unique and distinctive properties
- Biocompatibility
- Possibility to obtain consistent products with constant quality, molecular weight (Mw) and purity
- Widespread availability
- Elimination of geographical and climatic variations
- Feasibility of using regional products, agro-industrial by-products and industrial wastes as raw materials for polysaccharide production
- Low manufacture costs
- Shortened production times
- Utilization of genetic engineered microorganisms in order to increase polymer yields or to produce tailor-made polysaccharides with special properties and/or applications
- Ability to produce a wide variety of polysaccharides, some of them with great similarities to those of plant or animal origin

Although there are a great number of EPSs displaying outstanding potential for various industrial sectors, only very few of them are currently commercialized. This is because many of them can present the following **disadvantages**:

- Low yields or high production costs (being therefore not economically profitable)
- Difficulties to establish optimal conditions for the fermentation process
- Need for a strict control of fermentation conditions to obtain products with standardized characteristics

Based on all these pros and cons, and from a technological perspective, polysaccharides are products with a high added-value, whose range of applications in the most diverse fields is continuously growing. This wide applicability derives from the great diversity in their structural and functional properties. Accordingly, microbial polysaccharides with current or prospective commercial significance will be subsequently described.

General Properties of Microbial Polysaccharides

Chemical Composition

Among microbial polysaccharides, EPSs have gained greater attention because of their easier recovery from fermentation broths which facilitates further purification steps [7, 8]. They use to exhibit novel and unique physicochemical and biological properties, often even superior to those of traditionally employed polysaccharides from plant sources. For that reason, they have attracted worldwide interest. However, for a long time, their study became relegated due the complexity to successfully achieve this goal [9, 10]. Most of the initial research on biopolymers was mainly directed to biomacromolecules with more relevant cellular functions than polysaccharides, like nucleic acids or proteins. Meanwhile, polysaccharides were mainly seen as major food components or structural molecules in different organisms [7, 10].

Concerning the chemical composition of biopolysaccharides, it is widely recognized that despite being composed of monosaccharides with relatively simple structures (Table **2**), the possibility to combine them provides a complexity greater than that which can be obtained when combining amino acids in the case of proteins, for example. A high number of natural monomers are available in nature, which can additionally be modified, thus increasing the possible combinations to an indescribable count [11]. As for the main biomolecules, this remarkable variability might be translated to richness in information, being carbohydrates another vital biological reservoir of it. Nevertheless, not all the possible carbohydrate combinations are biologically found in nature [10]. The monosaccharides most commonly present in microbial polysaccharides are D-glucose, D-galactose and D-mannose. Other neutral sugars can be encountered, such as L-fucose and L-rhamnose, uronic acids and amino sugars. Pentoses are usually not found in prokaryotic polymers, this being a differential feature with respect to polysaccharides from eukaryotic origin [12], although some exceptions were reported in cyanobacteria [13] and *Sinorhizobium* [14].

Most microbial EPSs exhibit a polyanionic nature, with D-glucuronic acid as the more common uronic acid. The only recognized positively charged polymer is the one produced by *Staphylococcus epidermidis* [15, 16]. Usually, only one uronic acid type is contained in EPSs, with the well-known exception of bacterial alginates [12, 17], and polysaccharides from *Rhizobium* [18] and marine bacteria [19]. With respect to polysaccharide chemical modifications, some substituents frequently observed are enumerated in Table (**3**). The most commonly reported groups are acetate and pyruvate, but esterification with lactic or succinic acid was also reported [15]. Several amino acids have also been described as polysaccharide substituents, such as serine or L-glutamic acid in the case of the *Escherichia coli* K40 EPS [20], taurine [12], or tyrosine in the acidic heteropolysaccharide from *Paenibacillus* species [21].

Table 3. Main substituents of microbial polysaccharides [15].

Substituent	**Linkage**	**EPS source**
Organic acids		
Acetate	Ester	Frequent (*e.g. Azotobacter vinelandii, Klebsiella, etc.*)
Glycerate	Ester	*S. paucimobilis* (gellan)
Hydroxybutanoate	Ester	*R. trifolii, R. leguminosarum*
Malonate	Ester	*E. coli* K10
Propionate	Ester	*E. coli* K52
Pyruvate	Ketal	Frequent (*e.g. Kelbsiella, E. coli, etc.*)
Succinate	(half) Ester	*Rhizobium* spp., *Agrobacterium* spp.
Taurine	Ester	*Staphylococcus aureus*
Amino acids		
L-glutamate		*K. aerogenes* K82
L-serine		*E. coli* K40, *E. coli* K54
L-threonine		*E. coli* K54
Inorganic acids		
Phosphate		Common (mainly Gram-positives)
Sulfate		Cyanobacteria, *Halobacterium* spp., *Sulfobolus* spp.

Acetyl substituents do not normally contribute to the total net charge of the macromolecule, but might induce conformational changes and molecular interactions which affect the polysaccharide gelling ability without disturbing viscosity. In the case of gellan, for example, the acetylated form gives soft and elastic gels whilst the mechanical properties change to tough and brittle for the fully deacetylated form [22], being this latter the one commercially available [23]. Similarly, Shatwell *et al.* demonstrated the influence of acetyl substitution rate in the strength of thermoreversible gels of xanthan [24]. Deacetylation frequently allows adopting an adequate molecular conformation and an improved interaction with certain cations [12].

Inorganic substituents can also be found in microbial polysaccharide composition, generally attached through ester linkages [18]. The presence of phosphate groups is widely extended among Gram-positive bacterial polymers, particularly pathogen streptococci and industrially relevant milk-fermenting bacterial species. Some phosphorylated EPSs are similar to cell-wall teichoic acids of Gram-positive bacteria [12], whilst others are similar to lipopolysaccharides from Gram-negative bacteria [15]. On the other hand, sulfate groups are less common in prokaryotic than in seaweed EPSs, being more frequent in cyanobacterial and osmophilic bacterial specimens. Some examples are the high-viscosity, high-Mw EPS from the archaeal halophile *Haloferax mediterranei* [25] or an EPS from *Staphylococcus epidermidis* [26].

Some cations can also be involved in the EPS structure, particularly in anionic ones. For instance, alginates strongly bind to divalent cations such as Ca^{2+}, Ba^{2+} and Sr^{2+}, or the XM6 polysaccharide from *Enterobacter* attaches mono- and divalent cations like Na^{+} and Ca^{2+} [16, 27, 28]. These ions become part of the polysaccharide during cellular production, but can be subsequently displaced through ionic exchange, electrodialysis, *etc.*, and are sometimes related to the EPS gelation temperature [23].

Macromolecular Conformation

As for the rest of the well-known biopolymers (proteins, nucleic acids), polysaccharides properties are defined and governed by their primary, secondary and tertiary structure. This is the basis for classification and also, for the comprehension of their 3D-conformation under solid and solution states [29]. Some polysaccharides contribute to the structural integrity and mechanical strength of plant tissues by building a hydrated 3D-network (*e.g.* pectin in plants, carrageenan, agar and alginate in seaweeds). Others, like chitin, cellulose, and less frequently, xylans and mannans, because of a closer interchain packaging, are able to form solid structures or tough fibers. Within the animal kingdom, hyaluronate, chondroitin sulfate and other glycosaminoglycans play a fundamental role by determining the properties of physiological fluids and the intercellular matrix structure [29].

Polysaccharide primary structure relates to the glycosidic monomeric sequence and the linkage types that determine a variety of possible architectures. Meanwhile, secondary structure is defined by the regular geometrical arrangement in space and it depends on the interchain polysaccharide interaction [27]. The flexibility or rigidity of polysaccharide linear chains and the steric hindrances due to bond angles or their length determine the spatial association mode and the finally adopted secondary structure [12]. Laminar, double and triple helices polysaccharide structures (Fig. **1**) have been early described in the literature [30, 31].

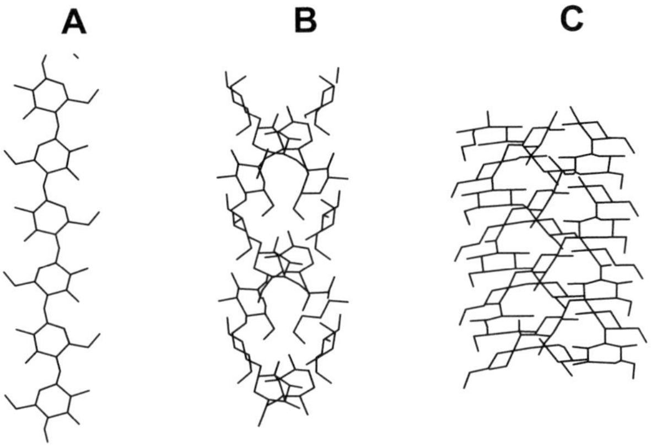

Fig. (1). Possible molecular conformations of microbial EPSs. **(A)** Laminar, **(B)** double helix, and **(C)** triple helix [31].

Laminar structure (Fig. **1A**) is typically found in cellulose, which possesses β-(1,4)-glycosidic linkages. This bond geometry gives place, once the chain is extended, to a slightly undulated ribbon where adjacent parallel chains interact through H-bonding to form connected layers with a three-dimensional appearance similar to that of β-sheet protein conformation. On the other hand, a classical example of polysaccharide double helix (Fig. **1B**) is constituted by amylose, one of the starch constituents. In this case, although (1,4)-glycosidic bonding is also found, the α-type linkage imposes an axial arrangement which affects the local stereochemistry and has a marked effect on the polymer conformation. The polysaccharide chain thus adopts a 3D-structure similar to a helical hollow tube which is stabilized through H-bonds, mimicking the protein α-helix conformation. These examples show how a simple difference in the condensation glycosidic linkages exerts a strong influence on the final polysaccharide secondary structure.

Finally, the third case (Fig. **1C**) is exemplified by curdlan, which bears β-(1,3)-glycosidic linkages. The macromolecule consists of three connected chains with the same torsion polarity around a common axis. As illustrated (Fig. **1C**), the three perfectly assembled chains form a triplex similar to a solid string. In the macromolecular core, three H-bonds take place at ~ 0.3 nm thus originating a very stable structure [31].

Tertiary structure represents a major-level organization resulting from energetically favored interactions between the polysaccharide component chains [29]. Macromolecular aggregates are also frequent, and the type of adopted arrangement will mainly depend on the molecular substituents and their spatial placement. These groups, according to their charge or polarity will determine the polysaccharide solubility in water. For many of the examined EPSs, the most favorable conformation exposes the charged fractions towards the periphery where they can interact with water molecules and different ions [32]. The main molecular interactions are represented by H-bonds, dipole-dipole and ionic forces, whilst hydrophobic ones, differently to protein or nucleic acid stabilization, are not as relevant. The final conformation stability will then result from the numerous cooperative interactions, which individually would be certainly weak [30].

Polysaccharides can also interact with other biomolecules, such as proteins, lipids, lectins or different polysaccharides, for example, when building biofilm complex structures [27]. Usually they are located in the cellular surface making contact with the surrounding environment, because of their hydrophilic character [10, 33]. Tertiary structures, in this case, become formed by a dense network of polysaccharides and further macromolecules, where cellular secreted products may eventually be trapped in the biofilm.

Based on all the aforementioned considerations, structural analyses of polysaccharides will offer crucial knowledge for the understanding of their functionality. Even though, this will never be a simple task because of their diversity and/or irregularity. Despite their partial organization, most of polysaccharides exhibit an amorphous structure, and up to date, only one crystalline form has been discovered [15]. Many polysaccharides with quite similar primary structures differ in their solubility and physicochemical properties. This fact emphasizes the relevance to elucidate polysaccharide tertiary or super-structure and the intermolecular interactions which support this behavior [12].

Gel Forming Ability

Although most polysaccharides have no crystalline structure, they assume in solution a non-random configuration which gives place to quasi-ordered domains [15]. Some polysaccharides form tridimensional networks or gels when a given chain goes through a tertiary structure conformation by contacting different regions of the same polysaccharide macromolecule. Interrupted regions play a relevant role for density control and reticulate distribution and so, they define the hydration degree and gel pore structure. These characteristics seem to be enzymatically regulated at a late stage during polysaccharide biosynthesis when, after polymerization, regions that must be associated become "unmasked" [30].

From the industrial point of view, gel-forming ability of microbial EPSs is a very attractive property. It can be either inherent to a given polysaccharide or induced by the presence of mono- or multivalent cations. Sometimes, gelling is the result of a synergic association with another polysaccharide, as in the case of xanthan [28, 34]. Among the most studied gelling microbial EPSs, gellan, curdlan and alginates could be highlighted. Scleroglucan exhibits the ability to form thermoreversible gels at low temperature, as a consequence of the weak crosslinking interaction of triplexes. However, more stable gels could be obtained in the presence of chromium salts and borax under alkaline conditions (pH 10-11) [8]. A similar situation occurs in the case of the EPS from *Sinorhizobium meliloti*, whose thermoreversible gels require the presence of mono, di o trivalent cations [35, 36].

Structural Analysis of Polysaccharides

It involves two aspects: 1) the characterization of the chain conformation, and 2) the analysis of the assembled chains (superstructure). The former one, as already mentioned, mainly relates to the identification of monomer composition, linkage types and substituents. The monomeric pattern can be evaluated through methylation analysis. After methylation of free hydroxyl groups and polysaccharide hydrolysis, monosaccharides are reduced to alditols and they are then acetylated to give partially methylated alditol acetates. The obtained low-Mw fragments can be analyzed by different methods, such as GC [37], GC-MS (coupled to mass spectra) [38], MALDI-TOF-MS [39] and NMR [38]. Alternatively, Smith degradation can be used to selectively fragment the studied polysaccharide. Three consecutive steps are involved: 1) periodate oxidation, 2) borohydride reduction (to polyalcohol) and, 3) mild acid hydrolysis [38, 40, 41].

Additionally, fragmentation (to low-Mw variants) or debranching might be achieved by means of the use of specific enzymes targeting particular glycosidic linkages. These techniques have been gradually replaced by high-pressure anion exchange chromatography with pulsed amperometric detection [42], which avoids derivatization steps. On the other hand, Nakano and co-workers reported carbohydrate characterization by means of capillary electrophoresis, without significant modification after hydrolysis [43].

Concerning superstructure analysis, polysaccharide diversity becomes a drawback. To understand the real situation it is enough to remind that whilst two identical monosaccharides can form eleven different disaccharides, two equal amino acids can only give place to a unique dipeptide [44]. The multiplicity of possible monosaccharide combinations along with the glycosidic linkage stereospecificity (α or β) leads to hard-to-solve complex structures, varying from homo- to heteropolysaccharides, either linear or highly branched.

The technique of X-ray diffraction has traditionally helped to elucidate the crystalline structure of polysaccharides. However, as for many other polymers, a polysaccharide usually never provides a unique and big enough crystal as to be analyzed by means of this technique. Sometimes, a unique adequate microscopic crystal can be obtained just for electron diffraction analysis [15]. For helical polysaccharides, with no individual crystals, polycrystalline fibers can be prepared so that most of the helices have their longitudinal axes parallel-oriented. The incidence of a monochromatic X-ray stream on the sample makes it diffract in concrete directions that originate typical signals which depend on the atom positioning within the polysaccharide [45]. The diffraction pattern thus provides information about the type and position of component atoms and with the help of mathematical modeling, atomic and molecular representation of the studied material can finally be obtained. One of the main shortcomings of this powerful analytical tool is first, the need to obtain crystalline systems, and further, the possibility that X-rays break some chemical bonds leading to a distorted molecular model.

On the other hand, significant contributions about polysaccharide structure have emerged from NMR spectroscopy, which can be used for either structural or dynamic conformational analysis of polymers under solid, solution or gel states [37]. Two-dimensional NMR and correlation spectroscopies, which can measure each C/H interaction with adjacent atoms and chemical groups, can provide their relative position within the molecule [15, 38, 46, 47]. Moreover, techniques based on laser light scattering (LLS) can also complement these data, not only allowing to determine the absolute polysaccharide Mw, but also providing information about the ordered/disordered state of the macromolecule or the transition temperature between these states [48 - 50].

Atomic force microscopy (AFM), transmission electron microscopy (TEM) and scanning electron microscopy (SEM) represent complementary powerful techniques for polysaccharide analysis [51 - 55]. These methodologies allow evaluating the tridimensional complex structures of EPSs and identifying individual molecular conformations and their assemblies (*e.g.* aggregation phenomena and gel systems) [56, 57].

Physical Properties

Molecular Mass

This is one of the main factors determining microbial EPS properties [58, 59]. This physical property has a direct influence on rheological behavior, particularly on viscosity [29, 60], and accordingly, it defines the feasibility of applying a given EPS for certain commercial or industrial uses [61]. Sutherland reported a Mw range for microbial polysaccharides between 1×10^3 up to 5×10^7 Da [12].

In the case of biopolymers it is important to consider that they do not usually exhibit a unique Mw, but in turn, they show Mw distribution. Accordingly, a set of Mw parameters is usually defined such as the number average molecular weight (Mn), which covers the widest distribution, the weight average molecular weight (Mw) and the polydispersity index (PDI), which indicates the Mw variation degree (Mw/Mn).

A traditionally and early applied technique for relative Mw and PDI determination is size exclusion chromatography (SEC), which makes use of Mw standards (generally dextran). Nevertheless, EPS sample viscosity or the tendency to form aggregates can represent a difficulty when applying this method. That may hinder proper determinations of Mw [12, 62]. For instance, in the case of xanthan, factors such as viscosity, aggregate and microgel formation may lead to Mw variability as large as from 1.1 up to 47×10^6 Da [12], a relatively inaccurate value.

To overcome many of these obstacles, techniques such as laser light scattering (LLS) or GPC coupled to multi- or low-angle LLS detectors (GPC-MALLS or LALLS) have shown to be efficient methods to accurately determine the absolute Mw with a great resolution as compared to traditional GPC [47, 63, 64]. A further advantage is the possibility to disregard from controls or column calibration [65].

In addition to these technical considerations, for most of the characterized microbial EPSs, such as xanthan, pullulan or scleroglucan, the Mw can vary depending on the culture conditions, downstream processing methodology and the producing strain [8, 12, 41, 66 - 68]. Moreover, fermentation time also showed some incidence on the Mw for certain EPSs like pullulan [69], while almost no differences were found in the case of scleroglucans [41, 68].

Rheological Behavior

Rheological properties of microbial EPS solutions determine their possible applications. At industrial level, EPSs are mainly chosen because of their gelling and viscosifying power [27, 34]. Most of these properties are affected by Mw, chemical structure, polysaccharide composition, presence of ionizable groups [47, 59], intrinsic viscosity, and the nature and number of intra- and intermolecular interactions [64, 70, 71]. Apart from these molecular parameters, which can also interact in between, other external factors such as the solvent quality can also influence the conformation adopted by the macromolecule as well as its supramolecular arrangements [29]. Usually most biopolymers are hydrosoluble, but water contribution to the system viscosity can be neglected. On the other hand, polymer concentration, temperature, pH, ionic nature and strength can actually influence the EPS solution viscosity, as will be later discussed.

Rheometry of concentrated polymer solutions constitutes a fundamental tool for evaluating viscosity and viscoelastic properties of polysaccharide systems [64, 72 - 74]. The rheological behavior of microbial EPS solutions allows their classification. Those fluids following the Newton's law, *i.e.* $\tau = \eta \cdot \gamma$ (τ, shear stress; η, apparent viscosity and γ, shear rate) are commonly called Newtonian. In that case, viscosity keeps constant while shear stress and shear rate change. It however depends on other parameters like temperature, pressure or fluid composition.

Conversely, non-Newtonian fluids show a non-linear relationship between τ and γ, so that η is not constant. Depending on the relationship between τ and γ, fluids are classified in pseudoplastic, dilatant, Bingham plastic and Casson or viscoplastic. In these cases, the reported η value should be accompanied by the shear rate at which the measurement was performed [28, 75].

Pseudoplastic and dilatant fluids obey to the Ostwald-de Waele model, in which $\tau = \kappa \cdot \gamma^n$, where κ is the consistency coefficient and n is the flow behavior index. The parameters κ and n characterize the fluid rheology, with $n < 1$ for pseudoplastic and $n > 1$ for dilatant fluids. On the other hand, Bingham and Casson (or viscoplastic) fluids require a threshold shear rate to start flowing, after which, Bingham plastics behave as Newtonian and Casson ones as pseudoplastic [76, 77]. Typical rheograms are depicted in Fig. (**2**).

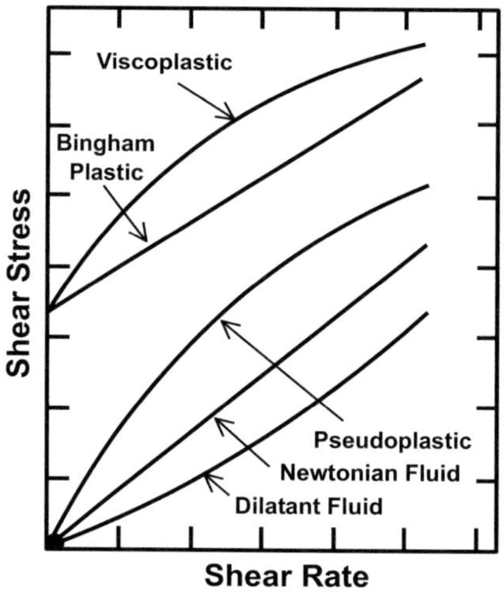

Fig. (2). Qualitative flow curves for different types of non-Newtonian fluids [77].

Most microbial EPS solutions exhibit a pseudoplastic behavior which is frequently taken advantage in the food industry. It additionally improves the texture and sensorial properties of the products obtained [29]. One of the typical and traditional cases is the *in situ* EPS production by lactic acid bacteria in dairy products, which thus naturally incorporates a thickening and anti-settling agent [78, 79]. Many pseudoplastic polymeric solutions also exhibit viscoelastic properties which mean that, after shear forces are stopped, the flow direction is reversible.

In some practical situations, apparent viscosity also depends on the time during which the shear force is applied. Based on this concept, fluids are classified as rheopectic, when apparent viscosity increases with the time of applying a shear force, or thixotropic for the opposite situation (Fig. **3**). The last behavior is the most common for microbial EPSs, in which case a given time is necessary to reach the equilibrium between the breaking structure and the new one that subsequently develops [75].

Another factor that can influence the EPS solution rheological behavior is its concentration. It relates to both the interactions between the EPS chains *per se*, as well as the interactions with the solvent molecules. A typical case was described for scleroglucan from *S. rolfsii* ATCC 201126 [41]. A marked exponential relationship between apparent viscosity and EPS concentration was demonstrated for this polymer, as witnessed by the notorious increase in κ and decrease in *n* as scleroglucan concentration augmented.

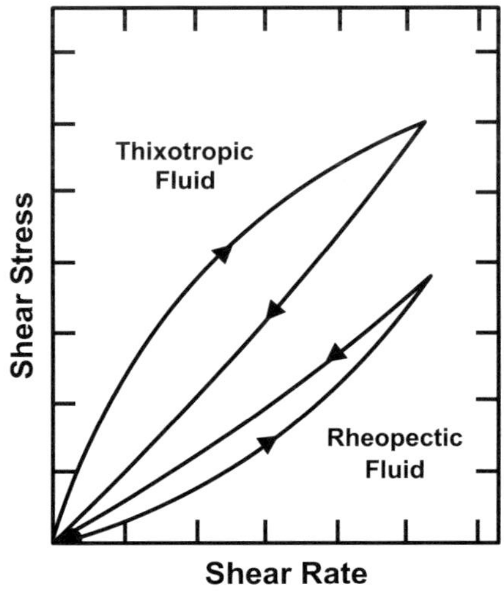

Fig. (3). Qualitative shear stress-shear rate behavior for thixotropic and rheopectic materials [77].

The reasons underlying high viscosity even at low scleroglucan concentration involve a high Mw and the chain structural rigidity [80]. Nevertheless, scleroglucan behavior can become progressively less pseudoplastic as EPS concentration drops below 2 g/L, when *n* index approaches to 1 [81]. In diluted solutions, intramolecular and solvent interactions become more significant than the intermolecular ones. On the other hand, high polymer concentration may induce supramolecular structural changes which also impact on the rheological behavior. Every chain will be surrounded or in close contact with other chains which thus imposes topological restrictions for chain configuration [29].

The influence of temperature on EPS viscosity relates to the induced conformational changes and the transition between ordered and disordered states [12, 82]. Thermal treatments normally affect the EPS aggregation state. When EPSs are subjected to heating and a critical temperature (Tm) is surpassed, they experience chain dissociation which leads to a viscosity drop. This behavior has been described for example for xanthan [83, 84] and also for scleroglucan [85]. Nevertheless, Capron *et al.* found for xanthan that above a critical EPS concentration, chain dissociation can be inhibited despite temperature exceeds Tm. After thermal denaturation, solutions can gradually recover their viscosity upon cooling as a consequence of macromolecular reorganization, which eventually may render more viscous solutions than prior to heating [85 - 88].

Viscosity of certain EPS solutions can additionally be affected by the presence of ions. Some polysaccharides bearing carboxyl groups can interact with certain ions (especially divalent) to form macromolecular networks with increased viscosity or even gel formation [27, 89]. For instance, the EPS produced by an *Alteromonas* strain has a selective interaction with divalent cations [90]. This particular chelating property has been proposed by Loaëc *et al.* for the treatment of wastewater effluents polluted with toxic metal ions [91].

The decrease in viscosity exerted by salt addition has been also described for some microbial EPSs. The lower availability of solvent molecules for the EPS stabilization in solution may account for this behavior. The excess of salts compromise solvent molecules for their own solvation. The reticulation degree of the macromolecule may also be affected by the interaction of EPS charged regions with the salt ions [92]. However, neutral polysaccharides such as scleroglucan might be quite stable even when confronted to high salt concentrations (up to 20% w/v NaCl) showing a negligible effect on viscosity [41].

Concerning pH effects, it can modify the net charge of EPSs thus altering attraction or repulsion forces between chains. This will then impact on apparent viscosity, although pH influence will be more pronounced for EPSs with ionic nature (*e.g.* xanthan) than for the neutral ones (*e.g.* scleroglucan). For instance, whilst scleroglucan exhibits a wide pH stability (0-13) [41], xanthan at $\leq 0.2\%$ w/v increases viscosity as pH augments but at higher EPS concentration (0.35-

0.4% w/v) acid pH favors viscosity [93, 94].

Chemical Structure and Physical Properties

Microbial EPS physical properties are determined by their chemical structure and conformation. One of these properties is the solubility in water, which is critically affected by the presence of lateral branching and ionizable groups in the EPS macromolecule. Those polysaccharides with linear chains of unsubstituted sugars tend to be insoluble [28].

The nature of glycosidic linkages also has a strong influence on the rigidity or flexibility of polymer chains. Linkages of β-(1,3) or β-(1,4) type render more rigid and less deformable EPSs, whilst α-(1,2) unions provide more flexibility [27]. The interaction with ions also determines important differences. One of the most representative examples is the case of alginates. Algal alginates form rigid, non-deformable gels due to the highly specific interaction with Ca^{2+} and Sr^{2+}, a property widely used with biotechnological purposes like cell or enzyme immobilization. In contrast, microbial alginate from *A. vinelandii* although structurally related, it is highly acetylated which inhibits the interaction with cations. However, the chemical removal of these groups significantly alters the physical properties and increases the union to divalent cations [27]. Similar situations have been described for other EPSs [95, 96].

Exopolysaccharides of Industrial or Commercial Interest

Acetan (Xylinan)

Xylinan (or acetan) is a complex anionic exopolysaccharide produced by the Gram-negative bacterium *Acetobacter xylinum*. It consists of a cellulosic backbone of β-(1,4)-D-glucopyranose where alternate glucopyranose residues are substituted at the 3-position with a pentasaccharide side chain, that protects the cellulosic backbone from enzymatic degradation by cellulases. The chemical structure of this side chain consists of: α-l-Rha*p*-(1,6)-β-D-Glc*p*-(1,6)-α-D-Glc*p*-(1,4)-β-D-GlcA*p*-(1,2)-α-D-Man*p*-(1,3), where branched glucopyranosyl residues together with the mannopyranosyl residue are partially O-acetylated at the 6-position (Fig. **4**). Xylinan is extracted as a water-soluble by-product during the commercial production of bacterial cellulose by *A. xylinum* [97].

Applications

Xylinan made its main place in the food industry as a viscosifying and gelling agent. It presents high gel strength and water-holding capacity and it is easy to mold. It is also a component in the coconut cream, a very popular confectionery in

Japan and Philippines [98].

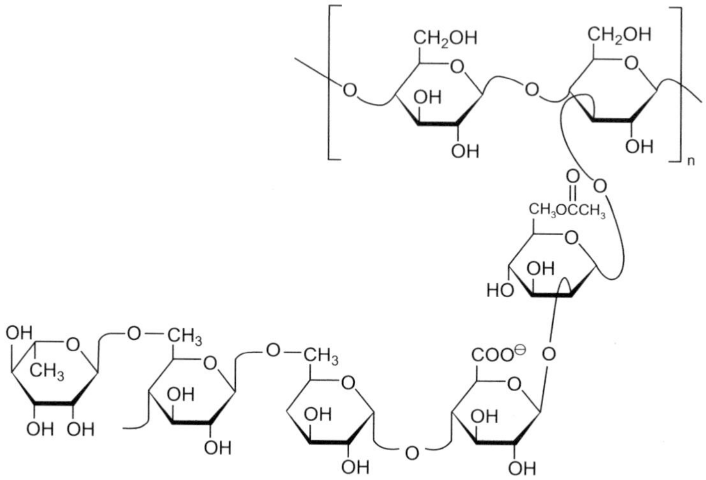

Fig. (4). Acetan basic structure [97].

Alginates

Alginates can be synthetized as structural components of various brown seaweed genera, but also as capsular polysaccharides in some bacteria of the genera *Pseudomonas* and *Azotobacter*. Alginates from these different sources show several 'pros and cons' that will be thereafter discussed. Seaweed alginates are complex mixtures of polymers that exhibit a wide range of molecular masses and heterogeneous β-D-mannuronate (M): α-L-glucuronate (G) (M:G) ratio (Fig. **5**).

These dissimilarities which lead to varied material properties occur because algae are harvested from the ocean where environmental conditions cannot be controlled and thus, climatic variations lead to disparities in algal production and composition.

Microbial Production

All current commercial alginates are isolated from farmed brown seaweeds, with over 30000 metric tons annually produced [99]. Nowadays, despite for the majority of industrial applications seaweed extraction is the main economic way to produce alginate, these polymers do not fulfill the requirements for some specific applications, especially in the pharmaceutical, chemical and biomedical fields, where polymers of well-defined composition are needed [100, 101].

As an alternative to the algal production process, alginate can also be produced by microorganisms, such as *Azotobacter vinelandii*, *Azotobacter chroococcum* and *Pseudomonas aeruginosa*. Conversely to *P. aeruginosa*, *Azotobacter* species are non-pathogenic soil bacteria and thus, they are more appropriate for the development of a biotechnological process for alginate large-scale industrial production. Moreover, the availability of the complete genome sequence of *Azotobacter vinelandii* turned this bacterium an ideal model of study, both from the technological and scientific point of view, because it can be genetically engineered in order to produce great amounts of tailor-made high-value alginates [101, 102].

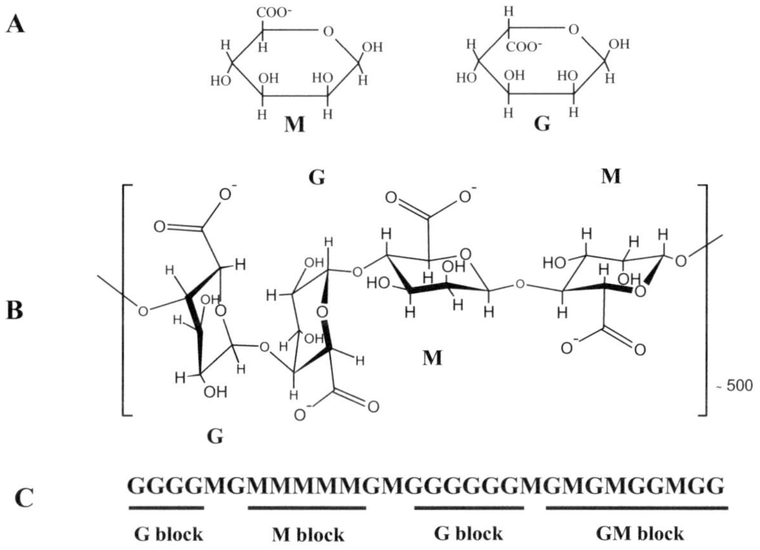

Fig. (5). Chemical structure of alginates. **(A)** Alginate monomers: **M** represents mannuronate residues, and **G**, glucuronate residues. **(B)** Chain conformation. **(C)** Block arrangement.

A key difference between algal and bacterial alginates is that the latter are O-acetylated. A higher degree of O-acetylation significantly increases viscosity and pseudoplastic behavior [103, 104]. The level of O-acetylation can be reasonably controlled by using specific strains/mutants, altering the growth media or controlling cultivation conditions such as aeration, pH and temperature [105 - 107]. In addition, live immobilized *P. syringae* cells have been successfully used to acetylate seaweed-derived alginates [108, 109].

General Properties

Alginates are a family of linear binary copolymers composed of variable amounts of β-D-mannuronate (M) and its C-5 epimer α-L-glucuronate (G), linked by (1,4) glycosidic bonds. Within the polymeric structure, monomers M and G are distributed in blocks of continuous mannuronate (M), glucuronate (G) or alternating residues (GM) (Fig. **5**). These EPSs under normal solution conditions (except in acidic environments), behave as highly charged polyanionic polymers [110].

The structure, monomeric composition and Mw of alginates closely depend on the EPS source, which also defines the physical properties [111]. Gel-forming ability is one of the most important features that make it attractive for the food industry. It depends on the block composition (M and G), acetyl substituents, and the selective binding to special cations like Ca^{2+} [112]. Additionally, numerous studies have reported the biocompatibility and low toxicity of alginate [6]. The most commercially interesting products are sodium alginate, propylene glycol alginate and alginic acid [113].

When dispersed in water, alginate may rapidly form harder-to-dissolve conglomerates if no caution is taken. As alginate concentration increases, the solution goes from a viscous liquid to a thick paste, which is difficult to handle. Monovalent cation alginate salts and its propylene glycol ester can be dissolved in water and organic solvents (*e.g.* alcohols), however, Ca^{2+} salts and alginic acid cannot. The presence of sugars, starches or proteins can compromise water hydration so that longer mixing times will then be required. Therefore, it is better to add these components, if necessary, after alginate dissolution [113].

Alginate solutions show non-Newtonian pseudoplastic behavior, and thixotropy is sometimes observed. Viscosity can be influenced by different factors; one highly relevant is alginate Mw which shows a direct relationship. The alginates from *A. vinelandii* may exhibit Mw between 80 and 4000 kDa [101]. Other influence is the typically derived from EPS concentration. Since a direct relationship between EPS concentration and viscosity does not exist, McDowell proposed an empirical equation:

$$\log_{10} \text{viscosity} = a \cdot (\text{EPS concentration} - b)^{1/2}$$

where, 'a' relates to the alginate DP, and 'b' is a constant for a given alginate [114]. Usually the manufacturers provide a [viscosity *vs.* alginate concentration] table for their particular products [115].

Alginate solution viscosity decreases as temperature rises. Heating above 50°C for several hours can induce depolymerization thus leading to an irreversible viscosity loss. However, freezing and thawing have almost no incidence on viscosity, provided that Ca^{2+} is below 0.5% [113].

Concerning pH effects, viscosity is stable within pH 5-11, but below 5, carboxyl groups of uronic acids become protonated thus reducing electrostatic repulsion between chains and increasing viscosity [116]. Between pH 3-4, the proton excess will favor linkages between G blocks which precipitate the polymer in the form of gel [17, 111], and inducible event at pH 5 if residual Ca^{2+} is present. Depolymerization can also take place above pH 11, with the consequent loss of viscosity [113].

Acetyl groups have a marked influence on bacterial alginate conformation since they alter the stereo regularity of the macromolecule. These substituents diminish the cation binding capacity of EPS which consequently affects gelation. On the other hand, their elimination allows intermolecular interaction which gives place to strong and rigid gels in the presence of cations [117]. Nevertheless, acetyl groups may be useful to improve for example the swelling ability and viscosity of alginates [105, 118].

Due to the polyanionic nature of alginates and their capacity to bind Ca^{2+} ions, this represents a way to raise viscosity without increasing alginate concentration. However, these solutions have a higher tendency to lose viscosity under stirring. The actual effect of Ca^{2+} is often experimentally determined. Normally, higher Mw and M/G proportions lead to higher viscosity changes [113, 114].

Alginate gels can be formed at temperatures up to 100°C and even under acidic conditions [113]. The direct mixture of alginate and polyvalent cations originates heterogeneous gels because the binding occurs very fast and irreversibly. The controlled cation introduction into the network can be performed by the method of adjusted diffusion [110]. Calcium is the standard selected cation for gel formation because of the cost, availability and non-toxicity. Normally, in alginate pearls the polymer is more concentrated on the surface, and decreases to the gel center [103]. This heterogeneous alginate distribution can be used to modify the final gel porosity or to control chemical diffusion into the alginate spheres [119].

Gel strength will depend on the number and length of crosslinking bridges between the blocks [120]. The higher the number of G blocks, the more porous and fragile the gel will be [17]. On the contrary, a higher number of M blocks give rise to a soft and elastic gel, of medium strength and smaller pores, but with high resistance to syneresis. Alginate can be enzymatically modified post-polymerization, which converts M to G blocks within a polymeric chain, thus altering the physicochemical properties [119]. The alginate gelation model commonly known as "egg box" is shown in Fig. (**6**). It starts with the dimer formation, followed by dimer aggregation [17].

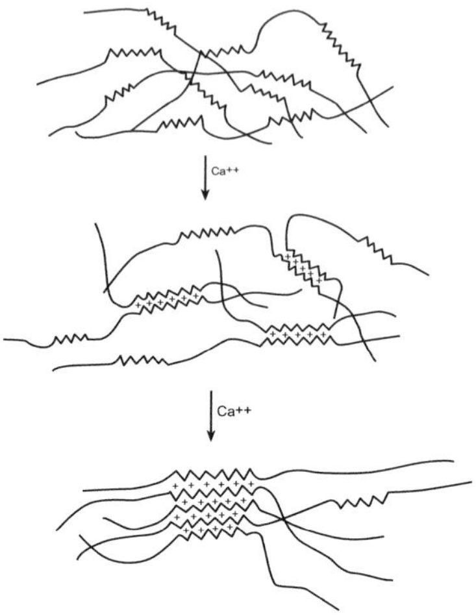

Fig. (6). Alginate gel formation from G blocks: The "egg-box" model [113].

Applications

Nowadays, bacterial alginates still represent only a small fraction of the current polymer market compared to alginates arising from algae. This is because the alginate extracted from marine brown algae and used for food and cosmetic industries is sold at prices as low as $5/kg, an unbeatable price for its bacterial counterpart. Nevertheless, biomedical industry requires a steady and consistent supply of alginates with defined properties, specific Mw and M:G proportions, which are requirements generally difficult to fulfil by algal alginates [102].

It is in this context where bacterial alginates occupy a significant place, because mechanisms of alginate production and modelling in bacteria can be controlled, thus enabling the manufacture of tailor-made and consistent bacterial alginates of high added-value which become relevant for biomedicine and extensive biotechnological applications [109]. Bacterial alginates (and to a lesser extent, seaweed alginate modified by *Azotobacter* epimerases) govern medical, pharmaceutical and biomedical markets due to their non-toxic, biocompatible, bio-inert, non-immunogenic, hydrophilic, conformational and biodegradable properties [102, 121, 122].

Alginates are utilized in a wide range of industries, because of their versatility, stabilizing, thickening, suspending, emulsifying, gel- and film-forming properties. In the food industry, prevails the use of alginates coming from algae. They can be applied to wide-ranging foods such as meat, fish, jams, soups, sauces, drinks, ice cream, dairy products and confectionery. As a surplus, it has been reported that some alginates used as food ingredients also possess anti-inflammatory and detoxifying properties [123, 124]. Currently, new claims are being reported for these polymers, such as their usage as a source of dietary fiber [125], as a matrix for controlled release systems (*e.g.* for conventional drugs, hormones, enzymes and specialized cells) [126, 127], in the prevention of thrombosis, and in fluid-absorbing wound dressings [128].

Within the tissue engineering field, alginate hydrogels are successfully implanted as matrices for *in vivo* and *in vitro* cell immobilization and also for cell growth stimulation [129]. For example, alginate has been used for co-immobilization of human osteoprogenitors and endothelial cells in bone regeneration therapy and has also been reported for neurological and cardiologic tissue regeneration [122]. Alginate gel beads are employed as entrapment devices of some hormone-producing cells before transplants to protect them from the host immune system.

In addition, some proteins and antineoplastic agents with therapeutic activities can be microencapsulated in alginate gels to increase their efficacy towards the maintenance of localized and controlled delivery without producing adverse side effects [121, 122, 130]. Alginate conjugates have also been proved as immunogens in vaccines to control cystic fibrosis [131] and further bacterial infections [132]. Likewise, calcium cross-linked alginate gels have been evaluated for the development of novel immunotherapy strategies for cancer treatments [112].

Chitin and Chitosan

Chitin and chitosan are co-polymers composed of a backbone of N-acetyl-D-glucosamine and D-glucosamine respectively, and linked together *via* β-(1,4)-glycosidic linkages, in a similar way to cellulose (Fig. **7**). The ratio of these two monomers is different in chitin and chitosan, and determines the physicochemical and biological differences in between.

Chitin

It is a naturally abundant linear mucopolysaccharide that was first isolated by H. Braconnot in 1811 (about 30 years earlier than cellulose) and became the first polysaccharide identified by man. Chitin can be found in innumerable natural sources, like crustacean shells (14-27% dry weight), fungal cell walls and septa of Ascomycetes, Zygomycetes, Basidiomycetes and Deuteromycetes, insect cuticles, and some algal (diatoms) support structures, among others [133, 134]. Chitin is the main source of chitosan, and can be distinguished from the latter by its higher proportion of N-acetyl-D-glucosamine residues over D-glucosamine. As an example, chitin derived from crustaceans is composed of more than 95% of N-acetyl-D-glucosamine and less than 5% D-glucosamine residues [135, 136].

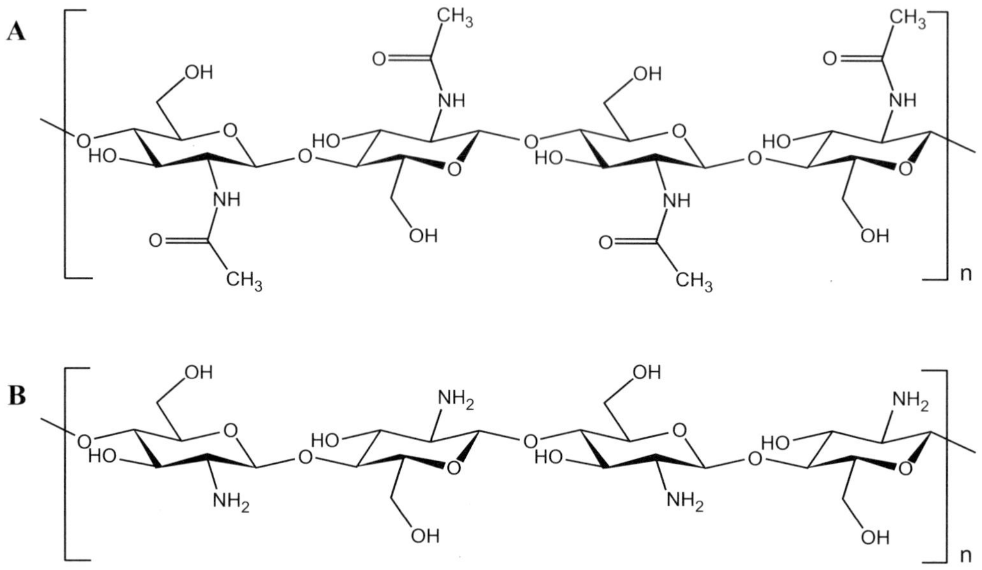

Fig. (7). Ideal structures of chitin **(A)** and chitosan **(B)**.

Because of its high content of acetyl groups, chitin is highly insoluble in water and most organic solvents. The average chitin DP can considerably vary depending on the source, with reported values from about 100 (chitin produced by some yeasts) to ~5000 (chitin obtained from crustaceans). In addition, two major types of chitin were reported, taking into account the polysaccharide arrangement in its native form: the α-chitin (chains arranged in a tightly compacted crystalline structure of antiparallel sheets and extensive intermolecular H-bonding) found in shrimp shells and fungal cell walls and; the β-chitin (chains arranged in a more mobile allomorph of parallel sheets) found in squid pens. These two chitin types were already well characterized [135].

Currently, in spite of its many available sources, commercial chitin is mainly obtained from shellfish [135, 137]. Nevertheless, recent advances in fermentation technology suggest that the cultivation of selected fungi can provide an advantageous alternative source of chitin, because of a wide number of reasons. For example, shellfish provisions depend on the season and are limited to the fishing industry, while fungal mycelium can be obtained by a simple fermentation process in any geographical location or season [138].

Likewise, processing of fungal mycelia for chitin extraction is cheaper and less laborious as compared to crustacean shells treatments [139]. In addition, shellfish-derived chitin is heterogeneous with respect to physicochemical properties [140].

Every year, thousands of tons of fungal-based waste materials are produced from mushroom biomass production and fermentation industries. These fungal wastes might be easily utilized for the production of chitin, thus simultaneously reducing costs of both polymer production and those derived from waste disposal [141]. In agreement, mycelium of several fungi, *i.e. Aspergillus, Penicillium, Fusarium, Allomyces, Rhizopus, Mucor, Zygorrhynchus* among others are being considered as possible commercial sources of chitin [142].

Chitosan

This term embraces a group of biodegradable, non-toxic, non-allergenic, polycationic, amino polysaccharides with varied Mw (from 10 to 2000 kDa) and a degree of deacetylation in the range 50-95%. It is generally composed of more than 80% β-(1,4) linked D-glucosamine and less than 20% N-acetyl-D-glucosamine residues [143]. It has been mainly described as a component of fungal cell walls, particularly in the Mucorales, although it was also detected in Basidiomycetes, like *Lentinula edodes* and *Pleurotus sajor-caju* [144, 145].

Unlike chitin, chitosan is not profusely found in nature and, it is mostly commercially obtained from chitin by means of a heterogeneous deacetylation process that involves strong alkali and high temperatures for long periods of time [146]. Alkaline deacetylation leads to the removal of over 80% of the acetyl groups from N-acetyl-D-glucosamine residues in the chitin backbone, thus converting them into D-glucosamine [147]. The varied nature of the production process renders to heterogeneous chitosan, with different Mw, polydispersity and deacetylation degree. The resulting polymers therefore exhibit different physical properties and biological functions. Methods aimed at obtaining more homogeneous polymers have been well documented [148].

Fungal chitosan polysaccharides, in addition to fungal chitin, have a considerable potential for their commercial use. Chitosan polymers are currently considered as one of the most promising materials for future applications due to their excellent biodegradability, biocompatibility, non-toxicity and extraordinary properties that can be outlined by selecting adequate producing strains, fermentation conditions and downstream processes. Moreover, it has been reported that fungal chitosan displays higher biological activity and economic advantages as compared to chitosan obtained from different sources [149].

General Properties of Chitin and Chitosan

These β-(1,4) linear copolymers of N-acetyl-D-glucosamine and D-glucosamine exhibit different monomeric proportions which consequently determine dissimilar physicochemical and biological properties. Chitosan is obtained by alkaline deacetylation of chitin thus releasing the amino groups [135, 150 - 152]. The degree of deacetylation (DD) will depend on temperature, time and NaOH concentration [153, 154]. The resulting chitosan is heterogeneous, linear and cationic [155, 156].

Chitin, due to the high acetyl content, is highly hydrophobic and insoluble in water and many organic solvents [6]. The low solubility of chitin has significantly restricted its application but, enzymatic or acid hydrolysate derivatives termed chitin oligosaccharides (COS), of lower Mw, are more easily dissolved in water [143]. Chitosan is insoluble in water, alkali and organic solvents. However, it is soluble in acidic solutions of acetic, nitric, hydrochloric, perchloric and phosphoric acids, because under these conditions amino groups become protonated thus favoring solubility [156].

Chitosan can be characterized according to Mw, DD, crystallinity and amino group distribution [157, 158]. Solubility, tensile strength, crystallinity, clotting and flocculant activities are normally affected by chitosan Mw [159, 160]. A high DD has been shown to favor chitosan dye adsorption capacity [161] and, along with heavy metal scavenging activity [162], may be applied for bioremediation purposes. The most biotechnologically relevant chitosan properties are probably the high-Mw, non-toxicity and biodegradability of this EPS [155, 156].

Concerning rheological properties, solutions of chitosan show pseudoplastic behavior with an excellent viscosifying power under acidic conditions. Viscosity can be affected by Mw, DD, pH, ionic strength, EPS concentration, and temperature. Generally, viscosity drops as temperature increases and it rises when EPS concentration becomes augmented. Meanwhile, the effect of pH is more dependent on the acid nature [149, 163].

Applications

Due to their wide spectrum of properties, chitin and chitosan have centered much interest in their potential applications in agriculture, biomedicine, biotechnology, and food industry [142, 164]. Chitin is a degradable biopolymer with low toxicity and, together with its derivatives, has been found useful in the prevention and treatment of various human and animal diseases. Among their properties, immunostimulating, antimicrobial, antitumor and hemostatic activities were reported [165]. Chitin-containing gels, films, membranes and fiber dressings have been currently developed for their application in controlled drug release and as wound dressing materials [166 - 168].

Chitin is also a promising material for tissue engineering due to its ability to stimulate connective tissue restoration, besides biodegradability, biocompatibility and antimicrobial properties. Sashiwa and Aiba summarized some of the emerging virtues of novel chitin derivatives, such as WSC (a partially deacetylated chitin), mercaptochitin, fluorinated chitin, N- and O-sulfated chitin, and DEAE-chitin. Properties such as anti-osteoporotic, anti-thrombogenic, anti-uremic and anti-metastatic have been already cited [169].

Interestingly, chitosan gives even more commercially interesting compounds than chitin. Its polycationic nature is considered essential for many useful biological activities. Additionally, its high nitrogen content makes chitosan a powerful chelating agent [149]. Chitosan also showed strong antimicrobial activity against Gram-positive and Gram-negative bacteria. This occurs because positively charged molecules of chitosan interact with negatively charged microbial surfaces, which results in the disruption of cellular membranes and the consequent microbial cell death [170].

Moreover, chitosan oligomers can penetrate into prokaryotic cells and interfere with the transcription of RNA and protein synthesis [171]. Because of this antimicrobial property, chitosan has been suggested as a food antimicrobial agent, showing promising results in the control of Gram-negative, foodborne pathogen *Salmonella typhimurium*, among others. Due to its non-toxic and bioadhesive properties to negatively charged cell surfaces, such as mucosal surfaces, proteins and other macromolecules (for example DNA), chitosan also behaves as an excellent matrix for oral and nasal drug delivery of proteins, vaccines and even genes [141].

Other reported biological activities of chitosan include antitumor, immunostimulatory, anticoagulant and hypocholesterolemic effects. It also promotes rapid dermal regeneration and accelerates wound healing. In addition, chitosan can form non-toxic biocompatible and biodegradable porous structures, tubes, blocks, films, sponges and gels, which are considered promising materials for cell-based transplantation and tissue engineering, especially for nerve, cartilage and bone regeneration [6].

Chitosan possesses numerous hydroxyl and highly reactive amino groups that allow preparing a variety of chitosan derivatives, which either improve its natural properties or open new possibilities for specific end uses. Chitosan derivatives with enhanced muco-adhesiveness, antimicrobial, anticancer, antioxidant and immuno-stimulant effects have been reported. The use of chitosan in blends with other natural polymers showed to be another promising strategy for producing tailor-made polymers for selected applications [6]. Chitin and chitosan were also proposed in the agricultural industry as biopesticides. Fungal chitins and chitosan (as well as commercially purified crustacean chitosan) were reported to induce defense responses in apple trees, and to significantly reduce apple lesions caused by plant pathogens such as *B. cinerea* and *P. expansum* [141].

Curdlan

Curdlan is a neutral, linear, non-branched homopolysaccharide of glucose. It is normally composed of 400-500 D-glucose residues joined by β-(1,3)-glycosidic linkages but can contain up to 12000 glucose units [172, 173], responding to the structure depicted in Fig. (**8**).

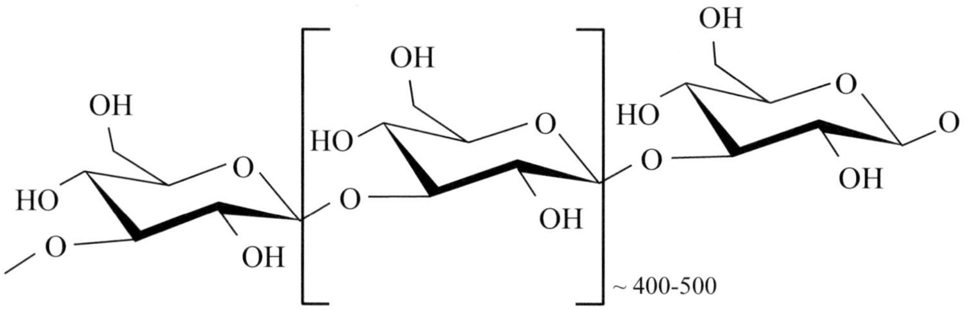

Fig. (8). Chemical structure of curdlan.

Microbial Production

Curdlan production was first detected in the alkaline tolerant mesophilic pathogen *Alcaligenes faecalis* var. *myxogenes* strain 10C3, coproduced with an acidic heteroglycan termed succinoglycan [174]. Thereafter, Harada and co-workers found a spontaneous high curdlan-yielding mutant of strain 10C3 (named 10C3K) that no longer produced succinoglycan [175]. This mutant became the predecessor of strains such as ATCC 31749 [176], ATCC 21680 [177], and ATCC 31750 [178], which are able to produce enhanced yields of curdlan with excellent gel-forming quality [179].

Currently, curdlan is commercially produced by non-pathogenic strains of *Agrobacterium*, like *Agrobacterium radiobacter* [98]. Curdlan is also produced as an extracellular storage polymer by species of *Cellulomonas,* including *C. flavigena* KU [180, 181]. Once synthetized, curdlan is excreted to the culture medium, but the secretory mechanisms of this EPS have not however been elucidated [182].

General Properties

Curdlan is insoluble in water, alcohols and most organic solvents, but can be dissolved in weakly alkaline solutions (0.25 M NaOH), DMSO and formic acid reagents [183]. The main peculiarity of curdlan is that, when heated in water, it forms thermally irreversible, tough gels. Alternatively, curdlan gels can also be obtained by means of neutralizing a polysaccharide solution in 0.1 M NaOH, followed by dialysis [184]. On the other hand, autoclaving at 120°C can induce curdlan triple helix formation but these gels are sensitive to contraction and syneresis, although resistant to most of β-(1,3)-glucanases. Curdlan gel properties are intermediate between gelatin elasticity and agar fragility [117].

It has been identified under three forms [183] and the structural differences are illustrated in Fig. (**9**). The anhydrous form, composed by three interlaced chains strongly connected by H-bonds, can be obtained by overheating under vacuum. When heating is performed in the presence of water, it is possible to obtain a crystalline hydrate with two water molecules *per* glucose residue. The hydrated form is constituted by triple helices, but symmetry is lost because of the water solvation molecules. A third form can be obtained by dialysis of alkaline curdlan solutions in water. The presence of either loose triple helices with ~ two water molecules *per* glycosidic residue or, the coexistence of individual helices with ~ 20 water molecules *per* glucose with the former ones is not clear [184].

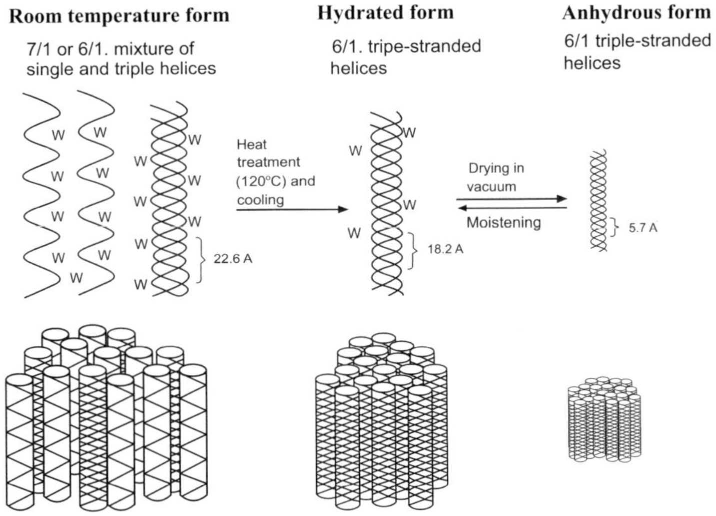

Fig. (9). Schematic representation of the three curdlan forms [184].

By means of X-ray diffraction, Marchessault and Deslandes [185] reported 30% curdlan crystallinity. Additionally, Harada *et al.* demonstrated that crystalline appearance depends on curdlan Mw; whilst low Mw curdlan (DP 49) forms hexagonal laminar crystals, higher-Mw curdlan (DP 131) gives pseudolaminar crystals [186].

Concerning the rheological behavior, viscoelastic properties were reported for curdlan gels and may be increased by temperature rise [172]. When a curdlan suspension is heated at 80-100°C and thereafter cooled down, it is possible to obtain a "high-set" thermoreversible gel, whereas the "low-set" gel corresponds to warming at 55°C [184]. Gelation implies the non-covalent association of triple helices. The underlying molecular mechanism can vary depending on the gel formation temperature. At high temperature, reticulation between curdlan micelles occurs through hydrophobic interactions, whilst H-bonds are the main responsible forces for low temperature reticulation [172].

Applications

Up to date, the use of this polysaccharide in the food industry is only permitted in Japan, where it is used to improve the texture of foodstuffs such as tofu, processed meats and fish pastes. It also provides viscosity to sauces, improves the rehydration of freeze-dried foods, acts as gelling agent of jellies and is used for the reconstitution and shaping of processed foods and confectionery [98].

Curdlan gels are tasteless, colorless and odorless and may present different strength, textural quality, physical stability and water-holding capacity, depending on the heating process. Because of their non-caloric value they became useful in calorie-reduced foods [187]. Other potential uses for curdlan relate to its film-forming ability, being thus useful for producing impermeable films for foods, as well as heat-resistant capsules [188, 189].

As scleroglucan, curdlan belongs to the group of the so-called BRM polymers, which have been attributed with numerous beneficial properties for human health. For instance, oral administration of curdlan helps to decrease blood glucose levels and accordingly, functional foods containing curdlan gels have been developed for the prevention or amelioration of diabetes [11]. In association to its non-digestible carbohydrate profile, curdlan behaves as prebiotic. Oral administration of partially hydrolyzed curdlan either in the form of an alkaline aqueous solution or as a gel showed to have immunopotentiating activity [190, 191]. Antitumor property of curdlan single helix and/or antiviral (anti-HIV) activity of sulfated derivatives, have been respectively reported. Additionally, it has been employed as antitumor drug delivery carrier, and as a support for enzyme immobilization [187].

Finally, curdlan applications in areas far apart from food and human health have also been described, for example, as an elicitor agent in stimulating natural protection in plants, and as an anti-segregating agent in concrete and mortar preparation [11].

Dextran

This is a broad group of neutral extracellular homopolysaccharides produced in substantial quantities by various bacteria, especially *Leuconostoc mesenteroides*, but also by other bacteria of the genera *Streptococcus*, *Leuconostoc*, *Acetobacter* and *Betabacterium* [187]. Dextrans are composed of α-D-glucopyranose residues mainly polymerized by α-(1,6) linkages (~ 50 to 95%) while the non-(1,6- -linkages may be either α-(1,2), α-(1,3) or α-(1,4) bonds (Fig. **10**).

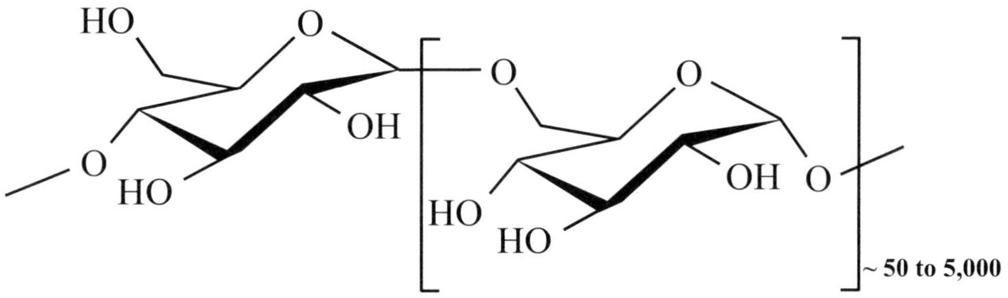

Fig. (10). Main backbone of dextran composed of α-(1,6)-D-glucopyranose units (95%). Branching points at α-(1,2), α-(1,3) or α-(1,4)- (5%) may be present.

At industrial level, some organisms of choice for dextran production are *Leuconostoc mesenteroides,* strains NRRL B-512 and B-512F [192, 193]. Dextrans produced by these strains have ~ 95% of α-(1,6) and ~ 5% of α-(1,3) linkages, and Mw between 40-50 × 10^6 Da [187]. In most cases, the side chain length is short, and the branched residues vary between 5 and 33% [97]. Differences in the linkage type and its extent, can determine important variations in solubility and rheological properties [9].

General Properties

Commercial dextran is usually produced by selected *Leuconostoc* strains. The recovered culture broth is acid hydrolyzed and enzymatically treated (dextranases) to reduce the final Mw to ~ 60000 g/mol [97]. Depending on the linkage types, their proportion and their location within the macromolecule, differences in solubility and rheological properties can be normally encountered among dextran preparations [187].

A high α-(1,6) linkage proportion provides high dextran solubility which is a desirable property for most industrial applications [193, 194]. Besides, dextrans exhibit low toxicity and immunogenicity, they are stable under acidic and mild alkaline conditions and are commercially available in a wide range of Mw with low polydispersity [6, 195]. Biocompatibility is a further advantage of this biopolymer, being well tolerated by the organism at moderate doses. As a plus, dextrans are also biodegradable through the action of dextranases of bacterial or fungal origin.

Depending on the sample preparation, structure and nature, the rheology of dextran solutions exhibit interesting features. Commercial dextrans have been the most intensively studied, such as the dextrans from Sigma Co. (Mw from 71500 to 531000 Da) whose solutions normally show a Newtonian behavior [196]. Carboxymethylated dextrans showed a viscosity increase as the substitution degree was raised, but viscosity becomes affected by the addition of salts [194, 197].

Applications

It is the first microbial polysaccharide that has been commercialized and approved for its use in foods. In the food industry it is used as thickener and gelling agent in gums and jelly candies. In pudding mixes, it provides texture and mouth feel. In confectionary it retains moisture, viscosity and inhibits sugar crystallization. However, its main field of application is in the pharmaceutical and chemical industry. The dextran produced by *L. mesenteroides* B-512 F is widely used because it is non-toxic, has low immunogenicity, high water solubility and high biological stability in the human blood stream. When parenterally administered, it can be completely metabolized in the body.

Fractions of dextran with lower and defined molecular masses (obtained by acid or enzymatic hydrolysis) are widely spread after parenteral administration. The product with Mw ~ 60000 g/mol is widely used in medicine as a blood-plasma expander, and the fraction of 40000 g/mol is used to enhance capillary blood flow [97]. On the other hand, dextrans of higher Mw and/or with greater proportions of non-(1,6)-linkages can cause allergic reactions [198].

Dextran sulfate is a derivative used as anticoagulant, anti-ulcerative and antilipemic drug. Because of the high proportion of (1,6)-linkages, dextran presents an unusual flexible backbone on which the 3-hydroxyl groups in consecutive positions are available for complexing metal ions. The complexes formed by dextran of Mw = 5000 g/mol with iron or calcium ions are also applicable to the treatment of iron-deficiency anemia and hypocalcemia, respectively. In recent years, novel dextran applications found a place in the biomedical field as a viscosity modifier and as a matrix for the immobilization of enzymes and drugs, like antibacterial nanoparticles [199].

In the oil industry, dextran is used as a component of the drilling mud and for secondary oil recovery. It has been also used in explosives, deflocculants, paper products, and as soil conditioner [200]. In the chemical industry, cross-linked dextran beads (Sephadex®) are widely used in gel filtration columns, for purification and fractionation of proteins. Dextran itself can also be used for purification, separation and concentration of metals.

Fractions of defined Mw (*e.g.*, the Pharmacia T500 Dextran, with an average Mw of 500000 g/mol) serve as polysaccharide standards in Mw calibrations [97]. The use of 'Blue dextran' is also widely reported as a void volume marker for gel filtration chromatography. Dextrans are also suitable in laboratory work because they do not show positive interactions with proteins. They can be used as part of incompatible phase separation systems, usually with polyethylene glycol. Their solubility in both aqueous and non-aqueous media renders them good membrane formers using electrospinning techniques [97].

Polyhydroxyalkanoates, (PHA)s

These polysaccharides, also known as bacterial bioplastics, contain repeating units of various hydroxyacids or mixtures thereof [201]. They are synthesized by several genera of bacteria, under an entirely biological process. Carbon sources are directly converted into PHAs which accumulate inside the bacteria in the form of intracellular granules. They can be produced as reserve materials and/or as energy storage backup when the C-source is in excess, or otherwise, when growth becomes limited due to the deprivation of other essential nutrients such as oxygen, nitrogen, phosphorus, sulfur, iron or magnesium [202 - 204].

The first PHA discovered was poly(3-hydroxybutyrate) (P3HB) produced by a *Bacillus megaterium* strain. It was characterized in 1920s by Maurice Lemoigne, a French microbiologist. Since then, more than 300 species of Gram-negative and Gram-positive bacteria able to accumulate PHA polymers have been described in the literature [205, 206]. Nevertheless, only a small number of them are able to synthesize enough PHA for industrial-scale production. Among them, *Cupriavidus necator*, *Alcaligenes latus*, *Methylobacterium organophillum*, *Pseudomonas oleovorans*, *Protomonas extorquens*, *Azotobacter vinelandii*, *Paracoccus denitrificans* and a recombinant *E. coli* are the most promising strains [207, 208]. The type of polymer produced depends on the bacterial strain, the phase of growth and the substrates supplied for cell culture [209, 210].

The PHA granules are composed by an amorphous, hydrophobic PHA core surrounded by proteins involved in PHA metabolism [211, 212]. Accumulation of PHAs is subjected to an extensive regulation by biosynthetic genes [213]. During PHAs industrial production, bacterial growth and polymer accumulation simultaneously occur in the bioreactor during the fermentation process [214].

In comparison to synthetic polymers, PHAs exhibit certain advantages. They represent attractive substitutes for petro-based polyesters, because they can be microbiologically degraded in nature under aerobic conditions to CO_2 and water, and to methane under anaerobic conditions, by the action of PHA depolymerases and PHA hydrolases. Additionally, PHA synthase (PhaC) presents broad substrate specificity: any organic molecule containing a carboxyl and a hydroxyl group able to be converted in the respective hydroxyacyl-CoA precursor can, in principle, be incorporated into a high-molecular-mass PHA [215]. That allows obtaining a great diversity of PHAs, with increased biocompatibility, improved end use properties and high biodegradability [216, 217].

As a plus, PHAs may be ecofriendly obtained from different raw materials such as agro-industrial wastes, which favorably impacts not only on biopolymer production costs but also on environmental health. Nevertheless, it should be mentioned that production costs of bioplastics are at present 5-10 times greater than the cost of petro-based plastics, and therefore this continues being the main limitation for the successful commercialization of these bioplastics.

Biosynthesis of PHAs basically takes place in two stages: generation of activated monomeric precursors (hydroxyacyl-CoAs), and polymerization of these monomers into PHAs. The polymerization of hydroxyacyl-CoAs is catalyzed by PHA synthases. These enzymes have a broad affinity towards various HA-CoAs, among which (R)-3-hydroxyacyl-CoAs are the most favorable substrates; although 4-, 5- and 6-hydroxyacyl-CoAs can also be used as substrates [102, 218]. More than 150 kinds of hydroxycarboxylic acids have been already identified as PHA monomers [219]. This gives the possibility of synthesizing an enormous variety of PHAs, with different composition and biotechnological properties (Fig. **11**).

The Mw of PHAs may vary between 2×10^5 and 3×10^6 Da, but as for many microbial polysaccharides, it depends on the producing microorganism, the substrate composition, the inoculum, culture conditions and downstream processing techniques [220]. The basic structure of PHAs, the main monomeric precursor and general lines of nomenclature for these polymers is depicted in Fig. (**11**).

Based on the chain length of their monomeric units PHAs can be classified in:

- Short-chain-length (scl) PHAs: contain 3-5 C-atoms and display thermoplastic material properties similar to polypropylene. They are produced by a wide range of bacteria and archaea, such as *A. latus* and *C. necator* [221].
- Medium-chain-length (mcl) PHAs: they consist of 6 to 14 C-atoms and have elastic material properties similar to rubber and elastomer. They are synthesized

by bacteria such as *Aeromonas caviae* and *Pseudomonas putida* [222].

On the other hand, PHA copolymers can be:

- scl-copolymers: *e.g.* P(3HB-co-4HB), P(3HB-co-3HV)
- mcl-copolymers: *e.g.* P(3HHx-xo-3HO)
- scl-mcl-copolymers: consisting both in scl- and mcl-monomers (*e.g.* P(3HB-co-3HHx) [223, 224].

A

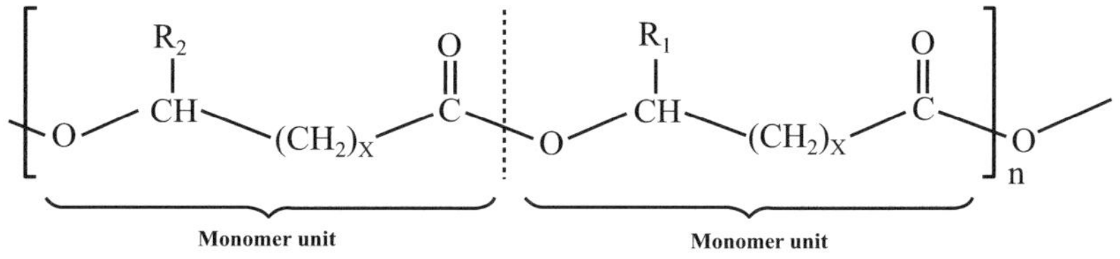

B

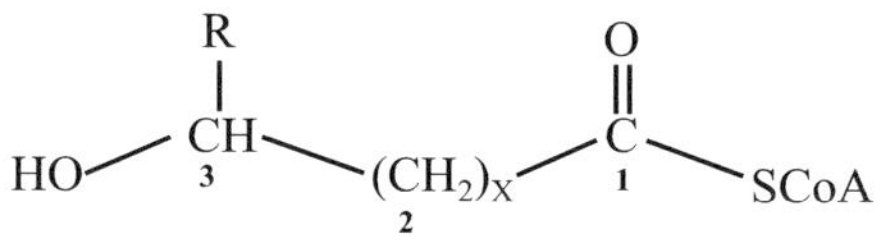

C

x	R_1	R_2	Polymer name	Symbol
1	H	H	Poly (3-hydroxypropionate)	PHP
1	CH_3	CH_3	Poly (3-hydroxybutyrate)	P3HB
2	H	H	Poly (4-hydroxybutyrate)	P4HB
3	H	H	Poly (5-hydroxyvalerate)	P5HV
1	CH_3	CH_2-CH_3	Poly (3-hydroxybutyrate-co-3-hydroxyvalerate)	PHBV

Fig. (11). (A) General structure of polyhydroxyalkanoates. R_1 and R_2 are alkyl groups, were R_1 can be equal (homopolymer) or different (copolymer) to R_2. Monomeric chains can be composed of 3 to 14 C in length. x = 1-4, and n = 50-15000. **(B)** Structure of main activated PHA precursor, (R)-3-hydroxyacyl-CoA. **(C)** Typical nomenclature for different PHAs according to the length and position of the side chain.

General Properties

Among PHAs, the most common ones are poly-hydroxy butyrate (PHB) and poly-hydroxy valerate (PHV). Short-chain PHAs exhibit thermoplastic properties whilst long-chain PHAs show elastomeric properties [204]. Monomer composition and Mw are the main influential parameters on PHA physical features. The PHB is highly crystalline and brittle and shows a Mw ~ 5×10^5 Da [225, 226]. On the other hand, PHAs of high Mw are useful for the production of films and strong fibers [226].

Currently, there is a great diversity of PHAs (> 150 types) due to the multiplicity of available monomeric constituents [216, 217]. Thermoplastic properties related to high polymerization degree, high crystallinity (60-80%) and melting temperature from 50 to 180°C, in addition to optical activity, isotactic behavior and insolubility in water make it competitive with polypropylene and other oil-derived plastics [227]. Beyond their use as plastics, PHAs can behave as chiral precursors for the chemical synthesis of biodegradable optically active components used in controlled delivery [210].

Applications

Polyhydroxyalkanoates are commercially relevant as renewable and biodegradable alternatives to oil-based materials [227]. Moreover, the broader substrate specificity of engineered PHA synthases provides versatile PHA biosynthetic platforms for the manufacture of tailor-made polyesters with required mechanical properties for a particular application [218].

Initially, PHAs were processed into films, fibers, molded products, extruded goods, coatings, adhesives, bottles, latex and several products of agricultural, commercial or packaging interest [228, 229]. Other uses include the manufacture of utensils, feminine hygienic products, cosmetic containers and shampoo bottles. One of the main PHAs currently commercialized is PHB, a biodegradable polymer that can be implanted into the human body and is widely used in biomedical applications. Recently, PHAs showing low density polyethylene (LDPE)-like material properties have also been synthesized [218].

Currently, more than 160 polyesters with diverse plastic properties have been described and this number is exponentially increasing by means of genetic and metabolic engineering techniques. Novel applications of these polyesters include medical purposes, such as orthopaedy (screws, scaffolds for cartilage engineering, bone graft substitutes), sutures, implants, dressings, urological stents, neural and cardiovascular-tissue engineering, fracture fixation and cell microencapsulation. They are used as osteo-synthetic materials to stimulate bone growth due to their piezoelectric properties in bone plates and surgical structures, and also as a replacement for blood vessels [230].

In addition, they can be used as biodegradable carriers for long-term controlled delivery of drugs, hormones, insecticides and herbicides, as supports of hypophyseal cells, or as precursors of molecules with anti-rheumatic, analgesic, radiopotentiator, chemopreventive, antihelmintic or antitumor properties. Finally, PHAs were also suggested for the treatment of narcolepsy and alcohol addiction [231 - 234].

Pullulan

The fundamental repeating unit of pullulan is the maltotriose trisaccharide (G3), which contains two (1,4)-linkages and no (1,6)-linkages. The main product of acid hydrolysis of pullulan is D-glucose [235]. In general lines, the basic structure of pullulan agrees with the following formula: [α-D-Glcp-(1,4)-α-D-Glcp-(1, 4)-α-D-Glcp-(1,6)]$_n$. Nevertheless, several variations along the polymer chain have been reported and so, there is no a unique structure of pullulan [69]. For instance, a 7% of maltotetraose subunits could be randomly found throughout the polymer molecule [236]. Moreover, α-(1,3), β-(1,3) and β-(1,6) linkages were also found in the main backbone of pullulan produced by certain strains [237]. Pullulan is essentially a linear polymer and its main general structure is shown in Fig. (**12**).

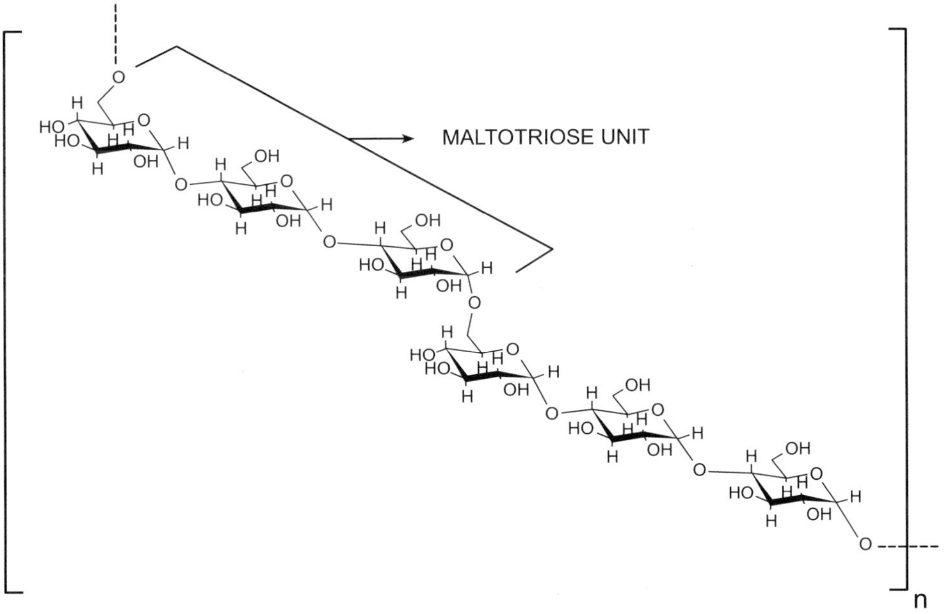

Fig. (12). Repeating maltotriose units of pullulan. Three glucoses linked by α-(1,4) bonds are polymerized to each other by α-(1,6) linkages, forming a linear chain.

Pullulan is a neutral, water-soluble EPS industrially produced by certain strains of *Aureobasidium pullulans* (formerly known as *Pullularia pullulans*). *A. pullulans* is a saprophytic and ubiquitous black fungus widely spread in diverse ecological niches, like soil, seawater, plant debris and animal tissues [238, 239], and it may exhibit two morphological forms, either yeast (high pullulan-producing state) or filamentous (low pullulan-producing state) states. *A. pullulans* is a strong pullulan producer, reaching yields up to 52.5 g/L of polysaccharide during batch processes [240, 241]. Apart from *A. pullulans*, a number of other microorganisms were also reported as pullulan producers, like the fungus *Phaeoacremonium aleophylum, Phaeomoniella chlamydospora, Tremella mesenterica* [242], *Cryphonectria parasitica* [243, 244], *Rhodotorula bacarum* [245] and *Teloschistes flavicans* [246].

Pullulan was discovered by R. Bauer in 1938, but it could not be isolated and characterized until 1958. In 1959, it was exhaustively studied by Bender, Lehmann and Wallenfels, who gave it the name "pullulan" [247]. This polysaccharide is produced since 1976 by Hayashibara Company Limited (Japan) [248]. This company has standardized different upstream and downstream processes for pullulan production and recovery, in such a way that the end product of each process holds a specific Mw and steady specific properties. Nowadays, this company is the chief commercial producer of pullulan [67].

The Mw of pullulan ranges between 362 to 480 kDa, whilst Mn is ~ 100-200 kDa [249]. Polydispersity (Mw/Mn) is usually between 2.1 and 4.1 [250, 251], being a relevant criterion for subsequent pharmaceutical applications [252]. The Mw value is significantly dependent on the producing strain, culture medium composition [253 - 255] or even cultivation time [256], usually increasing as fermentation proceeds [187]. However, there exists a critical point after which Mw can decrease because of the action of α-amylases, an undesirable effect that can be avoided by adding α-amylase inhibitors [255].

A higher Mw is frequently preferable for commercial applications, thus research focused into this commitment evaluated the influence of C-, N- or P-sources on the final pullulan Mw [250, 257]. Accordingly, in order to obtain high-Mw pullulan, NH_4^+ resulted better than NO_3^- as N-source.

General Properties

This neutral EPS exhibits structural flexibility and water solubility and has a characteristic film-forming ability [239]. Critical properties like biodegradability, impermeability to oxygen and non-toxicity are relevant for film applications [187, 258]. As a powder formulation, it is white and non-hygroscopic [239]. Its high solubility in water makes dissolution a fast step (either in hot or cold water) to form colorless, highly viscous and adhesive solutions.

Pullulan coatings are thermoresistent and a good barrier to oxygen thus increasing product shelf life. These properties make it useful for different pharmaceutical applications [187, 258]. Less soluble pullulan preparations can be obtained by chemical derivatization [258], either through esterification [259, 260] or etherification [261]. Additionally, Kato and Shiosaka reported pullulan hydrogenation as a mean to improve EPS thermal stability [262], whilst Tsuji and co-workers described a higher hydrosolubility by means of carboxylation [263].

Rheological properties of pullulan are closely related to Mw [252]. This EPS forms flexible random coils and produces low viscosity solutions similar to those of Arabic gum, with a Newtonian or nearly behavior. It is stable to heating, pH modifications and most of the added salts [239, 264]. Hydrodynamic properties of pullulan depend on the conformational flexibility and the solvent action, showing a high water solvation of 13 H_2O-molecules *per* glucose residue whereas this H-bond pattern can be abruptly disrupted in the presence of a chaotropic solvent like DMSO, which then imposes conformational changes [265].

Applications

Pullulan properties make it applicable in various fields. Purified pullulan is a biodegradable, highly water-soluble polymer. It has no mutagenic, carcinogenic, or toxicological activities [266]. When dissolved in water, it forms viscous solutions that are stable in the presence of many cations. Pullulan does not form gels. However, it forms transparent, non-hygroscopic, odorless, tasteless and non-toxic, thin films impermeable to oxygen. Films and solutions are widely used in food manufacturing and pharmaceutical industry [239, 267]. Significant efforts have focused in pullulan modification in order to improve its implementation in original fields. Therefore, modified and non-modified pullulan polymers currently have a wide range of applications in diverse industries [268 - 270].

One of the first successful uses of pullulan was as a blood plasma expander. For intravenous injection, purified pullulan must present a narrow Mw distribution (~ 30 to 90 kDa), and must be anionically modified with carboxyl and carbonyl groups which decrease its degradation rate in serum [248, 271 - 273].

The use of pullulan derivatives conjugated to viral antigens in vaccines was also patented, because of its effective adjuvant ability [274, 275]. Pullulan and acetylated pullulan can also be used for pharmaceutical and nutraceutical coatings for oral delivery in the form of tablets, pills or granules, as it prevents the oxidation and visual alterations of coatings and also allows a sustained drug release [258, 276].

Due to the pullulan ability to accumulate in the liver, a blend of pullulan and diethylenetriaminepentaacetic acid (DTPA) was successfully used to accumulate interferon (IFN) in human liver for the treatment of hepatitis C. In a similar way and because of its high liver tropism, pullulan derivatives were also utilized in the development of numerous non-viral carriers targeting the liver, for genes, proteins and cytokines therapy [277, 278]. Pullulan derivatives are also promising for the development of new biomaterials that can function as artificial extracellular matrix, able to attach cells and regulate their growth leading to three-dimensional tissue regeneration [258].

A heparin-conjugated carboxymethyl pullulan material was developed by Na *et al.* that can be used for the proliferation of vascular endothelial cells and is able to inhibit the proliferation of smooth muscle cells [279]. Carboxymethylated pullulan was also successfully used in chemotherapy [280]. In addition, derivatized pullulan nanoparticles (hydrogel nanoparticle of cholesterol-bearing pullulan) were designed for insulin controlled delivery in plasma [281]. Cholesterol/pullulan and amino-group-modified cholesterol/pullulan nanogels were proposed as promising fluorescent probes for biomedical imaging [282]. In the food industry, purified pullulan (with a white to off-white powder appearance) can be dissolved in water to give viscous odorless and tasteless solutions. It is nowadays used as thickener, viscosity stabilizer and also for the preparation of non-toxic, biodegradable, edible coatings. Pullulan can form oil-resistant, water-soluble and biodegradable films with high mechanical strength and low oxygen permeability, being suitable for packaging. It also contributes to flavor retention and to the preservation of fresh appearance in food. Colorants, flavoring, and other functional ingredients can be entrapped in these film matrices, which are very stable. In addition, pullulan films with antimicrobial properties were developed by the incorporation of lysozyme into the pullulan matrix [252]. Pullulan is also used in the manufacture of capsule shells and tablets of dietary supplements, and to make ham packaging films [98].

In food and beverages, pullulan imparts texture, structure, viscosity, dispersability and water retention. In food pastes and cookies it can be used as binder and stabilizer [259, 283]. Pullulan is also useful as a starch replacer in low-calorie food formulations. It is a non-digestible carbohydrate, and therefore it was proposed as a prebiotic with the ability to promote the growth of beneficial microbiota in the human intestine [284].

Among further pullulan applications, the use as an excipient in cosmetic emulsions has also been documented. It forms colorless solutions which can be employed for adhesives preparation, fibers preparation and as a sizing agent for paper [285]. Other usage involves the design of accurate molecular-mass standards for calibrating HPLC columns, which are used for size-exclusion chromatography of hydro-soluble polymers [117]. Its commercialization in the form of gel beads for chromatography [286] and for capillary electrophoresis allowed the purification of SDS-proteins and human histone H4 [287]. Pullulan was also successfully used as a preserving agent for bacteria like *E. coli* and *Bacillus subtilis*, by storing under different conditions. Finally, concerning environmental bioremediation, pullulan-producing microorganisms are able to remove heavy metals due to the adhesive nature of pullulan [252].

Scleroglucan

Scleroglucan is a high Mw, neutral, branched, weak-gelling glucan, produced mainly -but not exclusively- by filamentous fungi of the genus *Sclerotium* [288]. It consists in a backbone of β-(1,3)-D-glucopyranosyl residues bearing a single β-(1,6)-D-glucopyranosyl unit every three sugar residues of the main chain (Fig. **13**) [40, 289]. Scleroglucan displays a high degree of branching (DB ~ 0.33) responsible for its great water solubility. Depending on the microbial strain, fermentation process and the recovery method used, its Mw may vary between 1.3-3.2 × 10^5 and 0.3-6.0 × 10^6 Da [41, 81]. With regard to the degree of polymerization (DP, symbolized as "n" in Fig. **13**), reported values are highly variable, from values as low as 110 for the scleroglucan from *S. glucanicum*, 2400-2500 for *S. rolfsii* ATCC 201126 scleroglucan and up to 5600 for other cited scleroglucan [290].

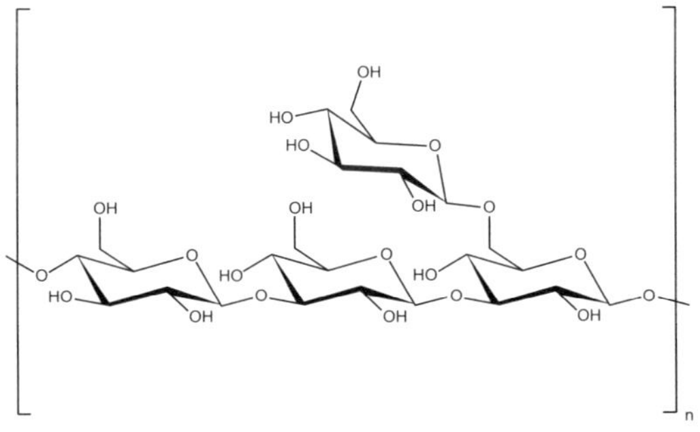

Fig. (13). Scleroglucan repeating unit showing the β-(1,3)-β-(1,6)-D-glucopyranosidic linkages.

First insights into the scleroglucan structure based on enzymatic approaches were provided by the Pioneering Research Division of the US Army Quartermaster Corps Research and Engineering Centre [291]. Later, Halleck patented the scleroglucan basic structure, including further information about producing strains, production process and preliminary data on the downstream processing [288]. Pillsbury Co. (Minneapolis, MN, USA) began the scleroglucan commercialization under the name Polytran® and up to date, other companies introduced it under different trademarks [292].

Scleroglucan production processes take place with selected producing strains and under submerged aerobic conditions in stirred-tank reactors. Production of EPS proceeds along with mycelial growth whilst the culture broth develops a gel-like consistency, and pH drops to ~ 2-2.5 normally during the first 12-24 h of cultivation [290, 293]. Much emphasis has been put on the need to standardize culture medium composition, process parameters or even the downstream processing in order to assure scleroglucan recovery and quality. This will allow obtaining high yields of a consistent EPS and will avoid eventual variations in chemical, physical and/or biological properties of scleroglucan [68, 290, 294].

Vast amounts of EPS (~ 20-25 g/L) could be obtained with *S. rolfsii* ATCC 201126. This usually involved batch processes at different working scales and/or with varying fermentation strategies [293]. Aspects such as strain preservation and inoculum preparation, even for different strains, have been critically assessed in order to stablish standardized operative protocols [293 - 295]. A high C/N source ratio in culture medium would be normally required to enhance EPS biosynthesis [8, 293], being preferential the use of sucrose as C-source and $NaNO_3$ as N-source. Ammonium-based N-sources would lead to decreased scleroglucan production, a fact likely related to a negative metabolic regulation of EPS biosynthesis by ammonium [293, 296].

Increased scleroglucan production under high-osmotic pressure conditions (*e.g.* 150 g/L sucrose) was described as a part of an osmoregulation mechanism [293]. Biosynthesis of EPS would occur in parallel with cell rescue processes and alternative energy-generating pathways, which may explain the EPS overproduction as a mechanism for survival under anoxic or other stressing conditions [297]. With regard to culture conditions, scleroglucan secretion is normally favored at 28-30°C, initial pH 4.5, aeration 0.5 vvm and stirring rate 400 rpm. Low dissolved oxygen levels would lead to restricted fungal growth and consequently, more C-source could be driven to scleroglucan biosynthesis [298]. This might also decrease by-product (oxalate) formation thus reducing the diversification of the C-source to undesirable products other than EPS [299]. Under these conditions, loose lenticular pellets of *S. rolfsii* ATCC 201126 are frequently associated to an optimal scleroglucan production [290].

Last improvements on this EPS production process are related to the employment of agro-wastes as suitable C-sources. This would represent a revalorization through the production of high added-value biotechnological products ("biorefinery" strategy) such as EPSs. At the same time, it might aid to cope with the depletion of natural resources and environmental concerns. Among alternative substrates, sugarcane molasses, treated lemon peel and commercial corn starch have been tested for scleroglucan production by *S. rolfsii* ATCC 201126 [300,

301].

Further challenges also relate to the scleroglucan downstream processing, a stage that may affect critical macromolecular features such as polymer Mw, DP, DB, purity and composition [68, 85, 290, 300]. A scleroglucan downstream processing protocol leading to an industrially acceptable recovery of EPS with high purity grade (~ 98%) involves a step of culture broth homogenization, followed by a 3-fold dilution with distilled water, and a final neutralization with NaOH. The diluted/neutralized broth is then heated (80 °C, 30 min) and centrifuged (27500 ×g, 15 °C, 20 min). The EPS from clear supernatant is precipitated with an equal volume of an organic solvent, commonly ethanol or isopropanol. Recovery of pure scleroglucan surrounds 30% (with sucrose or molasses media) up to 50% (with corn starch medium) [290].

General Properties

X-ray diffraction studies demonstrated that when dissolved in water at room temperature or at low concentrations of alkali, scleroglucan adopts a highly ordered, semi-rigid, triple helical tertiary structure. The existence of the triple helix for scleroglucan has been also verified by electron microscopy and light scattering measurements [97, 290, 302]. This conformation associated to a high DP account for its marked viscosifying ability, outstanding rheological behavior and the emerging scleroglucan successful applications [41, 303 - 308].

Scleroglucan is highly dispersible in water at room temperature mainly due to the presence of lateral β-1,6-glycosidic branching [187, 309]. Protruding groups on the macromolecular surface prevent the helices to close to each other which avoid aggregation [41, 310]. Sletmoen and Stokke reported that scleroglucan, and other β-1,3 substituted glucans, can complex organic molecules inside the central cavity of the triplex [311].

Some rheological parameters, salt tolerance, anti-syneresis effects, emulsifying and suspending properties of different *S. rolfsii* ATCC 201126 scleroglucan preparations showed certain variances [41, 306, 307]. Despite EPSs recovered at different cultivation times (48 or 72 h), showed identical primary structure [312], some conformational dissimilarities may account for this differential behavior [41, 68, 85].

Apparent viscosity of scleroglucan solutions exhibits a quite useful exponential relationship with EPS concentration, showing a non-Newtonian and non-thixotropic behavior. It displays high thermal stability (up to 100°C for 1 h) [41], but at ~ 7°C it forms thermoreversible gels [304, 313]. Among 140 polymers tested for EOR in the North Sea, scleroglucan was the more stable, being able to keep more than the 90% viscosity after 500 days at 90°C in seawater [304, 314]. Its neutral character may help this behavior, being hardly affected by very high salt concentrations or extreme pHs [41, 315, 316].

The highly ordered scleroglucan structure, can be however destroyed by conditions such as high alkali concentration (NaOH > 0.15 N; pH > 13) or chaotropic agents like dimethyl sulfoxide (DMSO > 87%). That may dissociate interchain H-bonding thus rendering random coil structures with the consequent abrupt decrease in viscosity [41, 317].

According to some references, pre-hydrated scleroglucan would be compatible with 5% w/v NaCl, 5% w/v Na_2SO_4, 20% w/v $CaCl_2$ and 10% w/v Na_2HPO_4 [309]. Compatibility of scleroglucan with some commercial thickeners such as alginates, xanthan, carrageenan, locust bean gum, and cellulose derivatives has been also reported [8, 304]. Scleroglucan from *S. rolfsii* ATCC 201126 showed compatibility and synergy with corn starch, xanthan, pectin and carboxymethylcellulose. Compatibility and a slight synergistic behavior were also observed with 30% (w/v) ethylene glycol [68].

Purified scleroglucan in solution (0.1-0.2% w/v) would efficiently stabilize fine powder suspensions (5-10% w/v) of ZnO, $CaCO_3$, and sulfamerazine [187, 304]. Compared to already known biopolymers, scleroglucan from *S. rolfsii* ATCC 201126 at 0.5 g/L showed a good dispersing ability against particulate suspensions of activated charcoal, bentonite, $CaCO_3$, celite and quartz powder [307]. Despite scleroglucan do not represent a primary emulsifier, as a surfactant does, it would allow the low-energy dispersion during emulsion formation of water-oil mixtures [187, 304]. Emulsifying ability of EPSs from *S. rolfsii* ATCC 201126 at 2 g/L was high (E = 56-60%) when tested with kerosene, moderate (~30%) with hexadecane, and negligible in the presence of olive oil-in-water emulsions [307].

Scleroglucan also exhibit biological properties which are the focus of intensive research. For instance, the administration of scleroglucan by diverse routes in rats and dogs did not induce toxicity, tissue pathology or blood abnormalities. Unsafe effects were neither found on eye nor skin in pigs, rabbits and humans. In addition, there are many reports supporting the immunostimulatory effects of scleroglucan and its potential for being applied as a non-digestible dietary fiber in humans [81, 308, 318]. Among health-promoting activities, some effects such as hypocholesterolemic, hypoglycemic, antioxidant and anti-obesity properties should be highlighted. It could thus be used for developing functional foods or nutraceuticals [308]. A general overview of relevant scleroglucan features and their actual or potential implications was recently reviewed [290].

Applications

The varied array of physico-chemical, nutritional, and biological properties of scleroglucan make it useful (or potentially useful) for a wide range of biotechnological products and processes. First uses of scleroglucan were reported in EOR, mainly as a xanthan competitor, due to the high heat stability, thickening power, high carrying capacity and lubricating power [319 - 321]. Use of scleroglucan aqueous fluids for petroleum recovery was reported since 1984, based on its viscosifying ability and salts tolerance [322, 323]. Likewise, Chaverot *et al.* reported the use of Actigum® CS-11 (85-90% scleroglucan) and Actigum® CS-6 (60-75% scleroglucan) for viscosity control of a bituminous binder used for road repair and construction, soil stabilization and sealing in civil engineering [324]. Because of its high flocculant capacity and fast settling rate, its addition to the liquor of a Bayer process fluid stream was also proposed to improve the recovery of alumina trihydrate [325, 326].

In the biomedical field, it is used as excipient in edible films, tablets and capsule granulates where it demonstrated to be a suitable matrix for controlled release of active substances, especially those sensitive to gastric acids. It also showed bioadhesive properties and good biodegradability [327, 328]. Its antitumor, antiviral, and antibacterial activity (of native or derivatized glucans) promoted the development of drugs, vaccines, and immunopotentiators [308, 329]. Similarly to curdlan, this β-glucan is a non-digestible fiber that behaves as a prebiotic. Further benefits include hypocholesterolemic, hypolipidemic, and hypoglycemic ability, and for that reason, it has been currently proposed for the design of nutraceuticals, low-calorie and functional foods [308].

Concerning further applications, scleroglucan can be used in the cosmetic industry as a dispersing, smoothing, lubricating, emulsifier or co-emulsifier agent, in products such as shampoos, conditioners, after-sun preparations and skin care compositions. A scleroglucan skin anti-inflammatory effect has also been reported [330, 331]. Its use as a saliva-substituting agent, in artificial tear waters, in mouth rinsing solutions and toothpastes was recently emphasized [290]. Additionally, in the environmental context, scleroglucans have been considered for bioremediation and agriculture, for the application of pesticides (to contribute their binding to leaves), and also in paper, painting and ceramic industries due to its particulate suspending properties and emulsifying activity [97, 307].

Sphingans

Sphingans are a family of structurally related EPSs having similar backbone arrangement except for the location of side chains. These include gellan, welan, rhamsan and diutan. Sphingans share a characteristic linear tetrasaccharide backbone structure, composed of two residues of β-D-glucose (β-D-Glcp), one of β-D-glucuronic acid (β-D-GlcpA), and one of α-L-rhamnose (α-L-Rha). Although the basic structure of the different sphingans is the same, differences in their sidechains impart dissimilar physical properties. General structure of these EPSs and their sidechains are depicted in Fig. (**14**).

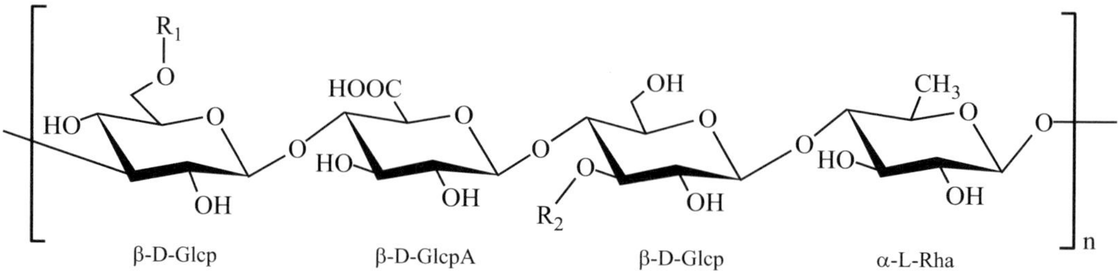

Name of sphingan	**R_1**	**R_2**
Gellan	H	H
Welan	H	← 1)-α-L-Rha or ← 1)-α-L-Manp
Rhamsan	← 1)-α-D-Glcp-(6 ← 1)-β-D-Glcp	H
Diutan	H	← 1)-α-L-Rha-(4 ← 1)-α-L-Rha

Fig. (14). Basic structure of the sphingan family (modified from [332]).

Gellan Gum

Gellan is the generic name for a partially O-acetylated, linear, anionic, gel-forming EPS, produced by the Gram-negative bacteria *Sphingomonas paucimobilis*, formerly known as *Pseudomonas elodea,* a microorganism first isolated from the elodea plant tissue [97, 333]. It was previously referred by the codenames S-60 and PS-60. Its approximate composition is glucose 60%, rhamnose 20% and glucuronic acid 20% and it also contains significant amounts of non-polysaccharide material such as cell protein and ash, which can be separated by filtration or centrifugation [334, 335].

Native form of gellan contains two types of acyl substituents, namely L-glyceryl and acetyl, bound to the glucose residue adjacent to glucuronic acid [336]. These residues can be removed by alkaline hydrolysis, obtaining deacetylated (or low-acyl) gellan gum [333]. Clarified gellan gum is obtainable after filtration of hot, low-acyl gellan; process that efficiently removes cell protein residues [337].

Gellan was discovered in 1978 [338], and in 1992, the USFDA approved gellan gum to be used as a food additive [339]. Currently, it is mainly commercialized by KELCO in Japan and USA, under the tradename KELCOGEL. Other gellan gum trademarks are GELRITE, Gel-Gro and GELZAN. Among all the sphingans, gellan became the most attractive polysaccharide because of its novel properties and industrial applications. The main strain currently used for industrial production of gellan is the reference strain *Sphingomonas paucimobilis* ATCC 31461, which achieves polymer yields of 35.7 g/L [22, 340, 341].

General Properties

When dissolved in water and in the presence of mono-, di- and trivalent cations, native and deacetylated gellan are able to form hydrogels through a temperature-dependent process [342]. Native gellan gives soft and elastic gels, whilst the deacetylated form turns them into rigid and brittle. This phenomenon occurs because acetyl and glyceryl groups located on the periphery of the helix of native polymers hamper chain associations, whereas the removal of these sidechains lead to strong intermolecular interactions and effective packing [343, 344]. After sidechain hydrolysis, the cations can easily form bridges between polysaccharide chains, which leads to the establishment of a polymer branched network [345]. Finally, after prior removal of all multivalent cations and the addition of Ca^{2+}, gellan is able to form gels in a similar manner to alginates. Gellan gels exhibit good rheological properties and are stable at extreme temperature and pH.

Gellan possesses excellent water solubility and tends to cycle under these conditions thus increasing the macromolecule volume [346]. Double helices interlace in a "union zone" by means of salt bridges and H-bonds [23]. The transition temperature (Tm) is around 35°C (can vary between 30-50°C) and above this value a rigid structure is obtained [351]. Below this point, a higher number of EPS molecules concentrate in the "union zone". Certain cations increase the number of union zones making it more resistant to temperature and improving gelation potential [346].

Gellan keeps its strength at 90°C, whilst xanthan for example, lose 74% at this temperature [347]. Thermal treatment at high temperatures dissociates double helices to give single coils therefore reducing viscosity to a significant extent. The variability of Tm value can relate to the concentration of cations (like K^+) in the gel, which can increase the union zones thus, improving heat resistance [348].

Acetyl content of gellan strongly influences gel properties. Native gellan gives soft, weak, elastic and thermoreversible gels due to the presence of acetyl and glyceryl voluminous groups that block the close approach of polymer chains [349]. Deacetylation gives place to extensive intermolecular association between gellan chains consequently originating firm and brittle gels [117].

Apart from the native gellan, there exist three modified known gellan types with variable acetylation degree: high-acetyl gellan, low-acetyl gellan and low-acetyl / clarified gellan. The presence of certain ions can also modify gel behavior. The addition of salts of Ca^{2+}, Na^{+}, K^{+} and Mg^{2+} increase gel resistance and fragility [350]. Gels containing 1% w/v gellan when amended with 0.1% w/v $MgCl_2$ or $CaCl_2$ developed high resistance (~ 800 g/cm^2), whereas the addition of KCl or NaCl even at 1% w/v, produced less resistant gels (~500 g/cm^2). This effect accounts for the indispensability of salt addition when using natural hard water for gellan dissolution [349, 351].

Gel strength improves within a pH range between 4 and 7, which is in agreement with the pH of the majority of foodstuff [350]. Other researchers reported a pH range of 2-10, even wider than that described for xanthan (4-8) [347]. The effect of pH can also modify the Tm value of gellan, significantly increasing at acidic pH in the presence of monovalent cations [349].

Hydrophilic ingredients can also influence gel formation by gellan. The presence of sucrose (~ 10% w/v), for instance, reduces the requirement for ions to achieve optimal gellan resistance [350, 352]. Sucrose-amended gellan gels at low (0-20% w/v) and high (80-85% w/v) sugar concentration were studied by transmission electron microscopy and demonstrated a reduction of the gel reticulation degree at high sucrose concentration [349, 353].

Among the functional properties of gellan, it could be highlighted its versatile texture, heat and pH stability, Tm flexibility, high clarity, easy dissolution in water, compatibility with other gums and polymers and low required concentration (0.04-0.05% w/v). For instance, the addition of gellan to gelatin provides stability to gelatin, increasing the fusion point but retaining its desirable properties [351, 352].

Applications

Gellan was first promoted particularly for its use in culture media as a potential agar substitute. Then, after obtaining permission for its use in foods, gellan was proposed as a xanthan competitor. Gellan has applications as a food additive, being useful as thickener, stabilizer and suspending agent, and giving gel textures in food products [97]. It is used as a food additive in dairy products, bakery fillings, frostings, icings and glazes, microwavable foods, dessert gels, jams, jellies, low-fat spreads, puddings, sauces, structured foods, confections, and toppings [354].

Its main application in the pharmaceutical industry is as an excipient in pharmaceutical tablets, capsules and bead matrices developed for oral or rectal drug administration, where it allows a sustained release [342]. Other solid, semi-solid and liquid formulations have been developed for varied administration routes: oral (*i.e.* liquid low-acyl gellan that in acidic environments undergo sol-gel transition, reducing appetite in obesity treatments), ophthalmic (*in situ* gelling ophthalmic solutions that improve absorption of ocular drugs, delivery ophthalmic systems) and nasal (nasal gels, microparticles, microspheres, where polymer bioadhesivity provides close adherence to the mucus layer thus increasing absorption rate).

In addition, gellan-based biomaterials are being investigated for tissue regeneration (as a component of wound dressings that impairs postsurgical adhesion and prevent scar formation), dental care (dental cavity fillings after tooth extraction), bone repair, cartilage reconstruction, gene delivery or biosensor synthesis, with very promising results. Because of its film forming property it is also being studied to develop implants for insulin delivery in diabetic patients [342, 355, 356].

Further applications relate to the cosmetic industry, where gellan gum is used as an excipient in lotions and creams because of its effective stabilization of emulsions during temperature fluctuations, besides maintaining consistency over time and imparting a shelf 'light and silky' feel when rubbed on the skin. In sunscreens and bronzers, it stabilizes the oil phase and allows uniform delivery of the therapeutic ingredients onto the skin. In hair care products such as shampoos and conditioners, it is used because of its stabilizing and suspending capacity. In toothpaste formulations, gellan gum provides an excellent flavor release, which makes it possible to reduce sweeteners and flavor components in the formulae. Other cosmetic products that may include gellan in their composition are facial masks, makeups and gels.

Concerning the use in the biotechnology field, gellan gum can be used as an agar substituent in microbiological culture media. In this regard, gellan presents the following advantages: forms solid gels at low concentrations (0.1%), with equivalent strength than agar but using much lower concentrations of gellan (the fifth part). In addition, gellan gels are more pure and transparent, and are ideal for growing thermophilic microorganisms, due to its high stability at elevated temperatures throughout prolonged incubations [357]. Gellan gum is also useful for plant tissue culture due to its high purity and resistance to mold contamination. It can be easily washed from the plant tissue for transplantation, and allows clear observation of root and tissue development [358]. Other reported application for gellan is its use in the bioremediation of polluted soils and aquifers [359, 360].

Welan Gum

Welan gum (previously known as S-130) is a polianionic acidic heteropolysaccharide produced by the Gram-negative bacterium *Sphingomonas* sp. The main strain used for industrial production is *Sphingomonas* sp. ATCC 31555 (also known as *Alcaligenes* sp. ATCC 31555). Nevertheless, other strains like *Alcaligenes* sp. CGMCC 2428, *Alcaligenes* sp. NX-3 and *Alcaligenes* sp. NX-3-1 have been reported to produce similar yields of welan [361 - 363]. It has an approximate Mw of 1.0×10^6 g/mol and it has the same tetrasaccharide backbone as gellan, only differing in the side chains [364]. Contrarily to gellan, welan is a non-gelling polysaccharide [365] bearing polyelectrolyte property due to the presence of D-glucuronic acid in its chemical structure.

Applications

Low viscosity welan gum is used in cement compositions, where it acts as a thickening, suspending, binding and emulsifying agent, improving the workability of cement. When the welan gum is added into a superplasticizer and is agitated, it swells in the superplasticizer, resulting in a stable suspension without much viscosity increase. This mixture is useful for providing underwater concrete, post-tensioning grout, and self-consolidating concrete. It also has prospective applications in oil-well drilling because of its viscosity and thermostability [332].

Rhamsan, Diutan and Other Sphingan Gums

Although gellan, because of its physicochemical properties and versatility, is clearly the sphingan of main interest from an industrial perspective, other sphingans are also being characterized and gradually find novel industrial applications. For example, rhamsan (formerly known as S-194) is used as suspending agent in fertilizers, pharmaceutical compositions and food systems. Diutan (or S-657) and Sphingan S-88 have found a place in the field of petroleum and water well treating fluids and muds; Sphingan S-7 is used as a component in textile printing pastes and herbicides, and is also used in oil well drilling [366 - 369].

Xanthan Gum

Xanthan gum is a heteropolysaccharide with a primary structure consisting of a repeated pentasaccharide unit fashioned by two glucose units, two mannose units, and one glucuronic acid unit. The main chain consists of β-(1,4)-D-glucopyranose residues. This chemical structure is identical to that of cellulose. The side chain consists of a D-glucuronic acid unit between two D-mannopyranose units linked at the C3 position of the D-glucose residue of the main chain (Fig. **15**).

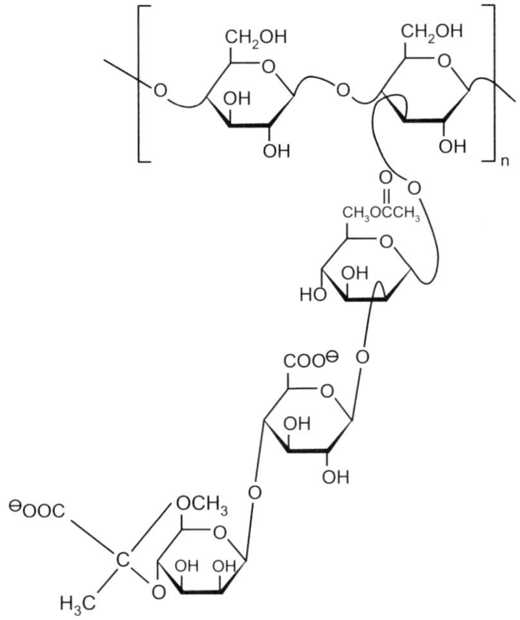

Fig. (15). Xanthan repeating unit [97].

Xanthan is an acidic EPS secreted by bacterial species of the genus *Xanthomonas*, a plant pathogen responsible of cabbage blight. *Xanthomonas campestris* is the bacterial species intensively used for xanthan gum production. This bacterium mainly grows in culture media with glucose, sucrose or starch as C-source, with high conversion efficiency (80%) to xanthan. Depending on the strain, growing conditions and media employed, they may have different ratios of O-acetyl substituents at the C6 position of the internal α-mannosyl residue, and pyruvate groups in the terminal β-D-mannosyl residue of the side chain. The presence of acetic and pyruvic acids produces a highly anionic polysaccharide [370].

This polysaccharide was discovered in the 1950s at the Northern Regional Research Laboratories (NRRL) of the United States Department of Agriculture [371] and its production was extensively studied in a pilot plant in the 1960s. Then, in 1963, it was first marketed by the Kelco Company under the tradename of Kelzan®. Since then, xanthan production increased uninterruptedly and its production is currently estimated around 9000 tons a year [97]. The main industries producing xanthan are Kelco, Merck, Pfizer (USA), Mero Rousselot Satia, Sanofi-Elf, Rhone-Poulenc (France), and Jungbunzlauer (Austria).

Although *Xanthomonas* is easy to grow in standard laboratory media, variations in the yield and quality of xanthan have been observed both in continuous and batch culture systems, suggesting the importance of the production process optimization

[372 - 374]. Industrial production of xanthan is commonly performed under batch mode, through an aerobic process. A typical xanthan production lasts for 60-90 h of cultivation and frequently yields a culture broth with 1.3% w/w xanthan, 0.1-0.3% w/w of cells and 0.1-1% w/w of unused sugar, salts and other residual medium components.

Production medium usually contains glucose or sucrose as a C-source, and NH_4NO_3, yeast extract, urea or peptone as N-source, plus phosphate, magnesium and other inorganic salts at lower concentrations (H_3BO_3, ZnO, $FeCl_3$, $CaCO_3$). Exponential-phase inoculum usually is in the order of 5-10% of the final culture volume. During production, dissolved oxygen tension (DOT) sharply drops in the first 15 h of culture due to the exponential growth, and thereafter because of the considerable increase in viscosity, which dramatically decreases the oxygen transfer rate. The DOT must be maintained at 10% saturation, which is commonly achieved by increasing agitation speed (~ 1000 rpm until the end of fermentation). The pH is automatically controlled at 7.0 (with NaOH) and temperature is normally set at 28°C.

Once fermentation is finished, the culture broth is heated (90°C, 15 min) and filtered (through 0.45-μm filter). The filtrate is first amended with NaCl (0.5 g/L) and once dissolved, 85% isopropanol (1:1, v/v) is added. The resultant precipitate is recovered by filtration, washed with isopropanol-water (3:1, v/v) and can be dried, milled and packaged in a suitable commercial form [7].

Several aspects should be specially controlled during xanthan industrial production, particularly to guarantee EPS consistency over time. For example, a major problem is the instability of the *X. campestris* strains. Three different strains were described: strain L (large), which produces high yields of xanthan, and Sm (small) and Vs (very small) strains which produce low or no yields of EPS. The two last strains are formed by deterioration of the L strain, normally due to aging, or as a result of deficient strain preservation. Variations in xanthan Mw, as well as in the acetylation degree and pyruvate content can also be observed as a consequence of ineffective strain conservation. Therefore, effective methods for long-term storage should be chosen in order to avoid a loss of the productive capacity and polymer properties. The construction of genetically modified strains of *X. campestris* able to use alternative low-cost raw materials for polymer production has been also under investigation [375].

General Properties

This is very likely the most studied EPS because of its interesting properties and relatively low production costs. Then, it constitutes a reference model for the study of further polysaccharides of microbial origin. The tertiary structure of xanthan is an extended rigid rod conformation, with Mw between 2-20 × 10^6 Da, depending on the association grade between chains, and the formation of aggregates of several individual chains [374]. It exhibits a narrow Mw distribution as compared to most polysaccharides.

In solution and under moderate temperatures, it principally exhibits a double-stranded helical structure [376]. According to X-ray diffraction analyses, dextro-rotatory single or double helices conformation can be identified [377]. Lateral chains protrude to the exterior of the macromolecule and are stabilized through H-bonds. Probably, this conformation accounts for the high xanthan stability against adverse conditions. Due to its anionic nature, the structure usually requires to be stabilized by salts in order to provide optimal functionality [378].

Xanthan gum imparts high viscosity to aqueous solutions, even at low concentrations, and can be maintained at varying temperature or pH. Its tridimensional structure also helps to stabilize emulsions and suspensions and gel formation is also achievable by mixing it with other polysaccharides (galactomannans) or trivalent metal ions. Xanthan gum is soluble in hot and cold water and, as the majority of hydrocolloids, it requires vigorous stirring when poured into the aqueous medium to avoid lump formation [379]. Factors such as stirring rate, particle size, and solvent composition have a critical effect on EPS hydration [380].

The native xanthan conformation can be denatured by thermal treatment or salt addition. The specific denaturing conditions will depend on the EPS composition, *e.g.* pyruvate content, which also affect rheological properties [187]. Thermally-induced denaturation implies a transition from an ordered (low temperature) to a disordered, more flexible state (high temperature). This change was early evidenced as change in viscosity [381]. As exposed, pyruvate and acetate contents as well as the medium ionic strength will influence the final conformational transition temperature (Tm) [378, 382].

Concerning the pH effect, xanthan is stable within a wide range, which is particularly useful for the manufacture of acid foodstuff, like salad dressings. Viscosity can decrease below pH 3, but as soon as the pH is neutralized, viscosity becomes recovered [380]. The effect of salts is closely related to xanthan concentration; salt addition decreases viscosity when EPS is below 0.3% w/v, but above this EPS concentration salts will have the opposite result. Likewise, the influence of salts is exerted when at low concentrations (*e.g.* 0.1% w/v NaCl). Higher salt values have an almost negligible incidence on viscosity [380].

Xanthan solutions exhibit a typical pseudoplastic behavior which can exalt sensory properties (taste release, food texture and mouthfeel) of some foodstuff.

This pseudoplasticity can vary depending on the xanthan concentration or salt addition [379, 380]. As for other semi-rigid macromolecules, xanthan solutions show high values of κ and low values of n. An additional feature of xanthan solutions is thixotropy, which implies the ability to recover the initial viscosity once the shear stress is removed, a reflection of the viscoelastic behavior [34, 117].

Synergistic interaction with other hydrocolloids can be used to improve xanthan properties. It is compatible with thickeners like starch, carrageenan, pectin, gelatin, agar, alginate and cellulose derivatives [380]. Synergism with galactomannans is also demonstrated, as well as the significant increase in viscosity of xanthan and guar gum. Xanthan, besides the synergistic interaction, is able to inhibit starch retrogradation and syneresis. Compatibility has been additionally tested with organic acids like acetic, citric, lactic and phosphoric, which is relevant for the food industry, being also possible to directly hydrate xanthan in the acid solution [380].

Applications

Some of the distinctive properties that make xanthan so suitable for several industries include [187]:

- Emulsifying and stabilizing properties
- Suspending capacity
- Ability to develop high-viscosity solutions at very low polymer concentrations (thickening agent)
- High acid and salt stability
- High shear-thinning (extremely large shear dependence on viscosity), and good pourability
- Little variations in viscosity under changing temperature conditions
- Gelling capacity *i.e.*, when mixed with certain other, non-gelling polysaccharides

Industrial use of xanthan as a food additive has been approved in 1969 by the US Food and Drug Administration (FDA) [187], and in 1980, the European Economic Community included xanthan to the food emulsifier/stabilizer list, as item E-415 [374]. This approval, together with its physicochemical properties, made this biopolymer useful for the design of packaging materials in contact with food and also for the design of oral drug delivery systems.

In the food industry, xanthan is extensively used in bakery products to help retention of water during baking, and because it increases food products shelf-life. Xanthan gum improves volume and texture to cold or hot processed bakery.

Because of its stabilizing properties and highly pseudoplastic behavior, it is used as an additive in most liquid and semi-liquid foods, such as syrups, sauce toppings, and salad dressings, where it improves emulsion stability.

Another utility is as a cholesterol-free egg substitute and also, as a gluten substitute. In drinks, it is used together with carboxymethylcellulose to add body to the liquids and, in the case of fruit pulp-containing drinks, it helps to maintain the suspension thus giving good appearance and texture. In low-fat foods, it increases the viscosity of the aqueous phase and stabilizes the food system. It is also used to stabilize and give body to dairy products and desserts such as ice creams, milk shakes and acidified yogurts [98]. In addition, xanthan suspensions are used as additives in frozen foods because of its high freeze-thaw stability. Likewise, xanthan gum stabilizes suspensions of vitamins and insoluble substances in liquid feed for pets, and acts as stabilizer and binder in the formulation of canned gravy based pet foods.

For pharmaceuticals, xanthan is used in emulsions and suspensions by providing excellent stability, uniformity in dosage formulations and good flow. In tablets, xanthan acts as a hydrophilic matrix carrier that enables the successfully controlled release of drugs such as theophylline and acetaminophen into the stomach fluid [97]. Concerning its application in the cosmetic industry, it is frequently used in toothpastes to afford easy pumpability, high viscosity and ease of teeth brushing [383]. In creams and lotions it provides creamy consistency and stabilizes emulsions, whilst its use in shampoo formulations keeps insoluble components suspended, favors uniform pigment dispersal and acts as thickener and stabilizer [374].

A largely described industrial use of xanthan relates to its early use for Enhanced Oil Recovery (EOR). Because of its viscosifying ability, shear thinning and stability to extreme conditions of temperature and pH, xanthan is also applicable in the tertiary recovery of petroleum. During the drilling operation, xanthan solutions are sufficiently fluid to facilitate the pumping of the mixture, but when the operation is stopped, the viscosity increases so that succeeds to keep the solids in suspension. With regard to further industrial uses, xanthan is used in textile printing and dyeing, to control the rheological properties of paste and to prevent dye migration. Application to the ceramic industry involves xanthan ability to prevent agglomeration during grinding of ceramic glazes. Additionally, in slurry explosive formulations, it is used as thickener and to improve heat stability (in combination with guar gum) [374].

CONFLICT OF INTEREST

The author (editor) declares no conflict of interest, financial or otherwise.

ACNOWLEDGEMENTS

Authors gratefully acknowledge the financial support from CONICET and MINCyT-DFG (grants PIP 0407 and PCBII 2015-2017).

REFERENCES

[1] D.J. Best, I.J. Higgins, and J. Jones, "Biotechnology: Principles and Applications", In: *Principles and Applications*Blackwell Scientific, 1985.

[2] M. Niaounakis, *Biopolymers: Reuse, Recycling, and Disposal.*1st ed Elsevier Science, 2013.

[3] M. Vert, Y. Doi, K-H. Hellwich, M. Hess, P. Hodge, P. Kubisa, M. Rinaudo, and F. Schué, "Terminology for biorelated polymers and applications (IUPAC Recommendations 2012)", *Pure Appl. Chem.,* vol. 84, pp. 377-410, 2012. [http://dx.doi.org/10.1351/PAC-REC-10-12-04]

[4] Z. Persin, K. Stana-Kleinschek, T.J. Foster, J.E. Van Dam, C.G. Boeriu, and P. Navard, "Challenges and opportunities in polysaccharides research and technology: The EPNOE views for the next decade in the areas of materials, food and health care", *Carbohydr. Polym.,* vol. 84, pp. 22-32, 2011. [http://dx.doi.org/10.1016/j.carbpol.2010.11.044]

[5] M.G. Rojas Cortés, B.M. Vallejo Díaz, and J.E. Perilla, "Biopolymers as materials for product development in pharmaceutical applications and biomedical uses", *Ing. Investig.,* vol. 28, pp. 57-71, 2008.

[6] A. Smelcerovic, Z. Knezevic-Jugovic, and Z. Petronijevic, "Microbial polysaccharides and their derivatives as current and prospective pharmaceuticals", *Curr. Pharm. Des.,* vol. 14, no. 29, pp. 3168-3195, 2008. [http://dx.doi.org/10.2174/138161208786404254] [PMID: 19075698]

[7] J.I. Fariña, "Biotecnología de polisacáridos microbianos", In: *Biotecnología microbiana I,* S.M. Tucumán, Ed., 1st ed Imprenta Central de la Universidad Nacional de Tucumán, 2005.

[8] S.A. Survase, P.S. Saudagar, I.B. Bajaj, and R.S. Singhal, "Scleroglucan: fermentative production, downstream processing and applications", *Food Technol. Biotechnol.,* vol. 45, pp. 107-118, 2007.

[9] I.W. Sutherland, "Biosynthesis of microbial exopolysaccharides", *Adv. Microb. Physiol.,* vol. 23, pp. 79-150, 1982. [http://dx.doi.org/10.1016/S0065-2911(08)60336-7] [PMID: 6180610]

[10] *Carbohydrate Biotechnology Protocols.*1st ed Humana Press, 1999.

[11] C. Laroche, and P. Michaud, "New developments and prospective applications for beta (1,3) glucans", *Recent Pat. Biotechnol.,* vol. 1, no. 1, pp. 59-73, 2007. [http://dx.doi.org/10.2174/187220807779813938] [PMID: 19075833]

[12] I.W. Sutherland, "Structure-function relationships in microbial exopolysaccharides", *Biotechnol. Adv.,* vol. 12, no. 2, pp. 393-448, 1994. [http://dx.doi.org/10.1016/0734-9750(94)90018-3] [PMID: 14545899]

[13] R. De Philippis, M. Margheri, E. Pelosi, and S. Ventura, "Exopolysaccharide production by a unicellular cyanobacterium isolated from a hypersaline habitat", *J. Appl. Pychol.,* vol. 5, pp. 387-494, 1993. [http://dx.doi.org/10.1007/BF02182731]

[14] M. Hisamatsu, S. Nomura, A. Shutsrirung, H. Obata, K. Teranishi, T. Yamada, S. Nuswantara, M. Yamashita, and Y. Murooka, "Structural characterization of a new acidic exopolysaccharide and cyclic (1→2) β-glucan produced by Rhizobium huakuii forming nodules on Astragalus sinicus", *J. Ferment. Bioeng.,* vol. 83, pp. 315-320, 1997. [http://dx.doi.org/10.1016/S0922-338X(97)80135-3]

[15] *Polysaccharides: Structural Diversity and Functional Versatility.*2 nd ed CRC Press, 2004.

[16] I.W. Sutherland, "Microbial polysaccharides from Gram-negative bacteria", *Int. Dairy J.,* vol. 11, pp. 663-674, 2001. [http://dx.doi.org/10.1016/S0958-6946(01)00112-1]

[17] G.C. Avendaño-Romero, A. López-Malo, and E. Paolu, "Propiedades del alginato y aplicaciones en alimentos", *Temas Selectos de Ingeniería de Alimentos,* vol. 7, pp. 87-96, 2013.

[18] B. Lindberg, "Components of bacterial polysaccharides", *Adv. Carbohydr. Chem. Biochem.,* vol. 48, pp. 279-318, 1990. [http://dx.doi.org/10.1016/S0065-2318(08)60033-5]

[19] J.G. Guezennec, P. Pignet, G. Raguenes, E. Deslandes, Y. Lijour, and E. Gentric, "Preliminary chemical characterization of unusual eubacterial exopolysaccharides of deep-sea origin", *Carbohydr. Polym.,* vol. 24, pp. 287-294, 1994. [http://dx.doi.org/10.1016/0144-8617(94)90073-6]

[20] T. Dengler, B. Jann, and K. Jann, "Structure of the serine-containing capsular polysaccharide K40 antigen from Escherichia coli O8:K40:H9", *Carbohydr. Res.,* vol. 150, pp. 233-240, 1986. [http://dx.doi.org/10.1016/0008-6215(86)80019-2] [PMID: 3093075]

[21] W.T. Seo, G.G. Kahng, S.H. Nam, S.D. Choi, H.H. Suh, S.W. Kim, and Y.H. Park, "Isolation and characterization of a novel exopolysaccharide-producing Paenibacillus sp. WN9 KCTC 8951P", *J. Microbiol. Biotechnol.,* vol. 9, pp. 820-825, 1999.

[22] K.S. Kang, and G.T. Veeder, Polysaccharide S-60 and bacterial fermentation process for its preparation, 1982.U.S. Patent 4,326,053

[23] H. Grasdalen, and O. Smidsrød, "Gelation of gellan gum", *Carbohydr. Polym.,* vol. 7, pp. 371-393, 1987. [http://dx.doi.org/10.1016/0144-8617(87)90004-X]

[24] K.P. Shatwell, I.W. Sutherland, S.B. Ross-Murphy, and I.C. Dea, "Influence of the acetyl substituent on the interaction of xanthan with plant polysaccharides — I. Xanthan-locust bean gum systems", *Carbohydr. Polym.,* vol. 14, pp. 29-51, 1990. [http://dx.doi.org/10.1016/0144-8617(90)90005-D]

[25] F. Rodriguez-Valera, J.G. Lillo, J. Antón, and I. Meseguer, "Biopolymer production by Haloferax mediterranei", In: *General and Applied Aspects of Halophilic Microorganisms*1st ed Springer US, 1991, pp. 373-380. [http://dx.doi.org/10.1007/978-1-4615-3730-4_45]

[26] A. Arvaniti, N.K. Karamanos, G. Dimitracopoulos, and E.D. Anastassiou, "Isolation and characterization of a novel 20-kDa sulfated polysaccharide from the extracellular slime layer of Staphylococcus epidermidis", *Arch. Biochem. Biophys.,* vol. 308, no. 2, pp. 432-438, 1994. [http://dx.doi.org/10.1006/abbi.1994.1061] [PMID: 8109972]

[27] I. Sutherland, "Biofilm exopolysaccharides: a strong and sticky framework", *Microbiology,* vol. 147, no. Pt 1, pp. 3-9, 2001. [http://dx.doi.org/10.1099/00221287-147-1-3] [PMID: 11160795]

[28] J.A. Mata Gómez, *"Caracterización de los exopolisacáridos producidos por microorganismos halófilos pertenecientes a los géneros Halomonas, Alteromonas, Idiomarina, Palleronia y Salipiger",* PhD thesis, Universidad de Granada, España, 2006.

[29] R. Lapasin, and S. Pricl, "Rheology of polysaccharide systems", In: *Rheology of Industrial Polysaccharides.*1st ed Theory and Applications: Springer, US, 1995, pp. 250-494. [http://dx.doi.org/10.1007/978-1-4615-2185-3_4]

[30] D.A. Rees, and E.J. Welsh, "Secondary and tertiary structure of polysaccharides in solutions and gels", *Angew. Chem. Int. Ed. Engl.,* vol. 16, pp. 214-224, 1977.

[http://dx.doi.org/10.1002/anie.197702141]

[31] E. Atkins, "Biomolecular structures of naturally occurring carbohydrate polymers", *Int. J. Biol. Macromol.,* vol. 8, pp. 323-329, 1986.
[http://dx.doi.org/10.1016/0141-8130(86)90050-4]

[32] E.D. Atkins, I.A. Nieduszynski, W. Mackie, K.D. Parker, and E.E. Smolko, "Structural components of alginic acid. II. The crystalline structure of poly-alpha-L-guluronic acid. Results of x-ray diffraction and polarized infrared studies", *Biopolymers,* vol. 12, no. 8, pp. 1879-1887, 1973.
[http://dx.doi.org/10.1002/bip.1973.360120814] [PMID: 4733713]

[33] W. Steiner, D. Haltrich, and R.M. Lafferty, "Production, Properties, and Practical Applications of Fungal Polysaccharides", In: *Biosurfactants. Production: Properties: Applications.,* N. Kosaric, F.V. Sukan, Eds., vol. 159. 1st ed CRC Press, 1993, pp. 175-230.

[34] I.W. Sutherland, "Microbial polysaccharide products", *Biotechnol. Genet. Eng. Rev.,* vol. 16, pp. 217-229, 1999.
[http://dx.doi.org/10.1080/02648725.1999.10647976] [PMID: 10819080]

[35] A. Heyraud, L. Dantas, J. Courtois, B. Courtois, W. Helbert, and H. Chanzy, "Crystallographic data on bacterial (1>4)-beta-D-glucuronan", *Carbohydr. Res.,* vol. 258, pp. 275-279, 1994.
[http://dx.doi.org/10.1016/0008-6215(94)84093-8] [PMID: 8039181]

[36] L. Dantas, A. Heyraud, B. Courtois, J. Courtois, and M. Milas, "Physicochemical properties of 'Exogel' exocellular β(1–4)-d-glucuronan from Rhizobium meliloti strain M5N1 C.S. (NCIMB 40472)", *Carbohydr. Polym.,* vol. 24, pp. 185-191, 1994.
[http://dx.doi.org/10.1016/0144-8617(94)90129-5]

[37] F. Freitas, V.D. Alves, J. Pais, N. Costa, C. Oliveira, L. Mafra, L. Hilliou, R. Oliveira, and M.A. Reis, "Characterization of an extracellular polysaccharide produced by a Pseudomonas strain grown on glycerol", *Bioresour. Technol.,* vol. 100, no. 2, pp. 859-865, 2009.
[http://dx.doi.org/10.1016/j.biortech.2008.07.002] [PMID: 18713662]

[38] L.F. Hallack, D.S. Passos, K.A. Mattos, O.A. Agrellos, C. Jones, L. Mendonça-Previato, J.O. Previato, and A.R. Todeschini, "Structural elucidation of the repeat unit in highly branched acidic exopolysaccharides produced by nitrogen fixing Burkholderia", *Glycobiology,* vol. 20, no. 3, pp. 338-347, 2010.
[http://dx.doi.org/10.1093/glycob/cwp181] [PMID: 19933228]

[39] S.S. Gauri, S.M. Mandal, K.C. Mondal, S. Dey, and B.R. Pati, "Enhanced production and partial characterization of an extracellular polysaccharide from newly isolated Azotobacter sp. SSB81", *Bioresour. Technol.,* vol. 100, no. 18, pp. 4240-4243, 2009.
[http://dx.doi.org/10.1016/j.biortech.2009.03.064] [PMID: 19403304]

[40] J.I. Fariña, *"Producción de escleroglucano por Sclerotium rolfsii"*, PhD thesis, Universidad Nacional de Tucumán, Argentina, 1997.

[41] J.I. Fariña, F. Siñeriz, O.E. Molina, and N.I. Perotti, "Isolation and physicochemical characterization of soluble scleroglucan from Sclerotium rolfsii. Rheological properties, molecular weight and conformational characteristics", *Carbohydr. Polym.,* vol. 44, pp. 41-50, 2001.
[http://dx.doi.org/10.1016/S0144-8617(00)00189-2]

[42] C. Panagiotopoulos, R. Sempéré, R. Lafont, and P. Kerhervé, "Sub-ambient temperature effects on the separation of monosaccharides by high-performance anion-exchange chromatography with pulse amperometric detection. Application to marine chemistry", *J. Chromatogr. A,* vol. 920, no. 1-2, pp. 13-22, 2001.
[http://dx.doi.org/10.1016/S0021-9673(01)00697-5] [PMID: 11452991]

[43] *Capillary Electrophoresis of Carbohydrates: From Monosaccharides to Complex Polysaccharides.* 1st ed Humana Press, 2010.

[44] R. Weiner, S. Langille, and E. Quintero, "Structure, function and immunochemistry of bacterial

exopolysaccharides", *J. Ind. Microbiol.,* vol. 15, no. 4, pp. 339-346, 1995. [http://dx.doi.org/10.1007/BF01569989] [PMID: 8605071]

[45] R. Chandrasekaran, and A. Radha, "Molecular architectures and functional properties of gellan gum and related polysaccharides", *Trends Food Sci. Technol.,* vol. 6, pp. 143-148, 1995. [http://dx.doi.org/10.1016/S0924-2244(00)89022-6]

[46] C. Liu, J. Lu, L. Lu, Y. Liu, F. Wang, and M. Xiao, "Isolation, structural characterization and immunological activity of an exopolysaccharide produced by Bacillus licheniformis 83701", *Bioresour. Technol.,* vol. 101, no. 14, pp. 5528-5533, 2010. [http://dx.doi.org/10.1016/j.biortech.2010.01.151] [PMID: 20199860]

[47] F. Freitas, V.D. Alves, and M.A. Reis, "Advances in bacterial exopolysaccharides: from production to biotechnological applications", *Trends Biotechnol.,* vol. 29, no. 8, pp. 388-398, 2011. [http://dx.doi.org/10.1016/j.tibtech.2011.03.008] [PMID: 21561675]

[48] I.W. Sutherland, "Bacterial Surface Polysaccharides: Structure and Function", In: *Int. Rev. Cytol.,* G.H. Bourne, K.W. Jeon, M. Friedlander, Eds., vol. 113. 1st ed Academic Press, 1989, pp. 187-231.

[49] P.J. Wyatt, "Light scattering and the absolute characterization of macromolecules", *Anal. Chim. Acta,* vol. 272, pp. 1-40, 1993. [http://dx.doi.org/10.1016/0003-2670(93)80373-S]

[50] B.J. Berne, and R. Pecora, *Dynamic Light Scattering: With Applications to Chemistry, Biology, and Physics.* 2nd ed Dover Publications Inc.: Mineola, New York, 1976.

[51] B.T. Stokke, A. Elgsaeter, D.A. Brant, T. Kuge, and S. Kitamura, "Macromolecular cyclization of (1>6)-branched-(1>3)-beta-D-glucans observed after denaturation-renaturation of the triple-helical structure", *Biopolymers,* vol. 33, no. 1, pp. 193-198, 1993. [http://dx.doi.org/10.1002/bip.360330118] [PMID: 8427935]

[52] S. Kitamura, T. Hirano, K. Takeo, H. Fukada, K. Takahashi, B.H. Falch, and B.T. Stokke, "Conformational transitions of schizophyllan in aqueous alkaline solution", *Biopolymers,* vol. 39, pp. 407-416, 1996. [http://dx.doi.org/10.1002/(SICI)1097-0282(199609)39:3<407::AID-BIP12>3.0.CO;2-8]

[53] A.K. Vuppu, A.A. Garcia, and C. Vernia, "Tapping mode Atomic Force Microscopy of scleroglucan networks", *Biopolymers,* vol. 42, no. 1, pp. 89-100, 1997. [http://dx.doi.org/10.1002/(SICI)1097-0282(199707)42:1<89::AID-BIP8>3.0.CO;2-Y] [PMID: 9209159]

[54] J-Y. Qian, W. Chen, W-M. Zhang, and H. Zhang, "Adulteration identification of some fungal polysaccharides with SEM, XRD, IR and optical rotation: A primary approach", *Carbohydr. Polym.,* vol. 78, pp. 620-625, 2009. [http://dx.doi.org/10.1016/j.carbpol.2009.05.025]

[55] T. Funami, "Atomic Force Microscopy Imaging of Food Polysaccharides", *Food Sci. Technol. Res.,* vol. 16, pp. 1-12, 2010. [http://dx.doi.org/10.3136/fstr.16.1]

[56] M. Iijima, M. Shinozaki, T. Hatakeyama, M. Takahashi, and H. Hatakeyama, "AFM studies on gelation mechanism of xanthan gum hydrogels", *Carbohydr. Polym.,* vol. 68, pp. 701-707, 2007. [http://dx.doi.org/10.1016/j.carbpol.2006.08.004]

[57] V.J. Morris, A. Gromer, A.R. Kirby, R.J. Bongaerts, and A. Patrick Gunning, "Using AFM and force spectroscopy to determine pectin structure and (bio) functionality", *Food Hydrocoll.,* vol. 25, pp. 230-237, 2011. [http://dx.doi.org/10.1016/j.foodhyd.2009.11.015]

[58] J. Cerning, C. Bouillanne, M. Landon, and M. Desmazeaud, "Isolation and Characterization of Exopolysaccharides from Slime-Forming Mesophilic Lactic Acid Bacteria", *J. Dairy Sci.,* vol. 75, pp. 692-699, 1992.

[http://dx.doi.org/10.3168/jds.S0022-0302(92)77805-9]

[59] D. van den Berg, G.W. Robijn, A.C. Janssen, M. Giuseppin, R. Vreeker, J.P. Kamerling, J. Vliegenthart, A.M. Ledeboer, and C.T. Verrips, "Production of a Novel Extracellular Polysaccharide by Lactobacillus sake 01 and Characterization of the Polysaccharide", *Appl. Environ. Microbiol.*, vol. 61, no. 8, pp. 2840-2844, 1995.
[PMID: 16535092]

[60] R. Rizzotti, "Les agents de texture épaississants, gélifiants, stabilisants", *Industries alimentaires et agricoles,* vol. 111, pp. 563-573, 1994.

[61] R.G. Gilbert, "Molecular weight distributions in free-radical polymerizations: their cause and cure", *Trends Polym. Sci. (Regul. Ed.),* vol. 3, pp. 222-226, 1995.

[62] I.W. Sutherland, *Biotechnology of microbial exopolysaccharides.*vol. 9. Cambridge University Press, 1990.
[http://dx.doi.org/10.1017/CBO9780511525384]

[63] I. Boukari, J.L. Putaux, B. Cathala, A. Barakat, B. Saake, C. Rémond, M. ODonohue, and B. Chabbert, "in vitro model assemblies to study the impact of lignin-carbohydrate interactions on the enzymatic conversion of xylan", *Biomacromolecules,* vol. 10, no. 9, pp. 2489-2498, 2009.
[http://dx.doi.org/10.1021/bm9004518] [PMID: 19655790]

[64] L. Hilliou, F. Freitas, R. Oliveira, M.A. Reis, D. Lespineux, C. Grandfils, and V.D. Alves, "Solution properties of an exopolysaccharide from a Pseudomonas strain obtained using glycerol as sole carbon source", *Carbohydr. Polym.,* vol. 78, pp. 526-532, 2009.
[http://dx.doi.org/10.1016/j.carbpol.2009.05.011]

[65] T.H. Mourey, and H. Coll, "Size-exclusion chromatography with two-angle laser light-scattering (SEC-TALLS) of high molecular weight and branched polymers", *J. Appl. Polym. Sci.,* vol. 56, pp. 65-72, 1995.
[http://dx.doi.org/10.1002/app.1995.070560109]

[66] F. García-Ochoa, V.E. Santos, and A. Alcón, "Xanthan gum production: An unstructured kinetic model", *Enzyme Microb. Technol.,* vol. 17, pp. 206-217, 1995.
[http://dx.doi.org/10.1016/0141-0229(94)00009-G]

[67] R.S. Singh, G.K. Saini, and J.F. Kennedy, "Pullulan: Microbial sources, production and applications", *Carbohydr. Polym.,* vol. 73, no. 4, pp. 515-531, 2008.
[http://dx.doi.org/10.1016/j.carbpol.2008.01.003] [PMID: 26048217]

[68] S.C. Viñarta, M.M. Yossen, J.R. Vega, L.I. Figueroa, and J.I. Fariña, "Scleroglucan compatibility with thickeners, alcohols and polyalcohols and downstream processing implications", *Carbohydr. Polym.,* vol. 92, no. 2, pp. 1107-1115, 2013.
[http://dx.doi.org/10.1016/j.carbpol.2012.10.065] [PMID: 23399135]

[69] B.J. Catley, "Pullulan, a relationship between molecular weight and fine structure", *FEBS Lett.,* vol. 10, no. 3, pp. 190-193, 1970.
[http://dx.doi.org/10.1016/0014-5793(70)80450-1] [PMID: 11945391]

[70] M.R. Kasaai, "Calculation of viscometric constants, hydrodynamic volume, polymer-solvent interaction parameter, and expansion factor for three polysaccharides with different chain conformations", *Carbohydr. Res.,* vol. 343, no. 13, pp. 2266-2277, 2008.
[http://dx.doi.org/10.1016/j.carres.2008.04.013] [PMID: 18511026]

[71] M.T. Nickerson, A.T. Paulson, and F.R. Hallett, "Pre-gel solution properties of gellan polysaccharides: Effect of potassium and calcium ions on chain associations", *Food Res. Int.,* vol. 41, pp. 462-471, 2008.
[http://dx.doi.org/10.1016/j.foodres.2007.12.009]

[72] S. Simsek, B. Mert, O.H. Campanella, and B. Reuhs, "Chemical and rheological properties of bacterial succinoglycan with distinct structural characteristics", *Carbohydr. Polym.,* vol. 76, pp. 320-324, 2009.

[http://dx.doi.org/10.1016/j.carbpol.2008.10.033]

[73] V.D. Alves, F. Freitas, N. Costa, M. Carvalheira, R. Oliveira, M.P. Gonçalves, and M.A. Reis, "Effect of temperature on the dynamic and steady-shear rheology of a new microbial extracellular polysaccharide produced from glycerol byproduct", *Carbohydr. Polym.,* vol. 79, pp. 981-988, 2010. [http://dx.doi.org/10.1016/j.carbpol.2009.10.026]

[74] M. Moosavi-Nasab, A.R. Taherian, M. Bakhtiyari, A. Farahnaky, and H. Askari, "Structural and Rheological Properties of Succinoglycan Biogums Made from Low-Quality Date Syrup or Sucrose Using Agrobacterium radiobacter Inoculation", *Food Bioproc. Tech.,* vol. 5, pp. 638-647, 2010. [http://dx.doi.org/10.1007/s11947-010-0407-4]

[75] P.M. Doran, *Principios de ingeniería de los bioprocesos* 1st ed Acribia: Zaragoza, España, 1998.

[76] H.A. Barnes, "Rheology: Principles, Measurements and Applications: by Christopher W. Macosko, published by VCH Publishers, Weinheim, Germany, 550 pp., £66.00, ISBN 1-56081-579-5", *Powder Technol.,* vol. 86, p. 313, 1996. [http://dx.doi.org/10.1016/S0032-5910(96)90008-X]

[77] R.P. Chhabra, "Non-Newtonian Fluids: An Introduction", In: *Rheology of Complex Fluids.,* M.J. Krishnan, P.A. Deshpande, S.P. Kumar, Eds., 1st ed Springer: New York, 2010, pp. 3-34. [http://dx.doi.org/10.1007/978-1-4419-6494-6_1]

[78] R. Tuinier, P. Zoon, M.A. Stuart, G.J. Fleer, and C.G. de Kruif, "Concentration and shear-rate dependence of the viscosity of an exocellular polysaccharide", *Biopolymers,* vol. 50, no. 6, pp. 641-646, 1999. [http://dx.doi.org/10.1002/(SICI)1097-0282(199911)50:6<641::AID-BIP8>3.0.CO;2-D] [PMID: 10508966]

[79] M.I. Torino, G. Font de Valdez, and F. Mozzi, "Biopolymers from lactic acid bacteria. Novel applications in foods and beverages", *Front. Microbiol.,* vol. 6, p. 834, 2015. [http://dx.doi.org/10.3389/fmicb.2015.00834] [PMID: 26441845]

[80] B.T. Stokke, A. Elgsaeter, E.Ø. Bjørnestad, and T. Lund, "Rheology of xanthan and scleroglucan in synthetic seawater", *Carbohydr. Polym.,* vol. 17, pp. 209-220, 1992. [http://dx.doi.org/10.1016/0144-8617(92)90006-C]

[81] N.E. Rodgers, "Scleroglucan", In: *Industrial Gums: Polysaccharides and Their Derivatives.,* R. Whistler, Ed., 2nd ed Academic Press Inc.: New York, U.S.A., 1973, pp. 499-511. [http://dx.doi.org/10.1016/B978-0-12-746252-3.50027-7]

[82] J. Ferguson, and Z. Kemblowski, *Applied fluid rheology.* 1st ed Springer Netherlands: London, 1991.

[83] H.W. Oviatt Jr, and D.A. Brant, "Thermal treatment of semi-dilute aqueous xanthan solutions yields weak gels with properties resembling hyaluronic acid", *Int. J. Biol. Macromol.,* vol. 15, no. 1, pp. 3-10, 1993. [http://dx.doi.org/10.1016/S0141-8130(05)80081-9] [PMID: 8443129]

[84] H. W. Oviatt, and D. A. Brant, "Viscoelastic behavior of thermally treated aqueous xanthan solutions in the semidilute concentration regime", *Macromolecules,* vol. 27, pp. 2402-2408, 1994. [http://dx.doi.org/10.1021/ma00087a007]

[85] S.C. Viñarta, O.D. Delgado, L.I. Figueroa, and J.I. Fariña, "Effects of thermal, alkaline and ultrasonic treatments on scleroglucan stability and flow behavior", *Carbohydr. Polym.,* vol. 94, no. 1, pp. 496-504, 2013. [http://dx.doi.org/10.1016/j.carbpol.2013.01.063] [PMID: 23544567]

[86] I. Capron, G. Brigand, and G. Muller, "Thermal denaturation and renaturation of a fermentation broth of xanthan: rheological consequences", *Int. J. Biol. Macromol.,* vol. 23, no. 3, pp. 215-225, 1998. [http://dx.doi.org/10.1016/S0141-8130(98)00050-6] [PMID: 9777709]

[87] E. Pelletier, C. Viebke, J. Meadows, and P.A. Williams, "A rheological study of the order-disorder conformational transition of xanthan gum", *Biopolymers,* vol. 59, no. 5, pp. 339-346, 2001.

[http://dx.doi.org/10.1002/1097-0282(20011015)59:5<339::AID-BIP1031>3.0.CO;2-A] [PMID: 11514937]

[88] N.A. Castillo, and J.I. Fariña, "Scleroglucans produced at bioreactor scale with alternative carbon sources: Differential rheological behavior against alkali, temperature and chaotrops", *II Workshop Polímeros Biodegradables y Biocompuestos. III Workshop of the project BIOPURFIL, Bio-based Polyurethane Composites with Natural Fillers,* 2015 Buenos Aires, Argentina

[89] E.R. Morris, D.A. Rees, G.R. Sanderson, and D. Thom, "Conformation and circular dichroism of uronic acid residues in glycosides and polysaccharides", *J. Chem. Soc., Perkin Trans. 2,* pp. 1418-1425, 1975.
[http://dx.doi.org/10.1039/p29750001418]

[90] L. Bozzi, M. Milas, and M. Rinaudo, "Characterization and solution properties of a new exopolysaccharide excreted by the bacterium Alteromonas sp. strain 1644", *Int. J. Biol. Macromol.,* vol. 18, no. 1-2, pp. 9-17, 1996.
[http://dx.doi.org/10.1016/0141-8130(95)01038-6] [PMID: 8852748]

[91] M. Loaëc, R. Olier, and J. Guezennec, "Uptake of lead, cadmium and zinc by a novel bacterial exopolysaccharide", *Water Res.,* vol. 31, pp. 1171-1179, 1997.
[http://dx.doi.org/10.1016/S0043-1354(96)00375-2]

[92] Y. Isobe, K. Endo, and H. Kawai, "Properties of a highly viscous polysaccharide produced by a bacillus strain isolated from soil", *Biosci. Biotechnol. Biochem.,* vol. 56, no. 4, pp. 636-639, 1992.
[http://dx.doi.org/10.1271/bbb.56.636] [PMID: 27280661]

[93] R. Clark, "Extensional viscosity of some food hydrocolloids", In: *Gums and stabilisers for the food industry.,* G.O. Phillips, D.J. Wedlock, P.A. Williams, Eds., vol. 6. 1st ed IRL Press: Oxford, 1992, pp. 73-85.

[94] M.V. Pastor, E. Costell, L. Izquierdo, and L. Durán, "Effects of concentration, pH and salt content on flow characteristics of xanthan gum solutions", *Food Hydrocoll.,* vol. 8, pp. 265-275, 1994.
[http://dx.doi.org/10.1016/S0268-005X(09)80339-1]

[95] R. Chandrasekaran, and V.G. Thailambal, "The influence of calcium ions, acetate and l-glycerate groups on the gellan double-helix", *Carbohydr. Polym.,* vol. 12, pp. 431-442, 1990.
[http://dx.doi.org/10.1016/0144-8617(90)90092-7]

[96] S.B. Ross-Murphy, K.P. Shatwell, I.W. Sutherland, and I.C. Dea, "Influence of acyl substituents on the interaction of xanthans with plant polysaccharides", *Food Hydrocoll.,* vol. 10, pp. 117-122, 1996.
[http://dx.doi.org/10.1016/S0268-005X(96)80062-2]

[97] G. Morris, and S. Harding, "Polysaccharides, Microbial", In: *Encyclopedia of Microbiology.,* M. Schaechter, Ed., 3rd ed Academic Press: Oxford, 2009, pp. 482-494.
[http://dx.doi.org/10.1016/B978-012373944-5.00135-8]

[98] C. Ramalingam, J. Priya, and S. Mundra, "Applications of microbial polysaccharides in food industry", *Int. J. Pharm. Sci. Rev. Res.,* vol. 27, pp. 322-324, 2014.

[99] K.I. Draget, O. Smidsrød, and G. Skjåk-Bræk, "Alginates from Algae", In: *Biopolymers Online,* A. Steinbüchel, Ed., 1st ed Wiley-VCH Verlag GmbH and Co. KGaA, 2005.
[http://dx.doi.org/10.1002/3527600035.bpol6008]

[100] W. Sabra, A-P. Zeng, and W-D. Deckwer, "Bacterial alginate: physiology, product quality and process aspects", *Appl. Microbiol. Biotechnol.,* vol. 56, no. 3-4, pp. 315-325, 2001.
[http://dx.doi.org/10.1007/s002530100699] [PMID: 11548998]

[101] E. Galindo, C. Peña, C. Núñez, D. Segura, and G. Espín, "Molecular and bioengineering strategies to improve alginate and polydydroxyalkanoate production by Azotobacter vinelandii", *Microb. Cell Fact.,* vol. 6, p. 7, 2007.
[http://dx.doi.org/10.1186/1475-2859-6-7] [PMID: 17306024]

[102] B.H. Rehm, "Bacterial polymers: biosynthesis, modifications and applications", *Nat. Rev. Microbiol.,*

vol. 8, no. 8, pp. 578-592, 2010.
[http://dx.doi.org/10.1038/nrmicro2354] [PMID: 20581859]

[103] G. Skjåk-Bræk, H. Grasdalen, and O. Smidsrød, "Inhomogeneous polysaccharide ionic gels", *Carbohydr. Polym.,* vol. 10, pp. 31-54, 1989.
[http://dx.doi.org/10.1016/0144-8617(89)90030-1]

[104] I. Donati, and S. Paoletti, "Material Properties of Alginates", In: *Alginates: Biology and Applications.,* B.H. Rehm, Ed., 1st ed Springer-Verlag Berlin Heidelberg: Berlin, Heidelberg, 2009, pp. 1-53.
[http://dx.doi.org/10.1007/978-3-540-92679-5_1]

[105] C. Peña, L. Hernández, and E. Galindo, "Manipulation of the acetylation degree of Azotobacter vinelandii alginate by supplementing the culture medium with 3-(N-morpholino)-propane-sulfonic acid", *Lett. Appl. Microbiol.,* vol. 43, no. 2, pp. 200-204, 2006.
[http://dx.doi.org/10.1111/j.1472-765X.2006.01925.x] [PMID: 16869905]

[106] A. Díaz-Barrera, P. Silva, J. Berrios, and F. Acevedo, "Manipulating the molecular weight of alginate produced by Azotobacter vinelandii in continuous cultures", *Bioresour. Technol.,* vol. 101, no. 23, pp. 9405-9408, 2010.
[http://dx.doi.org/10.1016/j.biortech.2010.07.038] [PMID: 20675122]

[107] I. Gaytán, C. Peña, C. Núñez, M.S. Córdova, G. Espín, and E. Galindo, "Azotobacter vinelandii lacking the Na(+)-NQR activity: a potential source for producing alginates with improved properties and at high yield", *World J. Microbiol. Biotechnol.,* vol. 28, no. 8, pp. 2731-2740, 2012.
[http://dx.doi.org/10.1007/s11274-012-1084-4] [PMID: 22806199]

[108] J.W. Lee, and D.F. Day, "Bioacetylation of seaweed alginate", *Appl. Environ. Microbiol.,* vol. 61, no. 2, pp. 650-655, 1995.
[PMID: 16534934]

[109] I.D. Hay, Z. Ur Rehman, M.F. Moradali, Y. Wang, and B.H. Rehm, "Microbial alginate production, modification and its applications", *Microb. Biotechnol.,* vol. 6, no. 6, pp. 637-650, 2013.
[PMID: 24034361]

[110] O. Smidsrød, and K.I. Draget, "Alginate gelation technologies", In: *Proc. from Food Colloids, Proteins, Lipids and Polysaccharides*, Ystad, Sweden, 1996.

[111] B.L. Pasin, C. Azón, and A. Garriga, "Microencapsulación con alginato en alimentos. Técnicas y aplicaciones", *Revista venezolana de Ciencia y Tecnología de Alimentos,* vol. 3, pp. 130-151, 2012.

[112] C. Peña, T. Castillo, C. Núñez, and D. Segura, "Bioprocess Design: Fermentation Strategies for Improving the Production of Alginate and Poly-β-Hydroxyalkanoates (PHAs) by Azotobacter vinelandii", In: *Progress in Molecular and Environmental Bioengineering - From Analysis and Modeling to Technology Applications,* A. Carpi, Ed., InTech, 2011, pp. 217-242.
[http://dx.doi.org/10.5772/20393]

[113] D.J. McHugh, "Production, properties and uses of alginates. Production and utilization of products from commercial seaweeds. FAO", *Fish. Tech. Pap,* vol. 288, pp. 58-115, 1987.

[114] R. McDowell, "Applications of alginates", *Rev. Pure Appl. Chem.,* vol. 10, pp. 1-15, 1960.

[115] *Kelco Algin: Hydrophilic Derivatives of Alginic Acid for Scientific Water Control.*2nd ed Kelco, Division of Merck and Company: San Diego, 1976.

[116] A. King, "Brown seaweed extracts (alginates)", *Food Hydrocoll.,* vol. 2, pp. 115-188, 1983.

[117] I.W. Sutherland, "Novel and established applications of microbial polysaccharides", *Trends Biotechnol.,* vol. 16, no. 1, pp. 41-46, 1998.
[http://dx.doi.org/10.1016/S0167-7799(97)01139-6] [PMID: 9470230]

[118] F. Clementi, "Alginate production by Azotobacter vinelandii", *Crit. Rev. Biotechnol.,* vol. 17, no. 4, pp. 327-361, 1997.
[http://dx.doi.org/10.3109/07388559709146618] [PMID: 9397533]

[119] K.I. Draget, and C. Taylor, "Chemical, physical and biological properties of alginates and their biomedical implications", *Food Hydrocoll.*, vol. 25, pp. 251-256, 2011. [http://dx.doi.org/10.1016/j.foodhyd.2009.10.007]

[120] M. Mancini, M. Moresi, and R. Rancini, "Mechanical properties of alginate gels: empirical characterisation", *J. Food Eng.*, vol. 39, pp. 369-378, 1999. [http://dx.doi.org/10.1016/S0260-8774(99)00022-9]

[121] A.D. Augst, H.J. Kong, and D.J. Mooney, "Alginate hydrogels as biomaterials", *Macromol. Biosci.*, vol. 6, no. 8, pp. 623-633, 2006. [http://dx.doi.org/10.1002/mabi.200600069] [PMID: 16881042]

[122] R.M. Hernández, G. Orive, A. Murua, and J.L. Pedraz, "Microcapsules and microcarriers for in situ cell delivery", *Adv. Drug Deliv. Rev.*, vol. 62, no. 7-8, pp. 711-730, 2010. [http://dx.doi.org/10.1016/j.addr.2010.02.004] [PMID: 20153388]

[123] B. McNeil, D. Archer, I. Giavasis, and L. Harvey, *Microbial production of food ingredients, enzymes and nutraceuticals.* 1st ed Elsevier, 2013. [http://dx.doi.org/10.1533/9780857093547]

[124] I. Giavasis, "Production of microbial polysaccharides for use in food", In: *Microbial Production of Food Ingredients, Enzymes and Nutraceuticals.*, B. McNeil, D. Archer, I. Giavasis, L. Harvey, Eds., 1st ed Woodhead Publishing: UK, 2013, pp. 413-468. [http://dx.doi.org/10.1533/9780857093547.2.413]

[125] I.A. Brownlee, A. Allen, J.P. Pearson, P.W. Dettmar, M.E. Havler, M.R. Atherton, and E. Onsøyen, "Alginate as a source of dietary fiber", *Crit. Rev. Food Sci. Nutr.*, vol. 45, no. 6, pp. 497-510, 2005. [http://dx.doi.org/10.1080/10408390500285673] [PMID: 16183570]

[126] B. Yao, C. Ni, C. Xiong, C. Zhu, and B. Huang, "Hydrophobic modification of sodium alginate and its application in drug controlled release", *Bioprocess Biosyst. Eng.*, vol. 33, no. 4, pp. 457-463, 2010. [http://dx.doi.org/10.1007/s00449-009-0349-2] [PMID: 19578877]

[127] K.Y. Lee, and D.J. Mooney, "Alginate: properties and biomedical applications", *Prog. Polym. Sci.*, vol. 37, no. 1, pp. 106-126, 2012. [http://dx.doi.org/10.1016/j.progpolymsci.2011.06.003] [PMID: 22125349]

[128] S. Thomas, "Alginate dressings in surgery and wound management Part 1", *J. Wound Care,* vol. 9, no. 2, pp. 56-60, 2000. [http://dx.doi.org/10.12968/jowc.2000.9.2.26338] [PMID: 11933281]

[129] T. Andersen, B.L. Strand, K. Formo, E. Alsberg, and B.E. Christensen, "Alginates as biomaterials in tissue engineering", In: *Carbohydrate Chemistry: Chemical and Biological Approaches,* T. L. Amelia Pilar Rauter, Ed., vol. 37. 1st ed Royal Society of Chemistry: Cambridge, UK, 2011, pp. 227-258.

[130] C.A. Hoesli, K. Raghuram, R.L. Kiang, D. Mocinecová, X. Hu, J.D. Johnson, I. Lacík, T.J. Kieffer, and J.M. Piret, "Pancreatic cell immobilization in alginate beads produced by emulsion and internal gelation", *Biotechnol. Bioeng.*, vol. 108, no. 2, pp. 424-434, 2011. [http://dx.doi.org/10.1002/bit.22959] [PMID: 20939004]

[131] N. Kashef, Q. Behzadian-Nejad, S. Najar-Peerayeh, K. Mousavi-Hosseini, M. Moazzeni, and G.E. Djavid, "Synthesis and characterization of Pseudomonas aeruginosa alginate-tetanus toxoid conjugate", *J. Med. Microbiol.*, vol. 55, no. Pt 10, pp. 1441-1446, 2006. [http://dx.doi.org/10.1099/jmm.0.46696-0] [PMID: 17005795]

[132] H. Sun, H. Pan, Z. Yang, and M. Shi, "The immune response and protective efficacy of vaccination with oral microparticle Aeromonas sobria vaccine in mice", *Int. Immunopharmacol.*, vol. 7, no. 9, pp. 1259-1264, 2007. [http://dx.doi.org/10.1016/j.intimp.2007.05.003] [PMID: 17630205]

[133] D.K. Arora, R.P. Elander, and K.G. Mukerji, *Handbook of Applied Mycology.* 1st ed Taylor & Francis, 1992.

[134] D.N. Hon, "Chitin and chitosan: Medical applications", In: *Polysaccharides in Medicinal Applications.,* S. Dumitriu, Ed., 1st ed Taylor & Francis: New York, 1996, pp. 631-649.

[135] M. Rinaudo, "Chitin and chitosan: Properties and applications", *Prog. Polym. Sci.,* vol. 31, pp. 603-632, 2006. [http://dx.doi.org/10.1016/j.progpolymsci.2006.06.001]

[136] F.A. Sagheer, M.A. Al-Sughayer, S. Muslim, and M.Z. Elsabee, "Extraction and characterization of chitin and chitosan from marine sources in Arabian Gulf", *Carbohydr. Polym.,* vol. 77, pp. 410-419, 2009. [http://dx.doi.org/10.1016/j.carbpol.2009.01.032]

[137] I. Younes, and M. Rinaudo, "Chitin and chitosan preparation from marine sources. Structure, properties and applications", *Mar. Drugs,* vol. 13, no. 3, pp. 1133-1174, 2015. [http://dx.doi.org/10.3390/md13031133] [PMID: 25738328]

[138] S.A. White, P.R. Farina, and I. Fulton, "Production and isolation of chitosan from Mucor rouxii", *Appl. Environ. Microbiol.,* vol. 38, no. 2, pp. 323-328, 1979. [PMID: 518086]

[139] W.L. Teng, E. Khor, T.K. Tan, L.Y. Lim, and S.C. Tan, "Concurrent production of chitin from shrimp shells and fungi", *Carbohydr. Res.,* vol. 332, no. 3, pp. 305-316, 2001. [http://dx.doi.org/10.1016/S0008-6215(01)00084-2] [PMID: 11376610]

[140] C. Crestini, B. Kovac, and G. Giovannozzi-Sermanni, "Production and isolation of chitosan by submerged and solid-state fermentation from Lentinus edodes", *Biotechnol. Bioeng.,* vol. 50, no. 2, pp. 207-210, 1996. [http://dx.doi.org/10.1002/bit.260500202] [PMID: 18626937]

[141] T. Wu, S. Zivanovic, F.A. Draughon, W.S. Conway, and C.E. Sams, "Physicochemical properties and bioactivity of fungal chitin and chitosan", *J. Agric. Food Chem.,* vol. 53, no. 10, pp. 3888-3894, 2005. [http://dx.doi.org/10.1021/jf048202s] [PMID: 15884813]

[142] J. Synowiecki, and N.A. Al-Khateeb, "Production, properties, and some new applications of chitin and its derivatives", *Crit. Rev. Food Sci. Nutr.,* vol. 43, no. 2, pp. 145-171, 2003. [http://dx.doi.org/10.1080/10408690390826473] [PMID: 12705640]

[143] P. Zou, X. Yang, J. Wang, Y. Li, H. Yu, Y. Zhang, and G. Liu, "Advances in characterisation and biological activities of chitosan and chitosan oligosaccharides", *Food Chem.,* vol. 190, pp. 1174-1181, 2016. [http://dx.doi.org/10.1016/j.foodchem.2015.06.076] [PMID: 26213092]

[144] J. Synowiecki, and N.A. Al-Khateeb, "Mycelia of Mucor rouxii as a source of chitin and chitosan", *Food Chem.,* vol. 60, pp. 605-610, 1997. [http://dx.doi.org/10.1016/S0308-8146(97)00039-3]

[145] M-T. Yen, and J-L. Mau, "“Physico-chemical characterization of fungal chitosan from shiitake stipes,” LWT -", *Food Sci. Technol. (Campinas.),* vol. 40, pp. 472-479, 2007.

[146] D. Knorr, "Recovery and utilization of chitin and chitosan in food processing waste management", *Food Technol.,* 1991.

[147] G. Muñoz, C. Valencia, N. Valderruten, E. Ruiz-Durántez, and F. Zuluaga, "Extraction of chitosan from Aspergillus niger mycelium and synthesis of hydrogels for controlled release of betahistine", *React. Funct. Polym.,* vol. 91–92, pp. 1-10, 2015. [http://dx.doi.org/10.1016/j.reactfunctpolym.2015.03.008]

[148] R. Pichyangkura, and S. Chadchawan, "Biostimulant activity of chitosan in horticulture", *Sci. Hortic. (Amsterdam),* vol. 196, pp. 49-65, 2015. [http://dx.doi.org/10.1016/j.scienta.2015.09.031]

[149] S. Ahmed, and S. Ikram, "Chitosan & its derivatives: a review in recent innovations", *Int. J. Pharm.*

Sci. Res., vol. 6, pp. 14-30, 2015.

[150] J. Cai, J. Li, C. Wang, J. Lin, Y. Hu, J. Yang, Y. Du, and H. Zheng, "Parametric Optimization of Extracellular Chitin Deacetylase Production by Scopulariopsis brevicaulis", *Journal of Biocatalysis and Biotransformation,* vol. 2, 2013.

[151] C.G. Li, F. Wang, W.G. Peng, and Y.H. He, "Preparation of chitosan and epichlorohydrin cross-linked adsorbents and adsorption property of dyes", *Applied Mechanics and Materials,* vol. 423, pp. 584-587, 2013.

[152] X. Li, M. Min, N. Du, Y. Gu, T. Hode, M. Naylor, D. Chen, R.E. Nordquist, and W.R. Chen, "Chitin, chitosan, and glycated chitosan regulate immune responses: the novel adjuvants for cancer vaccine", *Clin. Dev. Immunol.,* vol. 2013, p. 387023, 2013. [http://dx.doi.org/10.1155/2013/387023] [PMID: 23533454]

[153] A. Abdulkarim, M.T. Isa, S. Abdulsalam, A.J. Muhammad, and A.O. Ameh, "Extraction and characterisation of chitin and chitosan from mussel shell", *Civil and Environmental Research,* vol. 3, pp. 108-114, 2013.

[154] R. Hussain, M. Iman, and T.K. Maji, "Determination of degree of deacetylation of chitosan and their effect on the release behavior of essential oil from chitosan and chitosan-gelatin complex microcapsules", *International Journal of Advanced Engineering Applications,* vol. 6, pp. 4-12, 2013.

[155] R. Riva, H. Ragelle, A. Rieux, N. Duhem, C. Jérôme, and V. Préat, "Chitosan and Chitosan Derivatives in Drug Delivery and Tissue Engineering", In: *Chitosan for Biomaterials II,* R. Jayakumar, M. Prabaharan, A.R.A. Muzzarelli, Eds., Springer Berlin Heidelberg: Berlin, Heidelberg, 2011, pp. 19-44. [http://dx.doi.org/10.1007/12_2011_137]

[156] M. Vakili, M. Rafatullah, B. Salamatinia, A.Z. Abdullah, M.H. Ibrahim, K.B. Tan, Z. Gholami, and P. Amouzgar, "Application of chitosan and its derivatives as adsorbents for dye removal from water and wastewater: a review", *Carbohydr. Polym.,* vol. 113, pp. 115-130, 2014. [http://dx.doi.org/10.1016/j.carbpol.2014.07.007] [PMID: 25256466]

[157] P. Sorlier, A. Denuzière, C. Viton, and A. Domard, "Relation between the Degree of Acetylation and the Electrostatic Properties of Chitin and Chitosan", *Biomacromolecules,* vol. 2, pp. 765-772, 2001.

[158] S. Jana, A. Saha, A.K. Nayak, K.K. Sen, and S.K. Basu, "Aceclofenac-loaded chitosan-tamarind seed polysaccharide interpenetrating polymeric network microparticles", *Colloids Surf. B Biointerfaces,* vol. 105, pp. 303-309, 2013. [http://dx.doi.org/10.1016/j.colsurfb.2013.01.013] [PMID: 23399430]

[159] M. Jaworska, K. Sakurai, P. Gaudon, and E. Guibal, "Influence of chitosan characteristics on polymer properties. I: Crystallographic properties", *Polym. Int.,* vol. 52, pp. 198-205, 2003. [http://dx.doi.org/10.1002/pi.1159]

[160] I. Aranaz, M. Mengibar, R. Harris, I. Panos, B. Miralles, N. Acosta, G. Galed, and A. Heras, "Functional Characterization of Chitin and Chitosan", *Curr. Chem. Biol.,* vol. 3, pp. 203-230, 2009.

[161] J.S. Piccin, M.L. Vieira, J.O. Gonçalves, G.L. Dotto, and L.A. Pinto, "Adsorption of FD&C Red No. 40 by chitosan: Isotherms analysis", *J. Food Eng.,* vol. 95, pp. 16-20, 2009. [http://dx.doi.org/10.1016/j.jfoodeng.2009.03.017]

[162] Y. Ren, H.A. Abbood, F. He, H. Peng, and K. Huang, "Magnetic EDTA-modified chitosan/SiO2/Fe3O4 adsorbent: Preparation, characterization, and application in heavy metal adsorption", *Chem. Eng. J.,* vol. 226, pp. 300-311, 2013. [http://dx.doi.org/10.1016/j.cej.2013.04.059]

[163] M. Dash, F. Chiellini, R.M. Ottenbrite, and E. Chiellini, "Chitosan—A versatile semi-synthetic polymer in biomedical applications", *Prog. Polym. Sci.,* vol. 36, pp. 981-1014, 2011. [http://dx.doi.org/10.1016/j.progpolymsci.2011.02.001]

[164] M.N. Ravi Kumar, "A review of chitin and chitosan applications", *React. Funct. Polym.,* vol. 46, pp.

1-27, 2000.
[http://dx.doi.org/10.1016/S1381-5148(00)00038-9]

[165] K.V. Harish Prashanth, and R.N. Tharanathan, "Chitin/chitosan: modifications and their unlimited application potential—an overview", *Trends Food Sci. Technol.*, vol. 18, pp. 117-131, 2007. [http://dx.doi.org/10.1016/j.tifs.2006.10.022]

[166] P.R. Austin, and C.J. Brine, Chitin films and fibers, 1977. U.S. Patent 4,029,727

[167] B. Sagar, P. Hamlyn, and D. Wales, Wound dressing, 1991.EP. Patent 0,460,774 A2

[168] W-S. Hung, C-L. Fang, C-H. Su, W-F. Lai, Y-C. Chang, and Y-H. Tsai, "Cytotoxicity and immunogenicity of SACCHACHITIN and its mechanism of action on skin wound healing", *J. Biomed. Mater. Res.*, vol. 56, no. 1, pp. 93-100, 2001. [http://dx.doi.org/10.1002/1097-4636(200107)56:1<93::AID-JBM1072>3.0.CO;2-B] [PMID: 11309795]

[169] H. Sashiwa, and S-i. Aiba, "Chemically modified chitin and chitosan as biomaterials", *Prog. Polym. Sci.*, vol. 29, pp. 887-908, 2004. [http://dx.doi.org/10.1016/j.progpolymsci.2004.04.001]

[170] R.A. Muzzarelli, *Natural chelating polymers: alginic acid, chitin and chitosan.*vol. 55. Pergamon Press: Oxford, 1973.

[171] L.A. Hadwiger, D.F. Kendra, B.W. Fristensky, and W. Wagoner, "Chitosan Both Activates Genes in Plants and Inhibits RNA Synthesis in Fungi", In: *Chitin in Nature and Technology,* R. Muzzarelli, C. Jeuniaux, G.W. Gooday, Eds., Springer US: Boston, MA, 1986, pp. 209-214. [http://dx.doi.org/10.1007/978-1-4613-2167-5_28]

[172] T. Funami, M. Funami, H. Yada, and Y. Nakao, "Rheological and thermal studies on gelling characteristics of curdlan", *Food Hydrocoll.*, vol. 13, pp. 317-324, 1999. [http://dx.doi.org/10.1016/S0268-005X(99)00014-4]

[173] I-L. Shih, J-Y. Yu, C. Hsieh, and J-Y. Wu, "Production and characterization of curdlan by Agrobacterium sp", *Biochem. Eng. J.*, vol. 43, pp. 33-40, 2009. [http://dx.doi.org/10.1016/j.bej.2008.08.006]

[174] M. Hisamatsu, A. Amemura, T. Matsuo, H. Matsuda, and T. Harada, "Cyclic (1→2)-β-d-Glucan and the Octasaccharide Repeating-unit of Succinoglycan Produced by Agrobacterium", *Microbiology,* vol. 128, pp. 1873-1879, 1982. [http://dx.doi.org/10.1099/00221287-128-8-1873]

[175] T. Harada, K. Fujimori, S. Hirose, and M. Masada, "Growth and β-Glucan 10C3K Production by a Mutant of Alcaligenes faecalis var. myxogenes in Defined Medium", *Agric. Biol. Chem.*, vol. 30, pp. 764-769, 1966. [http://dx.doi.org/10.1080/00021369.1966.10858682]

[176] K.R. Phillips, and H.G. Lawford, "Curdlan: Its properties and production in batch and continuous fermentations", In: *Microbial Polysaccharides.,* M.E. Bushell, Ed., vol. 18. 1st ed Elsevier: Amsterdam, 1983, pp. 201-229.

[177] I. Nakanishi, K. Kimura, and T. Kanamaru, "Studies on curdlan-type polysaccharide. I. Industrial production of curdlan-type polysaccharide", *J. Takeda. Res. Lab.,* vol. 51, pp. 99-108, 1992.

[178] M-K. Kim, K-E. Ryu, W-A. Choi, Y-H. Rhee, and I-Y. Lee, "Enhanced production of (1 → 3)-β-d-glucan by a mutant strain of Agrobacterium species", *Biochem. Eng. J.,* vol. 16, pp. 163-168, 2003. [http://dx.doi.org/10.1016/S1369-703X(03)00032-9]

[179] E. Toksoy Öner, "Microbial Production of Extracellular Polysaccharides from Biomass", In: *Pretreatment Techniques for Biofuels and Biorefineries.,* Z. Fang, Ed., 1st ed Springer-Verlag Berlin Heidelberg, 2013, p. 457. [http://dx.doi.org/10.1007/978-3-642-32735-3_2]

[180] C.S. Buller, Water insoluble polysaccharide polymer and method thereof, 1990.U.S. Patent 4,908,310

[181] W.J. Kenyon, and C.S. Buller, "Structural analysis of the curdlan-like exopolysaccharide produced by Cellulomonas flavigena KU", *J. Ind. Microbiol. Biotechnol.,* vol. 29, no. 4, pp. 200-203, 2002. [http://dx.doi.org/10.1038/sj.jim.7000277] [PMID: 12355320]

[182] F. Donot, A. Fontana, J.C. Baccou, and S. Schorr-Galindo, "Microbial exopolysaccharides: Main examples of synthesis, excretion, genetics and extraction", *Carbohydr. Polym.,* vol. 87, pp. 951-962, 2012.
[http://dx.doi.org/10.1016/j.carbpol.2011.08.083]

[183] M. McIntosh, B.A. Stone, and V.A. Stanisich, "Curdlan and other bacterial (1>3)-β-D-glucans", *Appl. Microbiol. Biotechnol.,* vol. 68, no. 2, pp. 163-173, 2005.
[http://dx.doi.org/10.1007/s00253-005-1959-5] [PMID: 15818477]

[184] R. Zhang, and K.J. Edgar, "Properties, chemistry, and applications of the bioactive polysaccharide curdlan", *Biomacromolecules,* vol. 15, no. 4, pp. 1079-1096, 2014.
[http://dx.doi.org/10.1021/bm500038g] [PMID: 24552241]

[185] R.H. Marchessault, and Y. Deslandes, "Fine structure of (1→3)-β-d-glucans: curdlan and paramylon", *Carbohydr. Res.,* vol. 75, pp. 231-242, 1979.
[http://dx.doi.org/10.1016/S0008-6215(00)84642-X]

[186] T. Harada, A. Koreeda, S. Sato, and N. Kasai, "Electron Microscopic Study on the Ultrastructure of Curldan Gel: Assembly and Dissociation of Fibrils by Heating", *J. Electron. Microsc.,* vol. 28, pp. 147-153, 1979.
[http://dx.doi.org/10.1093/oxfordjournals.jmicro.a050168]

[187] P.A. Sandford, "Exocellular, microbial polysaccharides", *Adv. Carbohydr. Chem. Biochem.,* vol. 36, pp. 265-313, 1979.
[http://dx.doi.org/10.1016/S0065-2318(08)60238-3] [PMID: 380276]

[188] S. Toyoda, and M. Kimura, Meat products and processed meat products coated with curdlan gel film, 2004.U.S. Patent 20,040,047,949

[189] R. Kamaguchi, and T. Shiomi, Heat resistant capsule and process for producing the same, 2005.CA. Patent 2,533,471 A1

[190] H. Saitô, Y. Yoshioka, N. Uehara, J. Aketagawa, S. Tanaka, and Y. Shibata, "Relationship between conformation and biological response for (1----3)-beta-D-glucans in the activation of coagulation factor G from limulus amebocyte lysate and host-mediated antitumor activity. Demonstration of single-helix conformation as a stimulant", *Carbohydr. Res.,* vol. 217, pp. 181-190, 1991.
[http://dx.doi.org/10.1016/0008-6215(91)84128-2] [PMID: 1797400]

[191] W.M. Kulicke, A.I. Lettau, and H. Thielking, "Correlation between immunological activity, molar mass, and molecular structure of different (1>3)-beta-D-glucans", *Carbohydr. Res.,* vol. 297, no. 2, pp. 135-143, 1997.
[http://dx.doi.org/10.1016/S0008-6215(96)00273-X] [PMID: 9060180]

[192] J. Behravan, B.S. Bazzaz, and Z. Salimi, "Optimization of dextran production by Leuconostoc mesenteroides NRRL B-512 using cheap and local sources of carbohydrate and nitrogen", *Biotechnol. Appl. Biochem.,* vol. 38, no. Pt 3, pp. 267-269, 2003.
[http://dx.doi.org/10.1042/BA20030084] [PMID: 12927026]

[193] Ž. Petronijević, S. Ristić, D. Pešić, and A. Šmelcerović, "Immobilization of dextransucrase on regenerated benzoyl cellulose carriers", *Enzyme Microb. Technol.,* vol. 40, pp. 763-768, 2007.
[http://dx.doi.org/10.1016/j.enzmictec.2006.06.008]

[194] V. Tirtaatmadja, D.E. Dunstan, and D.V. Boger, "Rheology of dextran solutions", *J. Nonnewton. Fluid Mech.,* vol. 97, pp. 295-301, 2001.
[http://dx.doi.org/10.1016/S0377-0257(00)00226-3]

[195] R. Mehvar, "Dextrans for targeted and sustained delivery of therapeutic and imaging agents", *J. Control. Release,* vol. 69, no. 1, pp. 1-25, 2000. [http://dx.doi.org/10.1016/S0168-3659(00)00302-3] [PMID: 11018543]

[196] F. Carrasco, E. Chornet, R.P. Overend, and J. Costa, "A generalized correlation for the viscosity of dextrans in aqueous solutions as a function of temperature, concentration, and molecular weight at low shear rates", *J. Appl. Polym. Sci.,* vol. 37, pp. 2087-2098, 1989. [http://dx.doi.org/10.1002/app.1989.070370801]

[197] H. Nomura, S. Koda, and F. Hattori, "Viscosity of aqueous solutions of polysaccharides and their carboxylate derivatives", *J. Appl. Polym. Sci.,* vol. 41, pp. 2959-2969, 1990. [http://dx.doi.org/10.1002/app.1990.070411133]

[198] M. Naessens, A. Cerdobbel, W. Soetaert, and E.J. Vandamme, "Leuconostoc dextransucrase and dextran: production, properties and applications", *J. Chem. Technol. Biotechnol.,* vol. 80, pp. 845-860, 2005. [http://dx.doi.org/10.1002/jctb.1322]

[199] M. Moritz, and M. Geszke-Moritz, "The newest achievements in synthesis, immobilization and practical applications of antibacterial nanoparticles", *Chem. Eng. J.,* vol. 228, pp. 596-613, 2013. [http://dx.doi.org/10.1016/j.cej.2013.05.046]

[200] R. Whistler, and J. Daniel, "Functions of polysaccharides in foods", In: *Food Additives.,* A.L. Branen, P.M. Davidson, S.Y. Salminen, Eds., Marcel Dekker: New York, NY, 1990, pp. 395-424.

[201] H. Mutlu, and M.A. Meier, "Castor oil as a renewable resource for the chemical industry", *Eur. J. Lipid Sci. Technol.,* vol. 112, pp. 10-30, 2010. [http://dx.doi.org/10.1002/ejlt.200900138]

[202] S.Y. Lee, "Bacterial polyhydroxyalkanoates", *Biotechnol. Bioeng.,* vol. 49, no. 1, pp. 1-14, 1996. [http://dx.doi.org/10.1002/(SICI)1097-0290(19960105)49:1<1::AID-BIT1>3.3.CO;2-1] [PMID: 18623547]

[203] L.L. Madison, and G.W. Huisman, "Metabolic engineering of poly(3-hydroxyalkanoates): from DNA to plastic", *Microbiol. Mol. Biol. Rev.,* vol. 63, no. 1, pp. 21-53, 1999. [PMID: 10066830]

[204] J. Rincones, A.F. Zeidler, M.C. Grassi, M.F. Carazzolle, and G.A. Pereira, "The golden bridge for nature: the new biology applied to bioplastics", *Polym. Rev. (Phila. Pa.),* vol. 49, pp. 85-106, 2009. [http://dx.doi.org/10.1080/15583720902834817]

[205] S. Philip, T. Keshavarz, and I. Roy, "Polyhydroxyalkanoates: biodegradable polymers with a range of applications", *J. Chem. Technol. Biotechnol.,* vol. 82, pp. 233-247, 2007. [http://dx.doi.org/10.1002/jctb.1667]

[206] A. M. Gumel, M. S. M. Annuar, and Y. Chisti, "Recent advances in the production, recovery and applications of polyhydroxyalkanoates", *J. Polym. Environ.,* vol. 21, pp. 580-605, 2013. [http://dx.doi.org/10.1007/s10924-012-0527-1]

[207] S.P. Valappil, A.R. Boccaccini, C. Bucke, and I. Roy, "Polyhydroxyalkanoates in gram-positive bacteria: insights from the genera Bacillus and Streptomyces", *Antonie van Leeuwenhoek,* vol. 91, no. 1, pp. 1-17, 2007. [http://dx.doi.org/10.1007/s10482-006-9095-5] [PMID: 17016742]

[208] S. Chanprateep, "Current trends in biodegradable polyhydroxyalkanoates", *J. Biosci. Bioeng.,* vol. 110, no. 6, pp. 621-632, 2010. [http://dx.doi.org/10.1016/j.jbiosc.2010.07.014] [PMID: 20719562]

[209] B. Gupta, N. Revagade, and J. Hilborn, "Poly(lactic acid) fiber: An overview", *Prog. Polym. Sci.,* vol. 32, pp. 455-482, 2007. [http://dx.doi.org/10.1016/j.progpolymsci.2007.01.005]

[210] M.F. Valero-Valdivieso, Y. Ortegón, and Y. Uscategui, "Biopolímeros: avances y perspectivas", *DYNA,* vol. 80, pp. 171-180, 2013.

[211] K. Grage, A.C. Jahns, N. Parlane, R. Palanisamy, I.A. Rasiah, J.A. Atwood, and B.H. Rehm, "Bacterial polyhydroxyalkanoate granules: biogenesis, structure, and potential use as nano-/micr--beads in biotechnological and biomedical applications", *Biomacromolecules,* vol. 10, no. 4, pp. 660-669, 2009.
[http://dx.doi.org/10.1021/bm801394s] [PMID: 19275166]

[212] D. Jendrossek, "Polyhydroxyalkanoate granules are complex subcellular organelles (carbonosomes)", *J. Bacteriol.,* vol. 191, no. 10, pp. 3195-3202, 2009.
[http://dx.doi.org/10.1128/JB.01723-08] [PMID: 19270094]

[213] B. Kessler, and B. Witholt, "Factors involved in the regulatory network of polyhydroxyalkanoate metabolism", *J. Biotechnol.,* vol. 86, no. 2, pp. 97-104, 2001.
[http://dx.doi.org/10.1016/S0168-1656(00)00404-1] [PMID: 11245898]

[214] A. Rudin, and P. Choi, "Biopolymers", In: *The Elements of Polymer Science & Engineering.,* A.R. Choi, Ed., 3rd ed Academic Press: Boston, 2013, pp. 521-535.
[http://dx.doi.org/10.1016/B978-0-12-382178-2.00013-4]

[215] B.H. Rehm, and A. Steinbüchel, "Biochemical and genetic analysis of PHA synthases and other proteins required for PHA synthesis", *Int. J. Biol. Macromol.,* vol. 25, no. 1-3, pp. 3-19, 1999.
[http://dx.doi.org/10.1016/S0141-8130(99)00010-0] [PMID: 10416645]

[216] L.T. Lim, R. Auras, and M. Rubino, "Processing technologies for poly(lactic acid)", *Prog. Polym. Sci.,* vol. 33, pp. 820-852, 2008.
[http://dx.doi.org/10.1016/j.progpolymsci.2008.05.004]

[217] M.M. Reddy, S. Vivekanandhan, M. Misra, S.K. Bhatia, and A.K. Mohanty, "Biobased plastics and bionanocomposites: Current status and future opportunities", *Prog. Polym. Sci.,* vol. 38, pp. 1653-1689, 2013.
[http://dx.doi.org/10.1016/j.progpolymsci.2013.05.006]

[218] S.J. Park, T.W. Kim, M.K. Kim, S.Y. Lee, and S.C. Lim, "Advanced bacterial polyhydroxyalkanoates: towards a versatile and sustainable platform for unnatural tailor-made polyesters", *Biotechnol. Adv.,* vol. 30, no. 6, pp. 1196-1206, 2012.
[http://dx.doi.org/10.1016/j.biotechadv.2011.11.007] [PMID: 22137963]

[219] A. Steinbüchel, and H.E. Valentin, "Diversity of bacterial polyhydroxyalkanoic acids", *FEMS Microbiol. Lett.,* vol. 128, pp. 219-228, 1995.
[http://dx.doi.org/10.1016/0378-1097(95)00125-O]

[220] R. Rai, T. Keshavarz, J.A. Roether, A.R. Boccaccini, and I. Roy, "Medium chain length polyhydroxyalkanoates, promising new biomedical materials for the future", *Mater. Sci. Eng. Rep.,* vol. 72, pp. 29-47, 2011.
[http://dx.doi.org/10.1016/j.mser.2010.11.002]

[221] A.J. Anderson, G.W. Haywood, and E.A. Dawes, "Biosynthesis and composition of bacterial poly(hydroxyalkanoates)", *Int. J. Biol. Macromol.,* vol. 12, no. 2, pp. 102-105, 1990.
[http://dx.doi.org/10.1016/0141-8130(90)90060-N] [PMID: 2078525]

[222] B. Laycock, P. Halley, S. Pratt, A. Werker, and P. Lant, "The chemomechanical properties of microbial polyhydroxyalkanoates", *Prog. Polym. Sci.,* vol. 39, pp. 397-442, 2014.
[http://dx.doi.org/10.1016/j.progpolymsci.2013.06.008]

[223] E. Akaraonye, T. Keshavarz, and I. Roy, "Production of polyhydroxyalkanoates: the future green materials of choice", *J. Chem. Technol. Biotechnol.,* vol. 85, pp. 732-743, 2010.
[http://dx.doi.org/10.1002/jctb.2392]

[224] Y.K. Leong, P.L. Show, C.W. Ooi, T.C. Ling, and J.C. Lan, "Current trends in polyhydroxyalkanoates (PHAs) biosynthesis: insights from the recombinant Escherichia coli", *J. Biotechnol.,* vol. 180, pp. 52-

65, 2014.
[http://dx.doi.org/10.1016/j.jbiotec.2014.03.020] [PMID: 24698847]

[225] K. Sudesh, H. Abe, and Y. Doi, "Synthesis, structure and properties of polyhydroxyalkanoates: biological polyesters", *Prog. Polym. Sci.,* vol. 25, pp. 1503-1555, 2000. [http://dx.doi.org/10.1016/S0079-6700(00)00035-6]

[226] J. Agus, P. Kahar, H. Abe, Y. Doi, and T. Tsuge, "Molecular weight characterization of poly[(R)---hydroxybutyrate] synthesized by genetically engineered strains of Escherichia coli", *Polym. Degrad. Stabil.,* vol. 91, pp. 1138-1146, 2006. [http://dx.doi.org/10.1016/j.polymdegradstab.2005.07.006]

[227] B. Vergara, M. López, and F. Pérez, "Una perspectiva de los polihidroxialcanoatos bacterianos", *Avance y Perspectiva,* vol. 4, pp. 1-15, 2012.

[228] A. Steinbüchel, and B. Füchtenbusch, "Bacterial and other biological systems for polyester production", *Trends Biotechnol.,* vol. 16, no. 10, pp. 419-427, 1998. [http://dx.doi.org/10.1016/S0167-7799(98)01194-9] [PMID: 9807839]

[229] N. Angelova, and D. Hunkeler, "Rationalizing the design of polymeric biomaterials", *Trends Biotechnol.,* vol. 17, no. 10, pp. 409-421, 1999. [http://dx.doi.org/10.1016/S0167-7799(99)01356-6] [PMID: 10481173]

[230] C. Chatzidoukas, G. Penloglou, and C. Kiparissides, "Development of a structured dynamic model for the production of polyhydroxybutyrate (PHB) in Azohydromonas lata cultures", *Biochem. Eng. J.,* vol. 71, pp. 72-80, 2013. [http://dx.doi.org/10.1016/j.bej.2012.11.015]

[231] L.R. Castilho, D.A. Mitchell, and D.M. Freire, "Production of polyhydroxyalkanoates (PHAs) from waste materials and by-products by submerged and solid-state fermentation", *Bioresour. Technol.,* vol. 100, no. 23, pp. 5996-6009, 2009. [http://dx.doi.org/10.1016/j.biortech.2009.03.088] [PMID: 19581084]

[232] R. Jain, S. Kosta, and A. Tiwari, *Polyhydroxyalkanoates: A way to sustainable development of bioplastics.*vol. 1. 1st ed 2010.

[233] T. Keshavarz, and I. Roy, "Polyhydroxyalkanoates: bioplastics with a green agenda", *Curr. Opin. Microbiol.,* vol. 13, no. 3, pp. 321-326, 2010. [http://dx.doi.org/10.1016/j.mib.2010.02.006] [PMID: 20227907]

[234] J.M. Luengo, B. García, A. Sandoval, G. Naharro, and E.R. Olivera, "Bioplastics from microorganisms", *Curr. Opin. Microbiol.,* vol. 6, no. 3, pp. 251-260, 2003. [http://dx.doi.org/10.1016/S1369-5274(03)00040-7] [PMID: 12831901]

[235] B. Bernier, "The production of polysaccharides by fungi active in the decomposition of wood and forest litter", *Can. J. Microbiol.,* vol. 4, no. 3, pp. 195-204, 1958. [http://dx.doi.org/10.1139/m58-020] [PMID: 13536903]

[236] G. Carolan, B.J. Catley, and F.J. McDougal, "The location of tetrasaccharide units in pullulan", *Carbohydr. Res.,* vol. 114, pp. 237-243, 1983. [http://dx.doi.org/10.1016/0008-6215(83)88190-7]

[237] W. Sowa, A. Blackwood, and G. Adams, "Neutral extracellular glucan of Pullularia pullulans (de Bary) Berkhout", *Can. J. Chem.,* vol. 41, pp. 2314-2319, 1963. [http://dx.doi.org/10.1139/v63-336]

[238] E. Hermanides-Nijhof, "Aureobasidium and allied genera", *Stud. Mycol.,* vol. 15, pp. 141-177, 1977.

[239] T.D. Leathers, "Biotechnological production and applications of pullulan", *Appl. Microbiol. Biotechnol.,* vol. 62, no. 5-6, pp. 468-473, 2003. [http://dx.doi.org/10.1007/s00253-003-1386-4] [PMID: 12830329]

[240] S. Wu, Z. Jin, J.M. Kim, Q. Tong, and H. Chen, "Downstream processing of pullulan from

fermentation broth", *Carbohydr. Polym.,* vol. 77, pp. 750-753, 2009. [http://dx.doi.org/10.1016/j.carbpol.2009.02.023]

[241] L. Jiang, "Optimization of fermentation conditions for pullulan production by Aureobasidium pullulan using response surface methodology", *Carbohydr. Polym.,* vol. 79, pp. 414-417, 2010. [http://dx.doi.org/10.1016/j.carbpol.2009.08.027]

[242] C. Fraser, and H. Jennings, "A glucan from Tremella mesenterica NRRL-Y6158", *Can. J. Chem.,* vol. 49, pp. 1804-1807, 1971. [http://dx.doi.org/10.1139/v71-297]

[243] F. Delben, A. Forabosco, M. Guerrini, G. Liut, and G. Torri, "Pullulans produced by strains of Cryphonectria parasitica—II. Nuclear magnetic resonance evidence", *Carbohydr. Polym.,* vol. 63, pp. 545-554, 2006. [http://dx.doi.org/10.1016/j.carbpol.2005.11.012]

[244] A. Forabosco, G. Bruno, L. Sparapano, G. Liut, D. Marino, and F. Delben, "Pullulans produced by strains of Cryphonectria parasitica—I. Production and characterisation of the exopolysaccharides", *Carbohydr. Polym.,* vol. 63, pp. 535-544, 2006. [http://dx.doi.org/10.1016/j.carbpol.2005.10.005]

[245] Z. Chi, and S. Zhao, "Optimization of medium and cultivation conditions for pullulan production by a new pullulan-producing yeast strain", *Enzyme Microb. Technol.,* vol. 33, pp. 206-211, 2003. [http://dx.doi.org/10.1016/S0141-0229(03)00119-4]

[246] R.A. Reis, C.A. Tischer, P.A. Gorin, and M. Iacomini, "A new pullulan and a branched (1>3)-, (1>6)-linked beta-glucan from the lichenised ascomycete Teloschistes flavicans", *FEMS Microbiol. Lett.,* vol. 210, no. 1, pp. 1-5, 2002. [PMID: 12023070]

[247] H. Bender, J. Lehmann, and K. Wallenfels, "[Pullulan, an extracellular glucan from Pullularia pullulans]", *Biochim. Biophys. Acta,* vol. 36, pp. 309-316, 1959. [http://dx.doi.org/10.1016/0006-3002(59)90172-6] [PMID: 13798777]

[248] Y. Tsujisaka, and M. Mitsuhashi, "Pullulan", In: *Industrial Gums. Polysaccharides and Their Derivatives,* R. L. W. J. N. BeMiller, Ed., Academic Press: San Diego, CA, 1993, pp. 447-460.3rd ed [http://dx.doi.org/10.1016/B978-0-08-092654-4.50020-6]

[249] K. Okada, M. Yoneyama, T. Mandai, H. Aga, S. Sakai, and T. Ichikawa, "Digestion and fermentation of pullulan", *Journal of Japanese Society of Nutrition and Food Science (Japan),* 1990.

[250] B.J. Wiley, D.H. Ball, S.M. Arcidiacono, S. Sousa, J.M. Mayer, and D.L. Kaplan, "Control of molecular weight distribution of the biopolymer pullulan produced byAureobasidium pullulans", *J. Environ. Polym. Degrad.,* vol. 1, pp. 3-9, 1993. [http://dx.doi.org/10.1007/BF01457648]

[251] T. Roukas, and F. Mantzouridou, "Effect of the aeration rate on pullulan production and fermentation broth rheological properties in an airlift reactor", *J. Chem. Technol. Biotechnol.,* vol. 76, pp. 371-376, 2001. [http://dx.doi.org/10.1002/jctb.391]

[252] K.C. Cheng, A. Demirci, and J.M. Catchmark, "Pullulan: biosynthesis, production, and applications", *Appl. Microbiol. Biotechnol.,* vol. 92, no. 1, pp. 29-44, 2011. [http://dx.doi.org/10.1007/s00253-011-3477-y] [PMID: 21800030]

[253] L.H. Gibson, and R.W. Coughlin, "Optimization of high molecular weight pullulan production by Aureobasidium pullulans in batch fermentations", *Biotechnol. Prog.,* vol. 18, no. 3, pp. 675-678, 2002. [http://dx.doi.org/10.1021/bp0200043] [PMID: 12052094]

[254] H-P. Seo, C-W. Son, C-H. Chung, D-I. Jung, S-K. Kim, R.A. Gross, D.L. Kaplan, and J-W. Lee, "Production of high molecular weight pullulan by Aureobasidium pullulans HP-2001 with soybean pomace as a nitrogen source", *Bioresour. Technol.,* vol. 95, no. 3, pp. 293-299, 2004.

[http://dx.doi.org/10.1016/j.biortech.2003.02.001] [PMID: 15288272]

[255] S. Prasongsuk, M.A. Berhow, C.A. Dunlap, D. Weisleder, T.D. Leathers, D.E. Eveleigh, and H. Punnapayak, "Pullulan production by tropical isolates of Aureobasidium pullulans", *J. Ind. Microbiol. Biotechnol.,* vol. 34, no. 1, pp. 55-61, 2007. [http://dx.doi.org/10.1007/s10295-006-0163-7] [PMID: 16909265]

[256] T.J. Pollock, L. Thorne, and R.W. Armentrout, "Isolation of new aureobasidium strains that produce high-molecular-weight pullulan with reduced pigmentation", *Appl. Environ. Microbiol.,* vol. 58, no. 3, pp. 877-883, 1992. [PMID: 16348676]

[257] Y. Lin, Z. Zhang, and J. Thibault, "Aureobasidium pullulans batch cultivations based on a factorial design for improving the production and molecular weight of exopolysaccharides", *Process Biochem.,* vol. 42, pp. 820-827, 2007. [http://dx.doi.org/10.1016/j.procbio.2007.01.013]

[258] M.R. Rekha, and C.P. Sharma, "Pullulan as a promising biomaterial for biomedical applications: A perspective", *Trends Biomater. Artif. Organs,* vol. 20, pp. 111-116, 2007.

[259] H. Hijiya, and M. Shiosaka, Adhesives and pastes, 1975.U.S. Patent 3,873,333

[260] H. Hijiya, and M. Shiosaka, Shaped bodies of pullulan esters and their use, 1975.U.S. Patent 3,871,892

[261] F. Fujita, K. Fukami, and M. Fujimoto, Pullulan aminoalkyl ether, 1979.U.S. Patent 4,167,623

[262] K. Kato, and M. Shiosaka, Hydrogenated pullulan, 1976.U.S. Patent 3,931,146

[263] K. Tsuji, M. Fujimoto, F. Masuko, and T. Nagase, Carboxylated pullulan and method for producing same, 1978.U.S. Patent 4,090,016

[264] A. Lazaridou, C.G. Biliaderis, and V. Kontogiorgos, "Molecular weight effects on solution rheology of pullulan and mechanical properties of its films", *Carbohydr. Polym.,* vol. 52, pp. 151-166, 2003. [http://dx.doi.org/10.1016/S0144-8617(02)00302-8]

[265] K.I. Shingel, "Current knowledge on biosynthesis, biological activity, and chemical modification of the exopolysaccharide, pullulan", *Carbohydr. Res.,* vol. 339, no. 3, pp. 447-460, 2004. [http://dx.doi.org/10.1016/j.carres.2003.10.034] [PMID: 15013381]

[266] T. Kimoto, T. Shibuya, and S. Shiobara, "Safety studies of a novel starch, pullulan: chronic toxicity in rats and bacterial mutagenicity", *Food Chem. Toxicol.,* vol. 35, no. 3-4, pp. 323-329, 1997. [http://dx.doi.org/10.1016/S0278-6915(97)00001-X] [PMID: 9207894]

[267] P.A. Gibbs, and R.J. Seviour, "Pullulan", In: *Polysaccharides in Medicinal Applications,* S. Dumitriu, Ed., Marcel Dekker, Inc.: New York, USA, 1996, pp. 59-86.

[268] S. Alban, A. Schauerte, and G. Franz, "Anticoagulant sulfated polysaccharides: Part I. Synthesis and structure–activity relationships of new pullulan sulfates", *Carbohydr. Polym.,* vol. 47, pp. 267-276, 2002. [http://dx.doi.org/10.1016/S0144-8617(01)00178-3]

[269] G. Masci, D. Bontempo, and V. Crescenzi, "Synthesis and characterization of thermoresponsive N-isopropylacrylamide/methacrylated pullulan hydrogels", *Polymer (Guildf.),* vol. 43, pp. 5587-5593, 2002. [http://dx.doi.org/10.1016/S0032-3861(02)00415-9]

[270] P. Sivakumar, and K.P. Rao, "The use of cholesteryl pullulan for the preparation of stable vincristine liposomes", *Carbohydr. Polym.,* vol. 51, pp. 327-332, 2003. [http://dx.doi.org/10.1016/S0144-8617(02)00187-X]

[271] M. Yoshida, Process for the production of a pullulan composition, 1983.UK Patent GB2109391

[272] K. Shingel, and P. Petrov, "Behavior of γ-ray-irradiated pullulan in aqueous solutions of cationic (cetyltrimethylammonium hydroxide) and anionic (sodium dodecyl sulfate) surfactants", *Colloid*

Polym. Sci., vol. 280, pp. 176-182, 2002.
[http://dx.doi.org/10.1007/s00396-001-0599-2]

[273] W. Kulicke, and T. Heinze, "Improvements in polysaccharides for use as blood plasma expanders", *Macromolecular symposia,* pp. 47-59, 2005.
[http://dx.doi.org/10.1002/masy.200590024]

[274] R. Yamaguchi, H. Iwai, Y. Otsuka, S. Yamamoto, K. Ueda, M. Usui, Y. Taniguchi, and T. Matuhasi, "Conjugation of Sendai virus with pullulan and immunopotency of the conjugated virus", *Microbiol. Immunol.*, vol. 29, no. 2, pp. 163-168, 1985.
[http://dx.doi.org/10.1111/j.1348-0421.1985.tb00815.x] [PMID: 2989661]

[275] M. Mitsuhashi, and S. Koyama, Process for the production of virus vaccine, 1987.U.S. Patent 4,659,569 A

[276] R.F. Childers, P.L. Oren, and W.M. Seidler, *A copolymer of ethyl acrylate and methyl methacrylate and a polysaccharide pullulan for coating tablets to mask the taste.*1991. *Google Patents*1991.

[277] H. Hosseinkhani, T. Aoyama, O. Ogawa, and Y. Tabata, "Liver targeting of plasmid DNA by pullulan conjugation based on metal coordination", *J. Control. Release,* vol. 83, no. 2, pp. 287-302, 2002.
[http://dx.doi.org/10.1016/S0168-3659(02)00201-8] [PMID: 12363454]

[278] M.R. Rekha, and C.P. Sharma, "Synthesis and evaluation of lauryl succinyl chitosan particles towards oral insulin delivery and absorption", *J. Control. Release,* vol. 135, no. 2, pp. 144-151, 2009.
[http://dx.doi.org/10.1016/j.jconrel.2009.01.011] [PMID: 19331862]

[279] K. Na, E.S. Lee, and Y.H. Bae, "Adriamycin loaded pullulan acetate/sulfonamide conjugate nanoparticles responding to tumor pH: pH-dependent cell interaction, internalization and cytotoxicity in vitro", *J. Control. Release,* vol. 87, no. 1-3, pp. 3-13, 2003.
[http://dx.doi.org/10.1016/S0168-3659(02)00345-0] [PMID: 12618018]

[280] H. Nogusa, H. Hamana, N. Uchida, R. Maekawa, and T. Yoshioka, "Improved in vivo antitumor efficacy and reduced systemic toxicity of carboxymethylpullulan-peptide-doxorubicin conjugates", *Jpn. J. Cancer Res.*, vol. 91, no. 12, pp. 1333-1338, 2000.
[http://dx.doi.org/10.1111/j.1349-7006.2000.tb00922.x] [PMID: 11123434]

[281] K. Akiyoshi, S. Kobayashi, S. Shichibe, D. Mix, M. Baudys, S.W. Kim, and J. Sunamoto, "Self-assembled hydrogel nanoparticle of cholesterol-bearing pullulan as a carrier of protein drugs: complexation and stabilization of insulin", *J. Control. Release,* vol. 54, no. 3, pp. 313-320, 1998.
[http://dx.doi.org/10.1016/S0168-3659(98)00017-0] [PMID: 9766251]

[282] U. Hasegawa, S.M. Nomura, S.C. Kaul, T. Hirano, and K. Akiyoshi, "Nanogel-quantum dot hybrid nanoparticles for live cell imaging", *Biochem. Biophys. Res. Commun.*, vol. 331, no. 4, pp. 917-921, 2005.
[http://dx.doi.org/10.1016/j.bbrc.2005.03.228] [PMID: 15882965]

[283] Y. Hasa, H. Tazaki, M. Ohnishi, and Y. Oda, "Preparation of anti-sticking substance for cooked noodles by fungal hydrolysis of potato pulp", *Food Biotechnol.*, vol. 20, pp. 263-274, 2006.
[http://dx.doi.org/10.1080/08905430600904484]

[284] Y. Sugawa-Katayama, F. Kondou, T. Mandai, and M. Yoneyama, "Effects of pullulan, polydextrose and pectin on cecal microflora", *Oyo Toshitsu Kagaku,* vol. 41, pp. 413-418, 1994.

[285] M.S. Deshpande, V.B. Rale, and J.M. Lynch, "Aureobasidium pullulans in applied microbiology: a status report", *Enzyme Microb. Technol.*, vol. 14, pp. 514-527, 1992.
[http://dx.doi.org/10.1016/0141-0229(92)90122-5]

[286] Y. Motozato, H. Ihara, T. Tomoda, and C. Hirayama, "Preparation and gel permeation chromatographic properties of pullulan spheres", *J. Chromatogr. A,* vol. 355, pp. 434-437, 1986.
[http://dx.doi.org/10.1016/S0021-9673(01)97349-2]

[287] S. Olmo, R. Gotti, M. Naldi, V. Andrisano, N. Calonghi, C. Parolin, L. Masotti, and V. Cavrini, "Analysis of human histone H4 by capillary electrophoresis in a pullulan-coated capillary, LC-ESI-MS

and MALDI-TOF-MS", *Anal. Bioanal. Chem.,* vol. 390, no. 7, pp. 1881-1888, 2008. [http://dx.doi.org/10.1007/s00216-008-1903-5] [PMID: 18274734]

[288] F. E. Halleck, *Polysaccharides and methods for production thereof,* 1967.

[289] M. Rinaudo, and M. Vincendon, "13C NMR structural investigation of scleroglucan", *Carbohydr. Polym.,* vol. 2, pp. 135-144, 1982. [http://dx.doi.org/10.1016/0144-8617(82)90059-5]

[290] N.A. Castillo, A.L. Valdez, and J.I. Fariña, "Microbial production of scleroglucan and downstream processing", *Front. Microbiol.,* vol. 6, p. 1106, 2015. [http://dx.doi.org/10.3389/fmicb.2015.01106] [PMID: 26528259]

[291] E.T. Reese, and M. Mandels, "Beta-D-1, 3 Glucanases in fungi", *Can. J. Microbiol.,* vol. 5, no. 2, pp. 173-185, 1959. [http://dx.doi.org/10.1139/m59-022] [PMID: 13638895]

[292] T. Coviello, A. Palleschi, M. Grassi, P. Matricardi, G. Bocchinfuso, and F. Alhaique, "Scleroglucan: a versatile polysaccharide for modified drug delivery", *Molecules,* vol. 10, no. 1, pp. 6-33, 2005. [http://dx.doi.org/10.3390/10010006] [PMID: 18007275]

[293] J.I. Fariña, F. Siñeriz, O.E. Molina, and N.I. Perotti, "High scleroglucan production by Sclerotium rolfsii: Influence of medium composition", *Biotechnol. Lett.,* vol. 20, pp. 825-831, 1998. [http://dx.doi.org/10.1023/A:1005351123156]

[294] J. Schmid, *Genetics of scleroglucan production by Sclerotium rolfsii.*Doktor der Ingenieurswissenschaften, Von der Fakultät III – Prozesswissenschaften: Technischen Universität Berlin, Berlin, 2008.

[295] J.I. Fariña, F. Siñeriz, O.E. Molina, and N.I. Perotti, "Low-cost method for the preservation of Sclerotium rolfsii Proimi F-6656: Inoculum standardization and its use in scleroglucan production", *Biotechnol. Tech.,* vol. 10, pp. 705-708, 1996. [http://dx.doi.org/10.1007/BF00168484]

[296] J.I. Fariña, V.E. Santos, N.I. Perotti, J.A. Casas, O.E. Molina, and F. García-Ochoa, "Influence of the nitrogen source on the production and rheological properties of scleroglucan produced by Sclerotium rolfsii ATCC 201126", *World J. Microbiol. Biotechnol.,* vol. 15, pp. 309-316, 1999. [http://dx.doi.org/10.1023/A:1008999001451]

[297] J. Schmid, V. Meyer, and V. Sieber, "Scleroglucan: biosynthesis, production and application of a versatile hydrocolloid", *Applied Microbiology and Biotechnology,* vol. 91, pp. 937-947, 2011.

[298] Y. Wang, and B. McNeil, "Dissolved oxygen and the scleroglucan fermentation process", *Biotechnol. Lett.,* vol. 17, pp. 257-262, 1995. [http://dx.doi.org/10.1007/BF01190633]

[299] B. Schilling, A. Henning, and U. Rau, "Repression of oxalic acid biosynthesis in the unsterile scleroglucan production process with Sclerotium rolfsii ATCC 15205", *Bioprocess Eng.,* vol. 22, pp. 51-55, 2000. [http://dx.doi.org/10.1007/PL00009101]

[300] A.L. Valdez, "Producción de escleroglucano por Sclerotium rolfsii ATCC 201126: Influencia de las condiciones de cultivo y su relación con la secreción de oxalato", In: *Licenciatura en Biotecnología Tesina de Grado.*PROIMI-CONICET, Universidad Nacional de Tucumán, Archivos de la Facultad de Bioquímica, Química y Farmacia, 2013.

[301] C.E. Montes de Oca, "Caracteristicas morfo-fisiologicas de Sclerotium rolfsii ATCC 201126 productor de biopolímero. Prospecciones biotecnológicas", In: *Licenciada en Biotecnología.*PROIMI-CONICET, Universidad Nacional de Tucumán, 2014.

[302] S.C. Viñarta, "Biopolímero escleroglucano: preparación, caracterización fisico-química y molecular, actividad biológica, relación estructura-función y potencial aplicación fármaco-industrial", In: *Tesis Doctoral, PROIMI, Universidad Nacional de Tucumán.*Tesis Doctoral: Tucumán, 2009.

[303] T. Yanaki, T. Kojima, and T. Norisuye, "Triple Helix of Scleroglucan in Dilute Aqueous Sodium Hydroxide", *Polym. J.,* vol. 13, pp. 1135-1143, 1981. [http://dx.doi.org/10.1295/polymj.13.1135]

[304] G. Brigand, "Scleroglucan", In: *Industrial Gums: Polysaccharides and Their Derivatives*Academic Press: New York, USA, 1993, pp. 461-472. [http://dx.doi.org/10.1016/B978-0-08-092654-4.50021-8]

[305] B.H. Falch, T. Espevik, L. Ryan, and B.T. Stokke, "The cytokine stimulating activity of (1>3)-beta-D-glucans is dependent on the triple helix conformation", *Carbohydr. Res.,* vol. 329, no. 3, pp. 587-596, 2000. [http://dx.doi.org/10.1016/S0008-6215(00)00222-6] [PMID: 11128587]

[306] S.C. Viñarta, O.E. Molina, L.I. Figueroa, and J.I. Fariña, "A further insight into the practical applications of exopolysaccharides from Sclerotium rolfsii", *Food Hydrocoll.,* vol. 20, pp. 619-629, 2006. [http://dx.doi.org/10.1016/j.foodhyd.2005.05.006]

[307] S.C. Viñarta, N.J. François, M.E. Daraio, L.I. Figueroa, and J.I. Fariña, "Sclerotium rolfsii scleroglucan: the promising behavior of a natural polysaccharide as a drug delivery vehicle, suspension stabilizer and emulsifier", *Int. J. Biol. Macromol.,* vol. 41, no. 3, pp. 314-323, 2007. [http://dx.doi.org/10.1016/j.ijbiomac.2007.04.001] [PMID: 17586039]

[308] I. Giavasis, "Bioactive fungal polysaccharides as potential functional ingredients in food and nutraceuticals", *Curr. Opin. Biotechnol.,* vol. 26, pp. 162-173, 2014. [http://dx.doi.org/10.1016/j.copbio.2014.01.010] [PMID: 24518400]

[309] I. Cottrell, "Industrial potential of fungal and bacterial polysaccharides", In: *Fungal Polysaccharides.,* P.A. Sandford, K. Matsuda, Eds., vol. 126. 1st ed American Chemical Society, 1980, pp. 251-270. [http://dx.doi.org/10.1021/bk-1980-0126.ch013]

[310] A. Palleschi, G. Bocchinfuso, T. Coviello, and F. Alhaique, "Molecular dynamics investigations of the polysaccharide scleroglucan: first study on the triple helix structure", *Carbohydr. Res.,* vol. 340, no. 13, pp. 2154-2162, 2005. [http://dx.doi.org/10.1016/j.carres.2005.06.026] [PMID: 16043160]

[311] M. Sletmoen, and B.T. Stokke, "Higher order structure of (1,3)-beta-D-glucans and its influence on their biological activities and complexation abilities", *Biopolymers,* vol. 89, no. 4, pp. 310-321, 2008. [http://dx.doi.org/10.1002/bip.20920] [PMID: 18186085]

[312] J.I. Fariña, S.C. Viñarta, M. Cattaneo, and L.I. Figueroa, "Structural stability of Sclerotium rolfsii ATCC 201126 beta-glucan with fermentation time: a chemical, infrared spectroscopic and enzymatic approach", *J. Appl. Microbiol.,* vol. 106, no. 1, pp. 221-232, 2009. [http://dx.doi.org/10.1111/j.1365-2672.2008.03995.x] [PMID: 19054236]

[313] T.L. Bluhm, Y. Deslandes, R.H. Marchessault, S. Pérez, and M. Rinaudo, "Solid-state and solution conformation of scleroglucan", *Carbohydr. Res.,* vol. 100, pp. 117-130, 1982. [http://dx.doi.org/10.1016/S0008-6215(00)81030-7]

[314] P. Davison, and E. Mentzer, "Polymer Flooding in North Sea Reservoirs", *Society of Petroleum Engineers Journal,* vol. 22, pp. 353-362, 1982. [http://dx.doi.org/10.2118/9300-PA]

[315] C. Noïk, and J. Lecourtier, "Studies on scleroglucan conformation by rheological measurements versus temperature up to 150°C", *Polymer (Guildf.),* vol. 34, pp. 150-157, 1993. [http://dx.doi.org/10.1016/0032-3861(93)90298-O]

[316] T. M. McIntire, and D. A. Brant, "Observations of the (1→3)-β-d-Glucan Linear Triple Helix to Macrocycle Interconversion Using Noncontact Atomic Force Microscopy", *J. Am. Chem. Soc.,* vol. 120, pp. 6909-6919, 1998. [http://dx.doi.org/10.1021/ja981203e]

[317] R. Nardin, and M. Vincendon, "Isotopic exchange study of the scleroglucan chain in solution", *Macromolecules,* vol. 22, pp. 3551-3554, 1989.

[318] P. Rapp, "1, 3-β-Glucanase, 1, 6-β-Glucanase and β-Glucosidase Activities of Sclerotium glucanicum: Synthesis and Properties", *J. Gen. Microbiol.,* vol. 135, pp. 2847-2858, 1989.

[319] G. Holzwarth, "Xanthan and scleroglucan: structure and use in enhanced oil recovery", *Developments in industrial microbiology,* vol. 26, pp. 271-280, 1984.

[320] A. Donche, A. Vaussard, and P. Isambourg, "Application of scleroglucan muds to drilling deviated wells", , 1994.

[321] R. Pirri, J. Gadioux, and R. Riveno, Scleroglucan gel applied to the oil industry, 1996.U.S. Patent 5,555,936

[322] M.S. Doster, A.J. Nute, and C.A. Christopher, Method of recovering petroleum from underground formations, 1984.U.S. Patent 4,457,372

[323] A. Vaussard, A. Ladret, and A. Donche, Scleroglucan based drilling mud, 1997.U.S. Patent 5,612,294

[324] P. Chaverot, F. Demangeon, and Y. Huet, Scleroglucane-added viscosity-controlled bituminous binder emulsion, 2001.CA. Patent 2,044,592

[325] R. Chester, J.D. Kildea, J. Wang, H.E. Bode, and X.H. Li, Recovery of alumina trihydrate during the bayer process using scleroglucan, 2012. U.S. Patent 20,120,305,837 A1

[326] R. Chester, J.D. Kildea, J. Wang, H.E. Bode, and X.H. Li, Recovery of alumina trihydrate during the Bayer process using scleroglucan, 2013.U.S. Patent 8,394,290

[327] M.L. Lovrecich, and G. Riccioni, Pharmaceutical tablets and capsule granulates of scleroglucan and active substance, 1991.U.S. Patent 5,068,111

[328] M.L. Lovrecich, and G. Riccioni, Pharmaceutical tablets and capsule granulates of scleroglucan and active substance, 1993.U.S. Patent 5,215,752

[329] S.C. Jong, and R. Donovick, "Antitumor and antiviral substances from fungi", *Adv. Appl. Microbiol.,* vol. 34, pp. 183-262, 1989.
[http://dx.doi.org/10.1016/S0065-2164(08)70319-8] [PMID: 2672723]

[330] T. Maier, K. Huber, U. Rau, and B. Schilling, Scleroglucans and cosmetic composition containing the new compounds, 2000.U.S. Patent 6,162,449

[331] T. Maier, K. Huber, U. Rau, and B. Schilling, Scleroglucans and cosmetic compositions containing the new compounds, 2002.U.S. Patent 6,369,217

[332] V. Kaur, M.B. Bera, P.S. Panesar, H. Kumar, and J.F. Kennedy, "Welan gum: microbial production, characterization, and applications", *Int. J. Biol. Macromol.,* vol. 65, pp. 454-461, 2014.
[http://dx.doi.org/10.1016/j.ijbiomac.2014.01.061] [PMID: 24508918]

[333] K.S. Kang, G.T. Veeder, P.J. Mirrasoul, T. Kaneko, and I.W. Cottrell, "Agar-like polysaccharide produced by a pseudomonas species: production and basic properties", *Appl. Environ. Microbiol.,* vol. 43, no. 5, pp. 1086-1091, 1982.
[PMID: 16346007]

[334] P-E. Jansson, B. Lindberg, and P.A. Sandford, "Structural studies of gellan gum, an extracellular polysaccharide elaborated by Pseudomonas elodea", *Carbohydr. Res.,* vol. 124, pp. 135-139, 1983.
[http://dx.doi.org/10.1016/0008-6215(83)88361-X]

[335] M.A. O'Neill, R.R. Selvendran, and V.J. Morris, "Structure of the acidic extracellular gelling polysaccharide produced by Pseudomonas elodea", *Carbohydr. Res.,* vol. 124, pp. 123-133, 1983.
[http://dx.doi.org/10.1016/0008-6215(83)88360-8]

[336] R. Chandrasekaran, A. Radha, and V.G. Thailambal, "Roles of potassium ions, acetyl and L-glyceryl groups in native gellan double helix: an X-ray study", *Carbohydr. Res.,* vol. 224, pp. 1-17, 1992.

[http://dx.doi.org/10.1016/0008-6215(92)84088-A] [PMID: 1591755]

[337] E. Dreveton, F. Monot, J. Lecourtier, D. Ballerini, and L. Choplin, "Influence of fermentation hydrodynamics on gellan gum physico-chemical characteristics", *J. Ferment. Bioeng.*, vol. 82, pp. 272-276, 1996.
[http://dx.doi.org/10.1016/0922-338X(96)88819-2]

[338] T. Kaneko, and K. Kang, "Agar-like polysaccharide produced by a Pseudomonas species: Taxonomical studies", In: *79th annual meeting of the American Society for Microbiology*, Washington DC, 1979.

[339] D.E. Pszczola, "Gellan gum wins IFT's food technology industrial achievement award", *Food Technol.,* vol. 47, pp. 94-96, 1993.

[340] K.S. Kang, G.T. Colegrove, and G.T. Veeder, Deacetylated polysaccharide S-60, 1982. U.S. Patent 4,326,052 A

[341] I.B. Bajaj, P.S. Saudagar, R.S. Singhal, and A. Pandey, "Statistical approach to optimization of fermentative production of gellan gum from Sphingomonas paucimobilis ATCC 31461", *J. Biosci. Bioeng.,* vol. 102, no. 3, pp. 150-156, 2006.
[http://dx.doi.org/10.1263/jbb.102.150] [PMID: 17046526]

[342] T. Osmałek, A. Froelich, and S. Tasarek, "Application of gellan gum in pharmacy and medicine", *Int. J. Pharm.,* vol. 466, no. 1-2, pp. 328-340, 2014.
[http://dx.doi.org/10.1016/j.ijpharm.2014.03.038] [PMID: 24657577]

[343] E. Morris, M. Gothard, M. Hember, C. Manning, and G. Robinson, "Conformational and rheological transitions of welan, rhamsan and acylated gellan", *Carbohydr. Polym.,* vol. 30, pp. 165-175, 1996.
[http://dx.doi.org/10.1016/S0144-8617(96)00059-8]

[344] F. Mazen, M. Milas, and M. Rinaudo, "Conformational transition of native and modified gellan", *Int. J. Biol. Macromol.,* vol. 26, no. 2-3, pp. 109-118, 1999.
[http://dx.doi.org/10.1016/S0141-8130(99)00070-7] [PMID: 10517517]

[345] E.R. Morris, K. Nishinari, and M. Rinaudo, "Gelation of gellan–a review", *Food Hydrocoll.,* vol. 28, pp. 373-411, 2012.
[http://dx.doi.org/10.1016/j.foodhyd.2012.01.004]

[346] Y. Yuguchi, M. Mimura, S. Kitamura, H. Urakawa, and K. Kajiwara, "Structural characteristics of gellan in aqueous solution", *Food Hydrocoll.,* vol. 7, pp. 373-385, 1993.
[http://dx.doi.org/10.1016/S0268-005X(09)80233-6]

[347] A.A. Ashtaputre, and A.K. Shah, "Studies on a Viscous, Gel-Forming Exopolysaccharide from Sphingomonas paucimobilis GS1", *Appl. Environ. Microbiol.,* vol. 61, no. 3, pp. 1159-1162, 1995.
[PMID: 16534966]

[348] M. Watase, and K. Nishinari, "Effect of potassium ions on the rheological and thermal properties of gellan gum gels", *Food Hydrocoll.,* vol. 7, pp. 449-456, 1993.
[http://dx.doi.org/10.1016/S0268-005X(09)80240-3]

[349] I.B. Bajaj, S.A. Survase, P.S. Saudagar, and R.S. Singhal, "Gellan gum: fermentative production, downstream processing and applications", *Food Technol. Biotechnol.,* vol. 45, p. 341, 2007.

[350] G.R. Sanderson, and R.C. Clark, "Laboratory-produced microbial polysaccharide has many potential food applications as a gelling, stabilizing and texturizing agent", *Food Technol.,* vol. 37, pp. 63-70, 1983.

[351] I. Giavasis, L.M. Harvey, and B. McNeil, "Gellan Gum", *Crit. Rev. Biotechnol.,* vol. 20, pp. 177-211, 2000.

[352] M. Papageorgiou, and S. Kasapis, "The effect of added sucrose and corn syrup on the physical properties of gellan—gelatin mixed gels", *Food Hydrocoll.,* vol. 9, pp. 211-220, 1995.
[http://dx.doi.org/10.1016/S0268-005X(09)80218-X]

[353] S. Kasapis, R. Abeysekera, N. Atkin, M. Deszczynski, and J.R. Mitchell, "Tangible evidence of the tranformation from enthalpic to entropic gellan networks at high levels of co-solute", *Carbohydr. Polym.*, vol. 50, pp. 259-262, 2002. [http://dx.doi.org/10.1016/S0144-8617(02)00038-3]

[354] V.D. Prajapati, G.K. Jani, B.S. Zala, and T.A. Khutliwala, "An insight into the emerging exopolysaccharide gellan gum as a novel polymer", *Carbohydr. Polym.*, vol. 93, no. 2, pp. 670-678, 2013. [http://dx.doi.org/10.1016/j.carbpol.2013.01.030] [PMID: 23499110]

[355] J.T. Oliveira, T.C. Santos, L. Martins, M.A. Silva, A.P. Marques, A.G. Castro, N.M. Neves, and R.L. Reis, "Performance of new gellan gum hydrogels combined with human articular chondrocytes for cartilage regeneration when subcutaneously implanted in nude mice", *J. Tissue Eng. Regen. Med.*, vol. 3, no. 7, pp. 493-500, 2009. [http://dx.doi.org/10.1002/term.184] [PMID: 19598145]

[356] D.F. Coutinho, S. Sant, M. Shakiba, B. Wang, M.E. Gomes, N.M. Neves, R.L. Reis, and A. Khademhosseini, "Microfabricated photocrosslinkable polyelectrolyte-complex of chitosan and methacrylated gellan gum", *J. Mater. Chem.*, vol. 22, no. 33, pp. 17262-17271, 2012. [http://dx.doi.org/10.1039/c2jm31374j] [PMID: 23293429]

[357] D. Shungu, M. Valiant, V. Tutlane, E. Weinberg, B. Weissberger, L. Koupal, H. Gadebusch, and E. Stapley, "GELRITE as an Agar Substitute in Bacteriological Media", *Appl. Environ. Microbiol.*, vol. 46, no. 4, pp. 840-845, 1983. [PMID: 16346398]

[358] A. Rozier, C. Mazuel, J. Grove, and B. Plazonnet, "Gelrite®: A novel, ion-activated, in-situ gelling polymer for ophthalmic vehicles. Effect on bioavailability of timolol", *Int. J. Pharm.*, vol. 57, pp. 163-168, 1989. [http://dx.doi.org/10.1016/0378-5173(89)90305-0]

[359] P. Moslemy, S.R. Guiot, and R.J. Neufeld, "Production of size-controlled gellan gum microbeads encapsulating gasoline-degrading bacteria", *Enzyme Microb. Technol.*, vol. 30, pp. 10-18, 2002. [http://dx.doi.org/10.1016/S0141-0229(01)00440-9]

[360] A.R. Johnsen, and U. Karlson, "Evaluation of bacterial strategies to promote the bioavailability of polycyclic aromatic hydrocarbons", *Appl. Microbiol. Biotechnol.*, vol. 63, no. 4, pp. 452-459, 2004. [http://dx.doi.org/10.1007/s00253-003-1265-z] [PMID: 14716468]

[361] M.A. O'Neill, R.R. Selvendran, V.J. Morris, and J. Eagles, "Structure of the extracellular polysaccharide produced by the bacterium Alcaligenes (ATCC 31555) species", *Carbohydr. Res.*, vol. 147, pp. 295-313, 1986. [http://dx.doi.org/10.1016/S0008-6215(00)90638-4]

[362] H. Li, H. Xu, H. Xu, S. Li, H-J. Ying, and P-K. Ouyang, "Enhanced welan gum production using a two-stage agitation speed control strategy in Alcaligenes sp. CGMCC2428", *Bioprocess Biosyst. Eng.*, vol. 34, no. 1, pp. 95-102, 2011. [http://dx.doi.org/10.1007/s00449-010-0450-6] [PMID: 20640447]

[363] H. Li, H. Xu, S. Li, H. Xu, C. Guo, H. Ying, and P. Ouyang, "Strain improvement and metabolic flux modeling of wild-type and mutant Alcaligenes sp. NX-3 for synthesis of exopolysaccharide welan gum", *Biotechnol. Bioprocess Eng.*, vol. 15, pp. 777-784, 2010.

[364] P-E. Jansson, B. Lindberg, G. Widmalm, and P.A. Sandford, "Structural studies of an extracellular polysaccharide (S-130) elaborated by Alcaligenes ATCC 31555", *Carbohydr. Res.*, vol. 139, pp. 217-223, 1985. [http://dx.doi.org/10.1016/0008-6215(85)90022-9]

[365] P.A. Sandford, I.W. Cottrell, and D.J. Pettitt, "Microbial polysaccharides: new products and their commercial applications", *Pure Appl. Chem.*, vol. 56, pp. 879-892, 1984. [http://dx.doi.org/10.1351/pac198456070879]

[366] R. Moorhouse, and J.L. Shim, Use of either regular or low-calcium heteropolysaccharide S-7 in explosives and compositions thereby produced, 1980.U.S. Patent 4,238,255

[367] J.A. Peik, S.M. Steenbergen, and H.R. Hayden, Heteropolysaccharide S-194, a process for producing it, and compositions containing it, 1983.EP. Patent 0,077,680 A2

[368] K.S. Kang, and G.T. Veeder, Heteropolysaccharide S-88, 1985.U.S. Patent 4,535,153

[369] J.A. Peik, S.M. Steenbergen, and G.T. Veeder, Heteropolysaccharide S-657, 1992. U.S. Patent 5,175,278

[370] P.A. Sandford, and J. Baird, "7 - Industrial Utilization of Polysaccharides", In: *Academic Press,* G. O. Aspinall, Ed., 1983, pp. 411-490. [http://dx.doi.org/10.1016/B978-0-12-065602-8.50012-1]

[371] A. Margaritis, and J.E. Zajic, "Mixing, mass transfer, and scale□up of polysaccharide fermentations", *Biotechnol. Bioeng.,* vol. 20, pp. 939-1001, 1978. [http://dx.doi.org/10.1002/bit.260200702]

[372] R. Silman, and P. Rogovin, "Continuous fermentation to produce xanthan biopolymer: effect of dilution rate", *Biotechnol. Bioeng.,* vol. 14, pp. 23-31, 1972. [http://dx.doi.org/10.1002/bit.260140104]

[373] M.C. Cadmus, S.P. Rogovin, K.A. Burton, J.E. Pittsley, C.A. Knutson, and A. Jeanes, "Colonial variation in Xanthomonas campestris NRRL B-1459 and characterization of the polysaccharide from a variant strain", *Can. J. Microbiol.,* vol. 22, no. 7, pp. 942-948, 1976. [http://dx.doi.org/10.1139/m76-136] [PMID: 963616]

[374] F. García-Ochoa, V.E. Santos, J.A. Casas, and E. Gómez, "Xanthan gum: production, recovery, and properties", *Biotechnol. Adv.,* vol. 18, no. 7, pp. 549-579, 2000. [http://dx.doi.org/10.1016/S0734-9750(00)00050-1] [PMID: 14538095]

[375] S. Ghazal, W. Elsayed, U. Badr, H. Gebreel, and K. Khalil, "Genetically modified strains of Xanthomonas campestris higher xanthan producer and capable to utilize whey", *Curr. Res. Bacteriol.,* vol. 4, pp. 44-62, 2011. [http://dx.doi.org/10.3923/crb.2011.44.62]

[376] B.T. Stokke, O. Smidsrød, and A. Elgsaeter, "Electron microscopy of native xanthan and xanthan exposed to low ionic strength", *Biopolymers,* vol. 28, pp. 617-637, 1989. [http://dx.doi.org/10.1002/bip.360280207]

[377] R. Moorhouse, M. Walkinshaw, and S. Arnott, "Xanthan gum: molecular conformation and interactions", *ACS Symp. Ser. Amer. Chem. Soc.,* 1977 [http://dx.doi.org/10.1021/bk-1977-0045.ch007]

[378] E. Morris, "Molecular origin of xanthan solution properties", *ACS Symp. Ser. no. 45, Amer. Chem. Soc.,* 1977pp. 81-89 Washington, DC

[379] B. Katzbauer, "Properties and applications of xanthan gum", *Polym. Degrad. Stabil.,* vol. 59, pp. 81-84, 1998. [http://dx.doi.org/10.1016/S0141-3910(97)00180-8]

[380] G. Sworn, "Xanthan Gum", [http://dx.doi.org/10.1533/9781845695873.186]

[381] A. Jeanes, J.E. Pittsley, and F.R. Senti, "Polysaccharide B-1459: A new hydrocolloid polyelectrolyte produced from glucose by bacterial fermentation", *J. Appl. Polym. Sci.,* vol. 5, pp. 519-526, 1961. [http://dx.doi.org/10.1002/app.1961.070051704]

[382] N.A. Morrison, R. Clark, T. Talashek, and C.R. Yuan, "New forms of xanthan gum with enhanced properties", *Spec. Publ. R. Soc. Chem.,* vol. 294, pp. 124-130, 2004.

[383] S. Rosalam, and R. England, "Review of xanthan gum production from unmodified starches by

Xanthomonas comprestris sp", *Enzyme Microb. Technol.,* vol. 39, pp. 197-207, 2006. [http://dx.doi.org/10.1016/j.enzmictec.2005.10.019]

CHAPTER 3

Analysis of Biopolymers from Brazilian Seeds

Joana Léa Meira Silveira[1,*], **Neoli Lucyszyn**[2], **Clayton Fernandes de Souza**[2] and **Carmen Lúcia de Oliveira Petkowicz**[1]

[1] *Biochemistry and Molecular Biology Department, Federal University of Paraná, Curitiba-Paraná, Brazil*

[2] *Chemistry Undergraduate Program, School of Education and Humanities, Pontifical Catholic University of Paraná, Curitiba-PR, Brazil*

Abstract: Polysaccharides, which can be obtained from a variety of different sources, have important industrial applications owing to their properties in solution. Considering the wide variety of plant species found in Brazilian flora, the achievement of these biopolymers from plant seeds is highly promising. In view of this potential, various research groups from Brazil have been investigating several plant species as a source of these natural polymers. These studies include structural elucidation, modification, and industrial application using several approaches. Therefore, the aim of this chapter is to review the results from studies of native Brazilian sources of polysaccharides, especially galactomannans and xyloglucans, to inspire commercial exploitation of these unconventional sources.

Keywords: Brazilian seeds, Carbohydrate, Galactomannan, Hemicellulose, HPSEC-MALLS, Hydrocolloids, Interaction, Leguminosae, Macromolecular analysis, NMR, Nonstarch polysaccharides, Physicochemical analysis, Polysaccharide derivatives, Polysaccharide interaction, Rheology, Structural analysis, Thickener agent, Vegetal polysaccharide, Viscosity, Xyloglucan.

INTRODUCTION

In the seeds of higher plants, polysaccharides have structural and/or storage functions. Of polysaccharides with storage functions, starch is mostly studied and its economically important. However, not all plant seeds have starch as the main storage polysaccharide, and these "nonstarch" polysaccharides include hemicelluloses, amyloids, and mucilage [1]. Such polysaccharides are stored outside the plasmalemma and are therefore called reserve celluloses, and are

* **Corresponding author Joana Léa Meira Silveira:** Biochemistry and Molecular Biology Department, Federal University of Paraná, Curitiba-PR, Brazil; E-mail: jlms12@yahoo.com

Martin Masuelli & Denis Renard (Eds.)

found mainly in seeds, roots, rhizomes, tubers, and bulbs [2]. Studies on the physiology of plants have revealed that the seed germination process is dependent on the type and monosaccharide content of the polysaccharides present in the seeds [3]. To begin seed germination, these polysaccharides are degraded by enzymatic hydrolysis. Moreover, this process and all biosynthesis mechanisms are dependent on the sequence of the oligosaccharides [4].

In general, to explore these polysaccharides, the biopolymers in the plant are first isolated, followed by elucidation of their structures and evaluation of their biochemistry and physiological properties. One advantage of these storage polysaccharides compared with structural polysaccharides in seed cell walls is an easier extraction process [5, 6]. Polysaccharides with good water affinity can be isolated from seeds by aqueous extraction. A simple extraction procedure includes enzymatic inactivation, exhaustive water extraction (cold and/or hot), and precipitation using organic solvents [7 - 9]. In addition, some authors [10, 11] have used new methods to improve the polysaccharide isolation process, including pressurized liquid extraction, ultrasonic-assisted extraction, and microwave-assisted extraction.

The goal of these extraction procedures is to obtain the water-soluble polysaccharides (called hydrocolloids), free of contaminants and in high yields [7, 12 - 18]. Hydrocolloids form colloidal systems in water usually exhibit highly viscous behavior, forming gels consisting of three-dimensional networks of polysaccharide chains [18 - 22]. The absence of toxicity of these biopolymers favors their use in the textile, pharmaceutical, biomedical, cosmetics, and food industries [5, 6, 18, 23 - 25].

The properties of this biopolymers class in solution are dependent on the polysaccharide structure and to determine the primary structures of the polymers, it is necessary to identify the carbohydrate components, their monosaccharide ratios, the manner in which they are linked, and their monosaccharide sequence [26 - 35].

These biopolymers are normally subjected to several methodologies and techniques for structural and physicochemical elucidation, including chromatography high-performance liquid chromatography (HPLC) and gas-liquid chromatography (GLC), mass spectrometry (MS), and spectroscopic methods [nuclear magnetic resonance (NMR), UV-Vis, and Fourier transformed infrared spectroscopy (FTIR)] to gain information about molar mass, dispersity, internal linkages, composition, and monosaccharide sequences [30, 33 - 38].

Due to the complexity of polysaccharides, chromatographic methods are the most commonly used techniques to obtain physicochemical parameters. Purified polysaccharides can be analyzed by multi-angle laser light scattering (MALLS) coupled to high-performance size-exclusion chromatography with refractive index detection (HPSEC-RI-MALLS), which permits the determination of molar mass, the presence of aggregates, and the homogeneity of the polymers. This technique has been considered an absolute method for characterization of high molar mass polymers [39 - 51].

Light scattering measurements are one of the most important techniques for characterizing dilute polymer solutions [40 - 42]. Static and dynamic laser light scattering (SLS and DLS, respectively) methods, which measure the patterns of light scattered from a polymer in solution, can be used to evaluate the hydrodynamic dimensions of a polymer, which varies with its structure, and thus provide qualitative and quantitative information on the architecture of macromolecules [35, 46, 49, 51].

GLC and HPLC are commonly used for the determination of mono- and oligosaccharides with high efficiencies in separation, quantification, and speed of analysis. Normally, the monosaccharide composition is obtained by acid hydrolysis of the polysaccharides and subsequent evaluation by HPLC. For GLC analysis, the polysaccharides are converted into volatile derivatives, such as methylated alditol acetates, which can be used to predict the presence and position of branches in polysaccharide structures [35, 47, 49].

In addition, 1D and 2D NMR techniques can be employed to obtain information on the pattern of branching, monosaccharide ratios, anomeric configuration (α and β), and oligosaccharide sequences in the complex structures of polysaccharides. The advantage of this technique is that it is nondestructive and requires only a small amount of sample [35, 48].

Many studies have revealed that almost all polysaccharides obtained in high concentrations from leguminous seeds are xyloglucans (XGs) and galactomannans (GMs). The diversity of sources allows for a high diversity of side chains on the XG and GM backbone, resulting in variable physicochemical properties [12 - 14, 20, 46, 52 - 54].

Different research groups from Brazil have evaluated the characteristics, including structural and physicochemical properties of XGs and GMs from several native Brazilian sources. Considering the potential of these biopolymers compared with commercial sources, we present the structural characterization, chemical modification, and possible industrial applications of these polysaccharides.

Structural Features of Xyloglucan Seeds

XGs are major hemicellulosic polysaccharides found in the primary cell walls of dicotyledons and non-graminaceus monocotyledons, and are present as a storage polysaccharide in the seeds of many dicotyledons [1].

The chemical structure of XG consists of a cellulosic backbone of D-glucose units linked by β-(1→4) glycosidic bonds, which can be substituted at O-6 by α-D-xylopyranose units. These latter units may be substituted at O-2 by β-D-galactopyranose units (Fig. **1**). The basic structure of storage XG is quite similar to that of the cell wall polysaccharide structure (hemicellulose), which generally presents a D-glucose, D-xylose, and D-galactose ratio of about 4:3:1. However, in the primary cell wall, the β-D-galactopyranose units may be substituted at O-2 by α-L-fucose units [55 - 65].

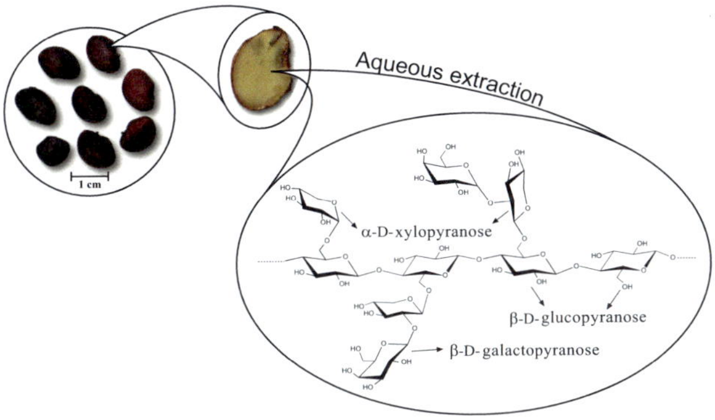

Fig. (1). Morphology and cross section of *Guibourtia hymenifolia* seeds and representation of a segment of a xyloglucan chain.

In Brazil, native species including *Hymenaea courbaril* [16, 22, 50, 51, 60 - 67], *Copaifera langsdorfii* [59, 68], *Mucuna sloanei* [69], and *Guibourtia hymenifolia* [48] are potential sources of XG for different applications. Tiné *et al.* [63] and Buckeridge *et al.* [59] studied the mobilization of XG cotyledon species from *H. courbaril* and *C. langsdorffii*, respectively. After germination, XG was degraded by hydrolases, such as β-galactosidase, endo-β-glucanase, α-xylosidase, and β-glucosidase, to produce oligosaccharides.

To facilitate these studies, representation of the monosaccharide sequences in the structure of XG has been standardized based on the degree of substitution on the main chain of the polysaccharide, allowing elucidation of the ratio of xylose, glucose, and galactose [70, 71]. The letter X specifies the glucose units of the main chain that are substituted by xylose, whereas the letter L designates the branches that contain galactose side chain terminal units and the letter G represents unsubstituted glucose [70]. For example, after enzymatic treatment of XG from *H. courbaril* seeds, Tiné *et al.* [65] demonstrated that the chain of this XG contains segments of type XXXG, XXXXG, and XXXXXG in a proportion of 2:1:0.2.

After exhaustive aqueous extraction from *C. langsdorffii* seeds at 25°C, Stupp *et al.* [68] obtained a yield of around 47% XG with a molar mass of 7.82×10^5 g/mol and radius of gyration (R_g) of 65 nm. The oligosaccharides obtained from this source had a composition similar to those from *Tamarindus indica*, but in different proportions and with differences in the fine structure of the biopolymers.

The seeds of *H. courbaril* (jatobá), also provided XG in high yields (around 45%), and methylation analysis indicated that these oligosaccharides had the same composition as that of XG from *T. indica* seeds. In water, this XG showed an intrinsic viscosity of 850 mL/g [60, 61]. In low concentrations, this XG presented near-Newtonian behavior (low viscosity), whereas at higher concentrations (>10 mg/mL), it behaved as a pseudoplastic [64].

Freitas *et al.* [46] demonstrated the presence of aggregates in solutions of XG from *H. courbaril* and found that heating the solution reduced aggregation. Moreover, the authors observed that the behavior of XG in solution is strongly affected by the presence of aggregates. A comparison of XG without preheating with that heated to 85°C for 2 h and cooled showed that heating reduced the molar mass for all samples tested, as determined by size-exclusion chromatography coupled with multi-angle laser light scattering (HPSEC-MALLS). The intrinsic viscosity [η], the absolute viscosity, and radius of gyration (R_g) were also lower for the heated and cooled XG solution. Within the physicochemical characteristic value ratio (C_∞), which is a measure of the short-range interactions that affect the internal rotation of the XG macromolecule, the presence of a unique series of XG oligosaccharides from jatobá seeds (XXXXG) probably generates a more rigid molecule compared with that from *T. indica.*

Using methylation and NMR analysis, Teixeira-Sá *et al.* [69] isolated and characterized a XG from *M. sloanei* seeds. The XG extracted with water accounted for 30% of the cotyledons. From the same seed, the author isolated a D-galactose binding lectin that interacts with the XG.

To study the structural features of the XG, many authors have used more extensively the NMR, providing information about the connectivity, molar ratio and distribution of the different carbohydrate unities that form the polysaccharide structure, with the advantage (in relation to others techniques) of being a non-destructive technique and requires small amounts of the sample. For example, the two-dimensional NMR technique was explored by Lucyszyn *et al.* [48] studied the structure of XG from *G. hymenifolia* seeds. After extraction procedures (yielding 54.2% wt), the mono-dimensional $^{1}H/^{13}C$ and the two-dimensional heteronuclear multiple-quantum correlation (HMQC) and homonuclear correlation spectroscopy (COSY) of purified XG was performed and the results are shown in Fig. (**2**).

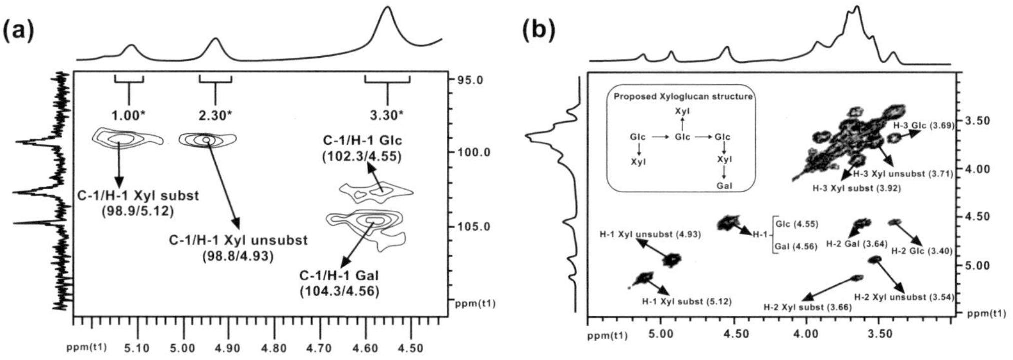

Fig. (2). (**a**) Two-dimensional NMR HMQC correlation spectrum of XG from *Guibourtia hymenifolia* seeds of anomeric region and (**b**) internal nucleous of XG obtained by COSY spectrum in D_2O at 70°C. (Glc: Glucose; Xyl: Xylose; Gal: Galactose). Note: the values indicate by * is relate of the Glc:Xyl:Gal molar ratio obtained by the GC-MS analysis of the alditol acetate of the methylated polysaccharide. Source: Lucyszyn *et al.* [48].

Studying the spectrum of ^{1}H RMN (data not shown), the integration of peaks of the anomeric region resulted in a glucose:xylose:galactose (Glc:Xyl:Gal) molar ratio of 3.3:2.3:1. The HMQC spectrum (Fig. **2a**) demonstrates $^{13}C/^{1}H$ signals with δ values of 104.3/4.56, 102.3/4.55, 98.9/5.12, and 98.8/4.93 ppm corresponding to β-D-galactopyranose, β-D-glucopyranose, and substituted and unsubstituted α-D-xylopyranose, respectively. The spectrum also confirmed the presence of four different linkages in the XG structure with chemical shifts at δ 104.3, 102.3, 98.9, and 98.8, which are assigned to the terminal β-D-galactopyranose and β-D-glucopyranose, and internal and terminal α-D- xylopyranose, respectively. The NMR assignments of sample *G. hymenifolia* present shifts similar to those obtained for other seed XGs such as *H. courbaril* [61] and *C. langsdorfii* [68]. These results of HMQC were confirmed by the authors [48], as observed in protons vicinal coupling obtained by COSY experiments (Fig. **2b**).

More recently, Arruda *et al.* [50] extracted polysaccharides from *H. courbaril* seeds with 0.1 M NaCl, followed by ethanol precipitation, to obtain XG with a yield of around 72% (*w/w*). The monosaccharide composition, as determined by GC/MS analysis, had a Glc:Xyl:Gal:Ara molar ratio of 40:34:20:6. 1D/2D NMR spectra confirmed that the structure of this XG has a central backbone composed of (1→4)-linked β-glucose units that partially branch at position 6 with non-reducing terminal units of α-xylose or β-galactose-(1→2)-α-xylose disaccharides.

Structural Features of Galactomannan Seeds

GMs are a water-soluble polysaccharides obtained from plant sources that are found mainly in the endosperm of plant seeds, mostly of the family Leguminosae Juss. or the Fabaceae Lindl., subfamilies Mimosaceae, Caesalpiniaceae, and Fabaceae [72, 73]. They are present in smaller amounts in species of the families Palmae, Annonaceae, Convolvulaceae, and Loganiaceae [73, 74], and are used as a reserve source of energy by plants during germination and as protection against dehydration [26].

With respect to their distribution in flora, GMs constitute the second largest group of storage polysaccharides [75].The generic GM structure consists of a main chain of D-mannopyranose linked by glycosidic β-(1→4) bonds that can besubstituted at O-6 by single units of α-D-galactopyranose (Fig. **3**). The ratio of mannose:galactose (Man:Gal) in GM ranges from 1.1:1 to 4:1 [73].

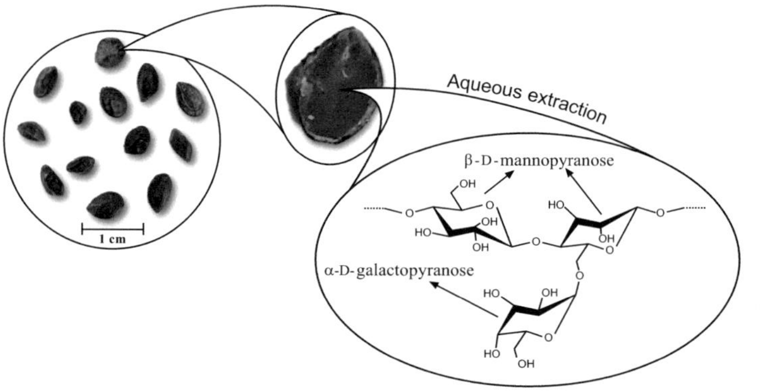

Fig. (3). Morphology and cross section of *Mimosa scabrella* var. *aspericarpa* seeds and representation of a segment of a galactomannan chain.

GM yields are very high in species of the Caesalpiniaceae and Mimosaceae subfamilies (15-38% of the seed dry mass) [73, 74]. The Man:Gal ratio, as well as the distribution of D-galactopyranose units along the main chain, may vary with the botanical source. The properties of these polysaccharides are related to their solubility in water, intermolecular interactions, and susceptibility to enzymatic degradation [30, 31, 73, 76 - 79].

The first Brazilian source of GM, described in the literature by the authors, was the seeds of *Mimosa scabrella* Bentham (bracatinga) [7 - 9, 28, 31, 76], which is a species of rapid growth, abundant in southern Brazil [80, 81]. The structure of the highly substituted GM obtained from these seeds was investigated (Man:Gal ~ 1.1:1, *i.e.*, for each 11 mannose units, 10 are replaced by galactose). Furthermore, solutions of the obtained GM exhibited high viscosities at relatively low concentrations, comparable to those of commercial GM (guar gum and carob) [30 - 32].

Considering the potential applications of this biopolymer, bracatinga seeds were chosen to obtain a pilot-scale product. The obtained GM had a yield of 20% of the seed weight, with 51% passing through a 100 mesh sieve, 10% moisture, a Man:Gal ratio of 1.1:1.0, and an intrinsic viscosity (η) of 740 mL/g, resulting in a product with characteristics similar to those of guar gum [8].

Vendruscolo *et al.* [9] have characterized the physicochemical and mechanical properties of GM from *M. scabrella* Bentham, obtained at pilot scale, using two different drying techniques: vacuum oven (GVO) and spray-drier (GSD). HPSEC-MALLS analysis suggested that both products behave as semi-flexible polymers, although the GSD sample showed more aggregation at the molecular level (10%) that was indicated for a higher chain stiffness (L_p= 9.1 nm) compared with GVO (L_p = 7.7 nm). Thermal analysis showed a weight loss event with peaks at 299.7 and 311.9°C for GVO and GSD, respectively, and a higher ash content in the GSD sample, in both inert and oxidant atmospheres. X-ray diffraction confirmed the amorphous nature of both galactomannans, although GVO showed high crystallinity.

In addition to *M. scabrella* Bentham, other sources, such as *M. scabrella* var. *aspericarpa* (bracatinga Argentina), *M. flocculosa*, *M. bimucronata*, and *M. taimbensis* were also investigated and found to provide high yields of GM. ^{13}CNMR spectra indicated that the obtained GM had similar chemical structures, with Man:Gal of ~1.1:1. Importantly, *M. taimbensis* and *M. flocculosa* are annual plants, like guar, which makes them, together with *M. scabrella*, viable alternatives for the production of GMs [82].

Stryphnodendron barbatiman, now classified as *Stryphnodendron adstringens* and known as barbartimão [83, 84], *Senna multijuga* known as sena [85], *Schizolobium amazonicum* known as pinho cuiabano [86, 87], *Schizolobium parahybae* known as guapuruvu [88, 89], *Cassia fastuosa* known as marimari or parica [23, 90], *Apuleia leiocarpa* known as grápia or jataí [18], *Senna macranthera* known as stick-bean or coffee senna [24], and *Caesalpinia ferrea* var. *ferrea* known as ironwood [35], are also part of the GM source group.

Galactomannan "Family" and the Distribution of Galactose

The occurrence of GM "families" in the same species was described for *S. amazonicum* and *S. parahybae*, both belonging to the Caesalpiniaceae family. Over five types of GM with different Man:Gal ratios were isolated from these species by aqueous extraction at different temperatures [33, 87]. In *S. multijuga*, different families were found in the aqueous and alkaline extractions (Man:Gal of 2.4:1 and 3.5:1, respectively) [85]. The same GM "families" were also observed in seeds of *Gleditsia triacanthos* [27]. Edwards *et al.* [29] suggested that these differences are due to the *in vivo* enzymatic activity of α-galactosidase.

Although there is agreement in the literature on the structure of GM [30], the distribution of galactose units along the main mannose chain is still debated [20, 36]. Degree of polydispersity (DP) of GM *S. amazonicum*, *S. parahybae*, and *C. fastuosa* were evaluated studying the oligosaccharide obtained by partial acid hydrolysis and ranged of the 2 to 6 [87]. Although the GM of *C. fastuosa* seeds had the same Man:Gal ratio as *Schizolobium* spp., the proportion of oligosaccharides (DP 2 to 6) was different, with a lower proportion of homo-oligosaccharides obtained from *C. fastuosa* when compared with *Schizolobium* spp. From these results, the authors proposed that fewer regions not substituted with galactose would be present along the main chain of the GM of *C. fastuosa*, suggesting a galactosyl group distribution along the mannose chain different from that observed for the two species of *Schizolobium*. Furthermore, the authors suggest that *S. parahybae* and *S. amazonicum* are the same species.

By evaluating the ^{1}H and ^{13}CNMR spectra, Souza *et al.* [35] obtained some structural information about the GM from *C. ferrea* var. *ferrea* seeds, as shown in Fig. (**4**). The integration of the peaks in anomeric region (Fig. **4b**) indicates that the GM from *C. ferrea* has a similar Man:Gal ratio as that from guar gum seeds (2.0:1), whereas from the HMQC spectrum (Fig. **4a**), based on literature data, the authors confirmed that the correlations correspond to those of a GM.

Based on the study of Dea and Morrison [73], the authors [35] estimated the distribution of α-D-galactopyranose in GM using the splitting of the *O*-substituted C-4 signals of β-mannopyranose in the ^{13}C NMR spectrum. The Man C-4 resonance region was split into three signals (Fig. **4c**). The first signal denoted I, at δ 77.5 ppm corresponds to two continuous *O*-substituted β-D-mannopyranose units; the second, denoted II, at δ 77.3 ppm corresponds to dyads in which only one of the two β-D-mannopyranose units is *O*-substituted; and the third, denoted III, at δ 77.0 ppm is assigned to non-substituted β-mannopyranose units that are adjacent to other units of the same monosaccharide. This result shows the dependence of the α-galactopyranose distribution along the β-mannan chain on the nearest neighbor probabilities. A comparison of the signal intensities shows that peak II in the spectrum is much more intense (54%) than peak III (24%) or peak I (22%) (Fig. **4d**), indicating that the galactose units are randomly distributed along the mannan chain.

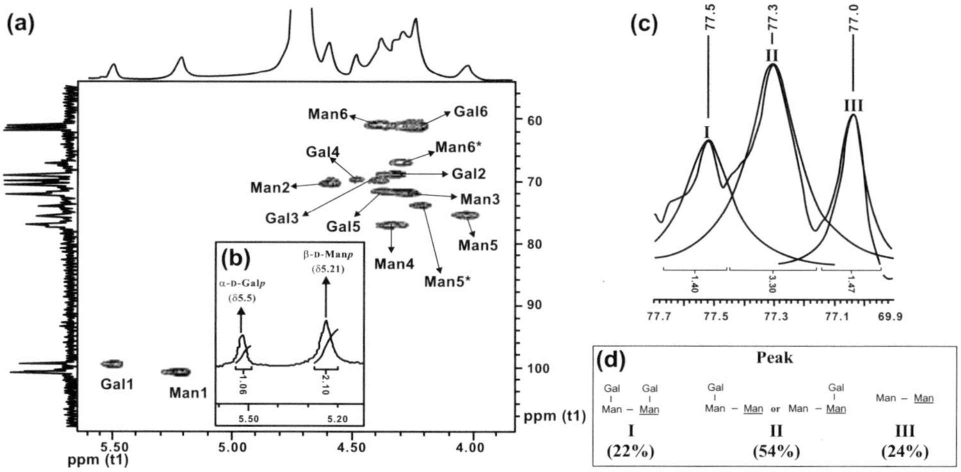

Fig. (4). (**a**) Two-dimensional NMR HMQC correlation spectrum, (**b**) anomeric region of the ^{1}H NMR spectrum, (**c**) splitting of the mannopyranose unities in C-4 region of the ^{13}C NMR spectrum and (**d**) distribution of α-D-galactopyranose on β-D-mannopyranose for galactomannan from *C. ferrea* var. *ferrea* seeds. Source: adapted from Souza *et al.* [35].

More recently, Bento *et al.* [91] subjected the endosperm of seeds from different subfamilies of Leguminosae to sequential aqueous and alkaline extractions. Seeds from the Caesalpinioideae subfamily provided GMs with different Man:Gal proportions. Moreover, a mannan was isolated from the immature endosperm of *Caesalpinia pulcherrima*, suggesting that there is no developmental relationship between GM and mannan. These results are consistent with the presence of a distinctive cell wall pattern in the endosperms of Leguminosae species.

Buckeridge and Dietrich [71] analyzed the yields and compositions of polysaccharides extracted with hot water from seeds of several Brazilian leguminous species. In 16 of the species studied, more than 10% of the dry weight of the seed was found to consist of polysaccharides, mainly mannose and galactose, with varying proportions of Man:Gal (1.4:1 to 4.4:1), which were detected by paper chromatography and GLC using the acid hydrolysates. Furthermore, high yields (~30%) of GM were observed from seeds of *Adenantera pavonina, Sesbania marginata*, *Sesbania punicea*, *Dimorphandra mollis*, *Stryphnodendron adstringens*, *Stryphnodendron polyphylum*, *Stryphnodendron obovatum*, *Senna alata*, *Senna bicapsularis*, *Senna spectabilis*, *Senna cernua*, and

Senna macranthera. This research has enabled studies on chemotaxonomy, where it was observed that the species of the Caesalpinaceae family had higher Man:Gal ratios compared with those of the species of the Mimosaceae and Fabaceae families, suggesting that GM is a factor for adaptation of these species to different climates [79]. The GM obtained from *Dimorphandra gardneriana* (fava danta) and *Dimorphandra mollis* (faveiro) seeds, which are Brazilian species that occur with high frequency in the cerrado and caatinga, were found to have Man:Gal ratios of around 1.8:1 and 2.7:1, respectively [34, 92].

The metabolism and degradation of several GMs from Brazilian seeds have been studied, for example, *S. virgate* (Fabaceae) [93, 94]. The seeds of *S. virgata* were observed to accumulate GM in the endosperm, which is degraded after germination and its degradation products are transferred to the growing embryo.

Galactomannan and Xyloglucan Properties in Solution

It is well known from the literature that there is a relation between the structural and physicochemical properties of natural polymers in aqueous solution. This relationship can be evaluated using theoretical models, as reported by Odijk [95] and Yamakawa and Fujii [96], who determined the radius of gyration (R_g) of polymeric chains and the intrinsic viscosity [η], and from these parameters, the persistence length (L_p).

Such parameters give information about stereo-regularity in the polymeric conformation, chain flexibility, and thermodynamic conditions, such as the presence of simple, double, or triple helices. These theoretically obtained parameters also can be obtained experimentally using MALLS and HPSEC-RI-MALLS techniques [9, 35, 42 - 44, 49, 51].

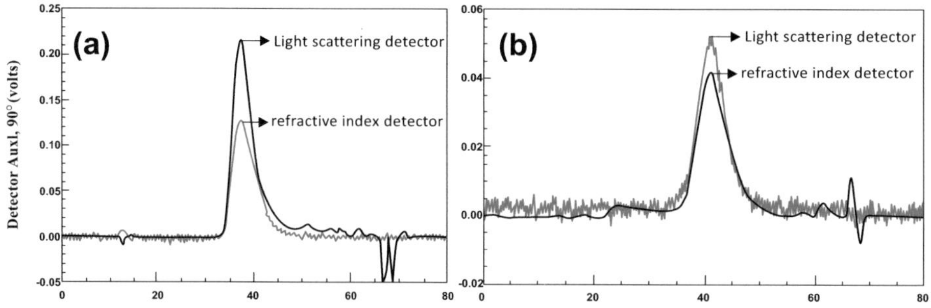

Fig. (5). Elution profiles for galactomannans from (**a**) *Mimosa scabrella* (bracatinga) and (**b**) *Schizolobium parahybae* (guapuruvu) using HPSEC-RI-MALLS with light scattering and refractive index detectors.

Fig. (**5**) shows the profiles of the GMs from *Mimosa scabrella* and *Schizolobium* *parahybae*, using HPSEC-RI-MALLS. Homogeneous, unimodal elution profiles are observed for both samples. Furthermore, the absence of aggregates can be observed from the mass distribution because these species are eluted in low elution time and a considerable increase in the light scattering is observed.

GMs from seeds are polysaccharides with high average molar masses ($\overline{M}_w$) on the order of 1 × 10^6 g/mol and dispersion indexes ($\overline{M}_w / \overline{M}_n$) greater than 1. Table **1** shows the chemical and physicochemical characteristics of some GMs and XGs obtained using HPSEC-RI-MALLS.

Table 1. Chemical and physicochemical properties of galactomannans (GMs) and xyloglucans (XGs) from some Brazilian sources after the extraction process.

GMs from Brazilian source							
Species	**Extraction process (solvent; time; temperature)**	**Yield (%)**	**Man:Gal Ratio**	**[η] (mL/g)**	**$\overline{M}_w$ (10^6 g/mol)**	**R_g (nm)**	**Ref.**
M. scabrella Bentham	water; 4 h; 25°C	30	1.1:1	740	1.15	73	82
M. scabrella var. *aspericarpa*	water; exhaustive extraction; 25°C	28	1.12:1	607	0.59	62	82
M. flocculosa	water; exhaustive extraction; 25°C	27	1.09:1	1000	1.48	85	82
M. taimbensis	water; exhaustive extraction; 25°C	30	1.16:1	490	0.50	54	82
M. bimucronata	water; exhaustive extraction; 25°C	30	1.05:1	890	1.08	87	82
S. adstringens	water; 24 h; 25°C	18	2.4:1	829	0.9	n.d.	17
M. floribunda	water; 4 h; 25°C	10	1.6:1	n.d.	1.18	n.d.	97
M. brauna	water; 4 h; 25°C	4	2.6:1	n.d.	1.46	n.d.	97
S. parahybae	water; 24 h; 25 °C	50	3.0:1	850	n.d.	n.d.	32
S. amazonicum	water; 24 h; 25 °C	50	3.0:1	950	n.d.	n.d.	32
C. fastuosa	water; exhaustive extraction; 25°C	27	4.0:1	n.d	0.56	n.d.	90
P. aculeata	water; 24 h; 10°C	26	3.1:1	558	0.77	n.d.	45

(Table 1) contd.....

XGs from Brazilian source							
Species	**Extraction process (solvent; time; temperature)**	**Yield (%)**	**Xyl:Glc:Gal Ratio**	**[η] (mL/g)**	**$\overline{M}_W$ (10^6 g/mol)**	**R_g (nm)**	**Ref.**
C. langsdorffii	water; exhaustive extraction; 25°C	47	1.7:1.6:1	804	0.78	65	68
H. courbaril	water; exhaustive extraction; 25°C	45	3.85:2.7:1	850	0.30	n.d.	60, 61
M. sloanei	water; 1 h; 25°C	30	1.8:1.7:1.0	n.d.	1.6	99.9	69
G. hymenifolia	water; exhaustive extraction; 25°C	54	3.3:2.3:1	665	0.84	97	48

n.d. - not determined.

Considering the polysaccharide solution systems, the intrinsic viscosities [η] values can be used to evaluate the viscosimetric molecular weight (Mw), through the Mark-Houwink-Sakurada equation, $[\eta] = K \cdot M_w^{\alpha}$, where K and α are constants for a given polymer-solvent-temperature system.

By this equation, it is possible to relate the Mw value with the average molecular weight ($\overline{M}_W$) values obtained by light scattering measurements, and [η]. It is possible to observe that higher values of $\overline{M}_W$ are related with higher intrinsic viscosities for polymers in solution. In addition, the values of R_g give information about the global dimension of the polymer chain and depend from the end-to-end distance of the polymer in solution [44].

A persistence length (L_p) of 9.3 nm was determined for *M. scabrella* GM using HPSEC-RI-MALLS. This value is in agreement with that predicted using molecular modeling studies (L_p = 9.6 nm), which was calculated using 4000 perfect conformations of [Gal α(1→6)] Man $\beta(1\rightarrow4)_x$ for a polymer with 10000 units. These results described, for the first time, the dimensions of a neutral, homogeneous polysaccharide. The persistence length obtained characterized the rigidity of a β-mannan chain in which all mannose units have been substituted at the 6 position by galactose units [98].

The static and hydrodynamic dimensions, are characteristics dependent of the structure of the macromolecules, and a combination of these two parameters provide qualitative information on the architecture of the macromolecules. Souza *et al.* [35] also studied the behavior of GM from *C. ferrea* var. *ferrea* using static light scattering (SLS) and dynamic light scattering (DLS), in terms of hydrodynamic radius (R_h, defined as the radius of an equivalent hard sphere diffusing at the same rate as the molecule under observation) and radius of gyration (R_g) parameters, to evaluate the molecular size and conformation in solution. In addition, the structure-sensitive parameter ρ (R_g/R_h), defined as the ratio of the radius of gyration R_g obtained from SLS to the hydrodynamic radius R_h from DLS provide the shape of the polymer in a good solvent. The DLS result of GM from *C. ferrea*, observed in Fig. (**6**), using a saline dilute solution shown a good correlation. Using the CONTIN method, the R_h of the GM at a specific angle of 90° was calculated as 53.5 nm. Using the R_g of 94.5 nm, the ρ parameter resulted in 1.77. According to these authors, the value of ρ decreases with increasing branching density, but an increase in polydispersity counteracts the effect of branching, and the result obtained for this GM is compatible with a random coil conformation in a good solvent and in accordance with HPSEC-RI-MALLS results (data not shown).

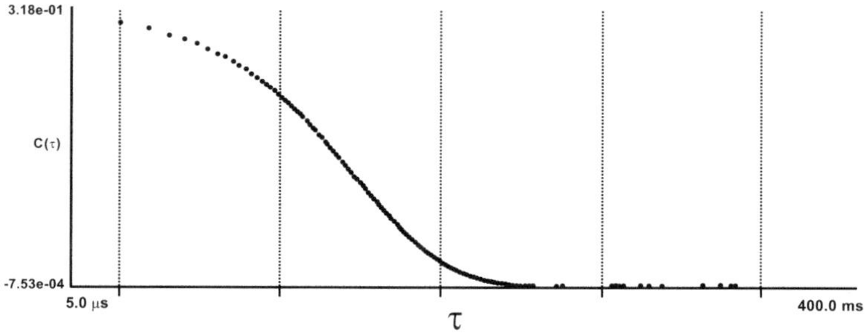

Fig. (6). Typical autocorrelation function obtained from dynamic light scattering measurements of the galactomannan from *C. Ferrea* seeds (1 mg/mL) in 0.1 M sodium nitrite solution containing 0.02% (*w/w*) sodium azide. Scattering angle θ= 90°. Source: Souza *et al.* [35].

New methodologies have been developed to evaluate the behavior of biopolymers in solution when subjected to extraction, solubilization, and chemical or enzymatic derivatization. Among these real-time analysis techniques, automatic continuous mixing (ACM) and time-dependent static light scattering (TDSLS) can be used during the processing of the polymers by coupling to the reactor and using various detection methods, including laser light scattering, viscometry, refractometry, and other spectroscopic methods, such as infrared and UV-Vis [44].

The enzymatic hydrolysis of GMs from native Brazilian species (*M. scabrella*, *Moldenhawera floribunda*, and *Melanoxylon brauna*) was monitored TDSLS in real time. The results demonstrated that the hydrolysis of isolated or combined β-mannanase and α-galactosidade is a random process and permitted an estimation of the lateral chain distribution. Moreover, this study demonstrated that there is a minimum sequence of unsubstituted mannose units needed for hydrolysis of β-mannanase to occur. Furthermore, the TDSLS results are in agreement with HPSEC-RI-MALLS and theoretical modeling studies [97].

Recently, Salvalaggio *et al.* [17] evaluated the aqueous extraction of GM from the seeds of *M. scabrella, S. adstringens*, and *S. parahybae* for 1, 2, 3, 4, 6, 24, and 48 h. The best extraction times for each species, based on the minimum extraction time, yield, and HPSEC-MALLS/RI/VIS results were 4, 6, and 2 h for *M. scabrella, S. adstringens*, and *S. parahybae*, respectively. Signs of degradation were verified by HPSEC-MALLS/RI/VIS, especially at 24 and 48 h for *S. adstringens* and *S. parahybae*. In most cases, the apparent persistence length, as determined by viscometry, indicated that the obtained GM contained aggregates, even after centrifugation and filtration. These results can be used to suggest an effective extraction time for each plant source of the biopolymer based on its performance and macromolecular behavior in solution.

Azero and Andrade [52] showed that the intrinsic viscosities [η] of the GMs from *Cassia javanica, C. pulcherrima*, and *Prosopis juliflora* seeds were dependent upon the purification process used.

Despite numerous studies on the chemical structure, conformation analysis, and physicochemical properties of GMs from Brazilian species, there are still many questions to be answered, such as the process of GM dissolution, which is dependent on the content of galactose substituents on the mannan main chain, the differences in molar mass distribution obtained during industrial processing of GM, as highlighted by Pollard *et al.* in recent studies with commercial GM [13 - 15], or the detection and quantification of partially hydrolyzed GM using high-performance anion-exchange chromatography (HPAEC) with pulsed amperometric detection (PAD) validation studies [36].

Rheology is the science of deformation and flow of materials, and an understanding of the rheology is essential for the rational use of biopolymers with high viscosities or gel-forming characteristics [99, 100]. From rheological analysis, it is possible to evaluate GM from seeds, which is generally present in solution in the semi-dilute regime, as having pseudoplastic behavior, *i.e.*, the viscosity decreases with increasing shear rate. This behavior was observed for GM obtained from the seeds of *M. scabrella* in aqueous solution at concentrations above 3.6 g/L. The critical concentration of this polysaccharide, c*[η], was 2.6, which corresponds to that of a randomly entangled biopolymer [7]. Pseudoplastic behavior was also observed for GM from the seeds of *C. fastuosa* [90] and *S. parahybae* [53].

Aqueous Binary System: Galactomannan or Xyloglucan

Synergistic interactions between polysaccharides are commercially attractive, as they allow for reduction of costs through replacement of expensive polysaccharides with less expensive products. Thus, new properties can be obtained by partial replacement of the polymers, whose characteristics can be measured using oscillatory and non-oscillatory rheology. With this goal, many systems have been formulated and are reported below.

GMs are soluble in water (hot or cold) and form high-viscosity solutions at low concentrations or induce the formation of gels when in synergy with other polymers, which in turn can control the structure and texture of commercial products [6, 101]. This property is directly related to the industrial interest of these polymers and can be evaluated by rheological analysis.

The synergistic effect observed in mixtures of xanthan (XT) and GM from Brazilian seed species (*M. scabrella* Bentham and *S. parahybae* (Vell) Blake) with Man:Gal ratios of 1.1:1 and 3:1, respectively, were studied by Bresolin *et al.* [53]. A strong interaction was observed for XT:GM (1:1, *w/w)* from *S. parahybae* seeds in water at 1 and 2 g/L. For the XT:GM mixture from *M. scabrella* (3:1, *w/w*; 2 g/L in water), there was a 32% increase in the apparent viscosity. The best interaction occurred when the system was previously heated to 80°C, independent of the XT conformation.

As the main chain of the GM from *M. scabrella* is almost completely substituted by galactose, the observed synergistic effect must have a different mechanism from that described by other authors in which xanthan interacts with galactomannan *via* the unsubstituted region of the latter [73, 102 - 104]. This interaction suggests that the presence of GM induces a conformational change in the ordered state of XT [19, 105].The rheological properties of the mixtures of XT and GM from *M. scabrella*, in the presence or absence of ascorbic acid, were analyzed in binary gels [106]; these results confirmed the synergy observed by Bresolin *et al.* [19, 53].

The synergistic properties of the gels formed by the mixture of GM from *C. fastuosa* seeds and XT were investigated by Amaral *et al.* [107] by measuring the viscoelastic properties. The values of the storage modulus (*G*') and loss modulus (*G*'') determined for 4 g/L solutions containing 50% (*w/w*) GM showed the formation of true gels.

The influence of XT conformation on the physicochemical behavior of mixtures with GM from the seeds of *S. parahybae* was studied by Bresolin *et al.* [108] using viscoelastic measurements, differential scanning calorimetry (DSC), and circular dichroism. The results suggest that the GM/XT interaction occurs between disordered segments, which are more abundant at low salt concentrations, but are present in low proportions at temperatures below the conformational transition temperature of XT (T_m). The observed dependence of elliptical light absorption with temperature, observed using circular dichroism, suggested that GM induced XT gel formation (T_g= 25°C), whereas XT alone exhibited a disordered conformation. The low T_g observed (25°C) compared with that usually reported in the literature (60°C) is related to the Man:Gal ratio and the distribution of galactosyl units along the mannose chain.

Mixtures of κ-carrageenan from *Hypnea musciformis* algae and GM from the seeds of *C. javanica*, a species found in southeastern Brazil, were studied by Andrade *et al.* [109] using dynamic rheological measurements and compression tests. Mixtures of 5 g/L κ-carrageenan and GM (2:1, *w/w*) with 0.1 mol/L KCl showed the viscoelastic region with the maximum synergy. The maximum synergy obtained by compression tests was at a total polymer concentration of 10 g/L with 0.25 mol/L KCl.

Recently, Soares *et al.* [110] developed a stable hydrogel composed of GM from *Cassia grandis* seeds and κ-carrageenan. This hydrogel showed temperature resistance and did not lose its shape until 42°C. Furthermore, thermo-rheological tests confirmed a higher organization of the polysaccharides in the hydrogel matrix over the storage period, with an elastic modulus that was always higher than the viscous modulus.

Synergistic effects have also been demonstrated in mixtures of GM and proteins. Gonçalves *et al.* [111] studied the effect of GM from *C. javanica* seeds in the thermal gelatinization of isolated whey protein at 80°C using rheological measurements. The gels were monitored at 80°C and after cooling to 20°C. At low concentrations, GM increased the rate of aggregation and gel strength of the protein, but did not intrinsically change the gelation process or the structure of the system. At high concentrations, the gel showed decreased values of the moduli and the mechanical spectra changed, indicating that the configuration of the system was modified. This result was interpreted to be a consequence of liquid–liquid phase separation during the aggregation process.

GM from *M. scabrella* and milk proteins showed synergy by rheological analysis after heating at 65°C. This phenomenon can be explained by a modification of the intermolecular network induced by GM, which became stronger with increasing thermal energy. These results were confirmed by HPSEC-RI-MALLS and calorimetry [112, 113].

Lucyszyn [114] analyzed the partial replacement of agar (A) by GMs (*Leucaena leucocephala*, *Cyamopsis tetragonolobus*, and *Cassia fastuosa*) with different Man:Gal ratio, namely 1.7:1, 1.9:1 and 3.0:1, respectively. Mixtures at 6 g/L, 50% (*w/w*) revealed that the best interaction. In terms of higher stiffness, the gel occurred with the mixture A:GM *Cassia fastuosa*, indicating that the best synergy occurs with less substituted GMs. Measurements of the breaking point showed that the gel showed a solid character up to a force of 3.8 N/m^2 applied to the surface.

Souza-Lima *et al.* [115] studied the interaction of XG from *H. courbaril* seeds (XGHC) with carboxymethylcellulose. The rheological properties of blends (2.5%) in water at 25°C showed synergy between the polysaccharides at concentrations of 2% and 0.5% for XGHC and carboxymethylcellulose, respectively.

Lima-Nishimura [116] produced semi-solid media, formed by interactions between agar and XGHC or Gelrite® and XGHC. Through rheological analysis, surface tension, and syneresis measurements, mixtures of agar (4 g/L)/XGHC (2 g/L) and Gelrite® (1.2 g/L)/XGHC (1.2 g/L) were found to show the best results in terms of gel strength.

Freitas *et al.* [117] investigated the structural features required for interactions between starch and XGHC. Mixtures of XGHC with maize starch, which has a high content of amylose (66%), and waxy maize starch showed increased viscosity of the slurry when compared with the isolated systems. Dynamic rheological analysis revealed the thermal stability XGHC-amylose mixtures when compared with amylose alone. The blends showed positive synergistic interactions when amylose-rich starches (maize starch and yam) were used, but not when starches with high levels of amylopectin (corn starch and cassava) were used.

Souza *et al.* [118] demonstrated the association between gellan gum and XGHC in saline media using small angle X-ray scattering (SAXS) measurements. The SAXS results (Fig. **7**) revealed the gel organization on of the voids/pores (R_g) on a nanometric scale. Furthermore, a dependence of R_g on the XGHC concentration used in the gellan gum blends was observed, with R_g generally increasing with increasing XGHC concentration.

Galactomannan and Xyloglucan Applications

The structural features and rheological properties of GMs and XGs show significantly increased water-retention capacities and paste viscosities, which promote their application as good thickening agents. Moreover, these macromolecules have good stability over a wide range of pH and temperature, and generate stable emulsions. The similar solution properties of these polysaccharides extracted from several sources, allows the exploration of these biopolymers as thickeners, binders, emulsifiers, and stabilizers in different industrial areas, such as food, pharmaceuticals, biotechnology, and medicine [18, 23, 24, 45, 47, 119 - 129].

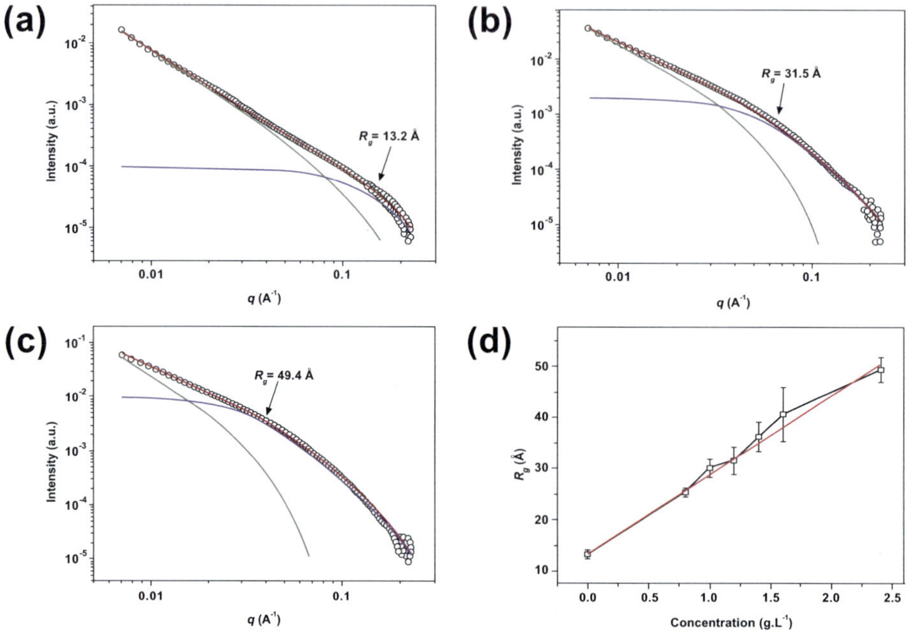

Fig. (7). SAXS profiles of (**a**) Pure gellan gum (2.4 g/L), (**b**) Mixture of gellan gum/xyloglucan from *Hymenaea courbaril* seeds (1.2 g/L/1.2 g/L), and (**c**) Pure xyloglucan in 0.15 mol/L NaCl at 25 ± 1°C. The unified fits are shown as solid red lines, and the two fit levels are represented by dashed green and dotted blue lines. (**d**) Dependence of the radius of gyration (R_g) of the voids/pores obtained from the unified fit on the concentration of xyloglucan present in the measured systems. Source: Souza *et al.* [118].

Studies to assess the interaction between GM from *M. scabrella* seeds with surfactants and proteins in cosmetic formulations were made [25, 119]. In other work, the same GM source was tested in formulations as a substitute for starch pudding [18], its interaction with milk proteins [112 - 114]. In studies conducted by Lima *et al.* [120], XGs obtained from seeds of *H. courbaril* L., *C. langsdorffii* Desf., *Tropaeolum majus* L., and *T. indica* L., and GM obtained from the seeds of *D. mollis* were used as additives in papermaking.

Bezerra *et al.* [47] used different GMs from seeds of *M. scabrella* to prepare chromatographic matrices for the isolation of jacalin lectin (galactose-linker). The gels obtained have been used for affinity chromatography, and in the work of Garros-Rosa *et al.* [45], modification of GM from *S. siamea* seeds was proposed for use as a chromatographic affinity matrix after crosslinking with epichlorohydrin.

Lima-Nishimura [116] tested XGHC mixed with agar as a new plant culture medium for the development of callus, somatic embryos, and regeneration of carrot roots, whereas this XG mixed with Gelrite® was tested for somatic embryogenesis of two genotypes of coffee. In other studies performed by Lucyszyn *et al.* [114, 121 - 125], mixtures of agar and GM from *C. fastuosa* were used in the micropropagation of apple rootstock Marubakaido (*Malus domestica* Borkh prunifolia), pear (*Pyrus communis* L.), strawberry (*Fragaria ananassa* Duchesne), and tobacco (*Nicotiana tabacum*).

Ribeiro *et al.* [126] developed a system for protecting enalaprilate (Enal), which is a potent intravenously administered angiotensin-converting enzyme inhibitor, intercalated in lamellar double hydroxides of aluminum and magnesium by coating the composite with XGHC.

Pires *et al.* [24] studied a heparinoid obtained by sulfation of a GM isolated from the seeds of *S. macranthera* and Ono *et al.* [23, 127] tested the *in vitro* and *in vivo* antiviral properties of sulfated GMs against yellow fever virus (YFV, BeH111 strain) and dengue 1 virus (DEN-1, Hawaii strain). More recently, the XGs from *C. langsdorffii*, *H. courbaril*, and *M. sloanei* seeds were used for biological activity tests using mouse peritoneal macrophages [128, 129].

Improvements in biological properties and/or products of commercial interest can be achieved through chemical surface modifications, and these polysaccharides have been used for the development of new materials. Souza *et al.* [130] achieved specific chemical modifications at C-6 in the GM from *C. ferrea* seeds using 2,2,6,6-tetramethylpiperidine-1-oxyl (TEMPO). New polyelectrolytes were produced with varying degrees of oxidation, and successive alkylations at C-6 were investigated to enhance the amphiphilic nature of the derivatives.

Lucyszyn *et al.* [131] oxidized XGHC with TEMPO to obtain derivatives with different degrees of oxidation (10, 23, and 100%). The HPSEC chromatogram and atomic force microscopy (AFM) images of the oxidized XGs are shown in Fig. (**8**). The HPSEC-RI-MALLS analysis (Fig. **8a**) suggests that the experimental conditions of the oxidation results in chain scission and degradation of the polymer during the synthesis. A comparison of the neutral polysaccharide and oxidized XG by AFM (Fig. **8b** and **8c**) showed that aggregation of the polysaccharide increased according to the degree of oxidation.

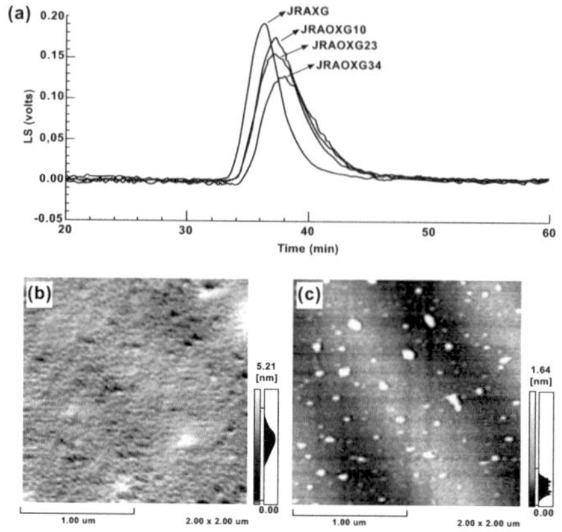

Fig. (8). (**a**) HPSEC-RI-MALLS analysis with light scattering (detector at 90°) for native (JRAXG) and oxidized (JRAOXG10, JRAOXG23, and JRAOXG34) xyloglucan samples. AFM dynamic mode (2.00×2.00 μm) images of xyloglucan from *Hymenaea courbaril* seeds deposited by drop incubation on mica: (**b**) JRAXG and (**c**) JRAOXG34. Source: Lucyszyn *et al.* [131].

CONCLUDING REMARKS

The structural and macromolecular properties of the "nonstarch" polysaccharides, specifically xyloglucan and galactomannan, from Brazilian sources were demonstrated. Considering the great diversity of seed polysaccharides (natural and modified) that can be applied in different industrial sectors and processes, it is extremely important to evaluate the chemical and physicochemical behavior of these biopolymers. Significant advances have been made by several research groups in the development of extraction process and new methodologies to evaluate the applicability of these hydrocolloids. These results show that the use of the seeds has the potential for truly sustainable production of natural resources.

CONFLICT OF INTEREST

The author (editor) declares no conflict of interest, financial or otherwise.

ACNOWLEDGEMENTS

The authors express their gratitude and deep admiration to Professors **Fany Reicher**, **João Batista Chaves Corrêa** (in memoriam), **Maria Rita Sierakowski**, and **Philip Albert James Gorin** (in memoriam), with whom they had the privilege to collaborate. They were the pioneers of plant carbohydrate chemistry in the Biochemistry and Molecular Biology Department of the Federal University of Paraná, Brazil, and examples of competence and dedication to science.

The authors would also like to acknowledge: (1) CNPq (Brazilian National Council for Scientific and Technological Development; process n°. 476950/2013-9 and n°.306949/2012); (2) Nanobiotec/Capes-Brazil Network; (3) Nanoglicobiotec

Network (Ministry of Science and Technology/CNPq; process n°564741/2010-8); (4) PRONEX-Carbohydrates; (5) Araucaria Foundation (Paraná, Brazil); and (6) CAPES (Higher Education Human Resources Development Agency of the Brazilian Ministry of Education) for providing the financial support needed to develop this project. J.L.M.S and C.L.O.P. are research members of CNPq.

REFERENCES

[1] G. Avigad, and P.M. Dey, "Carbohydrate Metabolism: Storage carbohydrates", In: *Plant Biochemistry.* Academic Press: London, UK, 1997, pp. 143-204. [http://dx.doi.org/10.1016/B978-012214674-9/50005-9]

[2] H. Meier, and J.S. Reid, "Reserve Polysaccharides Other Than Starch in Higher Plants", In: *Plant Carbohydrates I (Encyclopedia of Plant, Physiology - v.13A).* Springer-Verlag: Berlin, Heidelberg, 1982, pp. 418-461. [http://dx.doi.org/10.1007/978-3-642-68275-9_11]

[3] J.D. Bewley, and M. Black, *Seeds: Physiology of Development and Germination.* 2nd ed Plenum Press: New York, 1994. [http://dx.doi.org/10.1007/978-1-4899-1002-8]

[4] M.S. Buckeridge, "Seed cell wall storage polysaccharides: models to understand cell wall biosynthesis and degradation", *Plant Physiol.* , vol. 154, no. 3, pp. 1017-1023, 2010. [http://dx.doi.org/10.1104/pp.110.158642] [PMID: 20855518]

[5] D.J. Folkes, and M.A. Jordan, "Mono and Disaccharides: Analytical Aspects", In: *Carbohydrates in Food.* 2nd ed CRC Press, FL: Boca Raton, 2006, pp. 1-40.

[6] A.M. Stephen, and S.C. Churms, "Introduction", In: *Food Polysaccharides and Their Applications.* 2nd ed CRC Press, FL: Boca Raton, 2006, pp. 1-17. [http://dx.doi.org/10.1201/9781420015164.ch1]

[7] J.L. Ganter, M. Milas, J.B. Corrêa, F. Reicher, and M. Rinaudo, "Study of solution properties of galactomannan from the seeds of Mimosa scabrella", *Carbohydr. Polym.* , vol. 17, pp. 171-175, 1992. [http://dx.doi.org/10.1016/0144-8617(92)90001-7]

[8] J.L. Ganter, A.T. Cardoso, M. Kaminski, and F. Reicher, "Galactomannan from the seeds of Mimosa scabrella: a scale-up process", *Int. J. Biol. Macromol.* , vol. 21, no. 1-2, pp. 137-140, 1997. [http://dx.doi.org/10.1016/S0141-8130(97)00053-6] [PMID: 9283028]

[9] C.W. Vendruscolo, C. Ferrero, E.A. Pineda, J.L. Silveira, R.A. Freitas, M.R. Jiménez-Castellanos, and T.M. Bresolin, "Physicochemical and mechanical characterization of galactomannan from Mimosa scabrella: Effect of drying method", *Carbohydr. Polym.* , vol. 76, pp. 86-93, 2009. [http://dx.doi.org/10.1016/j.carbpol.2008.09.028]

[10] A.C. Soria, L. Ruiz-Aceituno, L. Ramos, and L.M. Sanz, "Microwave-assisted extraction of polysaccharides", In: *In.* Polysaccharides, 2014, pp. 1-18.

[11] C.P. Passos, and M.A. Coimbra, "Microwave superheated water extraction of polysaccharides from spent coffee grounds", *Carbohydr. Polym.* , vol. 94, no. 1, pp. 626-633, 2013. [http://dx.doi.org/10.1016/j.carbpol.2013.01.088] [PMID: 23544583]

[12] R.L. Whistler, "Solubility of Polysaccharides and Their Behavior in Solution", In: *Carbohydrates in Solution.* American Chemical Society: Washington, DC, 1973, pp. 242-255. [http://dx.doi.org/10.1021/ba-1971-0117.ch014]

[13] M.A. Pollard, B. Eder, P. Fischer, and E.J. Windhab, "Characterization of galactomannans isolated from legume endosperms of Caesalpinioideae and Faboideae subfamilies by multidetection aqueous SEC", *Carbohydr. Polym.* , vol. 79, pp. 70-84, 2010. [http://dx.doi.org/10.1016/j.carbpol.2009.07.028]

[14] M.A. Pollard, R. Kelly, C. Wahl, P. Fischer, E. Windhab, B. Eder, and R. Amadó, "Investigation of equilibrium solubility of carob galactomannan", *Food Hydrocoll.* , vol. 21, pp. 683-692, 2007. [http://dx.doi.org/10.1016/j.foodhyd.2006.08.010]

[15] M.A. Pollard, R. Kelly, P.A. Fischer, E.J. Windhab, and B. Eder, "B, and R. Amadó, "Investigation of molecular weight distribution of LBG galactomannan for flours prepared from individual seeds, mixtures, and commercial samples", *Food Hydrocoll.* , vol. 22, pp. 1596-1606, 2008. [http://dx.doi.org/10.1016/j.foodhyd.2007.11.004]

[16] K.C. Kai, and C.L. Petkowicz, "Influence of extraction conditions on properties of seed xyloglucan", *Int. J. Biol. Macromol.* , vol. 46, no. 2, pp. 223-228, 2010. [http://dx.doi.org/10.1016/j.ijbiomac.2009.12.012] [PMID: 20060409]

[17] M.O. Salvalaggio, R.A. Freitas, E.M. Franquetto, H.S. Koop, and J.L. Silveira, "Influence of the extraction time on macromolecular parameters of galactomannans", *Carbohydr. Polym.* , vol. 116, pp. 200-206, 2015. [http://dx.doi.org/10.1016/j.carbpol.2014.05.036] [PMID: 25458290]

[18] N. Lucyszyn, *"Galactomananas: novas fontes do biopolímero e aplicações na indústria alimentícia"*, Dissertation, Universidade Federal do Paraná, Brasil, 1994.

[19] T.M. Bresolin, M. Milas, M. Rinaudo, F. Reicher, and J.L. Ganter, "Role of galactomannan composition on the binary gel formation with xanthan", *Int. J. Biol. Macromol.* , vol. 26, no. 4, pp. 225-231, 1999. [http://dx.doi.org/10.1016/S0141-8130(99)00087-2] [PMID: 10569283]

[20] M. Rinaudo, "Relation between the molecular structure of some polysaccharides and original properties in sol and gel states", *Food Hydrocoll.* , vol. 15, pp. 433-440, 2001. [http://dx.doi.org/10.1016/S0268-005X(01)00041-8]

[21] J.P. Doyle, P. Giannouli, E.J. Martin, M. Brooks, and E.R. Morris, "Effect of sugars, galactose content and chain length on freeze-thaw gelation of galactomannans", *Carbohydr. Polym.* , vol. 64, pp. 391-401, 2006. [http://dx.doi.org/10.1016/j.carbpol.2005.12.019]

[22] A.P. Busato, F. Reicher, R. Domingues, and J.L. Silveira, ""Rheological properties of thermally xyloglucan gel from the seeds of Hymenaea courbaril," Mat. Sci. Eng. C-", *Bio. S.* , vol. 29, pp. 410-414, 2009.

[23] L. Ono, *"Atividade antiviral de galactomananas modificadas por oxidação e/ou sulfatação"*, PhD thesis, Universidade Federal do Paraná, 2004.

[24] L. Pires, P.A. Gorin, F. Reicher, and M-R. Sierakowski, "An active heparinoid obtained by sulphation of a galactomannan extracted from the endosperm of Senna macranthera seeds", *Carbohydr. Polym.* , vol. 46, pp. 165-169, 2001. [http://dx.doi.org/10.1016/S0144-8617(00)00298-8]

[25] R.P. Vianna-Filho, C.L. Petkowicz, and J.L. Silveira, "Rheological characterization of O/W emulsions incorporated with neutral and charged polysaccharides", *Carbohydr. Polym.* , vol. 93, no. 1, pp. 266-272, 2013. [http://dx.doi.org/10.1016/j.carbpol.2012.05.014] [PMID: 23465929]

[26] M. Edwards, P.V. Bulpin, I.C.M. Dea, and J.S.G. Reid, "Biosynthesis of legume-seed galactomannans *in vitro*. Cooperative interactions of a guanosine 5′-diphosphate-mannose-linked (1→4)-β-D-mann--syltransferase and a uridine 5′-diphosphate-galactose-linked α-D-galactosyltranferase in particulate enzyme preparations from developing endosperms of fenugreek (*Trigonella foenum-graecum* L.) and guar (*Cyamopsis tetragonoloba* [L.] Taub.)", *Planta,* vol. 78, pp. 41-51, 1989.

[27] A.E. Manzi, M.N. Mazzini, and A.S. Cerezo, "The galactomannan system from the endosperm of the seed of Gleditsia triacanthos", *Carbohydr. Res.* , vol. 125, pp. 127-143, 1984. [http://dx.doi.org/10.1016/0008-6215(84)85148-4]

[28] J.L. Ganter, J.B. Corrêa, F. Reicher, A. Heyraud, and M. Rinaudo, "Low molecular weight carbohydrates from Mimosa scabrella seeds", *Plant Physiol. Biochem.* , vol. 29, pp. 139-146, 1991.

[29] M. Edwards, C. Scott, M.J. Gidley, and J.S. Reid, "Control of mannose/galactose ratio during galactomannan formation in developing legume seeds", *Planta,* vol. 187, no. 1, pp. 67-74, 1992. [http://dx.doi.org/10.1007/BF00201625] [PMID: 24177968]

[30] J.L. Ganter, S.F. Zawadzki-Baggio, S.C. Leitner, M-R. Sierakowski, and F. Reicher, "Structural studies on galactomannans from Brazilian seeds", *J. Carbohydr. Chem.* , vol. 12, pp. 753-767, 1993. [http://dx.doi.org/10.1080/07328309308019005]

[31] J.L. Ganter, M-R. Sierakowski, and F. Reicher, "Evolution of polysaccharide research on Mimosa scabrella", *Cienc. Cult.* , vol. 45, pp. 14-18, 1993.

[32] J.L. Ganter, A. Heyraud, C.L. Petkowicz, M. Rinaudo, and F. Reicher, "Galactomannans from Brazilian seeds: characterization of the oligosaccharides produced by mild acid hydrolysis", *Int. J. Biol. Macromol.* , vol. 17, no. 1, pp. 13-19, 1995. [http://dx.doi.org/10.1016/0141-8130(95)93512-V] [PMID: 7772557]

[33] F. Reicher, and J.L. Ganter, "Composição química e caracterização estrutural de galactomananas de espécies brasileiras", In: *Fibra dietética en Iberoamérica: tecnologia y salud: obtención, caracterización, efecto fisiológico y aplicación en alimentos.* Varela: São Paulo, SP, 2001, pp. 27-41.

[34] P.L. Cunha, Í.G. Vieira, Â.M. Arriaga, R.C. de Paula, and J.P. Feitosa, "Isolation and characterization of galactomannan from Dimorphandra gardneriana Tul. seeds as a potential guar gum substitute", *Food Hydrocoll.* , vol. 23, pp. 880-885, 2009. [http://dx.doi.org/10.1016/j.foodhyd.2008.05.005]

[35] C.F. Souza, N. Lucyszyn, F.A. Ferraz, and M-R. Sierakowski, "Caesalpinia ferrea var. ferrea seeds as a new source of partially substituted galactomannan", *Carbohydr. Polym.* , vol. 82, pp. 641-647, 2010. [http://dx.doi.org/10.1016/j.carbpol.2010.05.031]

[36] G. Mercier, and C. Campargue, "Partially hydrolyzed guar gum characterization and sensitive quantification in food matrices by high performance anion exchange chromatography with pulsed amperometric detectionvalidation using accuracy profile", *J. Chromatogr. A,* vol. 1262, pp. 180-187, 2012. [http://dx.doi.org/10.1016/j.chroma.2012.09.027] [PMID: 23031439]

[37] S.F. Zawadzki-Baggio, M-R. Sierakowski, J.B. Corrêa, and F. Reicher, "A linear (1→5)-linked α-l-arabinofuranan from seeds of guapuruvu (Schizolobium parahybum)", *Carbohydr. Res.* , vol. 233, pp. 265-269, 1992. [http://dx.doi.org/10.1016/S0008-6215(00)90941-8]

[38] S. Zawadzki-Baggio, *"Arabinana e galactomanana de Schizolobium parahybum. Estudo da biossíntese de galactomanana"*, PhD thesis, Universidade Federal do Paraná, 1994.

[39] M. Rinaudo, "The Relation between Chemical Structure o Polysaccharides and their Physical properties", In: *Gums and Stabilisers for the Food Industry.* IRL Press: Oxford, 1992, pp. 51-61.

[40] P.J. Wyatt, "Light scattering and the absolute characterization of macromolecules", *Anal. Chim. Acta,* vol. 272, pp. 1-40, 1993. [http://dx.doi.org/10.1016/0003-2670(93)80373-S]

[41] W.F. Reed, "Coupled Multiangle Light-Scattering and Viscosimetric Detectors for Size Exclusion Chromatography with Application to Polyelectrolyte Characterization", In: *Strategies in Size Exclusion Chromatography (ACS Symposium Series 635)* American Chemical Society: Washington, DC, 1996. [http://dx.doi.org/10.1021/bk-1996-0635.ch002]

[42] L.K. Kostanski, D.M. Keller, and A.E. Hamielec, "Size-exclusion chromatography-a review of calibration methodologies", *J. Biochem. Biophys. Methods,* vol. 58, no. 2, pp. 159-186, 2004. [http://dx.doi.org/10.1016/j.jbbm.2003.10.001] [PMID: 14980789]

[43] A.M. Striegel, J.J. Kirkland, W.W. Yau, and D.D. Bly, *Modern Size-Exclusion Liquid Chromatography*. Wiley & Sons: New York, 2009. [http://dx.doi.org/10.1002/9780470442876]

[44] W.F. Reed, "Method for online monitoring of polydispersity during polymerization reactions", *Macromolecules,* vol. 33, pp. 7165-7172, 2000. [http://dx.doi.org/10.1021/ma0006023]

[45] I. Garros-Rosa, F. Reicher, C.L. Petkowicz, M-R. Sierakowski, and R.A. Moreira, "Characterization of the galactomannans from Parkinsonia aculeata seeds and their application on affinity chromatography", *Polímeros,* vol. 16, pp. 99-103, 2006. [http://dx.doi.org/10.1590/S0104-14282006000200008]

[46] R.A. Freitas, A.P. Busato, D.A. Mitchell, and J.L. Silveira, "Degalactosylation of xyloglucan: Effect on aggregation and conformation, as determined by time dependent static light scattering, HPSEC–MALLS and viscometry", *Carbohydr. Polym.* , vol. 83, pp. 1636-1642, 2011. [http://dx.doi.org/10.1016/j.carbpol.2010.10.021]

[47] G.P. Bezerra, D.M. Teixeira-Sá, D.L. Soares, and R.A. Moreira, "A comparative study of jacalin-galactomannans binding by affinity chromatography", *Proc. of the Thirty-Fifth Annual Meeting of SBBq*

[48] N. Lucyszyn, A.F. Lubambo, L. Ono, T.A. Jó, C.F. de Souza, and M-R. Sierakowski, "Chemical, physico-chemical and cytotoxicity characterisation of xyloglucan from Guibourtia hymenifolia (Moric.) J. Leonard seeds", *Food Hydrocoll.* , vol. 25, pp. 1242-1250, 2011. [http://dx.doi.org/10.1016/j.foodhyd.2010.11.012]

[49] K-L. Cheong, D-T. Wu, J. Zhao, and S-P. Li, "A rapid and accurate method for the quantitative estimation of natural polysaccharides and their fractions using high performance size exclusion chromatography coupled with multi-angle laser light scattering and refractive index detector", *J. Chromatogr. A,* vol. 1400, pp. 98-106, 2015. [http://dx.doi.org/10.1016/j.chroma.2015.04.054] [PMID: 25990349]

[50] I.R. Arruda, P.B. Albuquerque, G.R. Santos, A.G. Silva, P.A. Mourão, M.T. Correia, A.A. Vicente, and M.G. Carneiro-da-Cunha, "Structure and rheological properties of a xyloglucan extracted from Hymenaea courbaril var. courbaril seeds", *Int. J. Biol. Macromol.* , vol. 73, pp. 31-38, 2015. [http://dx.doi.org/10.1016/j.ijbiomac.2014.11.001] [PMID: 25450047]

[51] V.C. Spier, M-R. Sierakowski, A.T. Ibrahim, J.C. Scholze Baum, J.L. Silveira, and R.A. de Freitas, "Time-dependent viscometry study of endoglucanase action on xyloglucan: A real-time approach", *Int. J. Biol. Macromol.* , vol. 81, pp. 461-466, 2015. [http://dx.doi.org/10.1016/j.ijbiomac.2015.08.023] [PMID: 26297307]

[52] E.G. Azero, and C.T. Andrade, "Testing procedures for galactomannan purification", *Polym. Test.* , vol. 21, pp. 551-556, 2002. [http://dx.doi.org/10.1016/S0142-9418(01)00123-4]

[53] T.M. Bresolin, P.C. Sander, F. Reicher, M-R. Sierakowski, M. Rinaudo, and J.L. Ganter, "Viscometric studies on xanthan and galactomannan systems", *Carbohydr. Polym.* , vol. 33, pp. 131-138, 1997. [http://dx.doi.org/10.1016/S0144-8617(97)00051-9]

[54] N.K. Mathur, *Industrial Galactomannan Polysaccharides*. CRC Press: Boca Raton, FL, 2012.

[55] P. Kooiman, "Partial enzyme degradation of Tamarindus-amyloid", *Nature,* vol. 180, p. 201, 1957. [http://dx.doi.org/10.1038/180201a0] [PMID: 13451680]

[56] P. Kooiman, "On the occurrence of amyloids in plant seeds", *Acta Bot. Neerl.* , vol. 9, pp. 208-219, 1960. [http://dx.doi.org/10.1111/j.1438-8677.1960.tb00651.x]

[57] T. Hayashi, "Xyloglucans in the primary cell wall", *Annu. Rev. Plant Physiol.* , vol. 40, pp. 139-168, 1989.

[http://dx.doi.org/10.1146/annurev.pp.40.060189.001035]

[58] S.C. Fry, "The structure and functions of xyloglucan", *J. Exp. Bot.* , vol. 40, pp. 1-11, 1989. [http://dx.doi.org/10.1093/jxb/40.1.1]

[59] M.S. Buckeridge, D.C. Rocha, J.S. Reid, and S.M. Dietrich, "Xyloglucan structure and post-germinative metabolism in seeds of Copaifera langsdorfii from savanna and forest populations", *Physiol. Plant.* , vol. 86, pp. 145-151, 1992. [http://dx.doi.org/10.1111/j.1399-3054.1992.tb01323.x]

[60] N. Lima-Nishimura, "F. Reicher, J.B.C. Corrêa, J.L.M.S. Ganter, and M.-R. Sierakowski, “Partial structure of a xyloglucan from the seeds of Hymenaea courbaril var. stilbocarpa (jatobá)", *Cienc. Cult.* , vol. 45, pp. 22-26, 1993.

[61] N.N. Lima, C.G. Rechia, J.L. Ganter, F. Reicher, and M-R. Sierakowski, "Oligosaccharides derived from the xyloglucan isolated from the seeds of Hymenaea courbaril var. stilbocarpa", *Int. J. Biol. Macromol.* , vol. 17, no. 6, pp. 413-415, 1995. [http://dx.doi.org/10.1016/0141-8130(96)81855-1] [PMID: 8789349]

[62] M.S. Buckeridge, H.J. Crombie, C.J. Mendes, J.S. Reid, M.J. Gidley, and C.C. Vieira, "A new family of oligosaccharides from the xyloglucan of Hymenaea courbaril L. (Leguminosae) cotyledons", *Carbohydr. Res.* , vol. 303, no. 2, pp. 233-237, 1997. [http://dx.doi.org/10.1016/S0008-6215(97)00161-4] [PMID: 9352637]

[63] M.A. Tiné, A.L. Cortelazzo, and M.S. Buckeridge, "Xyloglucan mobilisation in cotyledons of developing plantlets of Hymenaea courbaril L. (Leguminosae-Caesalpinoideae)", *Plant Sci.* , vol. 154, no. 2, pp. 117-126, 2000. [http://dx.doi.org/10.1016/S0168-9452(99)00245-9] [PMID: 10729610]

[64] S. Martin, R.A. Freitas, E. Obayashi, and M-R. Sierakowski, "Physico-chemical aspects of galactoxyloglucan from the seeds of Hymenaea courbaril and its tetraborate complex", *Carboid. Pol.* , vol. 54, pp. 287-295, 2003. [http://dx.doi.org/10.1016/S0144-8617(03)00073-0]

[65] M.A. Tiné, C.O. Silva, D.U. Lima, N.C. Carpita, and M.S. Buckeridge, "Fine structure of a mixed-oligomer storage xyloglucan from seeds of Hymenaea courbaril", *Carboid. Pol.* , vol. 66, pp. 444-454, 2006. [http://dx.doi.org/10.1016/j.carbpol.2006.03.032]

[66] R.A. Freitas, S. Martin, G.L. Santos, F. Valenga, M.S. Buckeridge, F. Reicher, and M-R. Sierakowski, "Physico-chemical properties of seed xyloglucans from different sources", *Carboid. Pol.* , vol. 60, pp. 507-514, 2005. [http://dx.doi.org/10.1016/j.carbpol.2005.03.003]

[67] C. Vargas-Rechia, F. Reicher, A. Heyraud, H. Driguez, Y. Linart, and Y. Liénart, "Xyloglucan octasaccharide XXLGol derived from the seeds of hymenaea courbaril acts as a signaling molecule", *Plant Physiol.* , vol. 116, no. 3, pp. 1013-1021, 1998. [http://dx.doi.org/10.1104/pp.116.3.1013] [PMID: 9501133]

[68] T. Stupp, R.A. de Freitas, M-R. Sierakowski, F.C. Deschamps, A. Wisniewski Jr, and M.W. Biavatti, "Characterization and potential uses of Copaifera langsdorfii seeds and seed oil", *Bioresour. Technol.* , vol. 99, no. 7, pp. 2659-2663, 2008. [http://dx.doi.org/10.1016/j.biortech.2007.04.051] [PMID: 17574847]

[69] D.M. Teixeira-Sá, F. Reicher, R.C. Braga, L.M. Beltramini, and R. de Azevedo Moreira, "Isolation of a lectin and a galactoxyloglucan from Mucuna sloanei seeds", *Phytochemistry,* vol. 70, no. 17-18, pp. 1965-1972, 2009. [http://dx.doi.org/10.1016/j.phytochem.2009.08.011] [PMID: 19783018]

[70] S.C. Fry, W.S. York, P. Albersheim, A. Darvill, T. Hayashi, J-P. Joseleau, Y. Kato, E.P. Lorences, G.A. Maclachlan, M. McNeil, A.J. Mort, J.S. Reid, H.U. Seitz, R.R. Selvendran, A.G. Voragen, and A.R. White, "An unambiguous nomenclature for xyloglucan-derived oligosaccharides", *Physiol. Plant.*

, vol. 89, pp. 1-3, 1993.
[http://dx.doi.org/10.1111/j.1399-3054.1993.tb01778.x]

[71] M.S. Buckeridge, and S.M. Dietrich, "Galactomannans from Brazilian legume seeds", *Rev. Bras. Bot. Braz. J. Bot.*, vol. 13, pp. 109-112, 1990.

[72] A. Cronquist, *An Integraded System of Classification of Flowering Plants.* Columbia University Press: New York, 1981.

[73] I.C. Dea, and A. Morrison, "Chemistry and interactions of seed galactomannans", *Adv. Carbohydr. Chem. Biochem.*, vol. 31, pp. 241-312, 1975.
[http://dx.doi.org/10.1016/S0065-2318(08)60298-X]

[74] P.M. Dey, "Biochemistry of plant galactomannans", *Adv. Carbohydr. Chem. Biochem.*, vol. 35, pp. 341-376, 1978.
[http://dx.doi.org/10.1016/S0065-2318(08)60221-8]

[75] V.D. Shcherbukhin, "Galactomannans of native flora (Review)", *Appl. Biochem. Microbiol.*, vol. 29, pp. 599-606, 1993.

[76] J.L. Ganter, *"Galactomananas de sementes de Mimosa scabrella (bracatinga)"*, Dissertation, Universidade Federal do Paraná, Brasil, 1988.

[77] B.V. McCleary, R. Amado, R. Waibel, and H. Neukon, "Effect of galactose content on the solution and interaction properties of guar and carob galactomannans", *Carbohydr. Res.*, vol. 92, pp. 269-285, 1981.
[http://dx.doi.org/10.1016/S0008-6215(00)80398-5]

[78] I.C. Dea, "The role of structural modification in controlling polysaccharide functionality", In: *Industrial Polysaccharides: Genetic Engineering, Structure/Property Relations and Applications.* Elsevier Science Publishers: Amsterdam, 1987, pp. 207-216.

[79] M.S. Buckeridge, V.R. Panegassi, D.C. Rocha, and S.M. Dietrich, "Seed galactomannan in the classification and evolution of the leguminosae", *Phytochemistry,* vol. 38, pp. 871-875, 1995.
[http://dx.doi.org/10.1016/0031-9422(94)00717-8]

[80] P.E. Carvalho, "Bracatinga, uma alternativa para reflorestamento", In: *Anais do quarto Seminário sobre Atualidade e Perspectivas Florestais, Curitiba: Paraná*1981.

[81] P.E. Carvalho, *Espécies Florestais Brasileiras. Recomendações Silviculturais, Potencialidades e Uso da Madeira.* EMBRAPA-SPI: Colombo, 1994.

[82] J.L. Ganter, and F. Reicher, "Water-soluble galactomannans from seeds of Mimosaceae spp", *Bioresour. Technol.*, vol. 68, pp. 55-62, 1999.
[http://dx.doi.org/10.1016/S0960-8524(98)00078-9]

[83] F. Reicher, S.C. Leitner, M-R. Sierakowski, J.D. Fontana, and J.B. Corrêa, "Seed gum of Stryphnodendron barbatiman (barbatimão)", *Appl. Biochem. Biotechnol.*, vol. 28, pp. 353-361, 1991.
[http://dx.doi.org/10.1007/BF02922614]

[84] F. Reicher, S.C. Leitner, M-R. Sierakowski, J.D. Fontana, and J.B. Corrêa, "Properties of the seed gum of Strypnodendron barbatiman (barbatimao)", *Appl. Biochem. Biotechnol.*, vol. 34-35, pp. 349-357, 1992.
[http://dx.doi.org/10.1007/BF02920558] [PMID: 1622205]

[85] C.G. Rechia, M.R. Sierakowski, J.L. Ganter, and F. Reicher, "Polysaccharides from the seeds of Senna multijuga", *Int. J. Biol. Macromol.*, vol. 17, no. 6, pp. 409-412, 1995.
[http://dx.doi.org/10.1016/0141-8130(96)81854-X] [PMID: 8789348]

[86] C.L. Petkowicz, F. Reicher, H. Chanzy, F.R. Taravel, and R. Vuong, "Linear mannan in the endosperm of Schizolobium amazonicum", *Carboid. Pol.*, vol. 44, pp. 107-112, 2001.
[http://dx.doi.org/10.1016/S0144-8617(00)00212-5]

[87] C.L. Petkowicz, M-R. Sierakowski, J.L. Ganter, and F. Reicher, "Galactomannans and arabinans from

seeds of caesalpiniaceae", *Phytochemistry,* vol. 49, pp. 737-743, 1998. [http://dx.doi.org/10.1016/S0031-9422(98)00228-3]

[88] C.L. Petkowicz, S. Schaefer, and F. Reicher, "The mannan from Schizolobium parahybaeendosperm is not a reserve polysaccharide", *Carboid. Pol.* , vol. 69, pp. 659-664, 2007. [http://dx.doi.org/10.1016/j.carbpol.2007.02.002]

[89] V.C. Matos, *"Gomas endospérmicas de Delonix regia e Schizolobium parahybae"*, PhD thesis, Universidade Federal do Ceará, Brasil, 2000.

[90] G.A. Tavares, *"Estrutura e propriedades físico-químicas da galactomanana de sementes de Cassia fastuosa Willd. (cassia)"*, Dissertation, Universidade Federal do Paraná, Brasil, 1994.

[91] J.F. Bento, I. Mazzaro, L.M. de Almeida Silva, R. de Azevedo Moreira, M.L. Ferreira, F. Reicher, and C.L. Petkowicz, "Diverse patterns of cell wall mannan/galactomannan occurrence in seeds of the Leguminosae", *Carbohydr. Polym.* , vol. 92, no. 1, pp. 192-199, 2013. [http://dx.doi.org/10.1016/j.carbpol.2012.08.113] [PMID: 23218282]

[92] V.R. Panegassi, G.E. Serra, and M.S. Buckeridge, "“Seeds of faveiro (Dimorphandra mollis) as a potential source of galactomannan for the food industry,” Ciência Tecnol", *Alime.* , vol. 20, pp. 406-415, 2000.

[93] A. Potomati, and M.S. Buckeridge, "Effect of abscisic acid on the mobilisation of galactomannan and embryo development of Sesbania virgata (Cav.) Pers. (Leguminosae – Faboideae)", *Rev. Bras. Bot. Braz. J. Bot.* , vol. 25, pp. 303-310, 2002. [http://dx.doi.org/10.1590/S0100-84042002000300006]

[94] P.P. Tonini, C.G. Lisboa, L. Freschi, H. Mercier, S.C. Mazzoni-Viveiros, and M.S. Buckeridge, "Effect of abscisic acid on galactomannan degradation and endo-β-mannanase activity in seeds of Sesbania virgata (Cav.) Pers.(Leguminosae)", *Trees (Berl.),* vol. 20, pp. 669-678, 2006. [http://dx.doi.org/10.1007/s00468-006-0082-2]

[95] T. Odijk, and A.C. Houwaart, "On the theory of the excluded-volume effect of a polyelectrolyte in a 1-1 electrolyte solution", *J. Polym. Sci. Pol. Phys.* , vol. 16, pp. 627-639, 1978. [http://dx.doi.org/10.1002/pol.1978.180160405]

[96] H. Yamakawa, and M. Fujii, "Intrinsic viscosity of wormlike chains. Determination of the shift factor", *Macromolecules,* vol. 7, no. 1, pp. 128-135, 1974. [http://dx.doi.org/10.1021/ma60037a024] [PMID: 4837983]

[97] J.L. Ganter, J.C. Sabbi, and W.F. Reed, "Real-time monitoring of enzymatic hydrolysis of galactomannans", *Biopolymers,* vol. 59, no. 4, pp. 226-242, 2001. [http://dx.doi.org/10.1002/1097-0282(20011005)59:4<226::AID-BIP1020>3.0.CO;2-N] [PMID: 11473348]

[98] C.L. Petkowicz, M. Milas, K. Mazeau, T. Bresolin, F. Reicher, J.L. Ganter, and M. Rinaudo, "Conformation of galactomannan. Experimental and modelling approache", *Food Hydrocoll.* , vol. 13, pp. 263-266, 1999. [http://dx.doi.org/10.1016/S0268-005X(99)00008-9]

[99] G. Schramm, *Reologia e reometria: fundamentos teóricos e práticos.* Artliber: São Paulo, 2006.

[100] H.A. Barnes, J.F. Hutton, and K. Waters, *An Introduction to Rheology.* Elsevier: Amsterdam, 1989.

[101] H. Maier, M. Anderson, C. Karl, K. Magnuson, and R.L. Whistler, "Guar, locust bean, tara and fenegreek gums", In: *Industrial Gums: Polysaccharides and Their Derivatives.* Academic Press: New York, 1993, pp. 215-218. [http://dx.doi.org/10.1016/B978-0-08-092654-4.50012-7]

[102] I.C. Dea, E.R. Morris, D.A. Rees, E.J. Welsh, H.A. Barnes, and J. Price, "Associations of like and unlike polysaccharides: Mechanism and specificity in galactomannans, interacting bacterial polysaccharides, and related systems", *Carbohydr. Res.* , vol. 57, pp. 249-272, 1977. [http://dx.doi.org/10.1016/S0008-6215(00)81935-7]

[103] E.R. Morris, D.A. Rees, G. Young, M.D. Walkinshaw, and A. Darke, "Order-disorder transition for a bacterial polysaccharide in solution. A role for polysaccharide conformation in recognition between Xanthomonas pathogen and its plant host", *J. Mol. Biol.* , vol. 110, no. 1, pp. 1-16, 1977. [http://dx.doi.org/10.1016/S0022-2836(77)80095-8] [PMID: 845940]

[104] L. Lundin, and A-M. Hermansson, "Supermolecular aspects of xanthan-locust bean gum gels based on rheology and electron microscopy", *Carboid. Pol.* , vol. 26, pp. 129-140, 1995. [http://dx.doi.org/10.1016/0144-8617(94)00070-A]

[105] M. Rinaudo, M. Milas, T. Bresolin, and J. Ganter, "Physical properties of xanthan, galactomannan and their mixtures in aqueous solutions", *Macromol. Symp.* , vol. 140, pp. 115-124, 1999. [http://dx.doi.org/10.1002/masy.19991400113]

[106] H.S. Koop, C.E. Praes, F. Reicher, C.L. Petkowicz, and J.L. Silveira, "“Rheological behavior of gel of xanthan with seed galactomannan: Effect of hydroalcoholic–ascorbic acid,” Mat. Sci. Eng. C-", *Bio. S.* , vol. 29, pp. 559-563, 2009.

[107] J. Amaral, R.A. Freitas, J.L. Ganter, and M-R. Sierakowski, "The xanthan-galactomannan system: Rheological properties", *Technology and Ecology Symposium,* 1998pp. 110-112 Poznán

[108] T.M. Bresolin, M. Milas, M. Rinaudo, and J.L. Ganter, "Xanthan-galactomannan interactions as related to xanthan conformations", *Int. J. Biol. Macromol.* , vol. 23, no. 4, pp. 263-275, 1998. [http://dx.doi.org/10.1016/S0141-8130(98)00061-0] [PMID: 9849624]

[109] C.T. Andrade, E.G. Azero, L. Luciano, and M.P. Gonçalves, "Rheological properties of mixtures of kappa-carrageenan from Hypnea musciformis and galactomannan from Cassia javanica", *Int. J. Biol. Macromol.* , vol. 27, no. 5, pp. 349-353, 2000. [http://dx.doi.org/10.1016/S0141-8130(00)00139-2] [PMID: 10998493]

[110] P.A. Soares, J.R. de Seixas, P.B. Albuquerque, G.R. Santos, P.A. Mourão, W. Barros Jr, M.T. Correia, and M.G. Carneiro-da-Cunha, "Development and characterization of a new hydrogel based on galactomannan and κ-carrageenan", *Carbohydr. Polym.* , vol. 134, pp. 673-679, 2015. [http://dx.doi.org/10.1016/j.carbpol.2015.08.042] [PMID: 26428171]

[111] M.P. Gonçalves, D. Torres, C.T. Andrade, E.G. Azero, and J. Lefebvre, "Rheological study of the effect of Cassia javanica galactomannans on the heat-set gelation of a whey protein isolate at pH 7", *Food Hydrocoll.* , vol. 18, pp. 181-189, 2004. [http://dx.doi.org/10.1016/S0268-005X(03)00061-4]

[112] G.E. Perissutti, *"Galactomanana de Mimosa scabrella Bentham e proteínas do leite: análise da interação entre biopolímeros"*, PhD thesis, Universidade Federal do Paraná, Brasil, 2002.

[113] G.E. Perissutti, T.M. Bresolin, and J.L. Ganter, "Interaction between the galactomannan from Mimosa scabrella and milk proteins", *Food Hydrocoll.* , vol. 16, pp. 403-417, 2002. [http://dx.doi.org/10.1016/S0268-005X(01)00113-8]

[114] N. Lucyszyn, *"Aplicação de galactomananas como substitutas parciais do ágar no cultivo de tecidos vegetais"*, PhD thesis, Universidade Federal do Paraná, Brasil, 2004.

[115] M.M. Souza-Lima, and M-R. Sierakowski, "Xyloglucan-carboxymethyl cellulose as a model for synergistic interactions", *Technology and Ecology Symposium,* 1998 pp. 108-109 Poznán

[116] N. Lima-Nishimura, *"Utilização de xiloglucana extraída de sementes de Hymenaea courbaril em géis para micropropagação"*, PhD thesis, Universidade Federal do Paraná, Brasil, 2002.

[117] R.A. Freitas, P.A. Gorin, J. Neves, and M-R. Sierakowski, "A rheological description of mixtures of a galactoxyloglucan with high amylose and waxy corn starches", *Carbohydr. Polym.* , vol. 51, pp. 25-32, 2003. [http://dx.doi.org/10.1016/S0144-8617(02)00095-4]

[118] C.F. de Souza, I.C. Riegel-Vidotti, M.B. Cardoso, L. Ono, N. Lucyszyn, A.F. Lubambo, C.V. Sens, A. Grein-Iankovski, and M-R. Sierakowski, "Nanometric organisation in blends of gellan/xyloglucan

hydrogels", *Carbohydr. Polym.* , vol. 114, pp. 48-56, 2014. [http://dx.doi.org/10.1016/j.carbpol.2014.07.070] [PMID: 25263863]

[119] J.A. Sugui, *"Aplicação de galactomananas de Mimosa scabrella (bracatinga): interação com surfactantes e proteína"*, Dissertation, Universidade Federal do Paraná, Brasil, 1994.

[120] D.U. Lima, R.C. Oliveira, and M.S. Buckeridge, "Seed storage hemicelluloses as wet-end additives in papermaking", *Carboid. Pol.* , vol. 52, pp. 367-373, 2003. [http://dx.doi.org/10.1016/S0144-8617(03)00008-0]

[121] N. Lucyszyn, M. Quoirin, A. Anjos, and M-R. Sierakowski, "Blends of agar/galactomannan for Marubakaido apple rootstock shoot proliferation", *Polímeros,* vol. 15, pp. 146-150, 2005. [http://dx.doi.org/10.1590/S0104-14282005000200017]

[122] N. Lucyszyn, M. Quoirin, L.L. Ribas, and M-R. Sierakowski, "Effect of agar, galactomannan and indole-butyric acid on in vitro rooting of the pear cultivar 'Durondeau' and apple rootstock cultivar 'Marubakaido'", *J. Hortic. Sci. Biotechnol.* , vol. 81, pp. 310-314, 2006. [http://dx.doi.org/10.1080/14620316.2006.11512067]

[123] N. Lucyszyn, M. Quoirin, L.L.F. Ribas, H.S. Koehler, and M.-R. Sierakowski, "Micropropagation of `Durondeau' pear in modified-gelled medium", *in vitro Cell. Dev.-Pl.* , vol. 42, pp. 287-290, 2006. [http://dx.doi.org/10.1079/IVP2006753]

[124] N. Lucyszyn, M. Quoirin, H.S. Koehler, F. Reicher, and M-R. Sierakowski, "Agar/galactomannan blends for strawberry (Fragaria x ananassa Duchesne) cv. Pelican micropropagation", *Sci. Hortic. (Amsterdam),* vol. 107, pp. 358-364, 2006. [http://dx.doi.org/10.1016/j.scienta.2005.11.005]

[125] N. Lucyszyn, M. Quorin, M.M. Homma, and M-R. Sierakowski, "Agar/galactomannan gels applied to shoot regeneration from tobacco leaves", *Biol. Plant.* , vol. 51, pp. 173-176, 2007. [http://dx.doi.org/10.1007/s10535-007-0034-8]

[126] C. Ribeiro, G.G. Arizaga, F. Wypych, and M-R. Sierakowski, "Nanocomposites coated with xyloglucan for drug delivery: in vitro studies", *Int. J. Pharm.* , vol. 367, no. 1-2, pp. 204-210, 2009. [http://dx.doi.org/10.1016/j.ijpharm.2008.09.037] [PMID: 18938232]

[127] L. Ono, W. Wollinger, I.M. Rocco, T.L. Coimbra, P.A. Gorin, and M-R. Sierakowski, "in vitro and in vivo antiviral properties of sulfated galactomannans against yellow fever virus (BeH111 strain) and dengue 1 virus (Hawaii strain)", *Antiviral Res.* , vol. 60, no. 3, pp. 201-208, 2003. [http://dx.doi.org/10.1016/S0166-3542(03)00175-X] [PMID: 14638396]

[128] M.M. Rosário, G.R. Noleto, J.F. Bento, F. Reicher, M.B. Oliveira, and C.L. Petkowicz, "Effect of storage xyloglucans on peritoneal macrophages", *Phytochemistry,* vol. 69, no. 2, pp. 464-472, 2008. [http://dx.doi.org/10.1016/j.phytochem.2007.08.011] [PMID: 17888467]

[129] M.M. do Rosário, M.M. Kangussu-Marcolino, A.E. do Amaral, G.R. Noleto, and C.L. Petkowicz, "Storage xyloglucans: potent macrophages activators", *Chem. Biol. Interact.* , vol. 189, no. 1-2, pp. 127-133, 2011. [http://dx.doi.org/10.1016/j.cbi.2010.09.024] [PMID: 20888807]

[130] C.F. Souza, N. Lucyszyn, F.A. Ferraz, and M-R. Sierakowski, "Oxidation and N-alkylation at the C-6 position of galactomannan extracted from Caesalpinia ferrea var. ferrea seeds", *Macromol. Symp.* , vol. 299-300, pp. 66-73, 2011. [http://dx.doi.org/10.1002/masy.200900098]

[131] N. Lucyszyn, A.F. Lubambo, K.F. Matos, I. Marvilla, C.F. Souza, and M-R. Sierakowski, "“Specific modification of xyloglucan from Hymenaea courbaril seeds,” Mat. Sci. Eng. C-", *Bio. S.* , vol. 29, pp. 552-558, 2009.

CHAPTER 4

Chitosan, From Residue to Industry

María B. Bonecco[*], María G. Martínez Sáenz and **Lautaro M. Buffa**

National Institute of Industrial Technology, Mar del Plata, Buenos Aires, Argentina

Abstract: Industrial activity associated with processing of crustaceans generates large quantities of waste. The main residue corresponds to its exoskeletons constituted primarily of chitin. Through a modification on its structure, chitosan is derived, the only polycationic polymer in nature. This work aims to review a wide variety of aspects concerning chitosan. First it studies the chitosan properties: biodegradability, biocompatibility and non-toxicity. Its high density of positive charges makes it a versatile material, being able to be used in practically all the industrial fields. Agriculture, biotechnology, food, wastewater treatment and medicine are the main examples. Then it compares different chitin and chitosan obtaining methods being mostly used chemical treatments. The biological methods are presented as an alternative. Several techniques are necessary for chitosan´s characterization. Molecular weight and deacetylation degree are the most important characteristics which define its potential applications. This chapter analyzes different methods according to the necessity of each situation. An examination on the regulatory status and the global market of this biopolymer is made. A report in 2010 projected its market in 21.4 thousand tons by 2015, placing Japan as the biggest contributor. The main industries are placed in Asia and US. In Latin America, there are plants in Brazil, Chile and México. In Argentina, despite the crustaceans fishing industry continuously growing, there is no industrial chitosan production. This review concludes with a description of the work that is being carried out at the National Institute of Industrial Technology of Argentina in this regard.

Keywords: Antimicrobial, Biocompatibility, Biodegradability, Biopolymer, Characterization, Chitin, Chitosan, Deacetylation degree, Fishing industry, Market, Molecular weight, Non-toxicity, Obtaining process, Polycation, Regulations, Shrimp, Waste.

INTRODUCTION

In the last century, economic and industrial growth related to the development of societies has led to a corresponding increase in the volume of waste generated worldwide. As a consequence, its final destination has become one of the major

[*] **Corresponding author María B. Bonecco:** National Institute of Industrial Technology, Mar del Plata, Buenos Aires, Argentina; Tel/Fax: 54-0223-4802801; E-mail: bbonec@inti.gob.ar

Martin Masuelli & Denis Renard (Eds.)

problems that communities have to confront. Waste management, both locally and globally, is vital for the preservation of the environment and public health. Industrial activity associated with the processing of fishery products, crustaceans in particular (*i.e.* shrimp and crab), generates large quantities of waste. Despite being an excellent raw material, in many regions, it has become a pollutant of high environmental impact (Nuñez Rico, 2010 [1]). The problem is not only the volume of accumulated waste, but also its slow rate of degradation. This has led to the establishment of different lines of research which seek to find possible solutions to the environmental problem in addition to the potential economic exploitation of the resource.

One of the main residues which results from processing crustaceans corresponds to their exoskeletons; rigid structures that give support and protection to those animals. The most important structural component is chitin, which is the most abundant polysaccharide in nature after cellulose. Chitin does not have a wide range of uses in the chemical industry due to its insolubility in water, organic solvents and dilute acids. However, its importance resides in the possibility of changing its structure in order to obtain another polysaccharide, chitosan (Khor & Lim, 2003 [2]).

Chitosan is the only polycationic polymer in nature. The high density of its positive charges is responsible for its unique physical, biological and physiological properties. Chitosan in its solid state is a semi-crystalline polymer. It is biodegradable, biocompatible and non-toxic (Peh *et al.*, 2000 [3]). It is also well known for its antimicrobial and filmogenic properties (Skjak-Braek *et al.*, 1989 [4]). The molecular characteristics of chitosan make it an incredibly versatile material, allowing it to be used in several ways, such as solutions, films, hydrogels, fibers, porous structure scaffolds, pastes, sheets, tablets, nanoparticles, microspheres, among others. This versatility makes it suitable for being used in practically every industrial field with food and beverage, biotechnology, agriculture, wastewater treatment, medicine, pharmaceutical and textiles being the most important examples.

The discovery of chitosan dates back to 1859, with the studies of Professor C. Rouget, who reported deacetylated chitin forms. In 1894, researchers Hoppe and Seyler called it chitosan (Lárez Velásquez, 2006 [5]). Between the 30's and 40's, these polymers attracted a considerable amount of attention, registering a total of 50 patents. Chitosan has been considered the biopolymer of the 21st century because its study and resulting applications have grown exponentially during the last decades (Nuñez Rico, 2010 [1]). It is important to emphasize the growing activity developed around chitin and chitosan, which has resulted in the organization of congresses, conferences and symposiums dedicated exclusively to them. Their importance as polymeric materials was first recognized internationally during the 70s, when the book "Chitin" was published by Muzzarelli (1974 [6]) and the First International Conference of Chitin and Chitosan was held in 1978. Latin America has been also working with chitosan. In 2000, the First Latin American Symposium of Chitin and Chitosan was held in Havana (Cuba) during the XI Latin American Symposium on Polymers (Lárez Velásquez, 2003 [7]).

The global importance of chitosan and its industry has led to the authoring of this review, about its production, properties and characterization, as well as its main uses and regulations.

RAW MATERIAL

The raw material used in the production of chitosan are the exoskeletons of shellfish which come from the fish processing industry. The exoskeleton is a rigid non-cellular structure secreted by the epidermal cells of such animals, which are generally divided into three regions: head, thorax and abdomen. The exoskeleton is composed of chitin, proteins, minerals and pigments, where proportions vary depending on the species of crustacean, the backbone region, nutritional status and reproductive cycle of the animal from which they come from. Older specimens have a lot more calcified exoskeletons with a lower proportion of chitin (Tharanathan & Kittur, 2003 [8]). In general, the component proportions are: chitin (15%-40%), protein (20%-40%) and $CaCO_3$ (20%-50%). Carotenoid pigments such as lutein, astaxanthin and β-carotein are found in small quantities along with other mineral elements (Colina *et al.*, 2014 [9]).

In industrial processing, chitin is recovered from crustacean's exoskeletons by alkaline extraction to solubilize proteins followed by acid treatment to dissolve the $CaCO_3$, or *vice versa*. This treatment must be adapted to each chitin source due to the differences in the ultrastructure of the initial raw material (Rinaudo, 2006 [10]). As it will be seen below, this variability constitutes an obstacle to standardize a method for obtaining chitosan, so it is necessary to have different methodologies for the characterization of the obtained product. Table **1** shows the percentages of the major components present in some species of crustaceans.

Table 1. Proximate chemical composition of crustacean's exoskeletons by species (% v/v) on dry basis.

Origin of raw material	Protein	Chitin	Ash content	Lipid
Callinectes sapidus (Crab)[1]	25.1	13.5	58.6	2.1
Chionoecetes opilio (Crab)[1]	29.2	26.6	40.6	1.3
Paralithodes camtschaticus (Crab)[1]	22.0	31.0	46.0	1.0

(Table 1) contd.....

Origin of raw material	Protein	Chitin	Ash content	Lipid
Pandalus borealis (Shrimp)[1]	41.9	17.0	34.2	5.2
Penaeus monodon (Shrimp)[1]	47.4	40.4	23.0	1.3
Procambarus clarkii (Lobster)[1]	29.8	13.2	46.6	5.6
Euphausia superba (Krill)[1]	41.0	24.0	23.0	11.6
*Pleoticcus muelleri (S*hrimp)[2]	34.9	16.3	34.2	ND*

[1]Adaptation from Colina *et al.*, 2014 [9]
[2] Physical – Chemistry Laboratory INTI - Mar del Plata
*ND: Not determined

MOLECULAR CHARACTERISTICS

Molecular Structure

The precursor of chitosan, chitin, is a linear homopolysaccharide consisting of *N*-acetyl-β-D-Glucosamine units, a six carbon monosaccharide derived from glucose wherein the hydroxyl group (-OH) of carbon 1 (C1) is replaced by an amine group ($-NH_2$), where an acetyl group is added ($-CO-CH_3$) (Fig. **1**). The monomers are linked together by β (1-4) glycosidic bond. This type of linkage involves the C1 of the first monomer with the C4 of the following monomer, both linked by an oxygen atom.

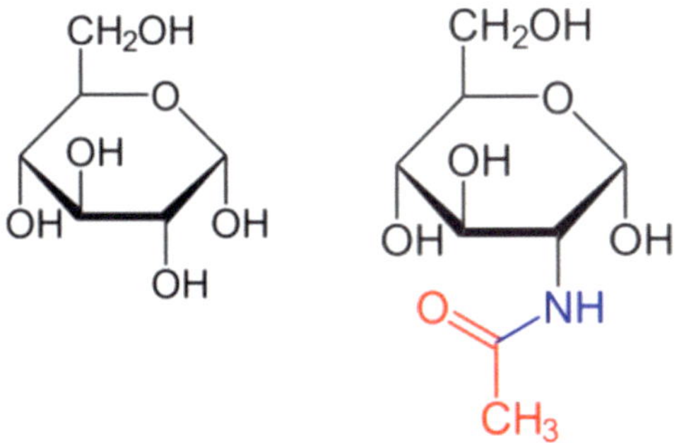

Fig. (1). Glucose and *N*-Acetil-Glucosamine structure.

Chitin in nature exists in two main allomorphs, α and β. A third one, γ-chitin, has also been described but it seems to be a variant of the α-family. The α-family is by far the most abundant, occuring in fungal and yeast cell walls, in krill, in lobster and crab tendons and shells, shrimp shells as well as in insect cuticles. The β-chitin can be found in association with proteins in the squid pen of cephalopods (Rudall & Kenchington, 1973 [11]). These allomorphs differ in the packing and polarities of adjacent chains in successive sheets (Fig. **2**). In the β-form, all chains are aligned in a parallel manner, which is not the case in α-chitin. In both structures, the chitin chains are organized in sheets where they are tightly held by a number of intra-sheet hydrogen bonds with the α and β chains packed in antiparallel arrangements (Yamaguchi *et al.*, 2005 [12]). The differences between both structures result in a β-chitin with better reactivity, swelling, and solubility than α-chitin (Pillai *et al.*, 2009 [13]). In spite of a lower reactivity, α-chitin is preferentially used by industrials due to its higher relative abundance.

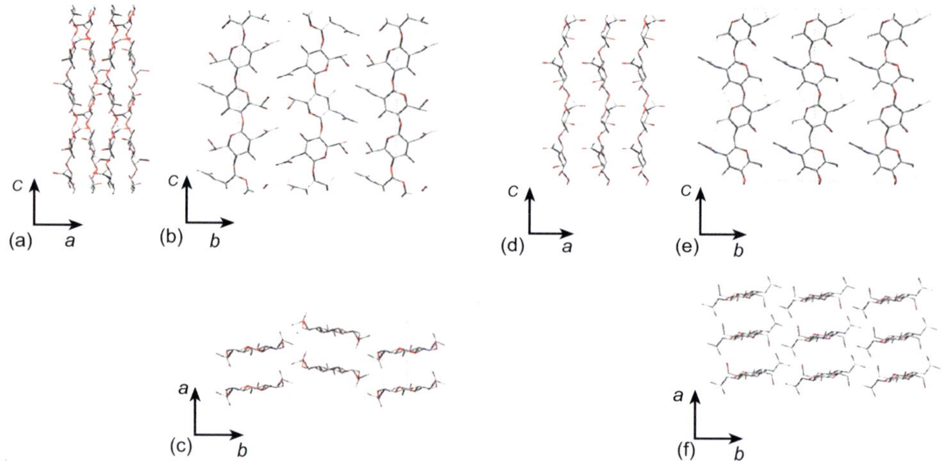

Fig. (2). Structure of α-chitin: (**a**) ac projection; (**b**) bc projection; (**c**) ab projection. Structure of anhydrous β-chitin: (**d**) ac projection; (**e**) bc projection; (**f**) ab projection. Extracted from Rinaudo, 2006 [10].

Chitosan is obtained by removing a large number of acetyl groups from the C2 of the chitin's monomers making the amine groups of those carbons to be exposed (Fig. **3**). Each monomer has three types of reactive functional groups, an amine (βNH_2) or acetamide group (-$NHCOCH_3$) in C2, and two hydroxyl groups (-OH) in C3 and C6. Most of chitosan functionalities involve specific reactions concerning the amine groups, which are responsible for making chitosan one of the most versatile materials being studied, giving it the ability to perform a wide variety of modifications, such as anchoring enzyme reactions, obtaining crosslinked films, *etc.* (Lárez Velásquez, 2003 [7]).

Deacetylation Degree and Molecular Weight

The deacetylation degree (DD) and the molecular weight (MW) are the most important characteristics of chitosan´s structure which determine its potential applications. The DD allows the differentiation of chitin from chitosan. It reflects the balance between the repeating units of 2-acetylamine-2-deoxy-β-D-glucosamine and 2-amine-2-deoxy-β-D-glucosamine (Khor & Lim, 2003 [2]). It is considered chitosan when the DD of the molecule is greater than 50%. When all the acetyl groups are removed, the resulting polymer is called chitan (Lárez Velásquez, 2006 [5]). On the other hand, chitosan's MW ranges from 5 kDa to 2000 kDa, from oligomers to larger structures (Ramírez *et al.*, 2002 [14]). Rheological and mechanical properties, such as friction coefficient and sedimentation and its hydrodynamic volume, can be estimated according to the MW and the conformation of the biopolymer molecules (Yamakawa, 2001 [15]). Both deacetylation degree and molecular weight depend on processing factors such as temperature, time and reactant concentrations used during chitosan production. Different determining techniques are mentioned further in this chapter. Table **2** shows different requirements of DD and MW according to each application.

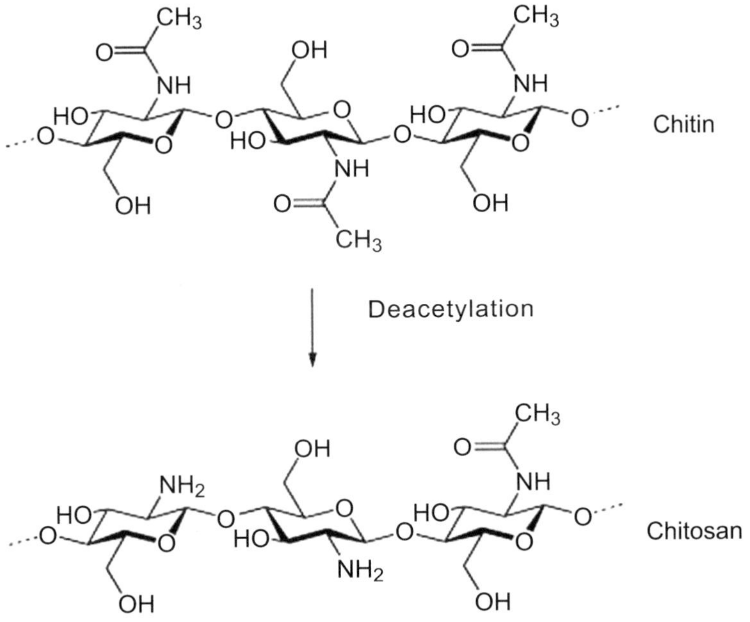

Fig. (3). Chitin and chitosan structure.

Table 2. General recommendations for the application of chitosan according its DD and MW.

Application	**Deacetylation Degree (DD)***	**Molecular Weight (MW)****
Wound healing	High	Low (oligomers)
Drug delivery systems	High	High
Gene Delivery	≤ 80%	Low (around 10kDa)

(Table 2) contd.....

Application	Deacetylation Degree (DD)*	Molecular Weight (MW)**
Scaffolds (tissue engineering)	Around 85% (Good proliferation and structure)	High (prolonged biodegradation)
Dietary ingredient	High	High (viscosity)
Food preservative	High	Medium-Low (5-80 kDa)
Emulsifying agent	Low (for emulsion stability)	(High viscosity)
Wastewater treatment	High (Depend on pollutant and water conditions: pH, ionic strength)	-
Metal reduction	High	Relationship between morphology and MW: Low-2D chains; Medium-single nanoparticles; High-nanoplates

*DD: Low:<60%, Medium:60%<DD<80%, High:>80%.
**MW: Low:<60 kDa, Medium:60 kDa<MW<100 kDa, High:>100 kDa.
Adaptation from Aranaz *et al.*, 2009 [16].

Solubility in Aqueous Media

Chitin is insoluble in water and organic solvents due to its intra and intermolecular hydrogen bonds. However, chitosan is soluble in organic acids such as formic, acetic, citric and tartaric acids, and also in mineral acids, except for sulfuric and phosphoric (Gauna & Nuñez, 2004 [17]). The solubility of chitosan depends mainly on three factors related to the properties of the molecule (Kubota *et al.*, 2000 [18]; Peniche & Argüelles-Monal, 2001 [19]):

- Deacetylation degree
- Molecular weight
- Distribution of acetoamide groups in the main chain

When introducing chitosan in its solid state to an acid solution, its amine groups are ionized with the available protons in solution giving the possibility of forming hydrogen bonds with water, remaining in solution. When chitosan is protonated, it becomes a soluble cationic polyelectrolyte in an aqueous medium, differentiating it from its original polymer, the chitin (Papineau *et al.*, 1991 [20]). What physically happens is that the ionized amines are unable to form hydrogen bonds with the carboxylic acids of the adjacent monomer, decreasing the impediment to rotation. Additionally, the acetyl group is heavier than the amine and by reducing its presence, it provides greater steric rotation, increasing the flexibility of the polymer chain and enabling chitosan to become a fluid (Wang *et al.*, 2006 [21]).

Formic acid is the best solvent for chitosan found so far, where solutions are obtained in aqueous systems containing 0.2%-100% of formic acid. On the other hand, as the pH increases above 6, chitosan's amines become deprotonated and the polymer loses its charge and becomes insoluble (Pillai *et al.*, 2009 [13]). However, depolymerization of chitosan contributes to its solubility in aqueous media at neutral pH. It can also be improved at neutral and/or slightly basic pH by the addition of glucosamine through the Maillard reaction which combines the amine group of the biopolymer with the carbonyl group of the reducing sugar, releasing a water molecule (Chung *et al.*, 2005 [22]; Tanaka *et al.*, 1993 [23]). The solubility is increased by the contribution of new hydroxyls groups and the amine group by the added glucosamine. The use of glucosamine *versus* other sugars is convenient because it does not alter the DD of the molecule, which is an important parameter to keep the biological, physiological and pharmacological properties of chitosan as mentioned earlier (Aranaz *et al.*, 2009 [16]).

MOLECULAR PROPERTIES

Toxicity and Biocompatibility

Chitosan is regarded as a non-toxic and biologically compatible polymer (Dash *et al.*, 2011 [24]). Its LD_{50} is 16 ml/kg (Paul & Sharma, 2000 [25]), a value similar to sucrose and less toxic than salt. However, modifications made to its molecular structure also modify this value.

The biocompatibility property of chitosan has made it a valuable polymer in the fields of medicine and pharmacy. It is expected that a biocompatible material shows minimal inflammatory reactions and toxicity at both local and systemic levels. To achieve this end, the tissues have to show a high tolerance to foreign agents. An effective approach for developing a clinically applicable chitosan, is to modify its physical and chemical surface properties through various derivatives and blending technologies. However, these modifications may affect chitosan biocompatibility and biodegradability (Sarmento & Das Neves, 2012 [26]).

Biodegradability

Biodegradation of chitosan refers to the breakdown of the polymer into smaller fractions and into its monomers (D-glucosamine, *N*-acetyl-glucosamine). This process is catalyzed by enzymes or chemicals (Sarmento & Das Neves, 2012 [26]). A variety of microorganisms synthesizes and/or degrades chitosan and chitin by chitinases. These enzymes are also present in higher plants, even though they do not have chitin in its structural components, reflecting a defense mechanism against microorganisms and insects that have structures composed of chitin (Collinge *et al.*, 1993 [27]). The degradation rate of these biopolymers depends on their chemical structure, MW, DD and the surrounding media to which they are applied (Xu *et al.*, 1996 [28]; Yang *et al.*, 2007 [29]). Higher DD shows lower degradation rates and the same behaviors are observed with the MW.

Chitosan's biodegradability has encouraged interest in research related to its application in medicine and pharmacology fields. In these fields, it is of vital importance to study its metabolic fate in the body once entered into it. The biodegradation rate is particularly crucial in the use of chitosan in drug-delivery systems and tissue engineering. The degradation kinetics may affect the performance of cell growth, tissue regeneration and host response (Halim *et al.*, 2012 [30]).

A number of studies have examined the *in vitro* and *in vivo* chitosan degradation, often in films or implants, while its fate upon oral administration is less reported (Baldrick, 2010 [31]). Lysozyme is one of the primary enzymes used for *in vitro* degradation studies. In human beings, besides being found in saliva and tears, lysozyme is also present in the spleen, lungs, leukocytes, plasma and cartilage. In the context of drug carrier resorption and wound healing, it is reported that chitooligomers and monomers are generated from chitosan biodegradation by lysozyme, *N*-acetylglucosaminidase and human chitinase (Muzzarelli, 1997 [32]).

It has also been shown in rabbits that serum lysozyme levels were elevated after an intravenous dose of chitooligomers (Hirano *et al.*, 1991 [33]). A proposed mechanism of this enzymatic degradation can be divided in the following stages: in the first stage, water and lysozyme permeate into the chitosan structure inducing swelling. The degradation occurs when the lysozyme acts on chitosan by cleavage of the β-(1-4) glycosidic bonds to form low MW chitosan, chitooligomers and *N*-acetyl-D-glucosamine residues. The low MW fragments remain in bulk until they are small enough to dissolve into the surrounding media. During the second stage, the swelling reaches the maximum point and the degradation proceeds as the chitosan structure continues to decrease in MW. At this point, small fragments are released into the media. Finally, chitosan is completely degraded to the smallest structure. However, this last stage depends on the DD of the chitosan; with higher DD it is more crystalline and, therefore, more difficult to be hydrolyzed requiring more time for its fragmentation. The hydrolyzed chitosan results in glucosamine, which is suitable for excretion from the body by urine (Halim *et al.*, 2012 [30]).

Further work following subcutaneous implantation of chitosan hydrogel in rats showed no detectable gel fragments in distant organs (brain, heart, lungs, liver, spleen and kidney) at times ranging from 4 to 30 days (Azab *et al.*, 2007 [34]). Gel degradation was suggested to be due to an oxidative process at the dose site. Intraperitoneal injection of fluorescin isothio-cyanate-labelled chitosan to mice resulted in rapid exposure to the kidney and urine, with scarce distribution to the liver, spleen, abdominal tissue and plasma (Onishi & Machida, 1999 [35]). The authors concluded that chitosan is highly biodegradable and easily excreted in urine, and showed no issues of accumulation/retention in the body.

Although the biodegradability of chitosan after oral administration to animals has not been well investigated, substantial amounts have been shown to be digested in the gut of rabbits (Hirano *et al.*, 1990 [36]). Indeed, chitinolytic enzymes are present in the intestinal mucosa of animals. An investigation of intestinal absorption of 14C-chitosan in rats showed that the material underwent digestion into low molecular weight substances within the gastrointestinal tract, and that it was distributed extensively in tissues (Nishimura *et al.*, 2003 [37]). Oral and intravenous studies in rats and dogs with radio-labelled glucosamine showed rapid clearance by the liver and kidney and excretion in the urine for a part of the dose, with the remainder ending up in skeletal muscle and articular cartilage (Anderson *et al.*, 2005 [38]). Chitooligomers produced by depolymerization of chitosan have been reported to have good intestinal absorption (Qin *et al.*, 2006 [39]). On the other hand, (Goycoolea *et al.* (2000 [40]), established that chitosan cannot be digested by humans, so it can be considered as a zero-calorie fiber. It has to be considered that *N*-substitution may affect enzymatic degradation. In some cases, chemical modification enhanced the biodegradation rate, while, in other cases it reduced it, so this should be considered when new derivatives are suggested for systemic administration (Kean & Thanou, 2010 [41]).

Antimicrobial Activity

The antimicrobial activity of chitosan has become of great interest, especially in the development of agricultural products which are effective in controlling pests and plant diseases. The mechanisms of action by which chitosan exerts its antibacterial activity have not been completely decoded. However, there are several postulated routes, some of them point to a direct effect on the pathogen, while others to a defense response on the host. (Helander *et al.* 2001 [42]) mentioned in their work that the electrostatic interaction between positively charged chitosan with negative residues of macromolecules present in the cell wall of bacteria, changes the permeability of the cell wall with the consequent alteration of its main function. Other authors have proposed that the formation of polyelectrolyte complexes physically block the outer cell membrane of the microorganism, preventing the normal flow of nutrients and waste, causing bacterial death (Chung *et al.*, 2004 [43]). A third postulated mechanism is that chitooligomers of low MW penetrate the cells of the microorganisms and prevent their growth inhibiting the action of several enzymes and interfering with protein synthesis by means of the inhibition of the transcription of DNA into mRNA (Tharanathan & Kittur, 2003 [8]).

Fungicidal activity of chitosan has also been studied both *in vitro* (El Ghaouth *et al.*, 1992 [44]) and *in vivo* (Li & Yu, 2001 [45]; Li *et al.*, 2007 [46]). In the same way, this property has been related with its positive charges, being less effective with those fungus who have chitosan in their cell walls (Allan & Hardwiger, 1979 [47]; Roller & Covill, 1999 [48]). This may be due to the fact that the electrostatic interactions of chitosan added (exogenous) with endogenous chitosan, both positively charged, is opposed to those cases in which the cell walls have macromolecules with negative charges. Chitosan also acts as a chelating agent that selectively binds trace metals and therefore inhibits mycotoxin production and mycelial growth (Cuero *et al.*, 1991 [49]).

The ability of chitosan to induce natural defense responses in plants has also been extensively studied. Some authors proposed that chitosan binds to fungal receptor sites simulating a fungal attack. The plant sends the information to the cell nucleus by a process of signal transduction causing the elicitation of numerous responses and biological processes that contribute to inhibit fungal infections. This response includes defense gene activation with subsequent synthesis of lignin and callose, synthesis of phytoalexins, protease inhibitors and the induction of enzymes related to pathogenicity: chitinase, glucanase, peroxidase and chitosanase (Ben-Shalom *et al.*, 2003 [50]; Hadwiger, 2013 [51]; Vander *et al.*, 1998 [52]).

Filmogenic Properties

The formation of intermolecular hydrogen bonds between the amine and hydroxyl groups in chitosan chains enables the production of chitosan films with good physical and mechanical properties (Salazar & Negr, 2013 [53]). Diverse techniques are used for their production, the most common being solvent evaporation and molding (Butler *et al.*, 1996 [54]). Diluted acids such as formic, acetic or propionic are used as the solvent in this process. Robust, flexible and transparent films can be obtained due to the linear structure of this polysaccharide (Kittur *et al.*, 1998 [55]).

The film forming ability of chitosan has been the object of many studies, some of which have led to industrial uses in various fields (Makino & Hirata, 1997 [56]). Photographic films, reverse osmosis membranes and cosmetics containing significant amounts of chitosan are now being marketed (Muzzarelli *et al.*, 1986 [57]). In food technology, the filmogenic property is complimentary to the biodegradability of chitosan and the low permeability to oxygen of its films (Kardas *et al.*, 2012 [58]).

Pure chitosan films are generally cohesive and compact, and the film surfaces have a smooth finish without pores or cracks (Coma *et al.*, 2002 [59]). They tend to exhibit resistance to fats and oils and have selective gas permeability, but they have a high water permeability because of their high hydrophilic character (Kittur *et al.*, 1998 [55]). To avoid this, many researchers seek to combine chitosan with other polymers to improve its properties (Buffa *et al.*, 2015 [60]).

Complex Formation with Metals

Amine groups present in the chitosan chain are highly reactive for chelating metals to a pH near neutrality (Acosta *et al.*, 2000 [61]). This property has become very important in the wastewater treatment industry (Muzzarelli, 1973 [62]). Heavy metals such as Hg, Cu, Pb, Zn, Ni, Cr and others are removed from wastewater by complexation with chitosan (Trimukhe & Varma, 2008 [63]). The speed and adsorption capacity of chitosan in wastewater depends on its state of implementation (powder, fiber or flakes) as well as specific reaction factors such as temperature, stirring speed and contact time among others (Rinaudo, 2006 [10]).

PRODUCTION METHODS

The method to produce chitosan from fishery industry waste consists of first isolating chitin from the crustacean's exoskeletons followed by a deacetylation reaction. Chemical and biological methods can be used in both stages and even a combination of these is also possible. As it was mentioned before, the composition of raw material varies from one species of crustacean to another or within the same species, making it necessary to adjust the chitin's isolation methods. In a similar way, chitosan production must be also adapted, but in this case, to give the required MW or DD for its desired application.

Chitin Isolation

The chemical method for the isolation of chitin involves two major steps: the extraction of protein matter in an alkaline medium (deproteinization) and the elimination of inorganic matter, mainly $CaCO_3$, in an acidic medium (demineralization). Generally, the raw material is crushed, washed with water and cut into small pieces prior to extraction (Abdou *et al.*, 2008 [64]).

The deproteinization process is performed with aqueous alkali solutions, traditionally of NaOH or KOH. The linkage between chitin and protein is through amide formation (Roberts, 1992 [65]). This bond occurs between amine groups in chitin with side chain carboxylate groups in protein. In presence of NaOH, the amide is hydrolyzed and the protein is removed (Lertsutthiwong *et al.,* 2002 [66]). It is important to emphasize that the use of NaOH on the deproteinization results in partial deacetylation of chitin, along with its hydrolysis, lowering its MW.

Demineralization of crustacean exoskeletons is carried out by aqueous acid solutions, normally by HCl. This process involves the decomposition of $CaCO_3$ into water soluble calcium salts with the release of CO_2. In the same way, other minerals present in the skeleton react similarly and give soluble salts in presence of acid. Prolonged decalcification time, even during 24 h, results in a very smooth drop in the ash content but can cause polymer degradation (No & Meyers, 1995 [67]).

The effectiveness of both deproteinization and demineralization depends on the temperature, time, reactant concentration, the ratio of raw material to solution, and the sequence in which they are carried out (Sagheer *et al.*, 2009 [68]). Within the range for these processes, it is feasible to obtain chitin with different ash contents, protein content and MW. On that basis, many scientific articles have been published using different combinations of these processing parameters. (Lertsutthiwong *et al.* 2002 [66]) concluded that when exoskeletons are decalcified first, the chitin obtained is more colorful than when the deproteinization is carried out first. Further, carrying out decalcification first gives higher ash contents, as well as a higher yield because the chitin is protected by the adhered protein resulting in less hydrolysis of the chitin's structure. In the opposed process, when deproteinization is carried out first, the protective layer of protein is removed and the unprotected chitin is exposed to the HCl, leading to efficient demineralization but also to greater hydrolysis of the polymer with the consequent loss of material. Table **3** shows the results of (Lertsutthiwong *et al.* 2002 [66]), where different times and reactant concentrations were compared as well.

Table 3. Characteristics of chitin isolated by different treatments.

Treatments	Moisture content (%)	Ash content (%)	Protein content (%)
0.25 M NaOH, 21 h & 1.1 M HCl, 2 h	7.93 ± 0.12	0.67 ± 0.06	9.64 ± 0.26
1.1 M HCl, 2 h & 0.25 M NaOH, 21 h	8.06 ± 0.23	2.18 ± 0.09	14.0 ± 0.21
0.5 M NaOH, 21 h & 1.1 M HCl, 2 h	7.38 ± 0.04	0.36 ± 0.03	3.19 ± 0.12
1.1 M HCl, 2 h & 0.5 M NaOH, 21 h	7.75 ± 0.19	1.82 ± 0.06	2.45 ± 0.00
1 M NaOH, 21 h & 1.1 M HCl, 2 h	8.08 ± 0.08	0.30 ± 0.02	0.80 ± 0.04
1.1 M HCl, 2 h & 1 M NaOH, 21 h	7.37 ± 0.10	1.71 ± 0.09	ND*
0.25 M NaOH, 21 h & 1.1 M HCl, 12 h	7.84 ± 0.06	0.47 ± 0.05	7.21 ± 0.08
1.1 M HCl, 12 h & 0.25 M NaOH, 21 h	7.84 ± 0.09	1.56 ± 0.09	11.1 ± 0.09
0.5 M NaOH, 21 h & 1.1 M HCl, 12 h	7.90 ± 0.06	0.36 ± 0.02	3.34 ± 0.09
1.1 M HCl, 12 h & 0.5 M NaOH, 21 h	7.95 ± 0.05	1.16 ± 0.02	1.95 ± 0.09
1 M NaOH, 21 h & 1.1 M HCl, 12 h	7.27 ± 0.17	0.20 ± 0.01	1.72 ± 0.10

(Table 3) contd.....

Treatments	Moisture content (%)	Ash content (%)	Protein content (%)
1.1 M HCl, 12 h & 1 M NaOH, 21 h	8.03 ± 0.10	1.05 ± 0.05	ND*

*ND: non-detectable
Adapted from Lertsutthiwong *et al.*, 2002 [66].

As an alternative to the traditional chemical process, biological methods are being evaluated to isolate chitin from marine sources (Sagheer *et al.*, 2009 [68]). The most common are the enzymatic and the fermentation methods.

The first attempts to use proteolytic enzymes to deproteinize crustacean waste were made by (Takeda & Abe 1962 [69]) and (Takeda & Katsuura 1964 [70]). In their works, tuna proteinase, papain and bacterial proteinase were used for the deproteinization of chitin with a remainder of 5%. Proteolytic enzymes are mainly derived from plants, microbial and animal sources and they can remove proteins under mild conditions. Alcalase, pepsin, papain, pancreatine and trypsin are some examples and they are able to be used with crustacean exoskeletons. Enzymatic deproteinization may be performed either after, or before the demineralization step of the solid material, which modifies the accessibility for the reactants (Younes & Rinaudo, 2015 [71]). A lower efficiency with the use of enzymes compared with chemicals is a disadvantage of this method. Thereby, after the treatment, more protein remains bound to the chitin, making a subsequent treatment with dilute NaOH necessary to increase its purity. As with the chemical method, demineralization of the shrimp shells is also necessary and a dilute solution of HCl is commonly used for this step. It must be noted that if enzymatic deproteinization is applied first, the minerals present in the cuticles may decrease the accessibility of the proteases and affect deproteinization efficiency. Thus, demineralization should be performed first (Younes & Rinaudo, 2015 [71]).

The fermentation process is an economically favorable alternative to the enzymatic method, employing proteases produced by endogenous microorganisms (called auto-fermentation) or by selected strains added intentionally. Lactic acid fermentation of crustacean exoskeletons can be performed by *Lactobacillus* sp. Lactic acid produced by these bacteria lowers the silage pH, suppressing the growth of spoilage microorganisms. At the same time, it reacts with the $CaCO_3$, leading to the formation of a precipitate of calcium lactate separated from lighter shells, which are recovered and rinsed with water. This process may be done either on purified crustaceous shells, or on complete shrimp waste (including heads and viscera). The efficiency of lactic acid fermentation depends on many factors, mainly the species and quantity of inoculums, carbon source and its concentration, initial pH and its change during fermentation, temperature and the duration of the process (Oh *et al.*, 2007 [72]; Prameela *et al.*, 2010 [73]; Rao *et al.*, 2000 [74]). Other works have isolated chitin from shellfish waste using whey and sucrose as substrate and carbonate source, respectively. However, a milder alkali treatment was needed to remove proteins and pigments that remained in the exoskeleton structure after the enzymatic treatment (Marcia *et al.*, 2011 [75]).

In addition, bacteria and fungi can be used for chitin isolation through non-lactic acid fermentation. (Ghorbel-Bellaaj *et al.* 2012 [76]) have used *Bacillus* strains, concluding that an excess of glucose has no influence on deproteinization degree and improves demineralization. The same author worked with *Pseudomonas aeruginosa* A2, reporting that the most important variables on the deproteinization and demineralization degree were shrimp shell concentration, glucose concentration, inoculum size and incubation time, while initial medium pH, temperature, speed of agitation and volume of culture had no effect on fermentation efficiency. (Khanafari *et al.* 2008 [77]) have isolated chitin from shrimp shells using chemical and microbial fermentation methods and the results showed that the biological process preserves the chitin structure better than the chemical alternative. Compared to chemical processes, longer processing times and the poorer accessibility of proteases caused by the presence of minerals (leading to higher residual protein), are drawbacks of microbial fermentation. However, fermentation methods are simple, more productive and environmentally friendly. It is possible to improve the deproteinization rate using simultaneous or successive processes such as two-step fermentations or co-fermentation of microorganisms. In order to obtain highly purified chitin, these types of processes must be completed by subsequent mild chemical treatment to remove the residual protein and minerals.

To choose between chemical and biological methods, (Heppe & Le Roux 2012 [78]) propose an integral analysis, where not only the physical and chemical characteristics of the product are considered, but also geographical and economical factors. In this presentation, Heppe and Le Roux compared chemical and enzymatic methods used on an industrial scale. According to them, the chemical process is convenient if there are low water costs, low chemical costs, low investment budget and low qualification of workers. In contrast, the enzymatic process is preferable when there are high water costs, high NaOH costs, profitable enzyme source available, good investment budget and high qualification of workers.

Chitosan Production Process

The most common method used for preparing chitosan is *via* the heterogeneous chemical deacetylation of chitin, employing highly concentrated alkaline solutions (Lamarque *et al.*, 2004 [79]). Time and temperature of the process influence the properties of the final product. To obtain a chitosan with high deacetylation degree, a highly concentrated aqueous solution of NaOH (at least 10 M) is commonly used at temperatures ranging from 50ºC to 130ºC for a time ranging from a few minutes up to several hours. With such severe reaction conditions, chitin undergoes significant alkaline hydrolysis responsible for extensive depolymerization. Several options were suggested to minimize this degradation. Some authors proposed changes to the reaction atmosphere by using an inert gas (N_2 or Ar), reduction of the volume of alkaline solution used, as well as the use of a diluting agent, such as acetone (Lamarque *et al.*, 2004 [79]). In addition, the chitin morphology influences the deacetylation reaction. β-chitin is more susceptible because of its weak intermolecular forces and accordingly, its greater accessibility of the chitin chains, giving a greater rate of reaction at all temperatures compared with α-chitin, although the activation energy is approximately the same for both polymorphic forms.

One of the disadvantages of chemical deacetylation is that the final product is a comprised of a group of polymers with a distribution of different MW, DD and functional groups distributions. In some cases, the presence of insoluble matter can be found after deacetylation and is related to chitosan particles with regions of insufficient deacetylation (Lertsutthiwong *et al.*, 2002 [66]). Another aspect to consider is that the wastewater generated needs to be treated due to its high NaOH concentration and by-products of the deacetylation reaction as well as pigments.

Along with chitin, chitosan can also be obtained by enzymatic treatment. Chitin deacetylase catalyzes the hydrolysis of *N*-acetamide bonds of chitin. It was first isolated and purified by (Araki & Ito 1974 [80]). There have been a number of studies devoted to the development of an enzyme-based method for converting chitin to chitosan using chitin deacetylases. The most frequently studied enzymes exhibit a very strong specificity for β-(1,4)-linked *N*-acetyl-D-glucosamine polymers and they are thermally stable at their optimal temperature (50ºC). They are extracted mainly from the fungi *Mucor rouxii*, *Absidia coerulea*, *Aspergillus nidulans* and *Colletotrichum lindemuthianum*. However, the enzymes vary significantly in their molecular weight and carbohydrate content and display a wide range of optimal pH values. It is interesting to notice that chitin deacetylases, produced by *C. lindemuthianum* and *A. nidulans*, are not inhibited by acetate (a product of deacetylation) which makes them suitable for potential biotechnological applications (Alfonso *et al.*, 1995 [81]; Tokuyasu *et al.*, 1996 [82]; Tsigos & Bouriotis, 1995 [83]).

An important drawback of this method is the very limited accessibility of chitin as a substrate for enzymes (Roberts, 1977 [84]). Several studies show that the pretreatment of chitin before enzyme addition improves the accessibility of the acetyl groups to the enzyme. In other studies, which took into account this consideration, the enzyme produced a deacetylated chitosan up to 97% in selected conditions. Some of the advantages of the enzymatic process are: that it is environmentally friendly, has lower chemical costs and is carried out at an ambient temperature. Also, the use of chitin deacetylases offers the possibility of a controlled, non-degradable process, resulting in the production of chitosan with a well-defined structure and composition, contrary to the chemical method (Younes & Rinaudo, 2015 [71]).

In practice, chitosan's oligomers are typically obtained *via* depolymerisation of the chitosan raw material. Chemical treatments with acid hydrolysis and enzymatic reactions using chitosanases, chitinases, hemicellulases, lizosimes, lipases, pectinases and amylases, are the most commonly used methods in practice (Kikkawa *et al.*, 1990 [85]; Hernández, 2004 [86]; Muzzarelli *et al.*, 2000 [87]). Enzymatic reaction differs from chemical hydrolysis in that during chemical hydrolysis, depolymerization occurs randomly as the amine groups of protonated chitosan form a kind of electrostatic shield that prevents the approach of acid due to charge repulsion. Depolymerization has also been achieved using oxidizing agents such as hydrogen peroxide (Qin *et al.*, 2002 [88]), microbiological methods using microorganisms producers of chitinolytic enzymes (Struszczyk *et al.*, 2006 [89]) and physical methods such as photo-degradation (Wang *et al.*, 2005 [90]) or ionizing radiation. Ionizing radiation is a simple and fast method, that can be performed in a wide temperature range, without introducing toxic residues into the final product and which is easy to control (International Atomic Energy Agency, 2002 [91]). Once the depolymerization reaction occurs, the resulting chitosan oligomers can be separated by chromatographic methods such as ion exchange and size exclusion chromatography according to their polymerization degree (Hernández, 2004 [86]).

CHARACTERIZATION

As it was mentioned earlier, the characteristics of the chitosan molecular structures are related to their chemical, biological and physiological properties, making the use of the appropriate chitosan vital to each application; with a particular DD, MW, and source, in some cases, even a specific distribution of the acetyl groups along the main chain. For this reason, the correct characterization of the biopolymer is necessary. Deacetylation degree and molecular weight are the most important parameters to be considered.

Deacetylation Degree

Different methods to determine the DD have been reported. Among them, the simplest, quickest and most inexpensive is the potentiometric method, as it can be done with readily accessible reagents and apparatus. There are some variations between procedures, but the most popular consists of the steps described below (Czechowska-biskup *et al.*, 2012 [92]; Hidalgo *et al.*, 2008 [93]; Parada *et al.*, 2004 [94]; Yuan *et al.*, 2011 [95]).

A sample of 0.3 g of chitosan is dissolved in 25 ml of 0.1 M HCl solution. The ionic strength of the medium is adjusted at 0.1 with KCl. The solution is stirred for 1 h at 55°C and finally a potentiometric evaluation is carried out with a 0.1 M NaOH solution, previously standardized, at 30°C in the presence of methyl orange. To avoid errors from precipitation of the biopolymer, the additions have to be made slowly with continuous stirring to homogenize the solution. Once the pH *vs.* added NaOH (ml) curve is finished, the volumes of the inflection point are obtained and with Ec. (1) the DD.

$$DD[\%] = \frac{16.1 * (V_{hip} - V_{lip}) * [NaOH]}{m_{chitosan}} \quad \textbf{(1)}$$

where: *16.1* is a constant related to the molecular weight of chitosan monomer and unit corrections; V_{hip} and V_{lip} are the higher and lower inflections points, respectively, expressed in ml; *[NaOH]* is the molarity of the solution of NaOH; and $m_{chitosan}$ is the mass in g of the sample.

In the search of more accurate results other techniques have been studied. The most explored techniques are infrared spectroscopy and liquid H^1 NMR (Czechowska-biskup *et al.*, 2012 [92]; Domszy & Roberts, 1985 [96]; Kasaai, 2010 [97]; Kasaai, 2008 [98]; Lavertu *et al.*, 2003 [99]; Nguyen *et al.*, 2009 [100]; Rinaudo *et al.*, 1993 [101]). The advantage of H^1 NMR is that there is no need for any calibration curve or reference sample of known DD. Another advantage, is that there is no need to know accurately the amount of chitosan used, or the purity of the sample, as long as the impurity peaks do not overlap with the relevant peaks of chitosan (Lavertu *et al.*, 2003 [99]). The limitation of this technique is the solubility of chitosan. To select a solvent, not only do the solubility properties for chitosan have to be considered, but also the resonance peaks of the biopolymer should not be obscured by resonance peaks of the solvent.

In order to calculate the DD by liquid H^1 NMR there are different equations which take into consideration peaks of different groups, as is shown in Ec. (2) (Hirai *et al.*, 1991 [102]) and Ec. (3) (Lavertu *et al.*, 2003 [99]). In the Ec. (2), the signal of peaks which corresponded to the three protons of the acetyl group (I_{HAc}) are used as reference of the acetylated monomers and the sum of the signals of the protons 2, 3, 4, 5 and 6 from both monomers as the total monomers ($I_{H6\text{-}H2}$) (Fig. **4**). In

Ec (3), I_{HAc} is used as reference of the acetylated monomers, and the signal of the proton 1 of deacetylated monomer (I_{H1D}) as the reference of the deacetylated monomer. In some cases, when there are contaminants such as acetate salt present, which overlap their signal with I_{HAc}, the signal of the proton 1 of acetylated monomer (I_{H1A}) can be used instead (Nguyen *et al.*, 2009 [100]). However, (Lavertu *et al.* 2003 [99]) do not recommend it for high DD because I_{H1A} is not visible in the spectrum.

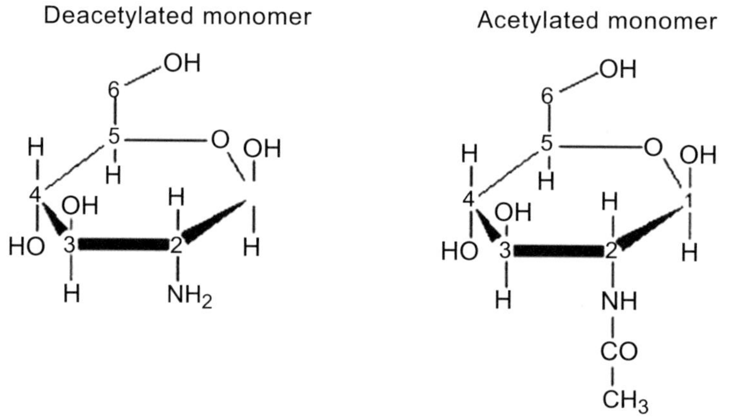

Fig. (4). Labeled carbons of deacetylated and acetylated monomers of chitosan. Extracted from Lavertu *et al.*, 2003 [99].

$$DD[\%] = \left[1 - \left(\frac{\frac{1}{3}I_{HAc}}{\frac{1}{6}I_{H6-H2}}\right)\right] \times 100 \quad (2)$$

$$DD[\%] = \left(\frac{I_{H1D}}{I_{H1D} + \frac{1}{3}I_{HAc}}\right) \times 100 \quad (3)$$

Several authors agree that the liquid H^1 NMR technique is adequate to use as a reference method (Czechowska-biskup *et al.*, 2012 [92]; Kasaai, 2010 [97]; Lavertu *et al.*, 2003 [99]). In addition, the technique is also known to be precise, reproducible and robust. In addition, it is stable and requires a small amount of chitosan.

Molecular Weight

In order to determine the molecular weight there are several methods which are employed, each having advantages and disadvantages. The viscosity method gives the viscosity average molecular weight ($\overline{M_v}$). Despite not being very precise, it is the most commonly used method due to its low cost and short analysis time. It is based on the fact that the intrinsic viscosity ($[\eta]$) of a polymer dissolved in solution is related to the polymer molecular weight according to the Mark-Houwink equation [Ec. (4)]:

$$[\eta] = K * \overline{M_v}^a \quad \textbf{(4)}$$

where K and a, are empirical constants that depend on the polymer-solvent pair and on temperature. The intrinsic viscosity is calculated using the Huggins equation Ec. (5) (Huggins, 1942 [103]):

$$\frac{\eta_{sp}}{c} = [\eta] + k * [\eta]^2 * c \quad \textbf{(5)}$$

In which η_{sp} is the specific viscosity, c is the concentration of the polymer and k is a constant depending on the polymer-solvent-temperature system.

The η_{sp} is calculated from a simple experiment that can be done with a relatively inexpensive glass viscometer. A series of dilute solutions of the polymer are prepared and the time it takes the volume of liquid between two points indicated on the viscometer to flow through a capillary is taken with a chronometer at a constant temperature. Then Ec. (6):

$$\eta_{sp} = \frac{t - t_o}{t_o} \quad \textbf{(6)}$$

where t_o is the flow time of the pure solvent and t the flow time of the polymer solution.

In order to obtain more accurate results, there are some important considerations: the viscometer must be clean to avoid changes in the viscosity, and this refers to, for example, previous solutions and dust particles; it is better to choose a viscometer where the flow time is greater than 100 s, so that it is permissible to ignore the kinetic energy terms, and use Ec. (6) to estimate η_{sp}.

Maghami & Roberts (1988 [104]) found that for chitosan with a range of 60% to

100% DD, the most accurate values currently available are K=1.81x10^{-3}cm^{3}g^{-1} and a=0.93, using 0.1 M acetic acid/0.2 M NaCl as solvent at 25°C.

There are other methods that have more specific results and allow the analysis of the MW distribution. Most of them have a first step of size exclusion chromatography (SEC) in common, combined with different detectors such as refractometers (Liu *et al.*, 2004 [105]; Muzzarelli *et al.*, 2003 [106]; Muzzarelli *et al.*, 1995 [107]). The combination of SEC with light scattering detectors to detect size and refractometers or UV detectors to detect concentration, generates absolute molecular weights and additional molecular size data.

The type of laser light scattering detector most frequently used is the multi-angle laser light scattering (MALLS) (Beri *et al.*, 1993 [108]; Brugnerotto *et al.*, 2001 [109]; Brugnerotto, 2001 [110]; Fernandez-Megia *et al.*, 2005 [111]; Rinaudo *et al.*, 1993 [101]). The difficulty of these cases is the selection of the column or sets of columns to use for chromatography due to the large polydispersity of chitosan (Nguyen *et al.*, 2009 [112]). (Nguyen *et al.* 2009 [100]) proposed a method to first fractionate chitosan and then characterize MW on SEC-MALLS, obtaining more accurate results for the MW and MW distribution as a consequence.

REGULATORY STATUS

As it is reported in this review, several safety and toxicological studies have been performed on chitosan. Some of them addressed to attend to its regulatory status, which varies from country to country. In 1983, chitosan received the "Generally Recognized as Safe" (GRAS) status by the Food and Drug Administration (FDA) in the United States for being used as an animal feed component (Shahidi & Abuzaytoun, 2005 [113]). For purification of potable water, chitosan was approved by the US Environmental Protection Agency (EPA) in 1986 (Knorr, 1986 [114]).

Nowadays, chitosan has the approval to be used as a food additive in Japan and South Korea (KFDA, 2015 [115]), and it is an approved wine processing aid in the European Union, Argentina, Chile, Australia and New Zealand (Legislación Chilena, 2012 [116]; EU, 2011 [117]; Food Standards Australia New Zealand, 2014 [118]; Instituto Nacional Vitivinícola Argentino, 2011 [119]). It is also listed as a processing aid in the Codex General Standard for Fruit Juices and Nectars (CODEX, 2005 [120]). The FDA has cleared several wound dressings made of chitosan and its derivatives (Dornish *et al.*, 2012 [121]). The European Community has granted the CE-mark to some similar products as well. In this area, Japan has several commercial wound dressing products made of chitosan and its derivatives for regeneration of the dermis. The FDA and the European Community have approved hemostatic topical agents based on this biopolymer

(Kardas *et al.*, 2012 [58]; Struszczyk & Struszczyk, 2007 [122]). For water treatment processes, the FDA has licensed the usage of chitosan for flocculation of protein (*Chitin & chitosan. A Global Strategic Business Report*, 2010 [123]). The EPA regulates chitosan applications for plants and crops. This agency approved biodegradable chitosan products for being used outdoors and indoors on plants and crops grown commercially and by consumers (EPA, 2008 [124]).

MARKET AND APPLICATIONS

As it can be noted from this review, chitosan as a biopolymer finds applications in almost all industrial sectors, making its market quite attractive. The high production cost of good quality chitosan is likely to be the major restraining factor. Scientists are working to make it cost-competitive, but, at the same time, this factor can be overcome by achieving economy of scale.

Transparency Market Research published in 2014 "Chitosan Market – Global Industry Analysis, Size, Share, Growth, Trends and Forecast, 2014-2020" in which the global market of chitosan in 2013 was valued at USD1.35 billon reaching USD4.22 billion in 2020, expanding at a Compound Annual Growth Rate (CAGR) of 17.7% between 2014 and 2020 (Sheela, 2014 [125]). Among the key factors that are driving market growth, are the increased demand from developing markets, such as Asia-Pacific, Latin America and the Middle East and the increasing applications and expansion on the agrochemicals and healthcare/medical end-use sectors (Strategro International, 2015 [126]).

A study made by Global Industry Analysts released in October 2010 (*Chitin & chitosan. A Global Strategic Business Report*, 2010 [123]) exhibits the volumes and projections for the chitosan market by geographic region and its end-use applications showed in Figs. (**5** and **6**). This report estimated the chitosan market for 2010 at 13.7 thousand tons and projected to reach 21.4 thousand tons by 2015, registering a CAGR of 9.7% over the period 2007-2015. As it can be seen in Fig. (**5**), Japan represents the biggest market of chitosan, reaching 5.8 thousand tons for 2010, followed by US with 3.4 thousand tons. The Asia-Pacific chitosan market represents the fastest growing region with a forecasted CAGR of 14.5% over the period 2007-2015. Latin America in this study is included in the rest of the world, which is the second fastest growing region with a CAGR of 12.2% over the same period.

Referring to end-use applications, the same report placed water treatment as the major end-use segment with 6.4 thousand tons estimated for 2010 (Fig. **6**). In Japan this application represents the major consumption of chitosan estimated at 4.4 thousand tons in 2010, 76% of the total consumption in that country. On the other hand, Cosmetics & Toiletries represent the major consumption of chitosan in US with an estimated volume of 900.4 t in 2010, 27% of total consumption there followed by Water treatment (23%), Food & Beverages (17%) and Healthcare/Medical (15%). The third place for end-use applications goes for Food & Beverages, 1.5 thousand tons estimated for 2010, in which the biggest contribution came from US with 40.2% in 2010.

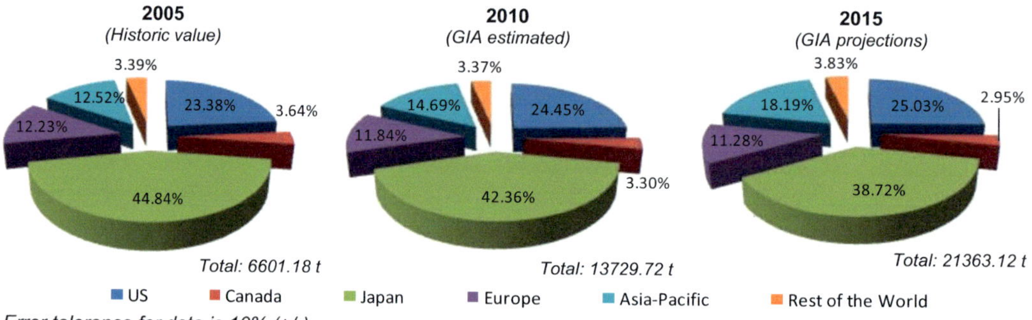

Error tolerance for data is 10% (+/-)
Data is reported al the Manufacturer´s level
Countries analyzed under Europe include Austria, Belgium, Bulgaria, Czech Republic, Denmark, Finland, France, Germany, Greece, Hungary, Ireland, Italy, The Netherlands, Norway, Poland, Portugal, Romania, Slovakia, Spain, Sweeden, Switzerland, Turkey and the UK
Countries analyzed under Asia Pacific include Australia, China, Hong Kong, India, Indonesia, Korea, Malaysia, New Zeland, Philippines, Singapure, Taiwan and Thailand.
Rest of the world includes Middle East (Iran, Irak, Israel, Kuwait, Saudi Arabia Syria and the UAE), Latin America (Argentina, Brazil, Chile, Colombia, Ecuador, México, Peru and Venezuela) and African countries.

Fig. (5). World perspective for chitosan by geographic region – percentage breakdown of volume sales for US, Canada, Japan, Europe, Asia-Pacific (excluding Japan), and rest of the world markets for years 2005, 2010 and 2015. Adapted from *Chitin & chitosan. A Global Strategic Business Report* 2010 [123].

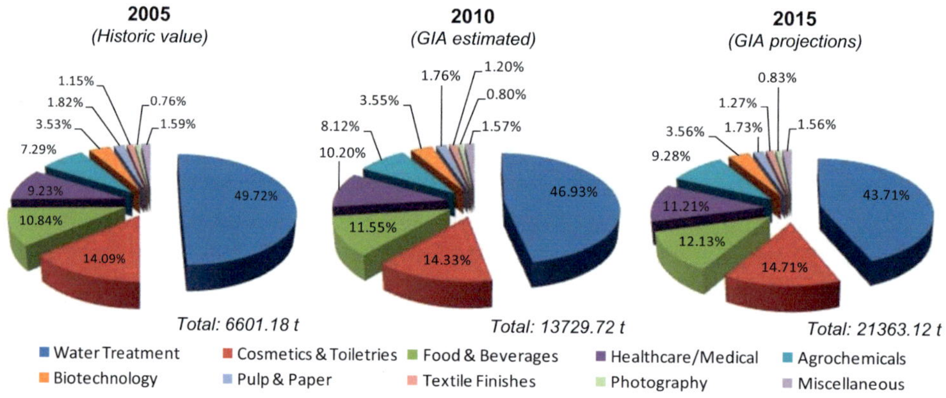

Error tolerance for data is 10% (+/-)
Data is reported al the Manufacturer´s level

Fig. (6). World perspective for chitosan by end-use application – water treatment. Cosmetics & toiletries, food & beverages, healthcare/medical, agrochemicals, biotechnology, pulp & paper, textile finishes, photography, and miscellaneous markets for years 2005, 2010 and 2015. Adapted from *Chitin & chitosan. A Global Strategic Business Report* 2010 [123].

Agrochemicals represents the fastest growing end-use segment, with a forecast CAGR of 12.6% over the period 2007-2015. The estimated consumption of chitosan in this particular segment was estimated in 1.1 thousand tons for 2010, and was projected to reach 2 thousand tons by 2015. In 2010 the major contribution to this segment was the Asia-Pacific region (54% of the total consumption). This segment is closely followed by the Healthcare/Medical end-use segment with an estimated chitosan consumption of 1.4 thousand tons for 2010 and a projected one of 2.4 thousand tons by 2015.

RESEARCH ON CHITOSAN AT INTI

As it is mentioned earlier in this chapter, the main chitosan industries are placed in Asia and US. In the case of Latin America, there are plants in Brazil, Chile and México. In Argentina, the crustaceans fishing industry is continuously growing. Its main commercially exploited decapod along the coast of Argentine Sea is the shrimp *Pleoticus muelleri* reaching 127.3 thousand tons in 2014 and 86.3 thousand tons until September 2015 (MAGyP, 2015 [127]). A large portion of this production is processed, generating a considerable amount of residues which becomes a serious environmental problem. As a consequence of the absence of a chitosan industry in Argentina, a potentially valuable material is being wasted.

Chitin isolation, chitosan production and characterization are being carried out at the National Institute of Industrial Technology of Argentine (INTI). At the same time, several applications of the biopolymer obtained are being tested in collaboration with other investigation groups from all around the country. The main objective of the project is the design and transfer of a plant of chitin and chitosan production from shrimp waste. A protocol to obtain this biopolymer was designed at laboratory scale. The procedure consists of a first stage of chemical chitin isolation using residues coming from a local fishery. In order to obtain chitosan, the chitin is treated with highly concentrated NaOH at high temperatures for a short period of time under continuous agitation. Time, temperature and NaOH concentration of the reaction are handled to obtain chitosan with different properties.

Chitin and chitosan characterization has been standardized. Chitin ash content and protein degree are evaluated to verify the process efficiency. Also, ash content is determined in chitosan as a control point of the washing stage. At the same time, the DD of this polymer is determined by two different methods. First, a potentiometric analysis is made to have a fast estimation and then eventually, a nuclear magnetic resonance (H^1 NMR) is carried out to give a precise result. MW of chitosan is determined by the viscosity method. In addition, the antibacterial property has been analyzed *in vitro* assays with positive results (Palacios *et al.*,

2012 [128]; Palacios *et al.*, 2012 [129]).

So far, the obtained chitosan has been tested on strawberries as coating to improve their shelf life (Bollini *et al.,* 2013 [130]; Martinez Saenz *et al.*, 2014 [131]), on strawberry fields to increase their yield (Bonecco *et al.*, 2015 [132]), in multilayer films with sodium caseinate for food preservation (Buffa *et al.*, 2015 [60]), as elicitors for resistance induction against pathogens in soybean (Vargas *et al.*, 2014 [133]) and as nanocarriers for insulin delivery (Prudkin Silva *et al.*, 2015 [134]). Other studies are still in process including its use for the treatment of fracking water and for the preservation of table grapes, among others. At present, a scale increase of the chitosan production process is being designed to continue with chitosan's applications research. In this regarding, INTI is also working on a clean chitosan production process, with the reutilization of the generated effluents, reducing the volume of used water and treating the final waste.

CONFLICT OF INTEREST

The author (editor) declares no conflict of interest, financial or otherwise.

ACKNOWLEDGEMENTS

The authors would like to the thanks Physico-Chemical Laboratory of INTI-Mar del Plata and Julieta Heba of INTI-Química for the contribution on the characterization analysis.

REFERENCES

[1] R. Nuñez Rico, *"Producción y aplicación de un biopolímero como una solución sustentable para el problema de abasto de agua en la Ciudad de México", Programa de becas ciudad de México-China.* Universidad Nacional Autónoma de México: México, 2010.

[2] E. Khor, and L.Y. Lim, "Implantable applications of chitin and chitosan", *Biomaterials,* vol. 24, no. 13, pp. 2339-2349, 2003. [http://dx.doi.org/10.1016/S0142-9612(03)00026-7] [PMID: 12699672]

[3] K. Peh, T. Khan, and H. Chng, "Mechanical, bioadhesive strength and biological evaluations of chitosan films for wound dressing", *J. Pharm. Pharm. Sci.* , vol. 3, no. 3, pp. 303-311, 2000. [PMID: 11177648]

[4] G. Skjak-Braek, T. Anthonsen, and P.A. Sandford, *Chitin and Chitosan: Sources, Chemistry, Biochemistry, Physical Properties and Applications.* Kluwer Academic Publishers Group, 1989.

[5] C. Lárez, "Quitina y quitosano : materiales del pasado para el presente y el futuro", *Av. en Química,* vol. 1, no. 2, pp. 15-21, 2006.

[6] R.A. Muzzarelli, *Chitin*Pergamon Press: Inglaterra, 1974.

[7] C. Lárez Velásquez, "Algunos usos del quitosano en sistemas acuosos", *Rev. Iberoam. Polímeros,* vol. 4, no. 2, pp. 91-109, 2003.

[8] R.N. Tharanathan, and F.S. Kittur, "Chitinthe undisputed biomolecule of great potential", *Crit. Rev. Food Sci. Nutr.* , vol. 43, no. 1, pp. 61-87, 2003. [http://dx.doi.org/10.1080/10408690390826455] [PMID: 12587986]

[9] M. Colina, A. Ayala, D. Rincón, J. Molina, J. Medina, J. Vargas, B. Montilla, E. Mixta, I. Ambiental, Q. Ca, I. Av, and S. Francisco, "Evaluación de los procesos para la obtención química de quitina y quitosano a partir de desechos de cangrejos", *Rev. Iberoam. Polímeros,* vol. 15, no. 1, pp. 21-43, 2014.

[10] M. Rinaudo, "Chitin and chitosan: Properties and applications", *Prog. Polym. Sci.* , vol. 31, pp. 603-632, 2006.
[http://dx.doi.org/10.1016/j.progpolymsci.2006.06.001]

[11] K.M. Rudall, and W. Kenchington, "The Chitin System", *Biol. Rev. Camb. Philos. Soc.* , vol. 48, no. 4, pp. 597-633, 1973.
[http://dx.doi.org/10.1111/j.1469-185X.1973.tb01570.x]

[12] Y. Yamaguchi, T.T. Nge, A. Takemura, N. Hori, and H. Ono, "Characterization of uniaxially aligned chitin film by 2D FT-IR spectroscopy", *Biomacromolecules,* vol. 6, no. 4, pp. 1941-1947, 2005.
[http://dx.doi.org/10.1021/bm0492172] [PMID: 16004431]

[13] C.K. Pillai, W. Paul, and C.P. Sharma, "Chitin and chitosan polymers: Chemistry, solubility and fiber formation", *Prog. Polym. Sci.* , vol. 34, pp. 641-678, 2009.
[http://dx.doi.org/10.1016/j.progpolymsci.2009.04.001]

[14] L. Ramírez, M. Plascencia, S. Huerta, H. Vázquez, and K. Shirai, "Obtención y caracterización de quitinas parcialmente desacetiladas mediante tratamineto biológico-químico", In: *VIII Simposio Latinoamericano de polímeros, II Simposio Iberoamericano de quitina, México*2002, pp. 593-594.

[15] H. Yamakawa, *Modern theory of polymer solutions.* Harper & Row Publishers Inc.: New York, 2001.

[16] I. Aranaz, M. Mengíbar, R. Harris, I. Paños, B. Miralles, N. Acosta, and G. Galed, "Functional characterization of chitin and chitosan", *Curr. Chem. Biol.* , vol. 3, pp. 203-230, 2009.

[17] M. Gauna, and N. Nuñez, *"Efecto ultrasonido en la desacetilación de la quitina de conchas de camarón"*, Thesis, Universidad del Zulia, Venezuela, 2004.

[18] N. Kubota, N. Tatsumoto, T. Sano, and K. Toya, "A simple preparation of half N-acetylated chitosan highly soluble in water and aqueous organic solvents", *Carbohydr. Res.* , vol. 324, no. 4, pp. 268-274, 2000.
[http://dx.doi.org/10.1016/S0008-6215(99)00263-3] [PMID: 10744335]

[19] C. Peniche, and W. Argüelles-Monal, "Overview on structural characterization of chitosan molecules in relation with their behavior in solution", *Macromol. Symp.* , vol. 168, pp. 1-20, 2001.
[http://dx.doi.org/10.1002/1521-3900(200103)168:1<1::AID-MASY1>3.0.CO;2-W]

[20] A.M. Papineau, D.G. Hoover, D. Knorr, and D.F. Farkas, "Antimicrobial effect of water-soluble chitosans with high hydrostatic pressure", *Food Biotechnol.* , vol. 5, pp. 45-57, 1991.
[http://dx.doi.org/10.1080/08905439109549790]

[21] Q.Z. Wang, X.G. Chen, N. Liu, S.X. Wang, C.S. Liu, X.H. Meng, and C.G. Liu, "Protonation constants of chitosan with different molecular weight and degree of deacetylation", *Carbohydr. Polym.* , vol. 65, pp. 194-201, 2006.
[http://dx.doi.org/10.1016/j.carbpol.2006.01.001]

[22] Y.C. Chung, C.L. Kuo, and C.C. Chen, "Preparation and important functional properties of water-soluble chitosan produced through Maillard reaction", *Bioresour. Technol.* , vol. 96, no. 13, pp. 1473-1482, 2005.
[http://dx.doi.org/10.1016/j.biortech.2004.12.001] [PMID: 15939275]

[23] T. Tanaka, M. Huang, J.R. Chiu, W.K. Ishizaki, and S. Taguchi, "Effect of the Maillard reaction on functional properties of chitosan", *Nippon Suisan Gakkai Shi,* vol. 59, pp. 1915-1921, 1993.
[http://dx.doi.org/10.2331/suisan.59.1915]

[24] M. Dash, F. Chiellini, R.M. Ottenbrite, and E. Chiellini, "Chitosan - A versatile semi-synthetic polymer in biomedical applications", *Prog. Polym. Sci.* , vol. 36, pp. 981-1014, 2011.
[http://dx.doi.org/10.1016/j.progpolymsci.2011.02.001]

[25] W. Paul, and C.P. Sharma, "Chitosan, a drug carrier for the 21 St century: a review", *STP Pharma Sci,* vol. 10, pp. 5-22, 2000.

[26] B. Sarmento, and J. Das Neves, *Chitosan-Based Systems for Biopharmaceuticals: Delivery, Targeting and Polymer Therapeutics.* John Wiley & Sons, Ltd, 2012. [http://dx.doi.org/10.1002/9781119962977]

[27] D.B. Collinge, K.M. Kragh, J.D. Mikkelsen, K.K. Nielsen, U. Rasmussen, and K. Vad, "Plant chitinases", *Plant J.* , vol. 3, no. 1, pp. 31-40, 1993. [http://dx.doi.org/10.1046/j.1365-313X.1993.t01-1-00999.x] [PMID: 8401605]

[28] J. Xu, S.P. Mccarthy, R.A. Gross, and D.L. Kaplan, "Chitosan film acylation and effects on biodegradability", *Macromolecules,* vol. 29, pp. 3436-3440, 1996. [http://dx.doi.org/10.1021/ma951638b]

[29] Y.M. Yang, W. Hu, X.D. Wang, and X.S. Gu, "The controlling biodegradation of chitosan fibers by N-acetylation in vitro and in vivo", *J. Mater. Sci. Mater. Med.* , vol. 18, no. 11, pp. 2117-2121, 2007. [http://dx.doi.org/10.1007/s10856-007-3013-x] [PMID: 17619982]

[30] A.S. Halim, L.C. Keong, I. Zainol, and A.H. Rashid, "Biocompatibility and Biodegradation of Chitosan and Derivatives", In: *Chitosan-Based Systems for Biopharmaceuticals: Delivery, Targeting and Polymer Therapeutics.* John Wiley & Sons, Inc., 2012, pp. 57-73. [http://dx.doi.org/10.1002/9781119962977.ch4]

[31] P. Baldrick, "The safety of chitosan as a pharmaceutical excipient", *Regul. Toxicol. Pharmacol.* , vol. 56, no. 3, pp. 290-299, 2010. [http://dx.doi.org/10.1016/j.yrtph.2009.09.015] [PMID: 19788905]

[32] R.A. Muzzarelli, "Human enzymatic activities related to the therapeutic administration of chitin derivatives", *Cell. Mol. Life Sci.* , vol. 53, no. 2, pp. 131-140, 1997. [http://dx.doi.org/10.1007/PL00000584] [PMID: 9118001]

[33] S. Hirano, M. Iwata, K. Yamanaka, H. Tanaka, T. Toda, and H. Inui, "Enhancement of serum lysozyme activity by injecting a mixture of chitosan oligosaccharides intravenously in rabbits", *Agric. Biol. Chem.* , vol. 55, no. 10, pp. 2623-2625, 1991.

[34] A.K. Azab, V. Doviner, B. Orkin, J. Kleinstern, M. Srebnik, A. Nissan, and A. Rubinstein, "Biocompatibility evaluation of crosslinked chitosan hydrogels after subcutaneous and intraperitoneal implantation in the rat", *J. Biomed. Mater. Res. A,* vol. 83, no. 2, pp. 414-422, 2007. [http://dx.doi.org/10.1002/jbm.a.31256] [PMID: 17455216]

[35] H. Onishi, and Y. Machida, "Biodegradation and distribution of water-soluble chitosan in mice", *Biomaterials,* vol. 20, no. 2, pp. 175-182, 1999. [http://dx.doi.org/10.1016/S0142-9612(98)00159-8] [PMID: 10022787]

[36] S. Hirano, C. Itakura, H. Seino, Y. Akiyama, I. Nonaka, N. Kanbara, and T. Kawakami, "Chitosan as an ingredient for domestic animal feeds", *J. Agric. Food Chem.* , vol. 38, no. 5, pp. 1214-1217, 1990. [http://dx.doi.org/10.1021/jf00095a012]

[37] Y. Nishimura, H-S. Kim, N. Ikota, H. Arima, H-S. Bom, Y-H. Kim, Y. Watanabe, M. Yukawa, and T. Ozawa, "Radioprotective effect of chitosan in sub-lethally X-ray irradiated mice", *J. Radiat. Res. (Tokyo),* vol. 44, no. 1, pp. 53-58, 2003. [http://dx.doi.org/10.1269/jrr.44.53] [PMID: 12841600]

[38] J.W. Anderson, R.J. Nicolosi, and J.F. Borzelleca, "Glucosamine effects in humans: a review of effects on glucose metabolism, side effects, safety considerations and efficacy", *Food Chem. Toxicol.* , vol. 43, no. 2, pp. 187-201, 2005. [http://dx.doi.org/10.1016/j.fct.2004.11.006] [PMID: 15621331]

[39] C. Qin, J. Gao, L. Wang, L. Zeng, and Y. Liu, "Safety evaluation of short-term exposure to chitooligomers from enzymic preparation", *Food Chem. Toxicol.* , vol. 44, no. 6, pp. 855-861, 2006. [http://dx.doi.org/10.1016/j.fct.2005.11.009] [PMID: 16442198]

[40] V. Goycoolea, "F. M., Argüelles-Monal, W., Peniche, C., Higuera-Ciapara, I., Doxastakis, G., & Kiosseoglou, “Chitin and chitosan", *Dev. Food Sci.* , vol. 41, pp. 265-308, 2000. [http://dx.doi.org/10.1016/S0167-4501(00)80013-8]

[41] T. Kean, and M. Thanou, "Biodegradation, biodistribution and toxicity of chitosan", *Adv. Drug Deliv. Rev.* , vol. 62, no. 1, pp. 3-11, 2010. [http://dx.doi.org/10.1016/j.addr.2009.09.004] [PMID: 19800377]

[42] I.M. Helander, E.L. Nurmiaho-Lassila, R. Ahvenainen, J. Rhoades, and S. Roller, "Chitosan disrupts the barrier properties of the outer membrane of gram-negative bacteria", *Int. J. Food Microbiol.* , vol. 71, no. 2-3, pp. 235-244, 2001. [http://dx.doi.org/10.1016/S0168-1605(01)00609-2] [PMID: 11789941]

[43] Y.C. Chung, Y.P. Su, C.C. Chen, G. Jia, H.L. Wang, J.C. Wu, and J.G. Lin, "Relationship between antibacterial activity of chitosan and surface characteristics of cell wall", *Acta Pharmacol. Sin.* , vol. 25, no. 7, pp. 932-936, 2004. [PMID: 15210068]

[44] A. El Ghaouth, J. Arul, J. Grenier, and A. Asselin, "Effect of chitosan and other polyions on chitin deacetylase in Rhizopus stolonifer", *Exp. Mycol.* , vol. 16, pp. 173-177, 1992. [http://dx.doi.org/10.1016/0147-5975(92)90025-M]

[45] H. Li, and Y. Yu, "Effect of chitosan on incidence of brown rot, quality and physiological attributes of postharvest peach fruit", *J. Sci. Food Agric.* , vol. 81, pp. 269-274, 2001. [http://dx.doi.org/10.1002/1097-0010(20010115)81:2<269::AID-JSFA806>3.0.CO;2-F]

[46] Y. Li, X.G. Chen, N. Liu, C.S. Liu, C.G. Liu, X.H. Meng, L.J. Yu, and J.F. Kennedy, "Physicochemical characterization and antibacterial property of chitosan acetates", *Carbohydr. Polym.* , vol. 67, pp. 227-232, 2007. [http://dx.doi.org/10.1016/j.carbpol.2006.05.022]

[47] A.H. Allan, and L.A. Hardwiger, "The fungicidal effect of chitosan on fungi of varying cell wall composition", *Exp. Mycol.* , vol. 3, pp. 285-287, 1979. [http://dx.doi.org/10.1016/S0147-5975(79)80054-7]

[48] S. Roller, and N. Covill, "The antifungal properties of chitosan in laboratory media and apple juice", *Int. J. Food Microbiol.* , vol. 47, no. 1-2, pp. 67-77, 1999. [http://dx.doi.org/10.1016/S0168-1605(99)00006-9] [PMID: 10357275]

[49] R.G. Cuero, G. Osuji, and A. Washington, "N-carboxymethylchitosan inhibition of aflatoxin production: Role of zinc", *Biotechnol. Lett.* , vol. 13, no. 6, pp. 441-444, 1991. [http://dx.doi.org/10.1007/BF01030998]

[50] N. Ben-Shalom, R. Ardi, R. Pinto, C. Aki, and E. Fallik, "Controlling gray mould caused by Botrytis cinerea in cucumber plants by means of chitosan", *Crop Prot.* , vol. 22, pp. 285-290, 2003. [http://dx.doi.org/10.1016/S0261-2194(02)00149-7]

[51] L.A. Hadwiger, "Multiple effects of chitosan on plant systems: solid science or hype", *Plant Sci.* , vol. 208, pp. 42-49, 2013. [http://dx.doi.org/10.1016/j.plantsci.2013.03.007] [PMID: 23683928]

[52] P. Vander, A. Domard, B.M. Moerschbacher, N. Eddine El Gueddari, and B.M. Moerschbacher, "Comparison of the ability of partially N-acetylated chitosans and chitooligosaccharides to elicit resistance reactions in wheat leaves", *Plant Physiol.* , vol. 118, no. 4, pp. 1353-1359, 1998. [http://dx.doi.org/10.1104/pp.118.4.1353] [PMID: 9847109]

[53] M.C. Salazar, and A.V. Negr, "Preparación y caracterización de películas de quitosano despolimerizado y reticulado con tripolifosfato de sodio", *Rev. la Soc. Química del Perú,* vol. 79, pp. 1-12, 2013.

[54] J. Butler, B. Vergano, P. Testin, R. Bunn, and J. Wiles, "Mechanical and barrier properties of edible chitosan films as affected by composition and storage", *J. Food Sci.* , vol. 61, no. 5, pp. 953-955, 1996.

[http://dx.doi.org/10.1111/j.1365-2621.1996.tb10909.x]

[55] F.S. Kittur, K.R. Kumar, and R.N. Tharanathan, "Functional packaging properties of chitosan films", *Zeitschrift für Leb. und -forsch. A,* vol. 206, no. 1, pp. 44-47, 1998.

[56] Y. Makino, and T. Hirata, "Modified atmosphere packaging of fresh produce with a biodegradable laminate of chitosan-cellulose and polycaprolactone", *Postharvest Biol. Technol.* , vol. 10, no. 3, pp. 247-254, 1997.
[http://dx.doi.org/10.1016/S0925-5214(96)01402-0]

[57] R.A. Muzzarelli, S. Aiba, Y. Fujiwara, T. Hideshima, C. Hwang, M. Kakizaki, M. Izume, N. Minoura, C.K. Rha, T. Shouij, A.J. Sinskey, and T. Tsutsumi, *Filmogenic Properties of Chitin / Chitosan.* Chitin Nat. Technol, 1986, pp. 389-402.

[58] I. Kardas, M.H. Struszczyk, L.A. van den Broek, J.E. van Dam, and D. Cienchanska, "Chitin and chitosan as functional biopolymers for industrial applications", In: *The European Polysaccharide Network of Excellence (EPNOE).* , P. Navard, Ed., Springer, 2012, pp. 329-373.
[http://dx.doi.org/10.1007/978-3-7091-0421-7_11]

[59] V. Coma, A. Martial-Gros, S. Garreau, A. Copinet, F. Salin, and A. Deschamps, "Edible antimicrobial films based on chitosan matrix", *J. Food Sci.* , vol. 67, no. 3, pp. 1162-1169, 2002.
[http://dx.doi.org/10.1111/j.1365-2621.2002.tb09470.x]

[60] L.M. Buffa, M.R. Ansorena, N.E. Marcovich, J.M. Herry, and S. Domenek, "Películas multicapa basadas en biopolímeros para su uso en contacto con alimentos", In: *12° Jornadas Abiertas de Desarrollo, Innovación y Transferencia Tecnológica. TECNO INTI (Póster)*San Martín: Argentina, 2015.

[61] N. Acosta, I. Aranaz, and A. Hera, "Chitin and chitosan in waste wáter treatment", In: *Primer simposio Latinoaericano de quitina y quitosán*2000, p. 482.

[62] R.A. Muzzarelli, *Natural chelating polymers: alginic acid, chitin and chitosan.* Pergamon Press: New York, 1973.

[63] K.D. Trimukhe, and A.J. Varma, "A morphological study of heavy metal complexes of chitosan and crosslinked chitosans by SEM and WAXRD", *Carbohydr. Polym.* , vol. 71, pp. 698-702, 2008.
[http://dx.doi.org/10.1016/j.carbpol.2007.07.010]

[64] E.S. Abdou, K.S. Nagy, and M.Z. Elsabee, "Extraction and characterization of chitin and chitosan from local sources", *Bioresour. Technol.* , vol. 99, no. 5, pp. 1359-1367, 2008.
[http://dx.doi.org/10.1016/j.biortech.2007.01.051] [PMID: 17383869]

[65] G.A. Roberts, *Chitin chemistry.* Macmillan: London, 1992.
[http://dx.doi.org/10.1007/978-1-349-11545-7]

[66] P. Lertsutthiwong, N.C. How, and S. Chandrkrachang, "Effect of chemical treatment on the characteristics of shrimp chitosan", *J. Met. Mater. Miner.* , vol. 12, no. 1, pp. 11-18, 2002.

[67] H.K. No, and S.P. Meyers, "Preparation and Characterization of Chitin and Chitosan — A Review", *J. Aquat. Food Prod. Technol.* , vol. 4, no. 2, pp. 24-52, 1995.
[http://dx.doi.org/10.1300/J030v04n02_03]

[68] F.A. Al Sagheer, M.A. Al-sughayer, S. Muslim, and M.Z. Elsabee, "Extraction and characterization of chitin and chitosan from marine sources in Arabian Gulf", *Carbohydr. Polym.* , vol. 77, no. 2, pp. 410-419, 2009.
[http://dx.doi.org/10.1016/j.carbpol.2009.01.032]

[69] M. Takeda, and E. Abe, "Isolation of crustgcean chitin. Decalcification by disodium ethylenediaminetetraacetate and enzymic hydrolysis of incidental proteins", *Norinsho Suisan Koshusho Kenkyu Hokoku,* vol. 11, pp. 339-406, 1962.

[70] M. Takeda, and H. Katsuura, "Purification of king crab chitin", *Suisan Daigakko Kenkyu Hokoku,* vol. 13, pp. 109-116, 1964.

[71] I. Younes, and M. Rinaudo, "Chitin and chitosan preparation from marine sources. Structure, properties and applications", *Mar. Drugs,* vol. 13, no. 3, pp. 1133-1174, 2015. [http://dx.doi.org/10.3390/md13031133] [PMID: 25738328]

[72] K.T. Oh, Y.J. Kim, V.N. Nguyen, W.J. Jung, and R.D. Park, "Demineralization of crab shell waste by Pseudomonas aeruginosa F722", *Process Biochem.* , vol. 42, no. 7, pp. 1069-1074, 2007. [http://dx.doi.org/10.1016/j.procbio.2007.04.007]

[73] K. Prameela, C.M. Mohan, P.V. Smitha, and K.P. Hemalatha, "Bioremediation of shrimp biowaste by using natural probiotic for chitin and carotenoid production an alternative method to hazardous chemical method", *Int. J. Appl. Biol. Pharm. Technol.* , vol. I, no. 3, pp. 903-910, 2010.

[74] M.S. Rao, J. Muñoz, and W.F. Stevens, "Critical factors in chitin production by fermentation of shrimp biowaste", *Appl. Microbiol. Biotechnol.* , vol. 54, no. 6, pp. 808-813, 2000. [http://dx.doi.org/10.1007/s002530000449] [PMID: 11152073]

[75] E. Marcia, J. Malespín, M. Sánchez, and M. Benavente, "Estudio de la fermentación láctica para la extracción de quitina a partir de desechos de crustáceos", *Nexo,* vol. 24, no. 1, pp. 33-42, 2011.

[76] O. Ghorbel-Bellaaj, I. Younes, H. Maâlej, S. Hajji, and M. Nasri, "Chitin extraction from shrimp shell waste using Bacillus bacteria", *Int. J. Biol. Macromol.* , vol. 51, no. 5, pp. 1196-1201, 2012. [http://dx.doi.org/10.1016/j.ijbiomac.2012.08.034] [PMID: 22981824]

[77] A. Khanafari, R. Marandi, and S. Sanatei, "Recovery of chitin and chitosan from shrimp waste by chemical and microbial methods", *Iran. J. Environ. Health Sci. Eng.* , vol. 5, no. 1, pp. 19-24, 2008.

[78] A. Heppe, and K. Le Roux, "Enzymatic and chemical production of chitin in comparison - problems in the realisation of production plants with 100 tons capacity", *6th Iberoamerican Chitin Symposium & 12th International Conference on Chitin and Chitosan.* , 2012p. 19 Fortaleza, Brazil

[79] G. Lamarque, C. Viton, and A. Domard, "Comparative study of the first heterogeneous deacetylation of alpha- and beta-chitins in a multistep process", *Biomacromolecules,* vol. 5, no. 3, pp. 992-1001, 2004. [http://dx.doi.org/10.1021/bm034498j] [PMID: 15132692]

[80] Y. Araki, and E. Ito, "A pathway of chitosan formation in Mucor rouxii: enzymatic deacetylation of chitin", *Biochem. Biophys. Res. Commun.* , vol. 56, no. 3, pp. 669-675, 1974. [http://dx.doi.org/10.1016/0006-291X(74)90657-3] [PMID: 4826874]

[81] C. Alfonso, O.M. Nuero, F. Santamaría, and F. Reyes, "Purification of a heat-stable chitin deacetylase from Aspergillus nidulans and its role in cell wall degradation", *Curr. Microbiol.* , vol. 30, no. 1, pp. 49-54, 1995. [http://dx.doi.org/10.1007/BF00294524] [PMID: 7765883]

[82] K. Tokuyasu, M. Ohnishi-Kameyama, and K. Hayashi, "Purification and characterization of extracellular chin deacetylase from Colletotrichum lindemuthianum", *Biosci. Biotechnol. Biochem.* , vol. 60, no. 10, pp. 1598-1603, 1996. [http://dx.doi.org/10.1271/bbb.60.1598] [PMID: 8987657]

[83] I. Tsigos, and V. Bouriotis, "Purification and characterization of chitin deacetylase from Colletotrichum lindemuthianum", *J. Biol. Chem.* , vol. 270, no. 44, pp. 26286-26291, 1995. [http://dx.doi.org/10.1074/jbc.270.44.26286] [PMID: 7592838]

[84] G.A. Roberts, "Thirty years of progress in chitin and chitosan", In: *Prog. Chem. Appl. chitin its Deriv.* vol. 13. 1977, pp. 7-15.

[85] T. Kikkawa, Y. Kawada, T. Furukawa, and I. Sakuno, "A convenient preparation method of chito-oligosaccharides by acid hydrolysis", *J. Fac. Agric. Tottori Univ.* , vol. 29, pp. 9-17, 1990.

[86] I. Hernández, "La quitosana: un producto bioactivo de diversas aplicaciones", *Cultiv. Trop.* , vol. 25, no. 3, pp. 97-110, 2004.

[87] R.A. Muzzarelli, V. Stanic, and V. Ramos, "Enzymatic depolymerization of chitins and chitosans",

Methods Biotecnol. , vol. 10, pp. 197-211, 2000. [http://dx.doi.org/10.1007/978-1-59259-261-6_16]

[88] C.Q. Qin, Y.M. Du, and L. Xiao, "Effect of hydrogen peroxide treatment on the molecular weight and structure of chitosan", *Polym. Degrad. Stabil.* , vol. 76, no. 2, pp. 211-218, 2002. [http://dx.doi.org/10.1016/S0141-3910(02)00016-2]

[89] K. Struszczyk, M. Szczęsna-Antczak, T. Antczak, M. Rzyska, and S. Bielecki, *Study on catalytic activity of mucor in the process of chitosan biodegradation.* Polish Chitin Soc., 2006, pp. 153-158.

[90] S.M. Wang, Q.Z. Huang, and Q.S. Wang, "Study on the synergetic degradation of chitosan with ultraviolet light and hydrogen peroxide", *Carbohydr. Res.* , vol. 340, no. 6, pp. 1143-1147, 2005. [http://dx.doi.org/10.1016/j.carres.2005.02.009] [PMID: 15797129]

[91] *Radiation synthesis and modification of polymers for biomedical applications.* International Atomic Energy Agency: Vienna, 2002, p. 202.

[92] R. Czechowska-biskup, D. Jarosińska, B. Rokita, P. Ulański, and J.M. Rosiak, "Determination of degree of deacetylation of chitosan - comparaison of methods", *Prog. Chem. Appl. chitin its Deriv.* , vol. 17, pp. 5-20, 2012.

[93] C. Hidalgo, Y. Suárez, and M. Fernandez, "Validación de una técnica potenciométrica para determinar el grado de desacetilación de la quitosana", *Ars Pharm.* , vol. 49, no. 3, pp. 245-257, 2008.

[94] L.G. Parada, R. Miranda, and S. Salvador, "Caracterización de quitosano por viscosimetría capilar y valoración potenciométrica", *Rev. Iberoam. polímeros,* vol. 5, no. 1, pp. 1-16, 2004.

[95] Y. Yuan, B.M. Chesnutt, W.O. Haggard, and J.D. Bumgardner, "Deacetylation of Chitosan: Material Characterization and in vitro Evaluation via Albumin Adsorption and Pre-Osteoblastic Cell Cultures", *Materials (Basel),* vol. 4, no. 12, pp. 1399-1416, 2011. [http://dx.doi.org/10.3390/ma4081399]

[96] J.G. Domszy, and G.A. Roberts, "Evaluation of infrared spectroscopic techniques for analysing chitosan", *Makromol. Chem.* , vol. 186, no. 8, pp. 1671-1677, 1985. [http://dx.doi.org/10.1002/macp.1985.021860815]

[97] M.R. Kasaai, "Determination of the degree of N-acetylation for chitin and chitosan by various NMR spectroscopy techniques: A review", *Carbohydr. Polym.* , vol. 79, no. 4, pp. 801-810, 2010. [http://dx.doi.org/10.1016/j.carbpol.2009.10.051]

[98] M.R. Kasaai, "A review of several reported procedures to determine the degree of N-acetylation for chitin and chitosan using infrared spectroscopy", *Carbohydr. Polym.* , vol. 71, no. 4, pp. 497-508, 2008. [http://dx.doi.org/10.1016/j.carbpol.2007.07.009]

[99] M. Lavertu, Z. Xia, A.N. Serreqi, M. Berrada, A. Rodrigues, D. Wang, M.D. Buschmann, and A. Gupta, "A validated 1H NMR method for the determination of the degree of deacetylation of chitosan", *J. Pharm. Biomed. Anal.* , vol. 32, no. 6, pp. 1149-1158, 2003. [http://dx.doi.org/10.1016/S0731-7085(03)00155-9] [PMID: 12907258]

[100] S. Nguyen, S. Hisiger, M. Jolicoeur, F.M. Winnik, and M.D. Buschmann, "Fractionation and characterization of chitosan by analytical SEC and 1H NMR after semi-preparative SEC", *Carbohydr. Polym.* , vol. 75, no. 4, pp. 636-645, 2009. [http://dx.doi.org/10.1016/j.carbpol.2008.09.002]

[101] M. Rinaudo, M. Milas, and P. Le Dung, "Characterization of chitosan. Influence of ionic strength and degree of acetylation on chain expansion", *Int. J. Biol. Macromol.* , vol. 15, no. 5, pp. 281-285, 1993. [http://dx.doi.org/10.1016/0141-8130(93)90027-J] [PMID: 8251442]

[102] A. Hirai, H. Odani, and A. Nakajima, "Determination of degree of deacetylation of chitosan by 1HNMR spectroscopy", *Polym. Bull.* , vol. 26, pp. 87-94, 1991. [http://dx.doi.org/10.1007/BF00299352]

[103] M.L. Huggins, "The Viscosity of Dilute Solutions of Long-Chain Molecules. IV. Dependence on Concentration", *J. Am. Chem. Soc.* , vol. 64, no. 11, pp. 2716-2718, 1942. [http://dx.doi.org/10.1021/ja01263a056]

[104] G.G. Maghami, and G.A. Roberts, "Evaluation of the viscometric constants for chitosan", *Makromol. Chem.* , vol. 189, no. 1, pp. 195-200, 1988. [http://dx.doi.org/10.1002/macp.1988.021890118]

[105] H. Liu, Y. Du, X. Wang, Y. Hu, and J.F. Kennedy, "Interaction between chitosan and alkyl β-d-glucopyranoside and its effect on their antimicrobial activity", *Carbohydr. Polym.* , vol. 56, no. 2, pp. 243-250, 2004. [http://dx.doi.org/10.1016/j.carbpol.2004.03.001]

[106] C. Muzzarelli, G. Tosi, O. Francescangeli, and R.A. Muzzarelli, "Alkaline chitosan solutions", *Carbohydr. Res.* , vol. 338, no. 21, pp. 2247-2255, 2003. [http://dx.doi.org/10.1016/S0008-6215(03)00373-2] [PMID: 14553988]

[107] R.A. Muzzarelli, W. Xia, M. Tomasetti, and P. Ilari, "Depolymerization of chitosan and substituted chitosans with the aid of a wheat germ lipase preparation", *Enzyme Microb. Technol.* , vol. 17, no. 6, pp. 541-545, 1995. [http://dx.doi.org/10.1016/0141-0229(94)00015-J]

[108] R.G. Beri, J. Walker, E.T. Reese, and J.E. Rollings, "Characterization of chitosans via coupled size-exclusion chromatography and multiple-angle laser light-scattering technique", *Carbohydr. Res.* , vol. 238, pp. 11-26, 1993. [http://dx.doi.org/10.1016/0008-6215(93)87002-A]

[109] J. Brugnerotto, J. Desbrières, L. Heux, K. Mazeau, and M. Rinaudo, "Overview on structural characterization of chitosan molecules in relation with their behavior in solution", *Macromol. Symp.* , vol. 168, no. 1, pp. 1-20, 2001. [http://dx.doi.org/10.1002/1521-3900(200103)168:1<1::AID-MASY1>3.0.CO;2-W]

[110] J. Brugnerotto, J. Desbrières, G. Roberts, and M. Rinaudo, "Characterization of chitosan by steric exclusion chromatography", *Polymer (Guildf.),* vol. 42, no. 25, pp. 09921-09927, 2001. [http://dx.doi.org/10.1016/S0032-3861(01)00557-2]

[111] E. Fernandez-Megia, R. Novoa-Carballal, E. Quiñoá, and R. Riguera, "Optimal routine conditions for the determination of the degree of acetylation of chitosan by 1H-NMR", *Carbohydr. Polym.* , vol. 61, no. 2, pp. 155-161, 2005. [http://dx.doi.org/10.1016/j.carbpol.2005.04.006]

[112] S. Nguyen, F.M. Winnik, and M.D. Buschmann, "Improved reproducibility in the determination of the molecular weight of chitosan by analytical size exclusion chromatography", *Carbohydr. Polym.* , vol. 75, no. 3, pp. 528-533, 2009. [http://dx.doi.org/10.1016/j.carbpol.2008.08.013]

[113] F. Shahidi, and R. Abuzaytoun, "Chitin, chitosan, and co-products: chemistry, production, applications, and health effects", In: *Advances in Food and Nutrition Research,* vol. 49. Elsevier Inc., 2005, pp. 93-135. [http://dx.doi.org/10.1016/S1043-4526(05)49003-8]

[114] D. Knorr, "Nutritional quality, food processing, and biotechnology aspects of chitin and chitosan: a review", *Process Biochem.* , vol. 21, no. 3, pp. 90-92, 1986.

[115] KFDA, *Korea Food Additives Code.* Available at: http://www.mfds.go.kr/fa/ebook/egongjeon_intro.jsp Accessed: Oct. 2015

[116] Legislación Chilena, *Decreto 59.* Biblioteca Electrónica del Congreso Nacional de Chile: Chile, 2012, pp. 37-38.

[117] EU, *Commission regulation (EU) No 53/2011.* Off. J. Eur. Union, 2011, pp. 1-6.

[118] Food Standards, *Application A1077, Food Standards Australia New Zeland.*Oct. 2015 Available at: https://www.comlaw.gov.au/Details/F2014L00033

[119] INV, *Instituto Nacional Vitivinícola, Resolución N°51/2011.*Oct. 2015 Available at: tp://infoleg.mecon.gov.ar/infolegInternet/anexos/190000-194999/190700/norma.htm

[120] CODEX, *CODEX General Standard for Fruit Juices and Nectars.* CODEX Alimentarius International Food Standards, 2005, pp. 1-19.

[121] M. Dornish, D.S. Kaplan, and S.R. Arepalli, *Regulatory Status of Chitosan and Derivatives.* Chitosan-Based Syst. Biopharm. Deliv. Target. Polym. Ther, 2012, pp. 463-481.

[122] M. Struszczyk, and K. Struszczyk, "Medical applications of chitin and its derivatives", *Polish Chitin Soc.* , vol. 93, pp. 139-147, 2007.

[123] *Chitin and chitosan. A Global Strategic Business Report,* .

[124] EPA, *Chitin and Chitosan Final Registration Review Decision Case 6063,* 2008.Oct. 2015 Available at: http://www.regulations.gov/

[125] A. Sheela, "Chitosan Market is Expected to Reach USD 4.22 Billion in 2020", *Press Releases. Transparecny Market Reserch,* 2014.Oct. 2015 Available at: htpp://www.transparencymarket research.com/chitosanmarket.html

[126] Strategro International L. L. C., "By Year 2015 : Global market for chitin derivatives expected to reach $ 63 billion / Global chitosan market could exceed $ 21 billion", *Aquqpreneur,* 2015.Oct. 2015 Available at: http://aquapreneur.com/issue-12/ by-year-2015-global-market-for-chitin-deri-atives-expected-to-reach-63-billion-global-chitosan-market-could-exceed-21-billion/

[127] MAGyP, Ministerio de Agricultura, Ganadería y Pesca, *Desembarques de Capturas Marítimas Totales - Por especie y mes,* 2015.Oct. 2015 Available at: http://www.minagri.gob.ar/site/pesca/ pesca_maritima/02-desembarques/lectura.php?imp=1&tabla=especie_mes_2015

[128] D. Palacios, M.B. Bonecco, M.G. Martinez Saenz, L.M. Buffa, A. Cuesta, and F. Bollini, "Effect of Culture Media on The Antimicrobial Activity of Chitosan Obtained from Fishin Industry Waste", *6th Iberoamerican Chitin Symposium & 12th International Conference on Chitin and Chitosan,* 2012p. 67

[129] D. Palacios, A. Cuesta, M.B. Bonecco, M.G. Martinez Saenz, L.M. Buffa, and F. Bollini, "Antimicrobial Effect of Chitosan Obtained From Fishing Industry Waste", *6th Iberoamerican Chitin Symposium & 12th International Conference on Chitin and Chitosan,* 2012p. 66

[130] F. Bollini, D. Palacios, L.M. Buffa, M.B. Bonecco, and M.G. Martinez Saenz, "Extensión de la vida útil de la frutilla en poscosecha utilizando quitosano como agente protector", In: *11° Jornadas Abiertas de Desarrollo, Innovación y Transferencia Tecnológica. TECNO INTI (Poster)*San Martín: Argentina, 2013.

[131] M.G. Martinez Saenz, Q.T. Ho, P. Verboven, and B.M. Nicolaï, "Efecto de cubiertas de quitosano en la tasa de respiración de frutillas", *V Congreso Internacional Ciencia y Tecnología de los Alimentos,* 2014p. 431

[132] M.B. Bonecco, L.M. Buffa, M.G. Martinez Saenz, and F. Bollini, "Quitosano y agricultura: Aplicación sobre cultivo de frutillas", In: *12° Jornadas Abiertas de Desarrollo, Innovación y Transferencia Tecnológica. TECNO INTI (póster)*San Martín: Argentina, 2015.

[133] L.I. Vargas, M.C. Minchiotti, J.O. Muñoz, and R.R. Madoery, "Chitosan Biopolymer and Oligomer Derivative as Elicitors for Resistance Induction Against Pathogens in Soybean (Glycine max)", *J. Chitin Chitosan Sci.* , vol. 2, no. 3, pp. 267-272, 2014. [http://dx.doi.org/10.1166/jcc.2014.1071]

[134] C. Prudkin Silva, M.B. Bonecco, and O.E. Pérez, "Nanocarriers biopoliméricos encapsulantes para el delivery de insulina", In: *XXIII Jornadas Jóvenes Investigadores. Grupo Montevideo (Poster)*La Plata: Argentina, 2015.

CHAPTER 5

Amphiphilic Chitosan-Polymer Derivatives and Albumin-Based Formulations for Drug Delivery Applications

João Pedro Martins, Andreia Almeida and **Bruno Sarmento***

i3S – Instituto de Investigação e Inovação em Saúde, Porto, Portugal

INEB – Instituto de Engenharia Biomédica, Porto, Portugal

IIFACTS – Instituto de Investigação e Formação Avançada em Ciências e Tecnologias da Saúde, Gandra, Portugal

Abstract: The use of biomaterials as drug carriers is being extensively investigated as a drug delivery strategy in pharmaceutical research. However, the selection of materials for such systems is rather complicated. Lately, the advent of biomaterials brought the use of polymers as drug vehicles into the spotlight.

Chitosan and albumin are two of the most promising biomaterials for the development of drug delivery systems. These are widely available in nature and can be used as scaffolds of unprecedented novel structure, presenting high biocompatibility and biodegradability and reduced toxicity. Moreover, these molecules promote optimized pharmacokinetics for targeted drug delivery and controlled release, and avoid accumulation or undesired side effects in healthy tissues

Chitosan is a linear polysaccharide obtained from chitin and, when used as drug carrier, demonstrates to improve drugs pharmacokinetic profiles when comparing with the drug alone.

Albumin is the most abundant protein in blood plasma, which has been receiving renewed interest in the development of drug delivery systems. Importantly, albumin is already an "off-the-shelf" product.

Advances in the study of engineered biomaterials represent a step forward in the exploitation, development and commercialization of new therapies for old, poorly served medical needs.

Keywords: Albumin, Biomaterials, Biopolymers, Chitosan, Drug delivery.

* **Corresponding author Bruno Sarmento:** i3S – Instituto de Investigação e Inovação em Saúde, Rua Alfredo Allen, 208, 4200-393 Porto, Portugal; E-mail: bruno.sarmento@ineb.up.pt

Martin Masuelli & Denis Renard (Eds.)

INTRODUCTION

The necessity of novel therapies for poorly served medical conditions is triggering the development and approval of new and enhanced drugs. Indeed, sub-optimal pharmacodynamics of many small molecules, peptides and proteins has been driving the development of innovative drug formulations, drug delivery technologies and medical devices [1]. The routes behind this strategy foresee the establishment of advanced formulations, characterized by high biocompatibility and biodegradability, reduced toxicity and optimized pharmacokinetics for targeted drug delivery and controlled release, avoiding accumulation or undesired side effects in healthy tissues [1].

The use of biomaterials as drug carriers is being extensively investigated as a drug delivery strategy in pharmaceutical research. In general, the main functions of a carrier include the protection of the drug from degradation, the enhancement of both drug release profile and absorption (by facilitating diffusion through epithelium) along with distribution and elimination [2]. To complement this research, carrier properties, such as chemistry, surface and functionalization can be precisely tuned to promote a better drug release pattern and biodistribution [2].

Selection of materials for such systems is rather complicated, but lately polymers are being widely used as biomaterials for drug transport.

Chitosan and albumin are two of the most promising biomaterials for the development of drug delivery systems. These are widely available in nature and can be used as scaffolds of unprecedented novel structure [3].

Chitosan is a polysaccharide, structurally similar to cellulose and has the desired properties for safe use as a pharmaceutical excipient [4]. Albumin is a natural polymer, physically and chemically very robust and stable, non-antigenic and metabolizable [1]. Importantly, albumin has an excellent binding capacity to various drugs [5]. The characteristics of these two biopolymers make them very suitable materials for developing drug delivery vehicles.

Advances in this field include the development of chitosan- and albumin-based platforms for drug delivery, aiming toward the treatment of high incident pathological states, as cancer or diabetes. Furthermore, approved therapies with superior therapeutic efficacy are being achieved, and indicate that, in the near future, chitosan- and albumin-based drug delivery systems will have an increased commercial status.

1. CHITOSAN AND AMPHIPHILIC DERIVATIVES

Polysaccharides are widely distributed in nature and, among them, chitosan is receiving increasing attention. Chitosan is a linear polysaccharide prepared from chitin, which is the second most abundant natural polymer in the world, after cellulose [6, 7]. Chitin can be found in the exoskeleton of insects, crustaceans and some cell wall of fungi. Chitosan, derivative of chitin can be obtained by alkaline deacetylation or enzymatic hydrolysis [7]. The reaction product is chitosan only when the average degree of acetylation (DA), is equal to or less than 50%. The DA is used to characterize the average content of *N*-acetyl-D-glucosamine units (acetylated unit) while degree of deacetylation (DD) is the number of D-glucosamine units (deacetylated unit) [6, 8]. Chitosan is constituted by *N*-acetyl-D-glucosamine and of D-glucosamine units linked by glycosidic linkages β-(1,4) (Fig. **1**) [6, 7, 9, 10].

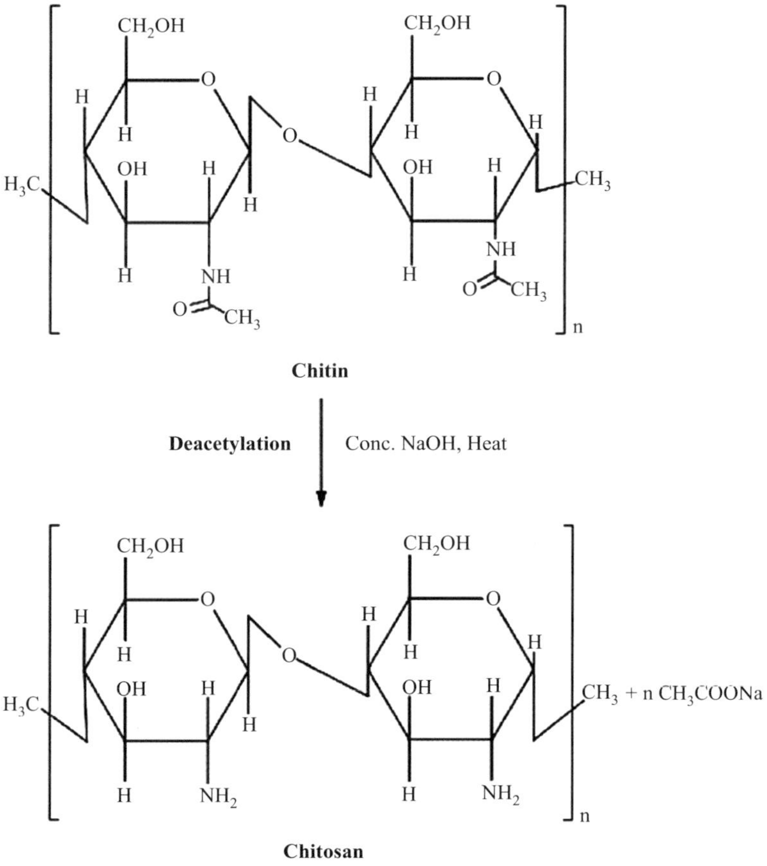

Fig. (1). Chitin deacetylation to chitosan.

These polymers show solubility in moderately acidic medium. Its solubility in acidic medium occurs due the protonation of the NH_2 functional group on the C-2 position of the D-glucosamine repeating unit (Fig. **1**) [6, 9, 11].

This polymer has one amino group and two hydroxyl groups in the repeating glucosidic residue, providing reactive sites for a variety of reactions, which is an advantage over other polymers [12]. Several types of chitosan can be obtained and with different physicochemical properties such as solubility, pK_a and viscosity [7, 13, 14].

The DD and the molecular weight (M_w) are two of the most important physicochemical properties of this polymer. An increase on the chitosan DD, leads to an increase of its viscosity, due to high and low deacetylated chitosan having different conformations in aqueous solution. Additionally, the chitosan solution viscosity can be also affected by factors as temperature and concentration [8]. For example, when chitosan concentration increases, or the temperature of the solution decreases, the viscosity increases.

The chitosan is a polycation whose charge density depends on the DA and pH. Due to the progressive amine protonation of pendant groups, the solubility is pH dependent. Therefore, chitosan is only soluble at pH values of approximately 6 or less [10, 11, 15]. This feature turns chitosan into a biopharmaceutical of interest, along with its properties of mucoadhesiveness and ability to open epithelial tight junctions [16]. In fact, at low pH (<6), chitosan amine groups are protonated and become positively charged, leading to polycationic behavior [8, 17]. By contrast, at a higher pH value (above around 6), chitosan amines are deprontonated and reactive (Fig. **2**).

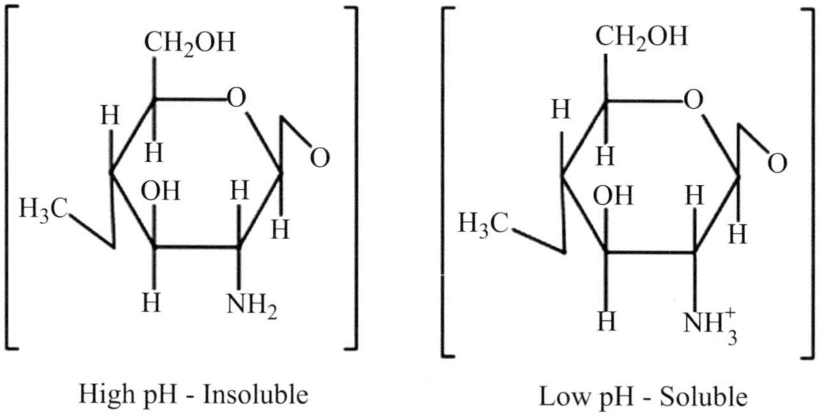

Fig. (2). The soluble-insoluble transition of chitosan occurs at about pH 6.

To improve the solubility of chitosan in physiological media, its biological properties are improved and its potential applications are expanded, while chitosan derivatives have been synthesized [11, 18]. Indeed, macromolecules with amphiphilic character have become a subject of great interest over the last years [19, 20].

Chitosan amphiphilic derivatives result from the attachment of hydrophobic structures to the hydrophilic chitosan backbone, and can be achieved by using different methodologies such as alkylation, acylation, graft polymerization, cross-linking, among others [11, 21, 22]. In general, the reaction products of an amphiphilic chitosan are, among others, *N*-alkyl chitosan, acyl-chitosan, graft derivatives, crown-ether derivatives or cyclodextrin derivatives [11]. Interestingly, the introduction of a hydrophobic molecule on chitosan backbone modifies its rheological behavior, being therefore used as rheological modifiers in pharmaceutical industries [19].

The ability of amphiphilic derivatives to self-assemble has triggered great interest in the biomedical field, particularly in the formation of micellar systems [11]. The applications of these chitosan amphiphilic derivatives in drug delivery will be discussed in the following sections.

1.1. Biological Properties and Biopharmaceutical Profile

Chitosan is a non-toxic, biodegradable and biocompatible polymer, of great biological interest, as shown in Table **1**. The presence of NH_2 groups in chitosan is the reason why it exhibits much higher potential as compared with chitin [8] and what makes it an excellent polymer for pharmaceutical applications.

Table **1** shows chitosan biological properties and the relationship between its structural characteristics (DD and M_w). For instance, the lower its DD and M_w, the higher the chitosan biodegradability [8, 23 - 25]. Moreover, chitosan plays a role in hemostasis, but independently of the coagulation cascade [26]. Okamoto *et al.* showed that chitosan reduces clotting time in a dose-dependently manner, with the ability to aggregate platelets [27].

Particular attention must be given to chitosan antimicrobial activity. This polymer inhibits the growth of microorganisms such as *Escherichia coli, Fusarium, Alternaria, Helminthosporium* [28] and *Candida* [29]. In fact, the growth of *E. coli* was inhibited in the presence of over 0.025% chitosan [28]. A possible reason for the growth inhibition of these bacteria might be the cationic chitosan amino groups binding to the anionic groups of these microorganisms [30]. The results obtained by Zheng and Zhu indicate that the effects of chitosan in the two types of bacteria differ from each other: in case of gram-positive bacteria, it is thought that high molecular weight chitosan form films around the cell, which ultimately inhibits the absorption of nutrients, while low molecular weight chitosan more easily penetrates in gram-negative bacteria, causing disturbances in the metabolism of these microorganisms [7, 31].

Table 1. Some biological and structural properties of chitosan.

Property	STRUCTURAL CHARACTERISTICS[a]	References
Biodegradability	↓ DD, ↓ Mw	[8, 23 - 25]
Biocompatibility	↑ DD	[8, 32, 33]
Cytocompatibility	↑ DD	[8, 32]
Mucoadhesion	↑ DD, ↑ Mw	[33 - 35]
Hemostatic action	↑ DD	[7, 26, 36]
Analgesic action	↑ DD	[37, 38]
Antimicrobial activity	↑ DD, ↑ Mw	[6, 7, 39 - 42]
Antioxidant activity	↑ DD, ↓ Mw	[6, 43 - 45]

[a] ↑ – Directly proportional to property; ↓ – inversely proportional to property.

The above mentioned chitosan properties make chitosan an exciting and promising excipient for the pharmaceutical industry, with a large margin for development. Due to chitosan bioactivity and intrinsic mucoadhesion properties, or the ability to open epithelial tight junctions, its formulations may double the therapeutic effects [10].

Chitosan amphiphilic derivatives have, indeed, attracted attention among researchers working in biomedical engineering, and particularly, in drug delivery, given the fact that they have improved physical and chemical properties compared with chitosan un-derivatized [13].

1.2. Chemical Conjugation between Chitosan and Hydrophobic Polymers

Water soluble polymers linked to a hydrophobic character is the new line of important biopharmaceutical macromolecules [19].

Chitosan hydrophobic derivatives can be easily obtained from its reaction with long chain acyl chlorides and anhydrides [46]. Additionally, polyester substituents provide the basis for hydrophobic interactions that contribute to the formation of chitosan hydrogels [46].

The stability of such formulations is attributed to hydrophobic interactions between the side chains [19, 46].

1.2.1. Hydrophobic Interactions

Self-assembly or aggregation of amphiphilic polymers has triggered growing interest in biological and pharmaceutical applications [47]. The aqueous solution properties of chitosan derivatives can be controlled by varying the degree of substitution with hydrophobic bunches for controlling the solubility properties [18]. These derivatives present hydrophilic sites (-NH_2 or -OH) and hydrophobic sites (alkyl chains) on the macromolecular chain [11, 19]. Chitosans that are substituted with alkyl chains having, at least, six carbon atoms, demonstrate hydrophobic interactions in solution [11, 13].

In order to observe hydrophobic interactions it is necessary to have apolar chains, which have a minimum length, but the degree of substitution does not need to be extremely high to have a compromise between solubility and interactions [19]. Therefore, a minimum level of hydrophobicity is dependent on the combination of the degree of substitution and the length of the chitosan chain.

Another aspect that influences the hydrophobic interactions is the temperature: a small increase of the temperature from 20ºC to 22°C promotes the hydrophobic interaction [19]. Likewise, the presence of an electrolyte favors the formation of hydrophobic interactions, since the alkyl chain has a sufficient length [19]. Hence, an increase in the ionic strength of the medium strengthens hydrophobic interactions.

Hydrophobic interactions differ from electrostatic interactions, hydrogen or Van der Waals bonds by the fact that they are not caused by direct cohesive forces between molecules but due to the specific structure of water molecules close to the polymer chains [18]. The hydrophobic associations lead to a three dimensional polymer network [19].

1.3. Chitosan-derivatives in Drug Delivery

In recent years, significant effort has been given to the development of biodegradable materials for drug delivery systems. Chitosan and its derivatives are suitable for drug delivery applications, for their biodegradability, biocompatibility, antibacterial activity, high affinity to *in vivo* macromolecules and ease of chemical modification, as already mentioned [6, 34, 48]. Furthermore, chitosan is a polyelectrolyte with reactive functional groups, absorption capacity and gel formation [8, 13].

1.3.1. Anti-cancer Drugs

Amphiphilic chitosan derivatives have already proven to be an excellent system for drug delivery when loaded with anticancer agents such as paclitaxel or doxorubicin.

In a study made by Mo *et al.*, *N*-octyl-*O*-sulfate chitosan micelles have been used to demonstrate the enhancing oral absorption of paclitaxel [49]. In fact, the results showed that these micelles significantly improved the oral bioavailability of paclitaxel compared with commercially available product (Taxol®) *in vivo* [49]. Micelles of amphiphilic chitosan may also be linked to peptides to effectively deliver anti-cancer drugs. In a recent study, an amphiphilic copolymer *N*-succinyl-palmitoyl-chitosan loaded with paclitaxel was synthetized and cRGDyK peptide was used as a ligand to target tumors *via* specific binding integrin receptor, which is overexpressed on tumor neovascularization and cells [50]. These micelles demonstrated an effective cellular uptake and low toxicity in mice, and therefore a great potential system to be used in the treatment of orthotopic lung cancer, rather than the exclusive use of Taxol® [50].

In another study, self-assembled nanoparticles were obtained by oleoyl chitosan, where the toxicity profile was in the normal range. Additionally, when loaded with doxorubicin, nanoparticles showed inhibitory rates in different human cancer cells (A549, Bel-7402, HeLa and SGC-7901) that were significantly higher than when using only the drug [48]. Hu *et al.*, formed polymeric micelles of stearic acid grafted chitosan oligosaccharide to encapsulate paclitaxel [51]. Prolonged and controlled drug release of paclitaxel was achieved [51]. A similar study also used stearic acid grafted chitosan oligosaccharide to form polymeric micelles, but encapsulating doxorubicin [52]. When using these micelles, the results showed *in vivo* tumor suppression in MCF-7 cells and reduced toxicity against the body of the animals in comparison with adriamycin (doxorubicin commercial), evidencing its potential in the treatment of tumors [52]. Studies of these micelles in others cells types showed that these micelles were rapidly internalized into tumor cells and the uptake in LLC and SKOV3 cells was higher than in A549 cells [53].

Linoleic acid-grafted chitosan oligosaccharide micelles was synthesized as a carrier for doxorubicin for treating cancer [54]. The results showed that the micelles have a high drug uptake capacity and controlled release profiles, thus being an excellent candidate for drug deliver [54].

Other drugs are also used in amphiphilic chitosan derivatives aiming cancer treatment. For example, nanoparticles formed by hydrophobically modified glycol chitosan were designed by chemical conjugation of hydrophobic 5β-cholanic acid moieties to the hydrophilic glycol chitosan backbone, loaded with camptothecin [55]. These nanoparticles exhibited high loading efficiency, longer blood-circulation time, better targeting of drug to tumor tissue towards MDA-MB231 human breast cancer xenografts subcutaneously implanted in nude mice [55]. This chitosan derivative is suggested as a drug delivery system for effective chemotherapy against human breast cancer [55]. Cisplatin-loaded chitosan microspheres were prepared with an effective incorporation rate [56]. A notable decrease in the number of necrosis of nodes and liver cells was observed [56].

1.3.2. Proteins/Peptides

Amphiphilic nanoparticles have many potential applications for the administration of therapeutic molecules, particularly proteins and peptides [28].

Chitosan-polyethylene oxide nanoparticles have large protein load capacity and may provide continuous release of the protein retained during 1 week [57]. Also, nanoparticles based on polymethacrylic acid-chitosan-polyethylene glycol for oral peptide delivery application were proven to be good candidates for this applications [48].

In a recent study, a bio-inspired nanocarrier was developed for protein delivery, based on biodegradable amphiphilic chitosan derivative with hydrophilic cell membrane mimic phosphorylcholine and hydrophobic deoxycholic acid moieties [21]. The nanoparticles showed excellent cytocompatibility and hemo-compatibility and a sustained-release behavior, being therefore considered a promising approach for the efficient delivery of therapeutic proteins [21].

The protein receiving the greatest attention as a candidate for nanoparticle based drug delivery has been, by far, insulin [16, 58]. It was found that insulin-loaded chitosan nanoparticles enhanced nasal absorption of proteins [58, 59]. These nanoparticles have good loading capacity (65–80%) and fast insulin release [60].

Chemical modifications of chitosan can increase the mucoadhesive properties, consequently increasing insulin absorption [16]. Another conjugation consists in chitosan and carboxymethyl-β-cyclodextrin nanoparticles [61]. These nanoparticles have great capacity for combination with insulin, good release profile, and are considered good carriers for fast or slow release of macromolecules for oral or nasal administration [61].

Insulin-loaded polyethylene glycol-grafted chitosan nanoparticles can be used to improve the systemic absorption of insulin after nasal administration [62]. In fact, these nanoparticles visibly enhance insulin nasal absorption in comparison with the same particles in suspension or control solution of insulin, being therefore a clearly indicated polymer to the nasal absorption of insulin [62]. Many other chitosan-based nanostructures have been designed for testing their ability to deliver insulin and other peptides and proteins.

1.3.3. Growth Factors

Chitosan derivatives for growth factors release are strongly present in tissue engineering, particularly in bone regeneration. For instance, chitosan hydrogels, together with bone morphogenetic protein (BMP) demonstrate ability to enhance the repair of an injury [23].

Furthermore, chitosan coated poly(L-lactide) matrices were prepared and examined for bone regeneration [63]. This study showed an improvement in bone forming ability, increasing its mechanical stability and biocompatibility [63]. Studies on the release of platelet-derived growth factor-BB from these matrices demonstrated a significant osteoinductive effect together with a high osteoconductive capacity [63]. Chitosan-modified DL-lactide/glycolide nanosp-heres were performed to improve the pulmonary delivery of elcatonin by nebulization [64]. The results showed that the nanospheres enhance drug absorption by opening the intercellular tight junctions in the lung epithelium [64].

Chitosan-poly-ε-caprolactone copolymer microspheres loaded with transforming growth factor-β1 were fabricated to study the properties of loading, swelling and release [65]. They concluded that the fast release rate of transforming growth factor-β1 from different microspheres could be significantly decreased with increasing poly-ε-caprolactone content. In a simulated acidic environment like cartilage lesions, the release of microspheres was modulated by pH and the cross-linker could act as a limited factor [65].

1.3.4. Antibiotics

Currently, there is still a large clinical failure on the administration of antibiotics [66]. The antibiotics encapsulated in microparticles or nanoparticles formed by a biodegradable polymer have shown great potential for replacing antibiotics in free form [66, 67]. Chitosan is one of the polymers used successfully for this purpose.

Ungaro *et al.*, developed a pulmonary delivery system for antibiotics such tobramycin based on poly(lactic-co-glycolic acid) (PLGA) nanoparticles in the presence of different types and amounts of hydrophilic polymers like chitosan [67]. The results showed a faster tobramycin release and a higher tendency to interact with mucin as compared to other polymers, such as poly(vinyl alcohol) [67]. The results suggest a great potential of these nanoparticles to act *in vivo* as a drug reservoir that slowly releases the active agent to the target. Chitosan-engineered nanoparticles were found in great quantities in the upper airways, lining lung epithelial surfaces, unlike other polymers [67].

In another study, chitosan has been blended with varying amounts of poly-ε-caprolactone for control delivery of ofloxacin [68]. The release of the antibiotic depends on the polymeric matrix being suitable for the basic environment of the large bowel, colon and rectal mucosa, as well as the pH of the media because the interactions are more easily broken at physiological pH than at low pH, making ofloxacin more quickly released at pH 7.4 than 3.4 [68]. Another example are chitosan salts coated with stearic acid prepared to release vancomycin. These salts supply pH-dependent systems to assure low vancomycin release in the stomach environment and adequate release in the colon as required [69].

1.3.5. Anti-inflammatory Drugs

The purpose of targeting anti-inflammatory drugs is to achieve the desired pharmacological response at a selected site without undesirable interactions [70]. The use of chitosan with Tween-80 and oleic acid to form a hydrogel as drug carrier for controlled drug delivery of anti-inflammatory is an example [70]. This hydrogel incorporated with three anti-inflammatory drugs (diclofenac, acetyl-salycilic acid and hydrocortisone acetate) with success, of which aspirin and diclofenac sodium are more included in the chitosan matrix than hydrocortisone acetate because the connection between the chitosan matrix and the guest molecule depend upon the molecular weight of the latter due to stereochemistry reasons [70].

Kumbar *et al.,* prepared microspheres of polyacrylamide-grafted-chitosan to encapsulate indomethacin, a non-steroidal anti-inflammatory drug used in the treatment of arthritis [71]. This study showed that there is a release of indomethacin, mainly by molecular diffusion from the fully swollen polymer and could be useful in the delivery of indomethacin and other similar drugs [71].

In a recent study, deoxycholic acid-phosphorylcholine-chitosan was synthesized using quercetin as a hydrophobic model drug [72]. The resulting micelles had low cytotoxicity and good hemocompatibility, further the drug loading and release experiments for quercetin showed it to be a potential nanocarrier for delivering and releasing hydrophobic drugs in a sustained manner, avoiding unfavorable biological responses [72].

Marques *et al.*, used a non-steroidal drug as an antitumor agent to combat colorectal cancer through micellar nanocarriers composed of amphiphilic chitosan [73]. They used ibuprofen to test the anti-cancer activity in breast cancer cells, concluding that these micelles could be a promising way to treat breast cancer because they led to a remarkable reduction in cancer cell viability at a relatively low drug dosage [73]. The conjugation of methoxy polyethylene glycol with chitosan was performed in order to increase the solubility of chitosan in water and control the release of water insoluble drugs like ibuprofen [74]. The results showed that this combination, providing a slower release of the drug along with a higher encapsulation efficiency, could be a promising delivery system for such drugs [74].

1.3.6. Vaccines Delivery

Due to its low production cost and excellent biological properties, chitosan has become of interest for vaccine delivery. It can easily form nano- and microparticles, and is also able to carry high quantities of several antigens such albumin [75]. Chitosan nanocarriers can significantly enhance the immunogenic response of antigens through intramuscular administration [16]. Jaganathan *et al.*, developed a surface-modified in poly(lactic-co-glycolic acid) (PLGA) microspheres with chitosan for nasal immunization using recombinant Hepatitis B (HBsAg) [76]. The results showed that the final microspheres were able to protect integrity and activity of HBsAg and able to produce a humoral and a cellular immune response by nasal administration. PLGA microspheres modified by chitosan had significantly reduced clearance rate of the nasal cavity compared with the control unmodified microspheres, demonstrating the potential to be an adjuvant in nasal vaccines [76].

Polymeric microspheres of poly-ε-caprolactone-chitosan was used for adsorption of *Streptococcus equi (S. equi)* enzymatic extract antigens to verify *in vivo* immune responses in mice [77]. The results showed that 45 days after animal vaccination, all mice had seroconverted and their IgG responses were significantly superior to the free antigen immunization group [77]. Furthermore, 300 days vaccination the groups also showed a serum specific IgG antibody in response to *S. equi*, suggesting be a potential vaccine adjuvant to prevent infection by *S. equi* [77].

A similar study was performed with alginate-chitosan-PLGA composite microspheres for formulation of Hepatitis B vaccine [78]. The results showed a great capacity for encapsulation of these microspheres due to the hydrophilic chitosan-alginate core. Also, the release of the vaccine showed an improvement when compared with conventional PLGA microspheres [78]. These microspheres showed better efficiency and also high amounts of antibody levels with only one dose of vaccine when compared with conventional single PLGA microspheres [78].

2. ALBUMIN

Albumin is by far the most plentiful protein in plasma, constituting more than half of blood plasma protein content of all vertebrates [79]. Similarly to most other plasma proteins, albumin is synthesized in the liver, in polysomes bound to the endoplasmic reticulum of hepatocytes, at a production rate of approximately 0.7 mg/h per gram of liver [80, 81]. It is a highly soluble and stable molecule but, contrary to other plasma proteins, is non-glycosylated [82].

Among its extensive range of features, albumin is responsible for 80% of the intravascular oncotic pressure [83],is a major source of sulphydryl groups and acts as a scavenger of reactive oxygen and nitrogen species [79, 84]. Moreover, albumin has an antithrombotic and anticoagulant effect, which may result from its capacity to bind nitric oxide (NO) to form *S*-nitrosothiols [81]. It is partially involved in the maintenance of the intravascular volume, and in a homeostatic condition, reduces microvascular permeability. Albumin influences aspects of the inflammatory pathway, including neutrophil rolling, adherence and activation [85], and also acts as a toxic waste carrier, by binding unconjugated bilirubin, the breakdown product of normal heme catabolism, delivering it to the liver for hepatic excretion [86]. Additionally, albumin has the ability to bind several endogenous molecules, such as long-chain fatty acids and steroids [81], and facilitate the transport of metabolites, electrolytes and drugs in the circulatory system. Interestingly, albumin's ability to bind drugs highly influences their activity and half-life [87].

Despite being perceived as an intravascular protein, the concentration of albumin in the extravascular space exceeds the total amount encountered in the intravascular space [81]. In fact, some of this albumin is tissue-bound, mainly found in locations such as skin, gut, muscle and fluids as the cerebrospinal or pleural, and also secretions as sweat, tears or milk [79, 88]. The total interstitial mass of albumin appears in even lower concentrations and varies between different anatomical regions [81].

Interestingly, there is a cyclic exchange of albumin between the extravascular space and blood-circulation. When secreted from the hepatocytes, albumin enters blood circulation. Later on, it translocates pores of sinusoidal or fenestrated endothelium present within several organs, including liver, pancreas, bone marrow and small intestine, therefore reaching the interstitium [79, 88]. In fact, albumin escapes the intravascular space through the endothelial barrier at the level of the capillaries [89]. Subsequently, albumin within the extracellular space returns back into the circulation through lymphatic drainage, which happens approximately 28 times during its lifetime [79, 88]. Organs presenting a continuous endothelium have demonstrated to allow the transport of albumin through transcytotic mechanisms, such as receptor-mediated transcytosis [90]. These are currently being investigated due to their potential impact in the development of new drug delivery systems.

2.1. Albumin in Drug Delivery

In recent decades, the constant demand for therapies with improved drug targeting, along with the necessity of improving pharmacokinetic profiles of peptide- or protein-based drugs has triggered renewed appreciation of albumin, which is now considered a versatile protein carrier, with an increasing role in the clinical setting

Currently, albumin-based drug carrier systems constitute a subject of major interest, and its long circulatory half-life is one of the main reasons for the development of strategies based on this endogenous molecule. Additionally, coupled to its well-known lack of toxicity and antigenicity, albumin has the ability to be degraded into innocuous natural products [91, 92]. Due to the different binding sites displayed by the albumin molecule, significant amounts of drug can be incorporated in these systems [2, 92].

Albumin is easily purified and very soluble, but extremely robust. Indeed, albumin shows stability in the pH-range of 4-9, solubility in 90% ethanol, and can withstand temperatures of 60 °C for up to 10 h without harmful effects [80]. Furthermore, manufacture of albumin-based nanosystems is easy to scale-up, yielding predictable and reproducible sized products [93], and demonstrate full maintenance of its properties after storage [2]. Since albumin primary structure is already defined, and knowing the great amount of charged amino acids within this molecule, including aspartic and glutamic acids or lysine, albumin-based carrier systems could allow the electrostatic adsorption of either positively or negatively charged molecules alone, thus obviating the need for other compounds [94]. Importantly, albumin reveals preferential uptake in tumor and inflamed tissue [80], which can be further enhanced by the addition of specific targeting ligands.

Due to all the above mentioned advantageous features, albumin is rendered an ideal candidate for the development of the desirable drug-loading carriers. Albumin is already an "off-the-shelf" product, since it can be largely obtained from egg white, bovine and human serum [95]. Albumins can be also obtained from soybeans, milk and grains [96 - 98].

2.1.1. Types of Albumin

Ovalbumin (OVA)

OVA is a glycoprotein that comprises 60-65% of egg white total protein [99]. For being readily available in large quantities it is massively used in structural and characterization studies, and in experimental models of allergy [100]. Also, the isolation of OVA from hen oviduct and its regulation by steroid hormones allowed the settlement of a model system for studying protein synthesis and secretion [101].

OVA has a molecular weight of 47 KDa and an isoelectric point of 4.8 [95]. It consists of 385 amino acid residues, and its sequence includes six cysteines, with a single disulfide bond between Cys74 and Cys121 [102].

The availability and low cost of OVA, its pH- and temperature-sensitive properties, its ability to form gel networks upon heating [103], and to form and stabilize foams [104], highlight its versatile profile as a drug delivery platform.

Bovine Serum Albumin (BSA)

BSA is a natural ellipsoid protein capable to form complexes in various shapes [105]. With a molecular weight of 69 KDa and an isoelectric point of 4.7 in water [95], BSA is a biocompatible, biodegradable, non-toxic and non-immunogenic protein [105] and presents unusual ligand-binding properties [106]. Similarly to OVA, BSA is widely available, of low cost and is easy to purify [107].

BSA is widely accepted in both research and commercial/pharmaceutical fields [108], and is considered a good system for drug and antigen delivery. Interestingly, bovine and human serum albumins share 76% of the amino acid sequence [109, 110].

Human Serum Albumin (HSA)

HSA is one of the most extensively investigated proteins. Its extraordinary binding capacity provides a depot for a wide range of compounds and therefore allowing their availability in quantities well beyond their solubility in plasma [111].

HSA has a molecular weight of 66 500 Da and exhibits an average half-life of 19 days [80]. X-ray crystallography was used to observe its heart-shaped structure

[112], and to unravel its 585 amino acid sequence, comprising 17 pairs of disulfide bridges and one free thiol from an unpaired cysteine (Cys34) [113]. The molecule consists of 67% α-helices and no β-sheets, and is formed from three homologous domains (DI, DII and DIII), subdivided in two separate helical subdomains (A and B), which are connected *via* long flexible loops [1, 82]. When broken, HSA amino acids provide nutrition to peripheral tissue [2].

Due to its small size and great abundance, many metabolic compounds and therapeutic drugs can be transported by this protein [80]. HSA binding sites for metabolic substrates and diagnostic and for therapeutic drugs are currently under extensive investigation.

2.2. Ligand-binding Capacity

Due to its high abundance and associated physical, chemical and pharmaceutical properties, albumin holds an astonishing ligand-binding capacity. Beyond avoiding undesired side effects, albumin binds almost every small molecule, both of endogenous or exogenous origins [1]. Therefore, when bound to albumin, drugs will be available in larger amounts than their normal plasma concentration, presenting decreased toxicity, lowered clearance rate and, subsequently, increased circulatory half-life [114]. Hence, albumin is one of the proteins with higher potential to be used as molecular cargo/or nanovehicle for clinical, biophysical, and industrial purposes [114].

The binding of substances to albumin occurs at the binding pockets present within its three domains. Despite their similar structures, the relative position of the domains amongst themselves is different, creating different interfaces between them and the binding substances [115].

Albumin-bound ligands account for up to 35 proteins and peptides (*e.g.* hemoglobin and transferrin) [116], several small molecules, and transition metal ions (*e.g.* copper and zinc) [117]. The set formed by albumin and its bound proteins and peptides is referred as "albuminome" [116].

When it comes to binding of fatty acids, crystallographic studies identified seven binding sites, asymmetrically distributed throughout the tertiary structure of albumin with moderate to high affinity [118]. The binding sites available for ligand or drug binding can be named as fatty acid binding sites, drug binding sites 1 and 2 or Sudlow's sites 1 and 2 [82]. For instance, bulky heterocyclic anions such as warfarin bind to drug binding site 1, which is located in the core of DIIA, while aromatic carboxilates (*e.g.* ibuprofen, diazepam) bind to drug binding site 2, located in DIIIA [1, 119 - 121].

Under normal physiological conditions and upon metabolic need, albumin is the major plasma fatty acid delivery vehicle, carrying 0.1-2 moles of fatty acids per mol of protein [122], but can increase in certain states of disease [114]. This value can rise up to 6 upon intensive exercise, thereby resulting in increased amount of fatty acids in blood plasma [122]. Interestingly, significant conformational changes are observed in the albumin molecule when bound to a fatty acid [111]. These changes result in the creation of new sub-sites in the molecule, which are thought to be necessary for the albumin-drug interaction [123, 124].

The active concentration of a drug considerably affects the overall pharmacokinetic and pharmacodynamics profile of a drug carrier. Therefore, studies on albumin-drug association and dissociation represent an important matter from both theoretical and practical point of views. These studies might allow deeper insights on the processes underlying drug disposition and elimination when associated to albumin, and on the effect of a drug-albumin system on pathological states, or the effect of co-administered drugs on delivery and efficacy of an albumin-based carrier [125].

2.3. Albumin Receptors, Half-life and Interaction with FcRn

Over the years, a great number of albumin-binding proteins and receptors have been identified in several tissues and cell lines. In fact, albumin uptake, transport, degradation, salvage and recycling was demonstrated to be highly regulated by the interaction with its receptors.

Several albumin-binding proteins and receptors have been described, namely glycoprotein (gp) 18, gp30, gp60 (albondin), cubilin, megalin, calreticulin, heterogeneous nuclear ribonucleoproteins (hnRNPs), the secreted protein acidic and rich in cysteine (SPARC; also named osteonectin), and the neonatal Fc receptor [126 - 132].

Gp18 and gp30 are found on an extensive variety of cells, including macrophages, fibroblasts, smooth muscle cells and a range of endothelia [133]. Additionally, these proteins are ubiquitously found in heart, lung, kidney, brain, and liver, amongst many other organs [133]. Both proteins are known to bind with high avidity to albumin with a modified conformation, rather than with native albumin, resembling other known scavenger receptors [133, 134]. In fact, denatured or modified albumin is degraded efficiently and at a faster rate than native albumin [86]. Therefore, one can conclude that g18 and g30 are involved in the regulation of the high affinity binding, endocytosis and degradation of conformationally modified albumins [134].

In contrary to gp18 and gp30, the gp60 albumin receptor, or albondin, presents a much more narrowed distribution, and is selectively expressed both on the plasma membrane of vascular continuous endothelium and alveolar epithelium [126, 127, 130]. Interestingly, gp60 binds native albumins, facilitates its internalization and subsequent transcytosis, but not the modified ones [135]. However, the genes encoding gp18, gp30 and gp60 have not been cloned yet, and mapping of their interaction with albumin is yet to be performed [136, 137].

Cubilin is a multi-ligand receptor located at absorptive intestinal cells, placenta, visceral yolk-sac cells, and proximal tubules of kidneys [79, 138], and is involved in the albumin endocytosis and transcellular transport [139]. This glycoprotein is also known to bind the transmembrane, endocytic receptor megalin [139]. Megalin is expressed in choroid plexus, kidney proximal tubule cells, thyrocytes, amongst many others [79, 131]. Despite being much more widely expressed than cubilin, both cubilin and megalin bind either native or modified albumin [79]. Together, these receptors form a high affinity complex, which is responsible for receptor-mediated reabsorption of albumin by the proximal tubes of the kidneys [128]. The absence of these receptors leads to albuminuria (abnormal rates of albumin in the urine), caused by a deficient uptake of albumin in the proximal tubule [128, 138, 139].

Calreticulin is primarily localized in the endoplasmic reticulum, functioning as a chaperone. However, it has been shown that calreticulin is found at multiple subcellular localizations, and is involved in several cellular functions, including apoptotic cell clearance, cell adhesion, and cell migration [140].

Calreticulin is the fifth protein identified from plasma membranes of human cancer cells. The remaining four are members of the hnRNP family (including hnRNP A2/B1, hnRNP C1, hnRNP A1 and hnRNP A3) [129]. In spite of not having their functions fully characterized, the members of hnRNP family are thought to be nuclear RNA-binding proteins involved in pre-mRNA processing (*e.g.* splicing, export and stability) [141]. hnRNP are localized on the surface of tumour cells [129]. However, it is not yet know what is the biological significance of the albumin-binding property of hnRNP proteins and whether they are involved in albumin endocytosis [129, 141].

Another molecule with albumin-binding properties is SPARC. SPARC is a multifunctional glycoprotein secreted by several cell types, which functions to regulate cell-matrix interactions, influencing physiological and pathological processes [142]. Additionally, SPARC is known to be highly expressed in malignant cells, and surrounding stromal cells (*e.g.* fibroblasts, endothelial cells and inflammatory cells), and importantly, to interact with native albumin.

However, it is not clear whether SPARC mediates albumin uptake into tumors [79]. In fact, SPARC-albumin interactions in tumors are undergoing major investigations, since it has been suggested that SPARC could sequester albumin-bound paclitaxel, an anti-cancer drug, thereby enhancing the delivery of this substance into the tumor microenvironment [142]. Nonetheless, pathways involved in SPARC signaling need to be further elucidated.

One of the main reasons for studying albumin in a drug delivery context is its extraordinary long circulatory half-life. In fact, albumin's molecular weight (≈66,5 KDa) is above the kidney filtration threshold, the mechanism preventing clearance of proteins larger than 60-70 kDa from the body through the urine [143]. However, this reason does not stands alone, since albumin's longevity is also highly dependent on another salvage mechanism, mediated by the neonatal Fc receptor (FcRn).

FcRn is a member of the major histocompatibility complex (MHC) class I molecules, and consists of a heterodimer of a glycosylated heavy α-chain in non-covalent association with β2-microglobulin (β2m) [144, 145]. It is widely distributed in several mammalian organs and cell types, such as vascular endothelium, antigen-presenting cells, gut, blood-brain barrier, kidneys and lungs [132]. However, the pattern of intestinal FcRn expression among species is markedly different. For instance, while FcRn is expressed throughout human life in intestinal epithelial cells [146, 147], intestinal FcRn expression levels in rodents suffer a significant decrease immediately after weaning [148].

FcRn was originally found in murine neonatal intestine, working as a transporter of immunoglobulin G (IgG) from ingested breast milk through the epithelial cell-layer of the small intestine [149]. Currently, a crucial role for FcRn in extending the serum half-life of IgG and albumin is well-recognized, since it protects both molecules from intracellular degradation, in a pH-dependent cellular recycling mechanism [132, 150]. This mechanism has been increasingly exploited due to its foreseen potential in several albumin-based therapeutic strategies.

2.4. Albumin-based Drug Delivery Systems

Over the past decades, pharmaceutical companies have been investing massive amounts of money in the development of novel drugs. Yet, most drug candidates fail to enter the clinic due to poor bioavailability and/or impermeability of the mucosal or epithelial barriers of the intestine, along with the unwanted side effects caused by non-specific cell/tissue interactions. For this reason, nanotechnology researchers are intensively investigating ways to develop new and more efficient drug delivery systems, with the purpose of taking these new drugs to the bedside.

In fact, the development of drug delivery systems has been an evolving process, and albumin is playing an increasing role as a drug carrier. Principally, the technologies that are being used for the formulation of albumin-based drug delivery systems involve coupling of low-molecular weight drugs to exogenous or endogenous albumin, conjugation with bioactive proteins and encapsulation of drugs into albumin nanoparticles (NPs) [80]. Additionally, binding a therapeutic peptide or protein, either covalently or physically to albumin, aiming to enhance its stability and half-life can be also considered for this purpose [80].

Association, conjugation or fusion of therapeutics to albumin is, indeed, a well-accepted and established technology, which might facilitate the modulation of the circulatory half-life of albumin-enabled drugs, ameliorating their pharmacokinetics tailored to a specific disease state and increased patient compliance [87].

Currently, several albumin drug-delivery systems are under extensive research towards the exciting avenue of their medical application, including albumin nano- and microparticles, conjugates, microbubbles and hydrogels.

2.4.1. Nano- and Microparticles

Nanoparticles are dispersions or solid colloidal particles, extensively investigated as a drug delivery system in pharmaceutical research. The use of nanoparticles as drug carriers allows to control particle size, surface charge, loading capacity and release of pharmacologically active agents in order to achieve the site-specific action of the drug at the therapeutically optimal rate and dose regimen [151]. To be transported, the drug needs to be dissolved, entrapped, encapsulated or attached to a nanoparticle matrix [151]. Depending on the preparation method, different nanosystems can be obtained, namely nanoparticles, nanospheres or nanocapsules [151].

Albumin nanocarriers are biodegradable, non-toxic, have well-defined very small sizes (50-300 nm), and can be easily prepared under mild conditions by coacervation, controlled desolvation or emulsion formation [2, 152]. Additionally, albumin-based nanoparticles have reactive functional groups (thiol, amino, and carboxyl) on their surface that can be used for ligand binding and other surface modifications [153]. Drug release can be achieved physiologically by protease digestion [153].

Microspheres are solid and approximately spherical particles, usually composed by polymeric, waxy or other materials, including starches, gums, fats, and proteins, which have been increasingly used as matrices for drug transport and delivery [154].

Albumin microspheres have received particular attention due to their high specificity, biodegradability, non-toxicity and biocompatibility as drug carriers [154]. Two main procedures are used for the preparation of these particles: chemical cross-linking using glutaraldehyde; and addition of an organic solvent and heat stabilization [155].

The biodistribution of the microspheres is highly influenced by their size, which usually ranges from 1-100 μm [80]. Interestingly, microspheres with sizes ranging from 1-3 μm are taken up by the reticuloendothelial system, being therefore retained in the liver, spleen and, importantly, in solid tumors [80]. By contrast, microspheres bigger than 15 μm effectively target the capillary bed of the lungs [80]. Moreover, albumin microspheres can carry therapeutic or diagnosis agents [154].

2.4.2. Drug-albumin Conjugates and Derivatives

The first report on a drug-albumin conjugate dates back to 1967, in a study aiming the development of a protein-based vehicle for the preferential introduction of cytotoxic agents targeting cancer cells [156, 157].

Due to its long plasma half-life, albumin is the preferential molecule to be used as a vehicle to produce long active drug-albumin conjugates, in which a drug is permanently attached to circulating albumin, producing a bioconjugate with intrinsic activity for the receptors of the targeted tissue [158]. Therefore, this strategy is able to prolong the duration of the drug effect.

The methods for preparing albumin-drug conjugates can be: azo-coupling of amino-bearing drugs to tyrosine moieties of albumin; activation of a drug with cyanurchloride and subsequently coupling the activated drug to tyrosine moieties of albumin; or activation of carboxyl groups of the drug with *N,N'*-dicyclohexylcarbodiimide and *N*-hydroxysuccinimide (NHS) and covalently coupling the drug to lysine moieties of albumin [158].

However, several reasons are hindering these conjugates to advance to late clinical trials. In fact, the coupling methods not always result in well-defined products, and the commercially available albumin used, as a plasma product, is potentially pathogenic [158]. Moreover, the synthesis and isolation of the conjugates is an *ex vivo* process, thereby increasing the costs of their production [158].

2.4.3. Albumin Microbubbles

Microbubbles have been proposed as a new vehicle for drug and gene delivery, and consist of a shell stabilized gas core, usually prepared through agitation methods such as sonication or shaking [159]. Microbubbles have specific acoustic properties, which makes them an ideal ultrasound contrast agent for both diagnostic and therapeutic applications [160]. Interestingly, microbubble oscillations produce non-linear harmonics that induce cell membrane permeabilization [160].

As an amphiphilic protein, albumin is an ideal shell material, able to form a complete shell at an air-liquid interface [159]. An ideal formulation should be able to form monodisperse microbubbles, with determined diameters that do not coalesce, at high production rates. Albumin, as the main component of plasma, presents an excellent patient safety profile, and brings insights into the possibility of producing microbubbles from blood, thereby potentiating the development of a fully biocompatibility microbubble [159].

2.4.4. Albumin Hydrogels

Protein-based hydrogels are becoming increasingly used as controlled or sustained drug delivery systems. Hydrogels are cross-linked polymeric networks that can be either swollen with water, buffers, or with physiological solutions without degradation [161]. The release of a drug from a hydrogel occurs gradually, by exchanging the interior water with an aqueous solution in a milieu [161].

Consequently, albumin-based engineered hydrogels have been under investigation [162]. HSA or BSA hydrogels are generally prepared by thermal denaturation, chemical cross-linking (*e.g.* glutheraldehyde), or polymer-albumin conjugates, but recently, electrostatically triggered self-assembly has demonstrated to be an alternative for the formation of albumin hydrogels [163].

Due to its biocompatibility and intrinsic similarity to the extracellular matrix of certain tissues, when using albumin as a hydrogel constituent, a drug with affinity to albumin is expected to be preferably loaded on the hydrogel and to be slowly released, in comparison with what would happen with a hydrogel without albumin [161].

2.5. Update on Clinical Applications

For several years, the use of albumin for medical applications was restricted to intravenous administration of HSA as a blood substitute for the treatment of patients with severe burns or cachexia [164]. However, in the mid-1990s, and due to the advantageous features presented by this molecule, albumin started to be envisaged as a potential carrier for therapeutic agents [164].

Extensive efforts have been made ever since, with the main goal of creating albumin-based systems that could improve drugs pharmacokinetic profile or specifically target the pathogenic site. Following the increasing number of studies in the field, there are already a number of albumin based approved products, as well as ongoing clinical trials and novel applications and insights in the preclinical setting.

The main albumin-based technologies available in the market include Levimir®, Victoza®, Tresiba®, Abraxane® and ^{99m}Tc-aggregated albumin. Levimir®, the albumin-binding derivative of human insulin, was developed by Novo Nordisk, and approved in 2004 [165]. It is used for the treatment of types 1 and 2 diabetes mellitus, working as an insulin analog [166]. It is administered subcutaneously and remains soluble before and after injection, with a predictable and prolonged pharmacokinetic profile [167].

Still with regard to the treatment of diabetes, but by stimulating insulin secretion, the same company developed Victoza® (approved in 2009), an albumin-binding derivative of glucagon-like 1 peptide (GLP-1), stable against metabolic degradation due to albumin-binding, and with a plasma half-life of 11-15 h following subcutaneous administration [164]. This means that one daily administration must be suitable for the maintenance of adequate blood glucose levels.

By the end of 2012, Novo Nordisk introduced a new drug vehicle in the market, Tresiba®, a follow-up to Levimir®. Tresiba® is an ultra-long-acting basal insulin analog, with a duration of action over 42 hours [168].

Abraxane® is an albumin-paclitaxel nanoparticle and the first approved as albumin-based delivery system for treating solid tumors, in 2005 [169]. This complex accumulates in the tumor by passively crossing the fenestrations in the diseased vasculature and, as it has been suggested, the albumin transport pathway mediated by the gp60 uptakes and distributes Abraxane® through the tumor, until the release of paclitaxel [170].

Technitium^{99m} (^{99m}Tc) radiopharmaceuticals are triggering the growth of nuclear medicine ^{99m}Tc is a metastable nuclear isomer used in more than 80% of all diagnostic procedures [171]. ^{99m}Tc-aggregated albumin is a gamma-emitting radionuclide imaging agent with several applications and a place in the market for approximately 30 years [172]. It consists of a sterile aqueous suspension of ^{99m}Tc to pathogen-free HSA aggregated particles in a pH between 3.8 and 8.0, and can be administered intravenously, subcutaneously, intradermally or intratumorally at doses ranging from 10 to 200 MBq [171].

Beyond the albumin-based market approved technologies, many others are being investigated, and are extensively reviewed by Elsadek and Kratz (2012) [171]. Already in clinical trials are: INNO-206, a novel pro-drug against sarcoma and pancreatic and gastric cancer [173]; albumin-modified exendin-4 conjugate, for treating type 2 diabetes [174]; albuferon® for treating hepatitis C [175]; and AFL-HSA, a fluorescein labeled albumin conjugate for detecting brain tumors during surgery [176].

Another albumin-based strategy being explored is based on the use of albumin as a drug carrier for antibodies and therapeutic peptides. Among the carriers in preclinical and clinical trials are: Albumod™ technology, which extends the half-life of biopharmaceuticals through a high-affinity albumin-binding domain [177]; camelid anti-TNF-α-anti-HSA trivalent nanobodies for treating rheumatoid arthritis and cancer [171]; the albudAb™ platform, which consists of anti-serum albumin domain antibodies for extending the half-lives of short lived drugs [178]; and the MM-111, an albumin fusion antibody targeting epidermal growth factor receptors [179].

Studies on albumin-based systems for antiviral and antibacterial applications showed that the polarity of the peptide influences the efficacy of albumin-binding [171]. The systems are PC-1505, an albumin-conjugate of an antiviral C34 peptide that inhibits the immunodeficiency virus (HIV)-1 entry [180], and SAFETY™, a lipopeptide RH01-albumin conjugate with improved antibacterial activity [171, 181].

Another promising approach is the use of albumin-binding contrast agents for imaging. The most recent advances achieved for this purpose are Vasovist® and B-22956, which allow the acquisition of high-resolution MRI images with improved delineation of vessel structure and vascular diseases [182]. Another option is the GmSA-20ROX, obtained after reacting non-immunogenic galactosamine-conjugated serum albumin with 20 molecules of an amido-reactive rhodamineX-succinimidyl ester, forming a self-quenching complex [183].

Several cardiovascular abnormalities, stoke and other pathophysiological malignancies derive from defects in the nitric oxide (NO) production [171]. Some strategies to replace or supplement endogenous NO production through exogenous NO administration based on albumin are already being under investigation. For instance, *S*-Nitrosilated human serum albumin (SNO-HSA) has cytoprotective activity and acts as an anticancer agent [184]; and albumin Liprizzi

(R410C), a genetic variant of HSA that carries NO, which has both antibacterial and cytoprotective properties [185].

CONCLUDING REMARKS

The ability to efficiently deliver a drug at a desired location is dependent on a variety of limiting factors. Nanotechnology offers the possibility of creating drug delivery vehicles such as polymeric materials, allowing new approaches to reduce unwanted side effects, increasing drug efficiency and effectiveness. Both chitosan and albumin are widely available in nature. They have low purchase costs and have desirable physical, chemical and pharmacological properties, making them promising targets for drug delivery applications.

Amphiphilic derivatives of chitosan have received great interest since they increase chitosan solubility and allow the encapsulation of poorly soluble drugs, thereby controllably regulating drug release, as well as improving the absorption/encapsulation efficiency. Further, chitosan has mucoadhesive properties and is capable of opening epithelial tight junctions, evidence of its potential as a drug delivery system.

Albumin, is widely abundant, and absolutely safe. It has the ability to cross the entire circulatory system, preferably being captured by tumors or inflamed tissues and is able to bind to various proteins/peptides and drugs, enabling a more efficient target. There are already several albumin-based technologies available on the market for the treatment of various diseases.

It is expected that soon, amphiphilic chitosan derivatives and albumin-based commercial products will have a tremendous impact in medical technologies, and subsequently, in patients treatment and compliance.

CONFLICT OF INTEREST

The author (editor) declares no conflict of interest, financial or otherwise.

ACKNOWLEDGEMENT

João Pedro Martins thanks Fundação para a Ciência c a Tccnologia (FCT), Portugal for the BiotechHealth PhD scholarship. This work was also financed by European Regional Development Fund (ERDF) through the Programa Operacional Factores de Competitividade – COMPETE, by Portuguese funds through FCT in the framework of the project PEst-C/SAU/LA0002/2013, and cofinanced by North Portugal Regional Operational Programme (ON.2 – O Novo Norte) in the framework of Project SAESCTN-PIIC&DT/2011 under the National Strategic Reference Framework (NSRF).

REFERENCES

[1] D. Sleep, "Albumin and its application in drug delivery", *Expert Opin. Drug Deliv.* , vol. 12, no. 5, pp. 793-812, 2015. [http://dx.doi.org/10.1517/17425247.2015.993313] [PMID: 25518870]

[2] A.O. Elzoghby, W.M. Samy, and N.A. Elgindy, "Albumin-based nanoparticles as potential controlled release drug delivery systems", *J. Control. Release,* vol. 157, no. 2, pp. 168-182, 2012. [http://dx.doi.org/10.1016/j.jconrel.2011.07.031] [PMID: 21839127]

[3] G.A. Joanitti, and L.P. Silva, *The Emerging Potential of Byproducts as Platforms for Drug Delivery Systems.* vol. 14. Curr. Drug Targets, 2013.

[4] J. Shaji, V. Jain, and S. Lodha, "Chitosan: a novel pharmaceutical excipient", *Int. J. Pharm. Appl. Sci.* , vol. 1, pp. 11-28, 2010.

[5] Y. Shahzad, I.U. Khan, and T. Hussain, "Bioactive albumin-based carriers for tumour chemotherapy", *Curr. Cancer Drug Targets,* vol. 14, pp. 752-763, 2014.

[6] H. Honarkar, and M. Barikani, "“Applications of biopolymers I: chitosan”. Monatshefte Für Chemie - Chem", *Mon.* , vol. 140, pp. 1403-1420, 2009.

[7] H. Silva, K. Dos Santos, and E.I. Ferreira, "Quitosana: derivados hidrossolúveis, aplicações farmacêuticas e avanços", *Quim. Nova,* vol. 29, p. 776, 2006. [http://dx.doi.org/10.1590/S0100-40422006000400026]

[8] M. Dash, F. Chiellini, R.M. Ottenbrite, and E. Chiellini, "Chitosan—A versatile semi-synthetic polymer in biomedical applications", *Prog. Polym. Sci.* , vol. 36, pp. 981-1014, 2011. [http://dx.doi.org/10.1016/j.progpolymsci.2011.02.001]

[9] F. Araújo, N. Shrestha, P.L. Granja, J. Hirvonen, H.A. Santos, and B. Sarmento, "Safety and toxicity concerns of orally delivered nanoparticles as drug carriers", *Expert Opin. Drug Metab. Toxicol.* , vol. 11, no. 3, pp. 381-393, 2015. [http://dx.doi.org/10.1517/17425255.2015.992781] [PMID: 25495133]

[10] A. Sosnik, "das Neves J, Sarmento B. “Mucoadhesive polymers in the design of nano-drug delivery systems for administration by non-parenteral routes: A review", *Prog. Polym. Sci.* , vol. 39, pp. 2030-2075, 2014. [http://dx.doi.org/10.1016/j.progpolymsci.2014.07.010]

[11] I. Aranaz, R. Harris, and A. Heras, "Chitosan amphiphilic derivatives. Chemistry and applications", *Curr. Org. Chem.* , vol. 14, p. 308, 2010. [http://dx.doi.org/10.2174/138527210790231919]

[12] P. Agrawal, G.J. Strijkers, and K. Nicolay, "Chitosan-based systems for molecular imaging", *Adv. Drug Deliv. Rev.* , vol. 62, no. 1, pp. 42-58, 2010. [http://dx.doi.org/10.1016/j.addr.2009.09.007] [PMID: 19861142]

[13] N.M. Alves, and J.F. Mano, "Chitosan derivatives obtained by chemical modifications for biomedical and environmental applications", *Int. J. Biol. Macromol.* , vol. 43, no. 5, pp. 401-414, 2008. [http://dx.doi.org/10.1016/j.ijbiomac.2008.09.007] [PMID: 18838086]

[14] S.K. Shukla, A.K. Mishra, O.A. Arotiba, and B.B. Mamba, "Chitosan-based nanomaterials: a state-o--the-art review", *Int. J. Biol. Macromol.* , vol. 59, pp. 46-58, 2013. [http://dx.doi.org/10.1016/j.ijbiomac.2013.04.043] [PMID: 23608103]

[15] F Andrade, F Goycoolea, and DA Chiappetta, "Chitosan-grafted copolymers and chitosan-ligand conjugates as matrices for pulmonary drug delivery", *Int. J. Carbohydr. Chem.* , vol. 2011, 2011.

[16] M. Garcia-Fuentes, and M.J. Alonso, "Chitosan-based drug nanocarriers: where do we stand?", *J. Control. Release,* vol. 161, no. 2, pp. 496-504, 2012. [http://dx.doi.org/10.1016/j.jconrel.2012.03.017] [PMID: 22480607]

[17] Y-W. Cho, J. Jang, C.R. Park, and S-W. Ko, "Preparation and solubility in acid and water of partially deacetylated chitins", *Biomacromolecules,* vol. 1, no. 4, pp. 609-614, 2000.
[http://dx.doi.org/10.1021/bm000036j] [PMID: 11710189]

[18] M. Yalpani, and L.D. Hall, "Some chemical and analytical aspects of polysaccharide modifications. III. Formation of branched-chain, soluble chitosan derivatives", *Macromol.* , vol. 17, pp. 272-281, 1984.
[http://dx.doi.org/10.1021/ma00133a003]

[19] J. Desbrières, C. Martinez, and M. Rinaudo, "Hydrophobic derivatives of chitosan: characterization and rheological behaviour", *Int. J. Biol. Macromol.* , vol. 19, no. 1, pp. 21-28, 1996.
[http://dx.doi.org/10.1016/0141-8130(96)01095-1] [PMID: 8782715]

[20] R. Riva, H. Ragelle, A. des Rieux, N. Duhem, C. Jérôme, and V. Préat, "Chitosan and Chitosan Derivatives", *Drug Deliv. Tissue Eng.* , vol. 244, pp. 19-44, 2011.

[21] M. Wu, H. Dong, K. Guo, R. Zeng, M. Tu, and J. Zhao, "Self-assemblied nanocomplexes based on biomimetic amphiphilic chitosan derivatives for protein delivery", *Carbohydr. Polym.* , vol. 121, pp. 115-121, 2015.
[http://dx.doi.org/10.1016/j.carbpol.2014.12.049] [PMID: 25659679]

[22] Y. Zambito, "Nanoparticles Based on Chitosan Derivatives", In: *Adv. Biomater. Sci. Biomed. Appl.* , PR Pignatello, Ed., 2013, pp. 243-263.
[http://dx.doi.org/10.5772/54944]

[23] J-K. Suh, and H.W. Matthew, "Application of chitosan-based polysaccharide biomaterials in cartilage tissue engineering: a review", *Biomaterials,* vol. 21, no. 24, pp. 2589-2598, 2000.
[http://dx.doi.org/10.1016/S0142-9612(00)00126-5] [PMID: 11071608]

[24] H. Zhang, and S.H. Neau, "in vitro degradation of chitosan by a commercial enzyme preparation: effect of molecular weight and degree of deacetylation", *Biomaterials,* vol. 22, no. 12, pp. 1653-1658, 2001.
[http://dx.doi.org/10.1016/S0142-9612(00)00326-4] [PMID: 11374467]

[25] T.A. Khan, K.K. Peh, and H.S. Chng, "Reporting degree of deacetylation values of chitosan: the influence of analytical methods", *J. Pharm. Pharm. Sci.* , vol. 5, no. 3, pp. 205-212, 2002.
[PMID: 12553887]

[26] S.B. Rao, and C.P. Sharma, "Use of chitosan as a biomaterial: studies on its safety and hemostatic potential", *J. Biomed. Mater. Res.* , vol. 34, no. 1, pp. 21-28, 1997.
[http://dx.doi.org/10.1002/(SICI)1097-4636(199701)34:1<21::AID-JBM4>3.0.CO;2-P] [PMID: 8978649]

[27] Y. Okamoto, R. Yano, K. Miyatake, I. Tomohiro, Y. Shigemasa, and S. Minami, "Effects of chitin and chitosan on blood coagulation", *Carbohydr. Polym.* , vol. 53, pp. 337-342, 2003.
[http://dx.doi.org/10.1016/S0144-8617(03)00076-6]

[28] M.N. Kumar, "A review of chitin and chitosan applications", *React. Funct. Polym.* , vol. 46, pp. 1-27, 2000.
[http://dx.doi.org/10.1016/S1381-5148(00)00038-9]

[29] S.S. Koide, "Chitin-chitosan: properties, benefits and risks", *Nutr. Res.* , vol. 18, pp. 1091-1101, 1998.
[http://dx.doi.org/10.1016/S0271-5317(98)00091-8]

[30] Z.G. Wei, R. Sandstroröm, and S. Miyazaki, "Shape-memory materials and hybrid composites for smart systems: Part I Shape-memory materials", *J. Mater. Sci.* , vol. 33, pp. 3743-3762, 1998.
[http://dx.doi.org/10.1023/A:1004692329247]

[31] L-Y. Zheng, and J-F. Zhu, "Study on antimicrobial activity of chitosan with different molecular weights", *Carbohydr. Polym.* , vol. 54, pp. 527-530, 2003.
[http://dx.doi.org/10.1016/j.carbpol.2003.07.009]

[32] C. Chatelet, O. Damour, and A. Domard, "Influence of the degree of acetylation on some biological properties of chitosan films", *Biomaterials,* vol. 22, no. 3, pp. 261-268, 2001. [http://dx.doi.org/10.1016/S0142-9612(00)00183-6] [PMID: 11197501]

[33] N.G. Schipper, K.M. Vårum, and P. Artursson, "Chitosans as absorption enhancers for poorly absorbable drugs. 1: Influence of molecular weight and degree of acetylation on drug transport across human intestinal epithelial (Caco-2) cells", *Pharm. Res.* , vol. 13, no. 11, pp. 1686-1692, 1996. [http://dx.doi.org/10.1023/A:1016444808000] [PMID: 8956335]

[34] M. Huang, E. Khor, and L-Y. Lim, "Uptake and cytotoxicity of chitosan molecules and nanoparticles: effects of molecular weight and degree of deacetylation", *Pharm. Res.* , vol. 21, no. 2, pp. 344-353, 2004. [http://dx.doi.org/10.1023/B:PHAM.0000016249.52831.a5] [PMID: 15032318]

[35] C-M. Lehr, J.A. Bouwstra, E.H. Schacht, and H.E. Junginger, "in vitro evaluation of mucoadhesive properties of chitosan and some other natural polymers", *Int. J. Pharm.* , vol. 78, pp. 43-48, 1992. [http://dx.doi.org/10.1016/0378-5173(92)90353-4]

[36] J. Yang, F. Tian, Z. Wang, Q. Wang, Y.J. Zeng, and S.Q. Chen, "Effect of chitosan molecular weight and deacetylation degree on hemostasis", *J. Biomed. Mater. Res. B Appl. Biomater.* , vol. 84, no. 1, pp. 131-137, 2008. [http://dx.doi.org/10.1002/jbm.b.30853] [PMID: 17514656]

[37] Y. Ohshima, K. Nishino, Y. Yonekura, S. Kishimoto, and S. Wakabayashi, "Clinical application of chitin non-woven fabric as wound dressing", *Eur. J. Plast. Surg.* , vol. 10, pp. 66-69, 1987. [http://dx.doi.org/10.1007/BF00578375]

[38] Y. Okamoto, K. Kawakami, K. Miyatake, M. Morimoto, Y. Shigemasa, and S. Minami, "Analgesic effects of chitin and chitosan", *Carbohydr. Polym.* , vol. 49, pp. 249-252, 2002. [http://dx.doi.org/10.1016/S0144-8617(01)00316-2]

[39] X. Fei Liu, Y. Lin Guan, D. Zhi Yang, Z. Li, and K. De Yao, "Antibacterial action of chitosan and carboxymethylated chitosan", *J. Appl. Polym. Sci.* , vol. 79, pp. 1324-1335, 2001. [http://dx.doi.org/10.1002/1097-4628(20010214)79:7<1324::AID-APP210>3.0.CO;2-L]

[40] I.M. Helander, E-L. Nurmiaho-Lassila, R. Ahvenainen, J. Rhoades, and S. Roller, "Chitosan disrupts the barrier properties of the outer membrane of gram-negative bacteria", *Int. J. Food Microbiol.* , vol. 71, no. 2-3, pp. 235-244, 2001. [http://dx.doi.org/10.1016/S0168-1605(01)00609-2] [PMID: 11789941]

[41] T. Kean, and M. Thanou, "Biodegradation, biodistribution and toxicity of chitosan", *Adv. Drug Deliv. Rev.* , vol. 62, no. 1, pp. 3-11, 2010. [http://dx.doi.org/10.1016/j.addr.2009.09.004] [PMID: 19800377]

[42] E.I. Rabea, M.E. Badawy, C.V. Stevens, G. Smagghe, and W. Steurbaut, "Chitosan as antimicrobial agent: applications and mode of action", *Biomacromolecules,* vol. 4, no. 6, pp. 1457-1465, 2003. [http://dx.doi.org/10.1021/bm034130m] [PMID: 14606868]

[43] R. Xing, S. Liu, Z. Guo, H. Yu, P. Wang, C. Li, Z. Li, and P. Li, "Relevance of molecular weight of chitosan and its derivatives and their antioxidant activities in vitro", *Bioorg. Med. Chem.* , vol. 13, no. 5, pp. 1573-1577, 2005. [http://dx.doi.org/10.1016/j.bmc.2004.12.022] [PMID: 15698774]

[44] P-J. Park, J-Y. Je, and S-K. Kim, "Free radical scavenging activities of differently deacetylated chitosans using an ESR spectrometer", *Carbohydr. Polym.* , vol. 55, pp. 17-22, 2004. [http://dx.doi.org/10.1016/j.carbpol.2003.05.002]

[45] J.C. Fernandes, P. Eaton, H. Nascimento, M.S. Gião, Ó.S. Ramos, and L. Belo, "Antioxidant activity of chitooligosaccharides upon two biological systems: Erythrocytes and bacteriophages", *Carbohydr. Polym.* , vol. 79, pp. 1101-1106, 2010. [http://dx.doi.org/10.1016/j.carbpol.2009.10.050]

[46] M.N. Kumar, R.A. Muzzarelli, C. Muzzarelli, H. Sashiwa, and A.J. Domb, "Chitosan chemistry and pharmaceutical perspectives", *Chem. Rev.*, vol. 104, no. 12, pp. 6017-6084, 2004. [http://dx.doi.org/10.1021/cr030441b] [PMID: 15584695]

[47] M. Yokoyama, S. Inoue, K. Kataoka, N. Yui, T. Okano, and Y. Sakurai, "Molecular design for missile drug: Synthesis of adriamycin conjugated with immunoglobulin G using poly (ethylene glycol)-bloc--poly (aspartic acid) as intermediate carrier", *Makromol. Chem.*, vol. 190, pp. 2041-2054, 1989. [http://dx.doi.org/10.1002/macp.1989.021900904]

[48] S. Sajeesh, and C.P. Sharma, "Novel pH responsive polymethacrylic acid-chitosan-polyethylene glycol nanoparticles for oral peptide delivery", *J. Biomed. Mater. Res. B Appl. Biomater.*, vol. 76, no. 2, pp. 298-305, 2006. [http://dx.doi.org/10.1002/jbm.b.30372] [PMID: 16130147]

[49] R. Mo, X. Jin, N. Li, C. Ju, M. Sun, C. Zhang, and Q. Ping, "The mechanism of enhancement on oral absorption of paclitaxel by N-octyl-O-sulfate chitosan micelles", *Biomaterials,* vol. 32, no. 20, pp. 4609-4620, 2011. [http://dx.doi.org/10.1016/j.biomaterials.2011.03.005] [PMID: 21440934]

[50] Z.Q. Yuan, J.Z. Li, Y. Liu, W.L. Chen, S.D. Yang, C.G. Zhang, W.J. Zhu, X.F. Zhou, C. Liu, and X.N. Zhang, "Systemic delivery of micelles loading with paclitaxel using N-succinyl-palmitol-chitosan decorated with cRGDyK peptide to inhibit non-small-cell lung cancer", *Int. J. Pharm.*, vol. 492, no. 1-2, pp. 141-151, 2015. [http://dx.doi.org/10.1016/j.ijpharm.2015.07.022] [PMID: 26188316]

[51] F.Q. Hu, G.F. Ren, H. Yuan, Y.Z. Du, and S. Zeng, "Shell cross-linked stearic acid grafted chitosan oligosaccharide self-aggregated micelles for controlled release of paclitaxel", *Colloids Surf. B Biointerfaces,* vol. 50, no. 2, pp. 97-103, 2006. [http://dx.doi.org/10.1016/j.colsurfb.2006.04.009] [PMID: 16759840]

[52] F.Q. Hu, L.N. Liu, Y.Z. Du, and H. Yuan, "Synthesis and antitumor activity of doxorubicin conjugated stearic acid-g-chitosan oligosaccharide polymeric micelles", *Biomaterials,* vol. 30, no. 36, pp. 6955-6963, 2009. [http://dx.doi.org/10.1016/j.biomaterials.2009.09.008] [PMID: 19782395]

[53] F.Q. Hu, X.L. Wu, Y.Z. Du, J. You, and H. Yuan, "Cellular uptake and cytotoxicity of shell crosslinked stearic acid-grafted chitosan oligosaccharide micelles encapsulating doxorubicin", *Eur. J. Pharm. Biopharm.*, vol. 69, no. 1, pp. 117-125, 2008. [http://dx.doi.org/10.1016/j.ejpb.2007.09.018] [PMID: 17997293]

[54] Y.Z. Du, L. Wang, H. Yuan, X.H. Wei, and F.Q. Hu, "Preparation and characteristics of linoleic acid-grafted chitosan oligosaccharide micelles as a carrier for doxorubicin", *Colloids Surf. B Biointerfaces,* vol. 69, no. 2, pp. 257-263, 2009. [http://dx.doi.org/10.1016/j.colsurfb.2008.11.030] [PMID: 19131223]

[55] K.H. Min, K. Park, Y-S. Kim, S.M. Bae, S. Lee, H.G. Jo, R.W. Park, I.S. Kim, S.Y. Jeong, K. Kim, and I.C. Kwon, "Hydrophobically modified glycol chitosan nanoparticles-encapsulated camptothecin enhance the drug stability and tumor targeting in cancer therapy", *J. Control. Release,* vol. 127, no. 3, pp. 208-218, 2008. [http://dx.doi.org/10.1016/j.jconrel.2008.01.013] [PMID: 18336946]

[56] J. Akbuğa, and N. Bergişadi, "Effect of formulation variables on cis-platin loaded chitosan microsphere properties", *J. Microencapsul.*, vol. 16, no. 6, pp. 697-703, 1999. [http://dx.doi.org/10.1080/026520499288645] [PMID: 10575622]

[57] P. Calvo, C. Remunan-Lopez, J.L. Vila-Jato, and M.J. Alonso, "Novel hydrophilic chitosan-polyethylene oxide nanoparticles as protein carriers", *J. Appl. Polym. Sci.*, vol. 63, pp. 125-132, 1997. [http://dx.doi.org/10.1002/(SICI)1097-4628(19970103)63:1<125::AID-APP13>3.0.CO;2-4]

[58] F. Andrade, P. Fonte, M. Oliva, M. Videira, D. Ferreira, and B. Sarmento, "Solid state formulations composed by amphiphilic polymers for delivery of proteins: characterization and stability", *Int. J.*

Pharm. , vol. 486, no. 1-2, pp. 195-206, 2015.
[http://dx.doi.org/10.1016/j.ijpharm.2015.03.050] [PMID: 25818062]

[59] W. Boonyo, H.E. Junginger, N. Waranuch, A. Polnok, and T. Pitaksuteepong, "Chitosan and trimethyl chitosan chloride (TMC) as adjuvants for inducing immune responses to ovalbumin in mice following nasal administration", *J. Control. Release,* vol. 121, no. 3, pp. 168-175, 2007.
[http://dx.doi.org/10.1016/j.jconrel.2007.05.025] [PMID: 17644205]

[60] A. Grenha, B. Seijo, and C. Remuñán-López, "Microencapsulated chitosan nanoparticles for lung protein delivery", *Eur. J. Pharm. Sci.* , vol. 25, no. 4-5, pp. 427-437, 2005.
[http://dx.doi.org/10.1016/j.ejps.2005.04.009] [PMID: 15893461]

[61] A.H. Krauland, and M.J. Alonso, "Chitosan/cyclodextrin nanoparticles as macromolecular drug delivery system", *Int. J. Pharm.* , vol. 340, no. 1-2, pp. 134-142, 2007.
[http://dx.doi.org/10.1016/j.ijpharm.2007.03.005] [PMID: 17459620]

[62] X. Zhang, H. Zhang, Z. Wu, Z. Wang, H. Niu, and C. Li, "Nasal absorption enhancement of insulin using PEG-grafted chitosan nanoparticles", *Eur. J. Pharm. Biopharm.* , vol. 68, no. 3, pp. 526-534, 2008.
[http://dx.doi.org/10.1016/j.ejpb.2007.08.009] [PMID: 17881202]

[63] J-Y. Lee, S-H. Nam, S-Y. Im, Y-J. Park, Y-M. Lee, Y-J. Seol, C.P. Chung, and S.J. Lee, "Enhanced bone formation by controlled growth factor delivery from chitosan-based biomaterials", *J. Control. Release,* vol. 78, no. 1-3, pp. 187-197, 2002.
[http://dx.doi.org/10.1016/S0168-3659(01)00498-9] [PMID: 11772460]

[64] H. Yamamoto, Y. Kuno, S. Sugimoto, H. Takeuchi, and Y. Kawashima, "Surface-modified PLGA nanosphere with chitosan improved pulmonary delivery of calcitonin by mucoadhesion and opening of the intercellular tight junctions", *J. Control. Release,* vol. 102, no. 2, pp. 373-381, 2005.
[http://dx.doi.org/10.1016/j.jconrel.2004.10.010] [PMID: 15653158]

[65] H. Wu, S. Wang, H. Fang, X. Zan, J. Zhang, and Y. Wan, "Chitosan-polycaprolactone copolymer microspheres for transforming growth factor-β1 delivery", *Colloids Surf. B Biointerfaces,* vol. 82, no. 2, pp. 602-608, 2011.
[http://dx.doi.org/10.1016/j.colsurfb.2010.10.024] [PMID: 21115281]

[66] M.H. Xiong, Y. Bao, X.Z. Yang, Y.H. Zhu, and J. Wang, "Delivery of antibiotics with polymeric particles", *Adv. Drug Deliv. Rev.* , vol. 78, pp. 63-76, 2014.
[http://dx.doi.org/10.1016/j.addr.2014.02.002] [PMID: 24548540]

[67] F. Ungaro, I. dAngelo, C. Coletta, R. dEmmanuele di Villa Bianca, R. Sorrentino, B. Perfetto, M.A. Tufano, A. Miro, M.I. La Rotonda, and F. Quaglia, "Dry powders based on PLGA nanoparticles for pulmonary delivery of antibiotics: modulation of encapsulation efficiency, release rate and lung deposition pattern by hydrophilic polymers", *J. Control. Release,* vol. 157, no. 1, pp. 149-159, 2012.
[http://dx.doi.org/10.1016/j.jconrel.2011.08.010] [PMID: 21864595]

[68] S. Sahoo, A. Sasmal, R. Nanda, A.R. Phani, and P.L. Nayak, "Synthesis of chitosan–polycaprolactone blend for control delivery of ofloxacin drug", *Carbohydr. Polym.* , vol. 79, pp. 106-113, 2010.
[http://dx.doi.org/10.1016/j.carbpol.2009.07.042]

[69] F. Bigucci, B. Luppi, A. Musenga, V. Zecchi, and T. Cerchiara, "Chitosan salts coated with stearic acid as colon-specific delivery systems for vancomycin", *Drug Deliv.* , vol. 15, no. 5, pp. 289-293, 2008.
[http://dx.doi.org/10.1080/10717540802006468] [PMID: 18763159]

[70] S. Dreve, I. Kacso, A. Popa, O. Raita, F. Dragan, and A. Bende, "Structural investigation of chitosan-based microspheres with some anti-inflammatory drugs", *J. Mol. Struct.* , vol. 997, pp. 78-86, 2011.
[http://dx.doi.org/10.1016/j.molstruc.2011.05.001]

[71] S.G. Kumbar, K.S. Soppimath, and T.M. Aminabhavi, "Synthesis and characterization of polyacrylamide-grafted chitosan hydrogel microspheres for the controlled release of indomethacin", *J. Appl. Polym. Sci.* , vol. 87, pp. 1525-1536, 2003.

[http://dx.doi.org/10.1002/app.11552]

[72] M. Wu, K. Guo, H. Dong, R. Zeng, M. Tu, and J. Zhao, "“in vitro drug release and biological evaluation of biomimetic polymeric micelles self-assembled from amphiphilic deoxycholic acid-phosphorylcholine-chitosan conjugate”. Mater. Sci. Eng. C Mater", *Biol. Appl.* , vol. 45, pp. 162-169, 2014.

[73] J.G. Marques, V.M. Gaspar, E. Costa, C.M. Paquete, and I.J. Correia, "Synthesis and characterization of micelles as carriers of non-steroidal anti-inflammatory drugs (NSAID) for application in breast cancer therapy", *Colloids Surf. B Biointerfaces,* vol. 113, pp. 375-383, 2014. [http://dx.doi.org/10.1016/j.colsurfb.2013.09.037] [PMID: 24129330]

[74] A. Hassani Najafabadi, M. Abdouss, and S. Faghihi, "“Synthesis and evaluation of PEG-O-chitosan nanoparticles for delivery of poor water soluble drugs: ibuprofen”. Mater. Sci. Eng. C Mater", *Biol. Appl.* , vol. 41, pp. 91-99, 2014.

[75] I.M. van der Lubben, J.C. Verhoef, G. Borchard, and H.E. Junginger, "Chitosan for mucosal vaccination", *Adv. Drug Deliv. Rev.* , vol. 52, no. 2, pp. 139-144, 2001. [http://dx.doi.org/10.1016/S0169-409X(01)00197-1] [PMID: 11718937]

[76] K.S. Jaganathan, and S.P. Vyas, "Strong systemic and mucosal immune responses to surface-modified PLGA microspheres containing recombinant hepatitis B antigen administered intranasally", *Vaccine,* vol. 24, no. 19, pp. 4201-4211, 2006. [http://dx.doi.org/10.1016/j.vaccine.2006.01.011] [PMID: 16446012]

[77] H.F. Florindo, S. Pandit, L.M. Gonçalves, H.O. Alpar, and A.J. Almeida, "Streptococcus equi antigens adsorbed onto surface modified poly-epsilon-caprolactone microspheres induce humoral and cellular specific immune responses", *Vaccine,* vol. 26, no. 33, pp. 4168-4177, 2008. [http://dx.doi.org/10.1016/j.vaccine.2008.05.074] [PMID: 18599166]

[78] X. Zheng, Y. Huang, C. Zheng, S. Dong, and W. Liang, "Alginate-chitosan-PLGA composite microspheres enabling single-shot hepatitis B vaccination", *AAPS J.* , vol. 12, no. 4, pp. 519-524, 2010. [http://dx.doi.org/10.1208/s12248-010-9213-1] [PMID: 20577912]

[79] A.M. Merlot, D.S. Kalinowski, and D.R. Richardson, "Unraveling the mysteries of serum albumin-more than just a serum protein", *Front. Physiol.* , vol. 5, p. 299, 2014. [http://dx.doi.org/10.3389/fphys.2014.00299] [PMID: 25161624]

[80] F. Kratz, "Albumin as a drug carrier: design of prodrugs, drug conjugates and nanoparticles", *J. Control. Release,* vol. 132, no. 3, pp. 171-183, 2008. [http://dx.doi.org/10.1016/j.jconrel.2008.05.010] [PMID: 18582981]

[81] T.W. Evans, "Review article: albumin as a drugbiological effects of albumin unrelated to oncotic pressure", *Aliment. Pharmacol. Ther.* , vol. 16, no. 5, suppl. Suppl. 5, pp. 6-11, 2002. [http://dx.doi.org/10.1046/j.1365-2036.16.s5.2.x] [PMID: 12423448]

[82] K.M. Sand, M. Bern, J. Nilsen, H.T. Noordzij, I. Sandlie, and J.T. Andersen, "Unraveling the Interaction between FcRn and Albumin: Opportunities for Design of Albumin-Based Therapeutics", *Front. Immunol.* , vol. 5, p. 682, 2015. [http://dx.doi.org/10.3389/fimmu.2014.00682] [PMID: 25674083]

[83] A.V. Trow, "Rozanski E a, Armelle M, Chan DL. “Evaluation of use of human albumin in critically ill dogs: 73 cases (2003–2006)", *J. Am. Vet. Med. Assoc.* , vol. 233, pp. 607-612, 2008. [http://dx.doi.org/10.2460/javma.233.4.607] [PMID: 18710318]

[84] J. Hankins, "The role of albumin in fluid and electrolyte balance", *J. Infus. Nurs.* , vol. 29, no. 5, pp. 260-265, 2006. [http://dx.doi.org/10.1097/00129804-200609000-00004] [PMID: 17035887]

[85] G.J. Quinlan, G.S. Martin, and T.W. Evans, "Albumin: biochemical properties and therapeutic potential", *Hepatology,* vol. 41, no. 6, pp. 1211-1219, 2005.

[http://dx.doi.org/10.1002/hep.20720] [PMID: 15915465]

[86] T. Peters Jr, "2 - The Albumin Molecule: Its Structure and Chemical Properties", In: *All about albumin.* , T.B. Peters, Ed., Academic Press: San Diego, 1995, pp. 9-II. [http://dx.doi.org/10.1016/B978-012552110-9/50004-0]

[87] D. Sleep, J. Cameron, and L.R. Evans, "Albumin as a versatile platform for drug half-life extension", *Biochim. Biophys. Acta,* vol. 1830, no. 12, pp. 5526-5534, 2013. [http://dx.doi.org/10.1016/j.bbagen.2013.04.023] [PMID: 23639804]

[88] T. Peters Jr, "5 - Metabolism: Albumin in the Body", In: *All about albumin.* , T.B. Peters, Ed., Academic Press: San Diego, 1995, pp. 188-250. [http://dx.doi.org/10.1016/B978-012552110-9/50007-6]

[89] J.P. Nicholson, M.R. Wolmarans, and G.R. Park, "The role of albumin in critical illness", *Br. J. Anaesth.* , vol. 85, no. 4, pp. 599-610, 2000. [http://dx.doi.org/10.1093/bja/85.4.599] [PMID: 11064620]

[90] J. Monks, and M.C. Neville, "Albumin transcytosis across the epithelium of the lactating mouse mammary gland", *J. Physiol.* , vol. 560, no. Pt 1, pp. 267-280, 2004. [http://dx.doi.org/10.1113/jphysiol.2004.068403] [PMID: 15297572]

[91] L. Brannon-Peppas, "Recent advances on the use of biodegradable microparticles and nanoparticles in controlled drug delivery", *Int. J. Pharm.* , vol. 116, pp. 1-9, 1995. [http://dx.doi.org/10.1016/0378-5173(94)00324-X]

[92] G.V. Patil, "Biopolymer albumin for diagnosis and in drug delivery", *Drug Dev. Res.* , vol. 58, pp. 219-247, 2003. [http://dx.doi.org/10.1002/ddr.10157]

[93] K. Langer, S. Balthasar, V. Vogel, N. Dinauer, H. von Briesen, and D. Schubert, "Optimization of the preparation process for human serum albumin (HSA) nanoparticles", *Int. J. Pharm.* , vol. 257, no. 1-2, pp. 169-180, 2003. [http://dx.doi.org/10.1016/S0378-5173(03)00134-0] [PMID: 12711172]

[94] M.A. Mateescu, P. Ispas-Szabo, E.B. Assaad, Ed., *6 - Protein–protein associative interactions and their involvement in bioformulations.* Woodhead Publ Ser Biomed, 2015, pp. 225-241.

[95] S. Doppalapudi, S. Katiyar, A. Domb, and W. Khan, "Biodegradable Natural Polymers", In: *Adv. Polym. Med.* , F. Puoci, Ed., vol. 2. Springer International Publishing: SE, 2015, pp. 33-66.

[96] M. Galbas, F. Porzucek, A. Woźniak, R. Słomski, and M. Selwet, "Isolation of low-molecular albumins of 2S fraction from soybean (Glycine max (L.) Merrill)", *Acta Biochim. Pol.* , vol. 60, no. 1, pp. 107-110, 2013. [PMID: 23520576]

[97] T.R. Neyestani, M. Djalali, and M. Pezeshki, "Isolation of alpha-lactalbumin, beta-lactoglobulin, and bovine serum albumin from cows milk using gel filtration and anion-exchange chromatography including evaluation of their antigenicity", *Protein Expr. Purif.* , vol. 29, no. 2, pp. 202-208, 2003. [http://dx.doi.org/10.1016/S1046-5928(03)00015-9] [PMID: 12767810]

[98] W. Weiss, G. Huber, K.H. Engel, A. Pethran, M.J. Dunn, A.A. Gooley, and A. Görg, "Identification and characterization of wheat grain albumin/globulin allergens", *Electrophoresis,* vol. 18, no. 5, pp. 826-833, 1997. [http://dx.doi.org/10.1002/elps.1150180529] [PMID: 9194615]

[99] J.A. Huntington, and P.E. Stein, "Structure and properties of ovalbumin", *J. Chromatogr. B Biomed. Sci. Appl.* , vol. 756, no. 1-2, pp. 189-198, 2001. [http://dx.doi.org/10.1016/S0378-4347(01)00108-6] [PMID: 11419711]

[100] G. Martos, P. Contreras, E. Molina, and R. López-Fandiño, "Egg white ovalbumin digestion mimicking physiological conditions", *J. Agric. Food Chem.* , vol. 58, no. 9, pp. 5640-5648, 2010. [http://dx.doi.org/10.1021/jf904538w] [PMID: 20345099]

[101] J.G. Jung, W. Lim, T.S. Park, J.N. Kim, B.K. Han, G. Song, and J.Y. Han, "Structural and histological characterization of oviductal magnum and lectin-binding patterns in Gallus domesticus", *Reprod. Biol. Endocrinol.*, vol. 9, p. 62, 2011.
[http://dx.doi.org/10.1186/1477-7827-9-62] [PMID: 21548987]

[102] D.M. Webster, W.K. Fisher, D.D. Koureas, and E.O. Thompson, "Amino acid sequences containing cysteine or cystine residues in ovalbumin from eggs of the quail Coturnix coturnix japonica", *Aust. J. Biol. Sci.*, vol. 34, no. 5-6, pp. 505-514, 1981.
[http://dx.doi.org/10.1071/BI9810505] [PMID: 7340765]

[103] M. Weijers, P.A. Barneveld, M.A. Cohen Stuart, and R.W. Visschers, "Heat-induced denaturation and aggregation of ovalbumin at neutral pH described by irreversible first-order kinetics", *Protein Sci.*, vol. 12, no. 12, pp. 2693-2703, 2003.
[http://dx.doi.org/10.1110/ps.03242803] [PMID: 14627731]

[104] A.C. Alleoni, "Albumen protein and functional properties of gelation and foaming", *Sci. Agric.*, vol. 63, pp. 291-298, 2006.
[http://dx.doi.org/10.1590/S0103-90162006000300013]

[105] Z. Yu, M. Yu, Z. Zhang, G. Hong, and Q. Xiong, "Bovine serum albumin nanoparticles as controlled release carrier for local drug delivery to the inner ear", *Nanoscale Res. Lett.*, vol. 9, no. 1, p. 343, 2014.
[http://dx.doi.org/10.1186/1556-276X-9-343] [PMID: 25114637]

[106] D.C. Carter, and J.X. Ho, "Structure of serum albumin", *Adv. Protein Chem.*, vol. 45, pp. 153-203, 1994.
[http://dx.doi.org/10.1016/S0065-3233(08)60640-3] [PMID: 8154369]

[107] Y-J. Hu, Y. Liu, T-Q. Sun, A-M. Bai, J-Q. Lü, and Z-B. Pi, "Binding of anti-inflammatory drug cromolyn sodium to bovine serum albumin", *Int. J. Biol. Macromol.*, vol. 39, no. 4-5, pp. 280-285, 2006.
[http://dx.doi.org/10.1016/j.ijbiomac.2006.04.004] [PMID: 16707156]

[108] G.L. Francis, "Albumin and mammalian cell culture: implications for biotechnology applications", *Cytotechnology,* vol. 62, no. 1, pp. 1-16, 2010.
[http://dx.doi.org/10.1007/s10616-010-9263-3] [PMID: 20373019]

[109] B.X. Huang, H-Y. Kim, and C. Dass, "Probing three-dimensional structure of bovine serum albumin by chemical cross-linking and mass spectrometry", *J. Am. Soc. Mass Spectrom.*, vol. 15, no. 8, pp. 1237-1247, 2004.
[http://dx.doi.org/10.1016/j.jasms.2004.05.004] [PMID: 15276171]

[110] A. Arasteh, S. Farahi, M. Habibi-Rezaei, and A.A. Moosavi-Movahedi, "Glycated albumin: an overview of the in vitro models of an in vivo potential disease marker", *J. Diabetes Metab. Disord.*, vol. 13, p. 49, 2014.
[http://dx.doi.org/10.1186/2251-6581-13-49] [PMID: 24708663]

[111] M. Fasano, S. Curry, E. Terreno, M. Galliano, G. Fanali, and P. Narciso, *"The extraordinary ligand binding properties of human serum albumin". IUBMB Life.* vol. 57. International Union Biochem. Mol. Biol. Life, 2005, pp. 787-796.

[112] X.M. He, and D.C. Carter, "Atomic structure and chemistry of human serum albumin", *Nature,* vol. 358, no. 6383, pp. 209-215, 1992.
[http://dx.doi.org/10.1038/358209a0] [PMID: 1630489]

[113] S. Sugio, A. Kashima, S. Mochizuki, M. Noda, and K. Kobayashi, "Crystal structure of human serum albumin at 2.5 A resolution", *Protein Eng.*, vol. 12, no. 6, pp. 439-446, 1999.
[http://dx.doi.org/10.1093/protein/12.6.439] [PMID: 10388840]

[114] G. Fanali, A. di Masi, V. Trezza, M. Marino, M. Fasano, and P. Ascenzi, "Human serum albumin: from bench to bedside", *Mol. Aspects Med.*, vol. 33, no. 3, pp. 209-290, 2012.

[http://dx.doi.org/10.1016/j.mam.2011.12.002] [PMID: 22230555]

[115] P.A. Zunszain, J. Ghuman, T. Komatsu, E. Tsuchida, and S. Curry, "Crystal structural analysis of human serum albumin complexed with hemin and fatty acid", *BMC Struct. Biol.*, vol. 3, p. 6, 2003. [http://dx.doi.org/10.1186/1472-6807-3-6] [PMID: 12846933]

[116] R.L. Gundry, Q. Fu, C.A. Jelinek, J.E. Van Eyk, and R.J. Cotter, "Investigation of an albumin-enriched fraction of human serum and its albuminome", *Proteomics Clin. Appl.*, vol. 1, no. 1, pp. 73-88, 2007. [http://dx.doi.org/10.1002/prca.200600276] [PMID: 20204147]

[117] W. Bal, M. Sokołowska, E. Kurowska, and P. Faller, "Binding of transition metal ions to albumin: sites, affinities and rates", *Biochim. Biophys. Acta,* vol. 1830, no. 12, pp. 5444-5455, 2013. [http://dx.doi.org/10.1016/j.bbagen.2013.06.018] [PMID: 23811338]

[118] G.J. van der Vusse, "Albumin as fatty acid transporter", *Drug Metab. Pharmacokinet.*, vol. 24, no. 4, pp. 300-307, 2009. [http://dx.doi.org/10.2133/dmpk.24.300] [PMID: 19745557]

[119] K. Yamasaki, V.T. Chuang, T. Maruyama, and M. Otagiri, "Albumin-drug interaction and its clinical implication", *Biochim. Biophys. Acta,* vol. 1830, no. 12, pp. 5435-5443, 2013. [http://dx.doi.org/10.1016/j.bbagen.2013.05.005] [PMID: 23665585]

[120] C-E. Ha, and N.V. Bhagavan, "Novel insights into the pleiotropic effects of human serum albumin in health and disease", *Biochim. Biophys. Acta,* vol. 1830, no. 12, pp. 5486-5493, 2013. [http://dx.doi.org/10.1016/j.bbagen.2013.04.012] [PMID: 23602811]

[121] A. Varshney, P. Sen, E. Ahmad, M. Rehan, N. Subbarao, and R.H. Khan, "Ligand binding strategies of human serum albumin: how can the cargo be utilized?", *Chirality,* vol. 22, no. 1, pp. 77-87, 2010. [http://dx.doi.org/10.1002/chir.20709] [PMID: 19319989]

[122] S. Fujiwara, and T. Amisaki, "Identification of high affinity fatty acid binding sites on human serum albumin by MM-PBSA method", *Biophys. J.*, vol. 94, no. 1, pp. 95-103, 2008. [http://dx.doi.org/10.1529/biophysj.107.111377] [PMID: 17827235]

[123] L. Zhu, F. Yang, L. Chen, E.J. Meehan, and M. Huang, "A new drug binding subsite on human serum albumin and drug-drug interaction studied by X-ray crystallography", *J. Struct. Biol.*, vol. 162, no. 1, pp. 40-49, 2008. [http://dx.doi.org/10.1016/j.jsb.2007.12.004] [PMID: 18258455]

[124] F. Yang, Y. Zhang, and H. Liang, "Interactive association of drugs binding to human serum albumin", *Int. J. Mol. Sci.*, vol. 15, no. 3, pp. 3580-3595, 2014. [http://dx.doi.org/10.3390/ijms15033580] [PMID: 24583848]

[125] Z.D. Zhivkova, "Studies on drug-human serum albumin binding: the current state of the matter", *Curr. Pharm. Des.*, vol. 21, no. 14, pp. 1817-1830, 2015. [http://dx.doi.org/10.2174/1381612821666150302113710] [PMID: 25732557]

[126] J.E. Schnitzer, W.W. Carley, and G.E. Palade, "Albumin interacts specifically with a 60-kDa microvascular endothelial glycoprotein", *Proc. Natl. Acad. Sci. USA,* vol. 85, no. 18, pp. 6773-6777, 1988. [http://dx.doi.org/10.1073/pnas.85.18.6773] [PMID: 3413125]

[127] J.E. Schnitzer, and P. Oh, "Antibodies to SPARC inhibit albumin binding to SPARC, gp60, and microvascular endothelium", *Am. J. Physiol.*, vol. 263, no. 6 Pt 2, pp. H1872-H1879, 1992. [PMID: 1481911]

[128] S. Amsellem, J. Gburek, G. Hamard, R. Nielsen, T.E. Willnow, O. Devuyst, E. Nexo, P.J. Verroust, E.I. Christensen, and R. Kozyraki, "Cubilin is essential for albumin reabsorption in the renal proximal tubule", *J. Am. Soc. Nephrol.*, vol. 21, no. 11, pp. 1859-1867, 2010. [http://dx.doi.org/10.1681/ASN.2010050492] [PMID: 20798259]

[129] T. Fritzsche, M. Schnölzer, S. Fiedler, M. Weigand, M. Wiessler, and E. Frei, "Isolation and

identification of heterogeneous nuclear ribonucleoproteins (hnRNP) from purified plasma membranes of human tumour cell lines as albumin-binding proteins", *Biochem. Pharmacol.* , vol. 67, no. 4, pp. 655-665, 2004.
[http://dx.doi.org/10.1016/j.bcp.2003.09.027] [PMID: 14757165]

[130] N. Ghinea, A. Fixman, D. Alexandru, D. Popov, M. Hasu, L. Ghitescu, M. Eskenasy, M. Simionescu, and N. Simionescu, "Identification of albumin-binding proteins in capillary endothelial cells", *J. Cell Biol.* , vol. 107, no. 1, pp. 231-239, 1988.
[http://dx.doi.org/10.1083/jcb.107.1.231] [PMID: 2839518]

[131] X.Y. Zhai, R. Nielsen, H. Birn, K. Drumm, S. Mildenberger, R. Freudinger, S.K. Moestrup, P.J. Verroust, E.I. Christensen, and M. Gekle, "Cubilin- and megalin-mediated uptake of albumin in cultured proximal tubule cells of opossum kidney", *Kidney Int.* , vol. 58, no. 4, pp. 1523-1533, 2000.
[http://dx.doi.org/10.1046/j.1523-1755.2000.00314.x] [PMID: 11012887]

[132] D.C. Roopenian, and S. Akilesh, "FcRn: the neonatal Fc receptor comes of age", *Nat. Rev. Immunol.* , vol. 7, no. 9, pp. 715-725, 2007.
[http://dx.doi.org/10.1038/nri2155] [PMID: 17703228]

[133] J.E. Schnitzer, A. Sung, R. Horvat, and J. Bravo, "Preferential interaction of albumin-binding proteins, gp30 and gp18, with conformationally modified albumins. Presence in many cells and tissues with a possible role in catabolism", *J. Biol. Chem.* , vol. 267, no. 34, pp. 24544-24553, 1992.
[PMID: 1447200]

[134] J.E. Schnitzer, and J. Bravo, "High affinity binding, endocytosis, and degradation of conformationally modified albumins. Potential role of gp30 and gp18 as novel scavenger receptors", *J. Biol. Chem.* , vol. 268, no. 10, pp. 7562-7570, 1993.
[PMID: 8463286]

[135] C. Tiruppathi, A. Finnegan, and A.B. Malik, "Isolation and characterization of a cell surface albumin-binding protein from vascular endothelial cells", *Proc. Natl. Acad. Sci. USA,* vol. 93, no. 1, pp. 250-254, 1996.
[http://dx.doi.org/10.1073/pnas.93.1.250] [PMID: 8552615]

[136] M. Bern, K.M. Sand, J. Nilsen, I. Sandlie, and J.T. Andersen, "The role of albumin receptors in regulation of albumin homeostasis: Implications for drug delivery", *J. Control. Release,* vol. 211, pp. 144-162, 2015.
[http://dx.doi.org/10.1016/j.jconrel.2015.06.006] [PMID: 26055641]

[137] A.J. Milici, N.E. Watrous, H. Stukenbrok, and G.E. Palade, "Transcytosis of albumin in capillary endothelium", *J. Cell Biol.* , vol. 105, no. 6 Pt 1, pp. 2603-2612, 1987.
[http://dx.doi.org/10.1083/jcb.105.6.2603] [PMID: 3320050]

[138] E.I. Christensen, and H. Birn, "Megalin and cubilin: multifunctional endocytic receptors", *Nat. Rev. Mol. Cell Biol.* , vol. 3, no. 4, pp. 256-266, 2002.
[http://dx.doi.org/10.1038/nrm778] [PMID: 11994745]

[139] H. Birn, J.C. Fyfe, C. Jacobsen, F. Mounier, P.J. Verroust, H. Orskov, T.E. Willnow, S.K. Moestrup, and E.I. Christensen, "Cubilin is an albumin binding protein important for renal tubular albumin reabsorption", *J. Clin. Invest.* , vol. 105, no. 10, pp. 1353-1361, 2000.
[http://dx.doi.org/10.1172/JCI8862] [PMID: 10811843]

[140] Y. Jiang, S. Dey, and H. Matsunami, "Calreticulin: roles in cell-surface protein expression", *Membranes (Basel),* vol. 4, no. 3, pp. 630-641, 2014.
[http://dx.doi.org/10.3390/membranes4030630] [PMID: 25230046]

[141] A. Chaudhury, P. Chander, and P.H. Howe, "Heterogeneous nuclear ribonucleoproteins (hnRNPs) in cellular processes: Focus on hnRNP E1's multifunctional regulatory roles", *RNA,* vol. 16, no. 8, pp. 1449-1462, 2010.
[http://dx.doi.org/10.1261/rna.2254110] [PMID: 20584894]

[142] R.A. Brekken, and E.H. Sage, "SPARC, a matricellular protein: at the crossroads of cell-matrix

communication", *Matrix Biol.* , vol. 19, no. 8, pp. 816-827, 2001.
[http://dx.doi.org/10.1016/S0945-053X(00)00133-5] [PMID: 11223341]

[143] K.M. Sand, M. Bern, J. Nilsen, H.T. Noordzij, I. Sandlie, and J.T. Andersen, "Unraveling the Interaction between FcRn and Albumin: Opportunities for Design of Albumin-Based Therapeutics", *Front. Immunol.* , vol. 5, p. 682, 2015.
[http://dx.doi.org/10.3389/fimmu.2014.00682] [PMID: 25674083]

[144] L.N. Gastinel, N.E. Simister, and P.J. Bjorkman, "Expression and crystallization of a soluble and functional form of an Fc receptor related to class I histocompatibility molecules", *Proc. Natl. Acad. Sci. USA,* vol. 89, no. 2, pp. 638-642, 1992.
[http://dx.doi.org/10.1073/pnas.89.2.638] [PMID: 1530991]

[145] N.E. Simister, and K.E. Mostov, "An Fc receptor structurally related to MHC class I antigens", *Nature,* vol. 337, no. 6203, pp. 184-187, 1989.
[http://dx.doi.org/10.1038/337184a0] [PMID: 2911353]

[146] U. Shah, B.L. Dickinson, R.S. Blumberg, N.E. Simister, W.I. Lencer, and W.A. Walker, "Distribution of the IgG Fc receptor, FcRn, in the human fetal intestine", *Pediatr. Res.* , vol. 53, no. 2, pp. 295-301, 2003.
[http://dx.doi.org/10.1203/00006450-200302000-00015] [PMID: 12538789]

[147] E.J. Israel, S. Taylor, Z. Wu, E. Mizoguchi, R.S. Blumberg, and A. Bhan, "Expression of the neonatal Fc receptor, FcRn, on human intestinal epithelial cells", *Immunology,* vol. 92, pp. 69-74, 1997.

[148] R.K. Gill, S. Mahmood, C.P. Sodhi, J.P. Nagpaul, and A. Mahmood, "IgG binding and expression of its receptor in rat intestine during postnatal development", *Indian J. Biochem. Biophys.* , vol. 36, no. 4, pp. 252-257, 1999.
[PMID: 10650727]

[149] F.W. Brambell, "The transmission of immune globulins from the mother to the foetal and newborn young", *Proc. Nutr. Soc.* , vol. 28, no. 1, pp. 35-41, 1969.
[http://dx.doi.org/10.1079/PNS19690007] [PMID: 4182340]

[150] K.M. Sand, M. Bern, J. Nilsen, B. Dalhus, K.S. Gunnarsen, J. Cameron, A. Grevys, K. Bunting, I. Sandlie, and J.T. Andersen, "Interaction with both domain I and III of albumin is required for optimal pH-dependent binding to the neonatal Fc receptor (FcRn)", *J. Biol. Chem.* , vol. 289, no. 50, pp. 34583-34594, 2014.
[http://dx.doi.org/10.1074/jbc.M114.587675] [PMID: 25344603]

[151] V.J. Mohanraj, and Y. Chen, "Nanoparticles – A Review", *Trop. J. Pharm. Res.* , vol. 5, pp. 561-573, 2006.

[152] S. Sebak, M. Mirzaei, M. Malhotra, A. Kulamarva, and S. Prakash, "Human serum albumin nanoparticles as an efficient noscapine drug delivery system for potential use in breast cancer: preparation and in vitro analysis", *Int. J. Nanomedicine,* vol. 5, pp. 525-532, 2010.
[PMID: 20957217]

[153] W. Lohcharoenkal, L. Wang, Y.C. Chen, and Y. Rojanasakul, "Protein nanoparticles as drug delivery carriers for cancer therapy", *Biomed. Res. Int.* , vol. 2014, 2014.

[154] S.T. Sayyed Abolghassem Naheed Razavi, "Preparation and characterization of albumin microspheres encapsulated with propranolol HCl", *DARU J. Pharm. Sci.* , vol. 11, pp. 137-141, 2003.

[155] J. Chatterjee, Y. Haik, and C.J. Chen, "Synthesis and characterization of heat-stabilized albumin magnetic microspheres", *Colloid Polym. Sci.* , vol. 279, pp. 1073-1081, 2001.
[http://dx.doi.org/10.1007/s003960100523]

[156] R. Wade, M.E. Whisson, and M. Szerkerke, "Some serum protein nitrogen mustard complexes with high chemotherapeutic selectivity", *Nature,* vol. 215, no. 5107, pp. 1303-1304, 1967.
[http://dx.doi.org/10.1038/2151303a0] [PMID: 6052739]

[157] L. Fiume, M. Manerba, and G. Di Stefano, "Albumin-drug conjugates in the treatment of hepatic

disorders", *Expert Opin. Drug Deliv.* , vol. 11, no. 8, pp. 1203-1217, 2014. [http://dx.doi.org/10.1517/17425247.2014.913567] [PMID: 24773257]

[158] F. Kratz, "Drug conjugates with albumin and transferrin", *Expert Opin. Ther. Pat.* , vol. 12, pp. 433-439, 2002. [http://dx.doi.org/10.1517/13543776.12.3.433]

[159] J.L. Chen, A.H. Dhanaliwala, A.J. Dixon, A.L. Klibanov, and J. Hossack, "Synthesis of albumin microbubbles using a microfluidic device for real-time imaging and therapeutics", *IEEE Int. Ultrason. Symp. IUS,* 2013pp. 1150-3 [http://dx.doi.org/10.1109/ULTSYM.2013.0294]

[160] P.A. Dijkmans, L.J. Juffermans, R.J. Musters, A. van Wamel, F.J. ten Cate, W. van Gilst, C.A. Visser, N. de Jong, and O. Kamp, "Microbubbles and ultrasound: from diagnosis to therapy", *Eur. J. Echocardiogr.* , vol. 5, no. 4, pp. 245-256, 2004. [http://dx.doi.org/10.1016/j.euje.2004.02.001] [PMID: 15219539]

[161] D. Tada, T. Tanabe, A. Tachibana, and K. Yamauchi, "Drug release from hydrogel containing albumin as crosslinker", *J. Biosci. Bioeng.* , vol. 100, no. 5, pp. 551-555, 2005. [http://dx.doi.org/10.1263/jbb.100.551] [PMID: 16384795]

[162] V. Abbate, X. Kong, and S.S. Bansal, "Photocrosslinked bovine serum albumin hydrogels with partial retention of esterase activity", *Enzyme Microb. Technol.* , vol. 50, no. 2, pp. 130-136, 2012. [http://dx.doi.org/10.1016/j.enzmictec.2011.11.002] [PMID: 22226199]

[163] K. Baler, R. Michael, I. Szleifer, and G.A. Ameer, "Albumin hydrogels formed by electrostatically triggered self-assembly and their drug delivery capability", *Biomacromolecules,* vol. 15, no. 10, pp. 3625-3633, 2014. [http://dx.doi.org/10.1021/bm500883h] [PMID: 25148603]

[164] F. Kratz, "A clinical update of using albumin as a drug vehicle - a commentary", *J. Control. Release,* vol. 190, pp. 331-336, 2014. [http://dx.doi.org/10.1016/j.jconrel.2014.03.013] [PMID: 24637463]

[165] A.B. Mayerson, "Levemir Early Clinical Experience Case Series", *Endocrinol. Metab. Clin. North Am.* , vol. 36, no. 1, suppl. Suppl. 1, pp. 45-52, 2007. [http://dx.doi.org/10.1016/S0889-8529(07)80007-7] [PMID: 17881331]

[166] P. Raskin, "Efficacy and safety of insulin detemir", *Endocrinol. Metab. Clin. North Am.* , vol. 36, suppl. Suppl. 1, pp. 21-32, 2007. [http://dx.doi.org/10.1016/S0889-8529(07)80005-3] [PMID: 17881329]

[167] P. Home, and P. Kurtzhals, "Insulin detemir: from concept to clinical experience", *Expert Opin. Pharmacother.* , vol. 7, no. 3, pp. 325-343, 2006. [http://dx.doi.org/10.1517/14656566.7.3.325] [PMID: 16448327]

[168] G.M. Keating, "Insulin degludec and insulin degludec/insulin aspart: a review of their use in the management of diabetes mellitus", *Drugs,* vol. 73, no. 6, pp. 575-593, 2013. [http://dx.doi.org/10.1007/s40265-013-0051-1] [PMID: 23620200]

[169] I. Cucinotto, L. Fiorillo, S. Gualtieri, M. Arbitrio, D. Ciliberto, N. Staropoli, A. Grimaldi, A. Luce, P. Tassone, M. Caraglia, and P. Tagliaferri, "Nanoparticle albumin bound Paclitaxel in the treatment of human cancer: nanodelivery reaches prime-time?", *J. Drug Deliv.* , vol. 2013, no. 905091, p. 905091, 2013. [PMID: 23738077]

[170] E. Miele, G.P. Spinelli, E. Miele, F. Tomao, and S. Tomao, "Albumin-bound formulation of paclitaxel (Abraxane ABI-007) in the treatment of breast cancer", *Int. J. Nanomedicine,* vol. 4, pp. 99-105, 2009. [PMID: 19516888]

[171] B. Elsadek, and F. Kratz, "Impact of albumin on drug deliverynew applications on the horizon", *J. Control. Release,* vol. 157, no. 1, pp. 4-28, 2012.

[http://dx.doi.org/10.1016/j.jconrel.2011.09.069] [PMID: 21959118]

[172] Y-F. Wang, M-H. Chuang, J-S. Chiu, T-M. Cham, and M-I. Chung, "On-site preparation of technetium-99m labeled human serum albumin for clinical application", *Tohoku J. Exp. Med.* , vol. 211, no. 4, pp. 379-385, 2007.
[http://dx.doi.org/10.1620/tjem.211.379] [PMID: 17409678]

[173] F. Kratz, I. Fichtner, and R. Graeser, "Combination therapy with the albumin-binding prodrug of doxorubicin (INNO-206) and doxorubicin achieves complete remissions and improves tolerability in an ovarian A2780 xenograft model", *Invest. New Drugs,* vol. 30, no. 4, pp. 1743-1749, 2012.
[http://dx.doi.org/10.1007/s10637-011-9686-5] [PMID: 21590366]

[174] I. Kim, T.H. Kim, K. Ma, E.S. Lee, D. Kim, K.T. Oh, D.H. Lee, K.C. Lee, and Y.S. Youn, "Synthesis and evaluation of human serum albumin-modified exendin-4 conjugate via heterobifunctional polyethylene glycol linkage with protracted hypoglycemic efficacy", *Bioconjug. Chem.* , vol. 21, no. 8, pp. 1513-1519, 2010.
[http://dx.doi.org/10.1021/bc100143c] [PMID: 20715855]

[175] R. Flisiak, and I. Flisiak, "Albinterferon-alpha 2b: a new treatment option for hepatitis C", *Expert Opin. Biol. Ther.* , vol. 10, no. 10, pp. 1509-1515, 2010.
[http://dx.doi.org/10.1517/14712598.2010.521494] [PMID: 20828335]

[176] P. Kremer, M. Fardanesh, R. Ding, M. Pritsch, S. Zoubaa, and E. Frei, "Intraoperative fluorescence staining of malignant brain tumors using 5-aminofluorescein-labeled albumin", *Neurosurgery,* vol. 64, pp. 53-60, 2009.

[177] J. Zurdo, A. Arnell, O. Obrezanova, N. Smith, R. Gómez de la Cuesta, and T.R. Gallagher, "Early Implementation of QbD in Biopharmaceutical Development: A Practical Example", *Biomed. Res. Int,* vol. 2015, pp. 1-19, 2015.

[178] C Herring, and O. Schon, *AlbudAb^{TM} Technology Platform–Versatile Albumin Binding Domains for the Development of Therapeutics with Tunable Half-Lives*Ther. Proteins, Wiley-VCH Verlag GmbH and Co. KGaA, 2012, pp. 249-268.

[179] C.F. McDonagh, A. Huhalov, B.D. Harms, S. Adams, V. Paragas, S. Oyama, B. Zhang, L. Luus, R. Overland, S. Nguyen, J. Gu, N. Kohli, M. Wallace, M.J. Feldhaus, A.J. Kudla, B. Schoeberl, and U.B. Nielsen, "Antitumor activity of a novel bispecific antibody that targets the ErbB2/ErbB3 oncogenic unit and inhibits heregulin-induced activation of ErbB3", *Mol. Cancer Ther.* , vol. 11, no. 3, pp. 582-593, 2012.
[http://dx.doi.org/10.1158/1535-7163.MCT-11-0820] [PMID: 22248472]

[180] C.A. Stoddart, G. Nault, S.A. Galkina, K. Thibaudeau, P. Bakis, N. Bousquet-Gagnon, M. Robitaille, M. Bellomo, V. Paradis, P. Liscourt, A. Lobach, M.E. Rivard, R.G. Ptak, M.K. Mankowski, D. Bridon, and O. Quraishi, "Albumin-conjugated C34 peptide HIV-1 fusion inhibitor: equipotent to C34 and T-20 in vitro with sustained activity in SCID-hu Thy/Liv mice", *J. Biol. Chem.* , vol. 283, no. 49, pp. 34045-34052, 2008.
[http://dx.doi.org/10.1074/jbc.M805536200] [PMID: 18809675]

[181] R. Hussain, and G. Siligardi, "Novel drug delivery system for lipophilic therapeutics of small molecule, peptide-based and protein drugs", *Chirality,* vol. 22, suppl. Suppl. 1, pp. E44-E46, 2010.
[http://dx.doi.org/10.1002/chir.20897] [PMID: 20872606]

[182] S. Kelle, G. Greil, R. Razavi, and E. Nagel, "Vasovist® for Imaging Ischemic and Congenital Heart Disease", In: *Clin. Blood Pool MR Imaging SE - 14.* , T. Leiner, M. Goyen, M. Rohrer, S. Schönberg, Eds., Springer Berlin Heidelberg, 2008, pp. 159-168.
[http://dx.doi.org/10.1007/978-3-540-77861-5_14]

[183] Y. Hama, Y. Urano, Y. Koyama, A.J. Gunn, P.L. Choyke, and H. Kobayashi, "A self-quenched galactosamine-serum albumin-rhodamineX conjugate: a smart fluorescent molecular imaging probe synthesized with clinically applicable material for detecting peritoneal ovarian cancer metastases", *Clin. Cancer Res.* , vol. 13, no. 21, pp. 6335-6343, 2007.

[http://dx.doi.org/10.1158/1078-0432.CCR-07-1004] [PMID: 17975145]

[184] N. Katayama, K. Nakajou, Y. Ishima, S. Ikuta, J. Yokoe, F. Yoshida, A. Suenaga, T. Maruyama, T. Kai, and M. Otagiri, "Nitrosylated human serum albumin (SNO-HSA) induces apoptosis in tumor cells", *Nitric Oxide,* vol. 22, no. 4, pp. 259-265, 2010. [http://dx.doi.org/10.1016/j.niox.2009.09.003] [PMID: 19818866]

[185] Y. Ishima, T. Sawa, U. Kragh-Hansen, Y. Miyamoto, S. Matsushita, T. Akaike, and M. Otagiri, "S-Nitrosylation of human variant albumin Liprizzi (R410C) confers potent antibacterial and cytoprotective properties", *J. Pharmacol. Exp. Ther.* , vol. 320, no. 3, pp. 969-977, 2007. [http://dx.doi.org/10.1124/jpet.106.114959] [PMID: 17135341]

CHAPTER 6

Biopolymers as Microencapsulation Materials in the Food Industry

Tomy J. Gutiérrez[1,2,*] and **Kelvia Álvarez**[1,2]

[1] *Department of Analytical Chemistry, Faculty of Pharmacy, Central University of Venezuela (UCV), PO Box 40109, Caracas 1040-A, Venezuela*

[2] *Institute of Food Science and Technology, Faculty of Sciences, Central University of Venezuela (UCV), PO Box 47097, Caracas 1041-A, Venezuela*

Abstract: Microencapsulation is a technology that physically wraps sensitive ingredients in a protective matrix. This may be required for several reasons, such as: 1) to contain aromatic compounds that can be rapidly evaporated or aromas that may be lost during storage, 2) to avoid undesirable interactions between the matrix and the aromas, 3) to minimize interactions between volatile compounds (*flavor*/*flavor*), 4) to protect substances against oxidative reactions, 5) to control and prolong the release of probiotics, drugs and/or *flavor*. Materials that have been microencapsulated include: enzymes, probiotics and microorganisms, acids, bases, oils, vitamins, antioxidants, salts, gases, pharmacologically active peptides and amino acids, flavorings and colorings. Several materials have been employed as microencapsulants, such as: gums (carrageenan, alginate and gum arabic), carbohydrates (starch, maltodextrin, β-cyclodextrin and chitosan), celluloses (cellulose acetate phthalate) and proteins (gelatin and dairy protein isolates). In this chapter we discuss some of the variables, such as the concentrations of the reactants used, that affect the formation of microencapsulation materials.

Keywords: Biopolymers, Food, Microencapsulation.

INTRODUCTION

Microencapsulation is a technology that provides a physical wrapping for ingredients considered as sensitive, in order to protect them from adverse reactions, nutritional deterioration and even the loss of volatile compounds, as a consequence of moisture, heat, or other extreme conditions [1]. This technique has proved to be extremely promising for the development of new products and processes [2].

* **Corresponding author Tomy J. Gutiérrez:** Institute of Food Science and Technology, Faculty of Sciences, Central University of Venezuela, PO Box 47097, Caracas 1041-A, Venezuela; Tel: +58 212 7534403; Fax: +58 212 7533971; E-mails: tomy.gutierrez@ciens.ucv.ve, tomy_gutierrez@yahoo.es

Martin Masuelli & Denis Renard (Eds.)

Biopolymers are widely used in the food industry to solve formulation problems that may arise from the chemical or physical instability of the active ingredient, or incompatibility between the active ingredient and the matrix. They may also be used for the controlled release of a compound, or to improve the bioavailability of a particular nutrient [3]. Furthermore, encapsulation is often required to: 1) immobilize cells and enzymes, 2) protect against degradation processes (O_2 and light), 3) extend shelf life and protect the nutritional value of food, 4) prevent the degradation of compounds with high nutritional value, thus maintaining their bioavailability and 5) mask flavors and odors.

The materials most often used for the microencapsulation of food are: gums (carrageenan, alginate and gum arabic), carbohydrates (starch, maltodextrin, β-cyclodextrin and chitosan), cellulose (cellulose acetate phthalate) and protein (gelatin and dairy protein isolates).

Some of the substances that have been microencapsulated for their application in the food industry are: enzymes, probiotics and microorganisms, acids, bases, oils, vitamins, antioxidants, salts, gases, pharmacologically active peptides and amino acids, flavorings and colorings.

Techniques for encapsulation include: spray drying, extrusion, coacervation, ionic gelation, entrapment with liposomes, freeze-drying, co-crystallization and emulsion, and sonochemistry, amongst others. However, the choice of encapsulation method depends on the required average particle size, the physical properties of the encapsulant and the substance to be encapsulated, how the encapsulated material is to be utilized, the desired release mechanism and the cost. The microencapsulation method used should also be quick to apply and versatile.

As already mentioned briefly, selection of the microencapsulation method is based on the physico-chemical characteristics of the polymer and the material to be encapsulated, taking into account the following factors [4, 5]:

- The yield should be high for the desired microparticle size.
- Encapsulation efficiency must be high.
- Reproducibility between batches must be within precise limits.
- The microspheres should not form clusters, and their form should be maintained.
- The environmental impact caused by manufacturing conditions and materials should be minimized.

In addition, the internal morphology of the particle must be controlled as it determines how the active ingredients are released. Particle morphology is heavily dependent on the microencapsulation method and protocol chosen.

A crucial step in microparticle production is the generation of the drops. This determines the size distribution and structure of the particles, thus affecting their usefulness for each application. The methods currently used for drop preparation are classified into three groups based on their productivity and capacity to control drop size: 1) high productivity and low control of drop size, 2) low productivity and high control of drop size, and 3) high productivity and high control of drop size.

Finally, changes in pH, concentrations of reactants and the temperature of the system also influence on the stability and release mechanism of the encapsulated material.

In this chapter, we discuss basic and general aspects of microencapsulation as an innovative technique for the food industry.

Microencapsulation

Although the term microencapsulation is relatively old, its application is relatively recent. The earliest references describing a microencapsulation process (gelatin microspheres prepared by coacervation) date from 1931. The use of this technology has increased in recent decades due to the interest in the development of biodegradable micro- and nanoparticles.

Microencapsulation is a technology that allows the physical wrapping of sensitive ingredients in a protective matrix (a “wall type” material) in order to protect them from adverse reactions, nutritional deterioration and even the loss of volatile compounds [6] as a consequence of moisture, heat, or other extreme conditions. Microencapsulation thus enables the development or improvement of the stability of the ingredients, as well as maintaining the product viability [1]. This technique has proved to be extremely promising for the creation of new products and processes [2]. Within the food industry, microencapsulation is typically used to resolve formulation problems that may arise from the chemical or physical instability of the active ingredient, or incompatibility between the active ingredient and the matrix. They may also be used for the controlled release of the compound, or the bioavailability of a particular nutrient [3].

From a technological point of view, microencapsulation is defined as the process of coating a product with molecules, solid particles or liquid globules, derived from different kinds of materials, to produce micron-sized particles. There are three types of particles resulting from this technological process: “microparticles”, “microcapsules” or “microspheres”, which differ from each other in their morphology and internal structure. Nevertheless, all three share a common characteristic: particle size, which is always less than 1 mm. Particles with a size less than 1 mm (1000 μm) are known as "nanospheres", "nanoparticles" or "nanocapsules" [7, 8].

Fig. (**1**) shows the different types of particles that can be obtained, depending on the encapsulation method employed.

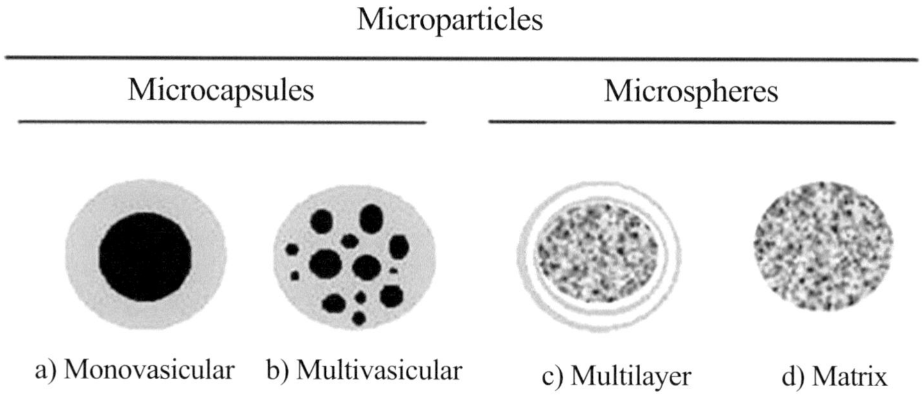

Fig. (1). Classification of microparticles according to their morphology.

The internal morphology of the particle must be controlled as it determines how the active ingredients are released. Particle morphology is heavily dependent on the microencapsulation method and protocol chosen.

There are currently several encapsulation techniques such as: 1) coacervation, a process brought about by the partial desolvation of a polymer which is deposited as drops of coacervate around the particles of the active ingredient. Coacervation thus represents an intermediate stage between dissolution and precipitation; 2) extraction-evaporation, a set of procedures in which an oil/water (o/w), water/oil (w/o) or water/water (w/w) emulsion is formed, with the surfactant compound incorporated into the outer phase and the coating material dissolved in the lipophilic inner phase. The active ingredients are then dissolved or dispersed into the inner phase; 3) interfacial polymerization, a process carried out within a w/o emulsion, in which the polymerization process develops in the interface, giving rise to the formation of the microcapsules; 4) atomization, whereby the active ingredient is dissolved or dispersed in a solution of the coating polymer in a suitable solvent. The polymer-active ingredient mixture is pulverized into a chamber within which hot air circulates (150-200 °C). This evaporates the solvent leaving the microcapsules behind; 5) atomization-freezing, in which the coating material is melted and mixed with the active ingredient, and then sprayed using sonication or air-atomization. This produces droplets which are then frozen into a chamber where cold air or liquid nitrogen circulates; 6) fluidized bed coating, which is used for the coating of solid particles containing active ingredients in a fluid bed system. In this system, the particles are maintained in suspension by the air stream entering through the base, whilst being sprayed by the polymer coating solution [9].

There are several reasons why the encapsulation of some compounds is advantageous, for example: 1) to contain flavorings associated with compounds that evaporate quickly or are lost from the food matrix during storage; 2) to avoid undesirable interactions between the flavors and the food matrix [10, 11]; 3) to minimize interactions between volatile compounds (*flavor*/*flavor*). With this in mind, it is worth mentioning that aldehydes and amines can form Schiff bases, with the result that the *flavor* is divided into two parts, one containing the aldehyde and the other the amine. Each of these may be encapsulated separately and applied to dry products; 4) to prevent the food matrix from reacting with light or oxygen [10]; 5) to control the release of *flavor* and/or *probiotics* [12]; and 6) to impart properties of a solid onto a liquid, for example, when incorporating the essential oils, omega 3 and omega 9, into powdered infant milk formulas.

Nanoemulsion forms part of the microencapsulation process, and consists of an emulsion with particles ranging from 50 to 200 nm in size. Emulsions can be prepared by low-energy emulsification methods or may arise spontaneously. Submicron emulsions can, however, be produced using high-energy techniques such as: ultrasound, high-pressure homogenization and microfluidization. In the particular case of the microencapsulation of aromatic liquids (*flavor*), emulsification occurs within an aqueous medium in the presence of the encapsulant, which is then dried under conditions that minimize its loss. All these methods are used in order to improve the stability of the encapsulated materials [12].

Within the food industry, several types of microencapsulation agents are used, such as: gums, natural and modified carbohydrates, lipids, proteins and synthetic polymers [1]. Other microencapsulation agents, characterized as "wall type" materials, are obtained from wheat protein isolates in combination with maltodextrins or corn syrup. These are effective in the microencapsulation of volatile compounds, such as ethyl caprylate, which would suffer great losses if they underwent spray drying [13]. The use of gum arabic is also common due to its solubility, viscosity and emulsification capacity [1]. The physico-chemical properties of the encapsulation material and the volatile compound must always be taken into account, however, retention of the compound depends on its functional group, molecular weight, polarity and the relative volatility of the volatile component [14].

Suslick [15] and Avivi and Gendaken [16] studied the coating of drugs. For this, they generated microcapsules, microspheres and microparticles between 1 μm and 1 mm in size from different types of materials such as: fats, carnauba wax, stearyl alcohol, proteins (gelatin and albumin), polysaccharides and alginates. Interfacial polymerization, fluidized bed coating and ionic gelation were the encapsulation techniques used. From these studies, the authors were able to observe: 1) the stabilization of the active ingredients; 2) the transformation of liquids to solids; 3) the compatibility of the active ingredient with the pharmaceutical formulation; 4) the masking of odors and tastes; 5) a reduction in the irritant effect on the gastric mucosa, and 6) the controlled release of the active ingredients. They concluded that the techniques they used have a wide range of biomedical applications such as the development of contrast agents for sonography and magnetic resonance imaging, as well as the controlled release of drugs and oxygen.

Drinks have also been developed with microencapsulated particles, using gum solutions and milk proteins with $CaCl_2$, by suspending the particles formed by stirring and emulsification in the presence of a stabilizer [17]. In this study, the particles were observed to take on various colors and flavors. This is due to structural differences between microcapsules, which may be formed by: 1) a single wall or membrane with spherical or irregular characteristics; 2) a multi-walled structure with either a homogeneous or a varied composition; and 3) numerous microcapsules covered with the same structural wall (Fig. **1**) [1, 18].

Suslick [19] developed protein microspheres (bovine whey albumin at 96% purity, dissolved in distilled water) using the ultrasound technique. It has also been observed that adding proteins and polysaccharides to carbohydrate mixtures at different concentrations improves the encapsulation properties [20]. Sheu and Rosenberg [21] prepared microcapsules by the spray drying technique using mixtures of whey protein isolates (WPI) and maltodextrin or corn syrup. These researchers found that the microencapsulation systems were affected both by the type of carbohydrate used and the relationship between the WPI and the carbohydrates. The mixture of whey protein with maltodextrins improved the surface of the microcapsules, making it smoother and decreasing the cavity depth. The relationship between the molecular weights of the wall systems and their surface characteristics also affected the microcapsule structure. The combination of WPI with carbohydrates of high molecular weight seemed to be the most effective as this formed microcapsules with shallower surface cavities. Thus, maltodextrins (high molecular weight) or corn syrup powder are useful for minimizing the formation of surface cracks. This is very important since the absence of cracks on the wall system prevents the deterioration and/or loss of the active ingredient during storage.

The high pressure homogenization technique has been used to encapsulate orange *flavor* in the presence of gum arabic, soy protein isolate, milk whey protein isolate and sodium caseinate, obtaining the highest yields with sodium caseinate proving as being the most effective. However, gum arabic has also been very efficient for obtaining smaller sized microcapsules [22].

Kaushik and Roos [20] evaluated the microencapsulation of limonene with the freeze-drying technique in various matrices including: gum arabic, gelatin and sucrose. The highest percentage retention of the encapsulated compounds (75.3 ± 0.3% of initial added quantity) was found for the 9:1 (wxw) emulsion with the gum arabic matrix. These researchers also compared ultra-high-pressure homogenization (50-250 MPa) to the freeze-drying technique for limonene encapsulation, concluding that freeze-drying was the more efficient method.

Finally, we wish to highlight the fact that the microencapsulation system must be water soluble, show good emulsifying and dehydrating properties, and have low viscosity at high concentrations of solids [10, 12].

Applications of Microencapsulation

The reasons for carrying out a microencapsulation process can be summarized into four groups:

- To immobilize cells, enzymes, *etc.*
- To protect against degradation processes (O_2, light, humidity, *etc.*) in order to: extend the useful life and protect the nutritional value of the food, avoid food degradation in the stomach due to its acidity, and mask tastes and odors.
- To control the release of drugs or other substances such as odors, *etc.*
- To remove the solvent in a slurry to: 1) facilitate the ease of handling of a product; 2) increase the half-life of the food; and 3) modify its properties (solubility, dispersion, hygroscopicity, rheology, *etc.*) [5, 23 - 25].

All the above have made the microencapsulation process popular in numerous industrial sectors. Of these, perhaps the best known application has been the controlled release of drugs by the pharmaceutical industry, but it is also used in this sector for masking flavors and odors. This technology has also enabled drugs to be directed to specific areas, even managing to prevent their degradation in the stomach. Within the food industry, microencapsulation has transformed food manufacturing processes, since it can intervene in the health of consumers through the encapsulation of probiotics or other ingredients that improve health. It has also led to the creation of new textures and products, thus allowing their differentiation and re-evaluation. One of the sectors where microencapsulation shows most promising results is the infant food sector with the addition of vitamins, minerals, *etc.* into baby foods. In the textile industry, microencapsulation has unleashed a revolution in the shape of "intelligent textiles", whereby particles incorporated into fibers can transport substances, for example deodorants, fragrances, insect repellents or materials with cosmetic value such as *aloe vera* [26]. Furthermore, microencapsulation technology has been used in the manufacture of self-repairable paints, which has been proven invaluable to the shipping industry.

Microparticles have also been used as model systems in medicine, biochemistry, colloidal chemistry, aerosols technology, *etc.* They have also been employed as a support for peptide synthesis, as well as for cell separation or the decontamination of water bodies polluted with heavy metals [27]. In addition, their use in the development of equipment calibration patterns for flow cytometers, particle analyzers and confocal microscopes, amongst others, has sparked much interest. A further application for microparticles is their use in the flow measurement of gases and liquids (*Laser Doppler Anemometry* (LDA); *Particle Dynamics* (PDA); *Particle Image Velocimetry* (PIV)).

The first commercial success of a product using microparticles was in the field of information science and imaging with the development of carbonless copy paper by the National Cash Register Company in the 1950s to overcome the problems of evaporation and diffusion of the ink on the paper. In recent years, the interest in developing new types of papers or displays with unique optical and electronic properties is increasing. The microparticles employed in these systems contain pigments such as titanium oxide and carbon black. These have electrical charges thereby allowing their orientation within an electric field, thus generating the desired image [28]. The advantages of these systems are that they can be read under faint or unsteady light, consume low amounts of energy and are easily transportable. Currently Sony Corp. has developed products in which the technology of E Ink Corp. electrophoretic ink has been incorporated.

Finally, microparticles have been crucial to advances in the use of medical diagnostic systems (multiplex systems) dramatically improving their efficiency and resolution [29].

Microencapsulated Substances

Some of the materials that can be encapsulated for their application in the food industry are: enzymes (bromelain obtained from pineapples, papain extracted from papayas, *etc.*), probiotics (*Lactobacillus Acidophilus*), acids, bases, oils (orange essential oil, oregano essential oil, *etc.*), vitamins (folic acid), antioxidants, salts, gases, pharmacologically active peptides and amino acids, flavorings and colorings [5, 30 - 36].

Microencapsulation Materials

Independent of the microencapsulation method chosen, a suitable matrix or wall system must be selected. Some of the encapsulating agents currently in use are listed in Table (**1**):

Table 1. Materials used for microencapsulation [5, 24, 40].

Type of coverage	Specific coverage
Gums	Agar, carrageenan, sodium alginate, gum arabic.
Carbohydrates	Starch, dextran, sucrose, corn syrup, chitosan.
Celluloses	Ethylcellulose, methylcellulose, acetylcellulose, nitrocellulose, carboxymethylcellulose (CMC), cellulose acetate phthalate.
Lipids	Waxes, paraffins, diglycerides, monoglycerides, oils, fats, stearic acid, trisetearina.
Proteins	Gluten, gelatin, casein, albumin, pectin.
Inorganic materials	Calcium sulphate, silicates.

Microencapsulation with Gums

Carrageenan

Carrageenan is a natural polysaccharide extracted from marine algae commonly used as a food additive. High temperatures are needed (60-80 °C) to dissolve this compound at concentrations ranging from 2 to 5% [37]. The gelation process of k-carrageenan is directly dependent on variations in temperature. Microcapsules are formed by dripping onto a polymer-potassium chloride (KCl) mixture. Studies conducted by Audet *et al.* [38, 39] suggest that a combination of k-carrageenan with other polymers, such as carob gum, can improve the strength of the encapsulant. This has been demonstrated by the tolerance of some carrageenan-encapsulated probiotics to stresses such as freezing, drying, hydrogen peroxide and simulated gastrointestinal conditions.

Alginates

Alginates are extracted primarily from three species of brown algae (*Laminaria hyperborea*, *Ascophyllum nodosum* and *Macroscystis pyrifera*) [40]. These compounds are linear anionic polymers of high molecular weight with both rigid and flexible regions composed of block structures formed by different configurations of the monomers, β-D-mannuronic acid (M) and α-L-guluronic acid (G), linearly bonded by glycosidic linkages (1, 4). The block structures may be formed by one or other of the monomers (M or G), or a mixture of them both (MG) [5, 41]. The composition, molecular weight and extent of the polymeric sequences of each alginate determine its physico-chemical properties [40]. One of the advantages of alginates is that they are not toxic to the immobilized probiotic bacteria, and are thus acceptable as a food additive [41].

Alginates have been used in the food industry as thickeners, gelling and colloid stabilizers and, due to their properties, in the microencapsulation and release of drugs and microorganisms [40]. Some of the advantages of alginates are mentioned below:

- Encapsulation may be carried out at room temperature.
- Toxic organic solvents are not required.
- They are extremely porous.
- They permit the rapid and controlled diffusion of macromolecules.
- They dissolve and degrade under normal physiological conditions.

Moreover, alginates in the presence of calcium form a structure known as an "egg crate" (Fig. **2**), whereby the calcium ions act as bridges between the negatively charged groups of the guluronic acid. The alginate chains can thus be joined in multiple ways, depending on the concentration of calcium present, thus giving greater rigidity to the gel [41]. The interfacial polymerization process is instantaneous, and particle size generally depends on the viscosity of the polymer solution, the bore diameter of the injector through which sodium alginate is injected into the system, and the distance between the injector and the gelling solution [41, 42]. In addition, calcium alginate capsules tend to be very porous, thus allowing the easy and rapid diffusion of water and other fluids through the capsule walls [43, 44].

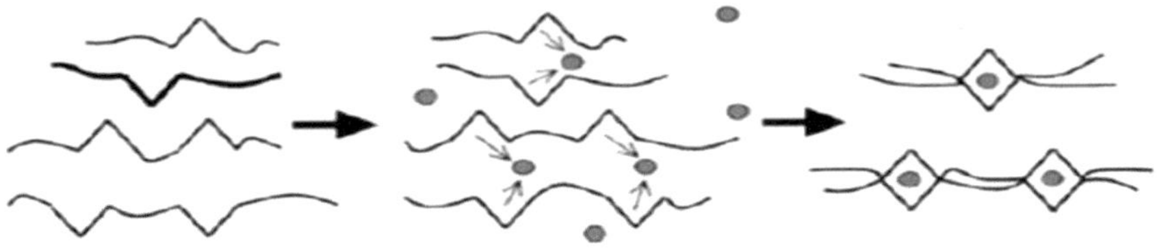

Fig. (2). Chemical reaction between alginate and calcium, leading to the formation of the "egg crate" structure.

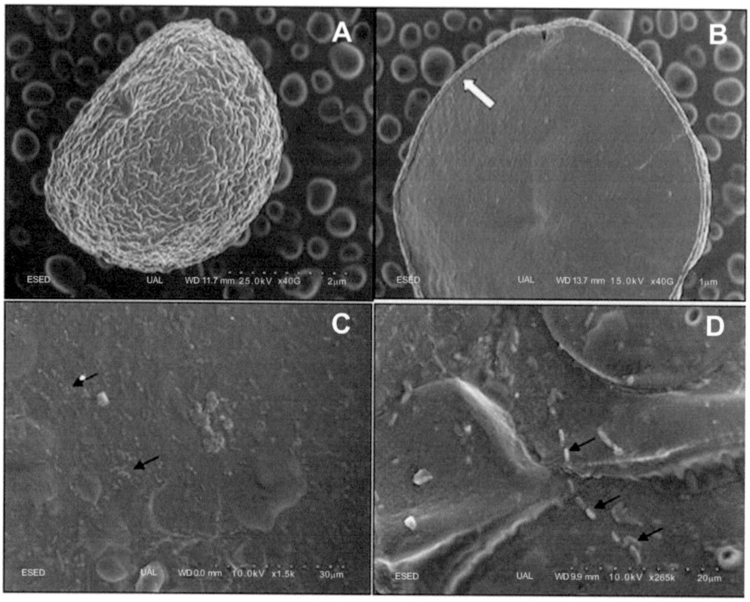

Fig. (3). **a**) Hydrated alginate microcapsule at 40X, **b**) Cross-section of an alginate-chitosan microcapsule with detail of the outer layer at 40X and, **c** and **d**) Detail of bacteria inside dehydrated capsules at 1.5 kX and 2.5 kX, respectively, metallized with gold [41].

As is well known, probiotics are inactivated at low pH values. However their encapsulation, using calcium chloride ($CaCl_2$) in the presence of sodium alginate, leads to the formation of calcium alginate microcapsules that protect the bacteria from the acid conditions of the digestive tract. This method has greatly increased the survival rates of the probiotic bacteria (*L. acidophilus*), compared to unencapsulated cells. Encapsulation percentages of above 80% have been reported for probiotics microencapsulated with alginates [45]. In addition, microencapsulation enables probiotic bacteria to be released into the intestine without any damage to them [46], and recent studies have shown that the viability of the microencapsulated probiotics in gastric fluid increases with capsule size [41]. However, other studies suggest that probiotic cultures contained in calcium alginate capsules greater than 1 mm in size are no more protected from gastric fluids than free cells [47].

$$2\ \text{Na-Alginate} + Ca^{2+} \rightarrow \text{Ca-Alginate} + 2\ Na^{+}$$

Furthermore, probiotic cells encapsulated with calcium alginate remain viable when maintained at 4 °C for a month, with a survival rate higher than 90% [45]. Encapsulation with calcium alginate thus increases the survival of probiotics under storage conditions and facilitates their administration at a specific site.

Fig. (**3a**) shows a freshly prepared alginate microcapsule, and Fig. (**3b**) a cross section of an alginate-chitosan microcapsule in which the thick outer chitosan coating can be clearly observed. In Figs. (**3c**) and (**d**), bacillus shaped probiotic bacteria trapped in a calcium alginate matrix may be distinguished [41].

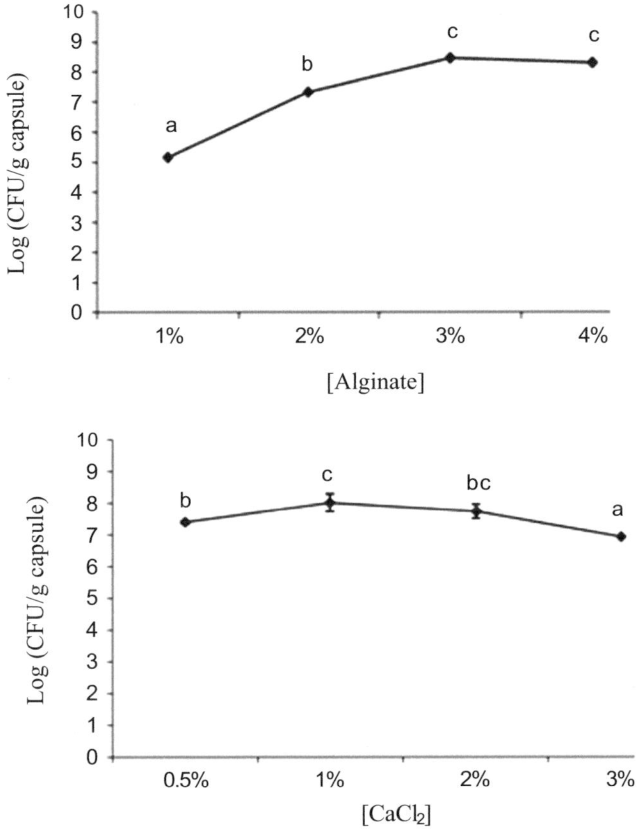

Fig. (4). **a**) Effect of sodium alginate concentration on the encapsulation yield and **b**) Effect of calcium chloride concentration on the encapsulation yield. The values show the mean ± s.d. (n = 3). Different letters (a, b and c) against values indicate significant differences between treatments ($p < 0.05$) [41].

Increasing the concentrations of alginate also has a favorable effect on the numbers of encapsulated bacteria, although there is a maximum concentration beyond which this does not increase. This phenomenon was highlighted in a study carried out by Martinez [41] where an increase in alginate concentration up to 3% produced a significant increase in the number of encapsulated bacteria ($p < 0.05$) after which the total number of encapsulated bacteria remained constant (Fig. **4a**).

Martinez [41] also found that the concentration of calcium chloride affects microencapsulation performance (Fig. **4b**). This author recorded the highest numbers of encapsulated bacteria at calcium chloride concentrations ranging from 1 to 2% (108 UFC). Performance decreased, however, at higher alginate concentrations ($p < 0.05$). Optimization of the process is thus necessary.

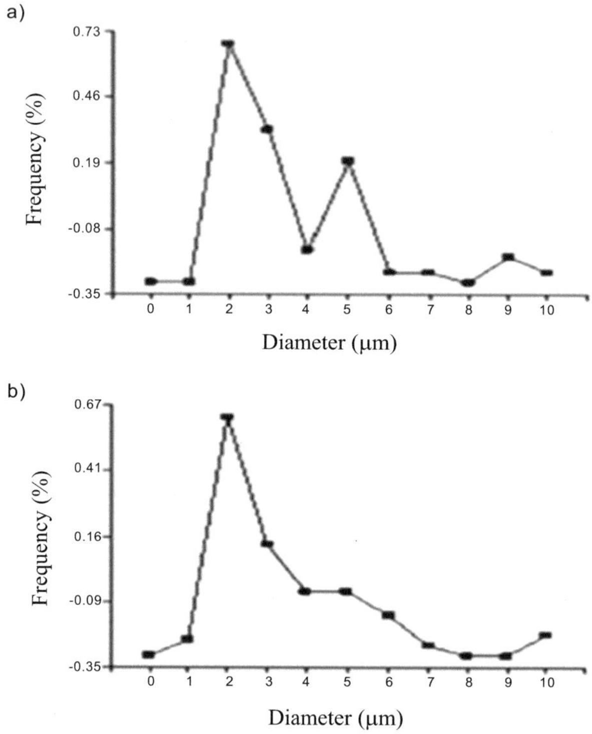

Fig. (5). Mean distribution of the particle sizes of w/o emulsion droplets as determined by optical microscopy in samples of **a**) 5% and **b**) 15% gum arabic [36].

Gum Arabic

Gum arabic has been used as a microencapsulation agent for phenolic compounds, with performance levels between 20 and 80% [48]. It has been reported that microcapsules with 5% gum arabic show a bimodal size distribution with two maximums, the first at approximately 2 μm, and the second at 5 μm. In contrast, microcapsules with 15% gum arabic have a monodisperse size distribution with a maximum of about 2 μm (Fig. **5**). Microcapsule size distribution may thus be attributed to the properties of the gum arabic, the concentration of the surfactant used and the degree of cross-linking between them to form the w/o emulsion. An increase in the concentration of gum arabic also results in a significant decrease in the encapsulation yield, and more agglomerates are produced when surfactant concentration increases [36].

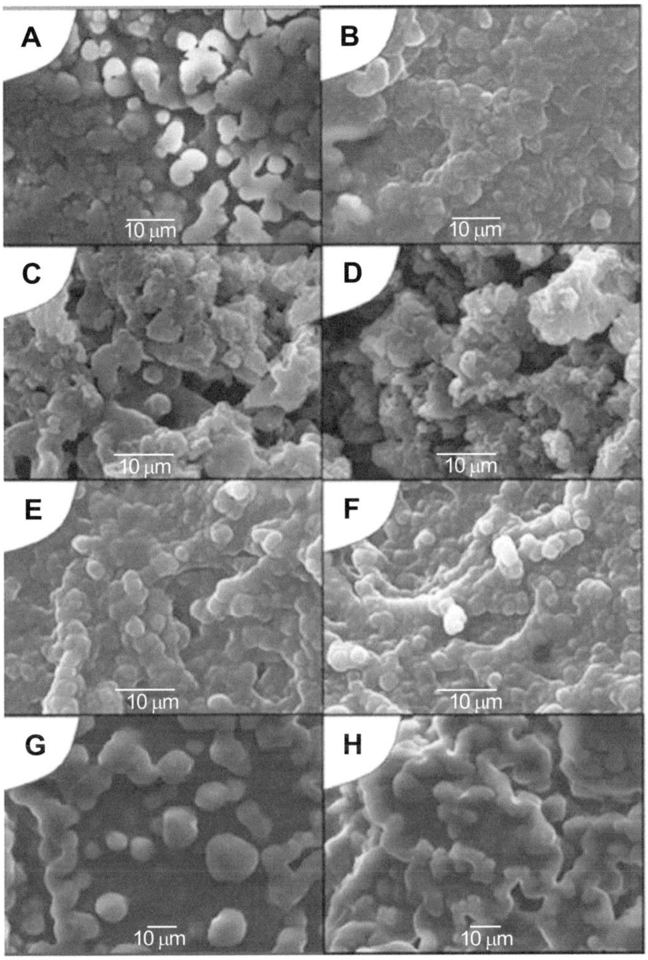

Fig. (6). SEM micrographs at 3.7kX of microcapsules in a water in oil (w/o) emulsion. Microcapsules prepared from 0.25% maqui leaf extract, microencapsulated with gum arabic (GA) and Tween 80 (T80) [36].

Vidal *et al.* [36] developed spherical microcapsules with a high degree of integrity, no surface ruptures or pores (Fig. **6**), and only a very few cracks. This can be attributed to the method used. The microcapsules produced by Vidal *et al.* [36] also show good sensory characteristics, as the small particle size increases emulsion stability thus avoiding gravitational separation, protecting the natural preservatives (phenolic compounds) and preventing the growth of deteriorative bacteria. In contrast, large size particles generate a gritty texture in the mouth, phase separation produces visual rejection, and the presence of deteriorative bacteria causes consumer rejection [36].

Microencapsulation with Carbohydrates

Starch

Starch is a biopolymer synthesized by plants, composed of two units (amylose and amylopectin). This material is cheap, readily available and resistant to thermal degradation [49]. Some studies have used modified starch (octenyl succinate) as an alternative, in order to fully or partially replace the more expensive gum arabic [50 - 53]. An example of this is Capsul® which is a chemically modified starch, derived from waxy corn, with a lipophilic component incorporated into its structure.

The emulsifying properties shown by Capsul® are useful for obtaining highly stable oil-in-water emulsions [10, 54 - 56]. In addition, its low viscosity has enabled the preparation of emulsions with a high solid content, thus obtaining microparticles with high retention values of the active compound [57, 58]. Capsul® has been used principally in the encapsulation of flavors, aromas, vitamins and spices [25].

Resistant starch may also be used for the microencapsulation of probiotic cultures. It also provides an ideal adhesive surface during processing, storage and transit through the gastrointestinal tract [59, 60].

Studies done by Iyer and Kailasapathy [61] and Sultana *et al.* [62] have demonstrated that lactic acid bacteria encapsulated with modified starches can survive for more than 6 months at room temperature under normal atmospheric and humidity conditions, and for at least 18 months when stored below freezing.

Microencapsulated probiotic cultures have been used in many fermented milk products such as: yogurt, cheese, creams and frozen desserts, as well as for biomass production. The encapsulated probiotics are protected against various agents such as bacteriophages, as well as adverse conditions produced by gastric juices and freezing. The manufacture of fermented milk products is thus greatly facilitated through the encapsulation of these microorganisms, as greater stability during storage of the product can be achieved [43].

Cyclodextrins can also be used as microencapsulation agents (Fig. **7a**). These are cyclic oligosaccharides formed by starch degradation processes and are made up of 6, 7 or 8 glucose units, linked by links α 1 $\rightarrow$ 4. Each cyclic oligosaccharide is given a specific name according to the number of glucose units contained within it (Table **2**) [24, 63].

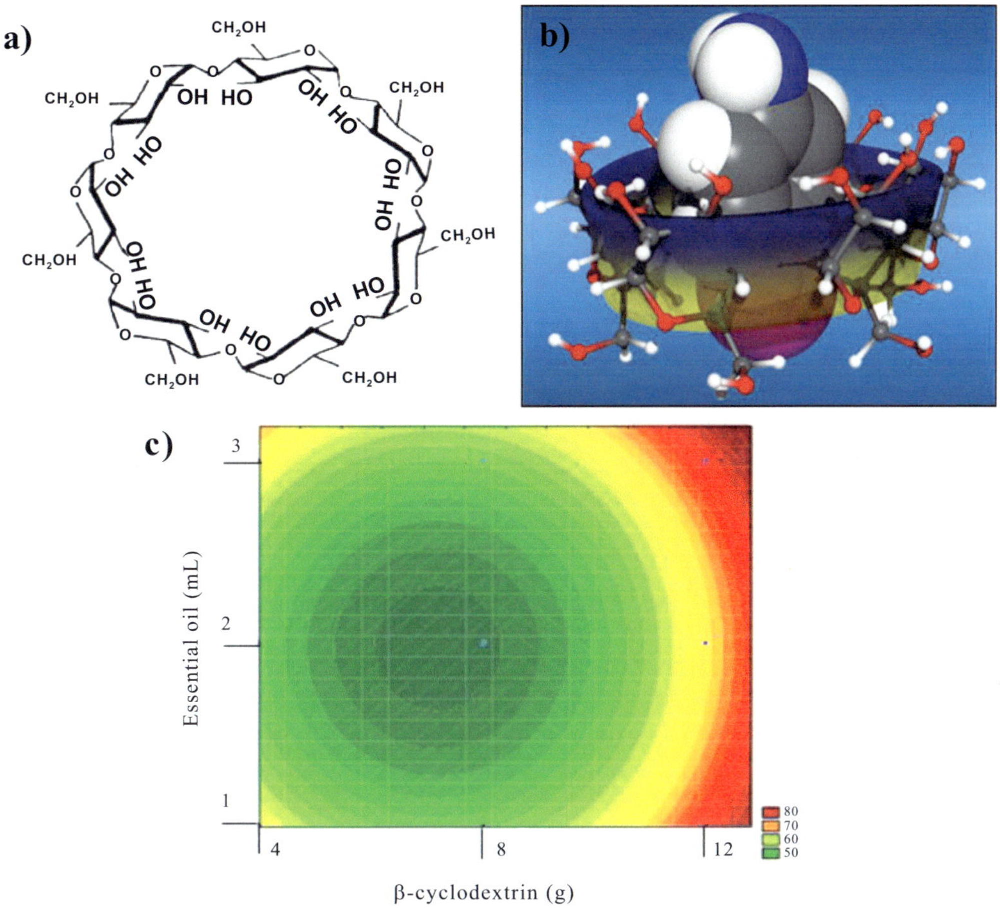

Fig. (7). a) Molecular structure of β-cyclodextrin [64], **b**) Microencapsulation of oregano essential oil (*Origanum vulgare*) with β-cyclodextrin [68] and **c**) Encapsulation efficiency of oregano essential oil (*Origanum vulgare*) using β-cyclodextrin [24].

Table 2. Characteristics and specifications of cyclodextrins [63].

Nomenclature	Glucose units (n)	Internal diameter (nm)	Height (nm)
α-cyclodextrin	6	0.57	0.78
β-cyclodextrin	7	0.78	0.78
γ-cyclodextrin	8	0.95	0.78

The volatile constituents of essential oils, which are also hydrophobic, may be enclosed in the lipophilic central cavity of the oligosaccharide ring, forming a complex molecular structure with a defined stoichiometry, whereby the substrate and the encapsulant are linked by (hydrophobic) physicochemical forces [65 - 68].

Leyva *et al.* [66] explain that during the formation of the β-cyclodextrin complex, the carboxyl and phenolic groups of the substrate (mainly polyphenols found in the essential oils) are arranged in a hydrophilic environment, whilst the aromatic region is accommodated in a hydrophobic region. This suggests that the β-cyclodextrin-essential oil complex acquires an equitable distribution of the substances (aromatic group and carboxyl group) present on the substrate (essential oil).

Complexes (microcapsules) made using β-cyclodextrin (Fig. **7b**) whilst providing protection to the compounds contained within them may also be released by controlling humidity levels. This is because the outer layer of the β-cyclodextrin is hydrophilic. This characteristic allows the interaction of water molecules with the outside of the complex, weakening the binding forces, and releasing the volatile components into the atmosphere [69].

Matos *et al.* [24] indicate that the highest microencapsulation efficiency (72-81%) occurred within a range of 8 to 12 g of β-cyclodextrin and 2 to 3 mL of essential oil (Fig. **7c**). This means that the concentration of β-cyclodextrin greatly influences microencapsulation efficiency, which is itself directly proportional to the microencapsulation percentage of the essential oil. Thus, an increases in the amount of β-cyclodextrin will increase microencapsulation efficiency.

Starch can be hydrolyzed using acids or enzymes in order to obtain maltodextrins, which can be used as microencapsulation agents. Guzmán [70] prepared maltodextrins of 20 dextrose equivalent from plantain starch (*Musa AAB*) and green "topocho" (*Musa ABB*) using α-amylase Termamyl 120L. He then used the maltodextrins obtained to achieve encapsulation yields of ascorbic acid of 76% and 72% with the plantain, and green topocho maltodextrins, respectively, using a spray dryer (Büchi Mini Spray Dryer, model B-290) under the following conditions: diameter of the spray nozzle 0.30 mm, temperature in the drying chamber 180 °C, feed flow rate 360 mL/h, pressure 0.41 bar, air drying speed 37.5 L/h, and product exit temperature 50°C. These conditions avoided the destabilization of the glass matrix of the maltodextrins that often occurs during the drying process, resulting in the high encapsulation yields obtained.

The morphology of ascorbic acid microencapsulated with maltodextrins from plantain and green "topocho" are shown in Fig. (**8**). It can be observed that most of the microcapsules are in the form of homogeneous globules with a smooth surface. Nevertheless, there were some cracks on the globules, which is characteristic of maltodextrins during drying. The diameters of the ascorbic acid microcapsules prepared using plantain-derived maltodextrin from were shorter than those made with green topocho-derived maltodextrin [70].

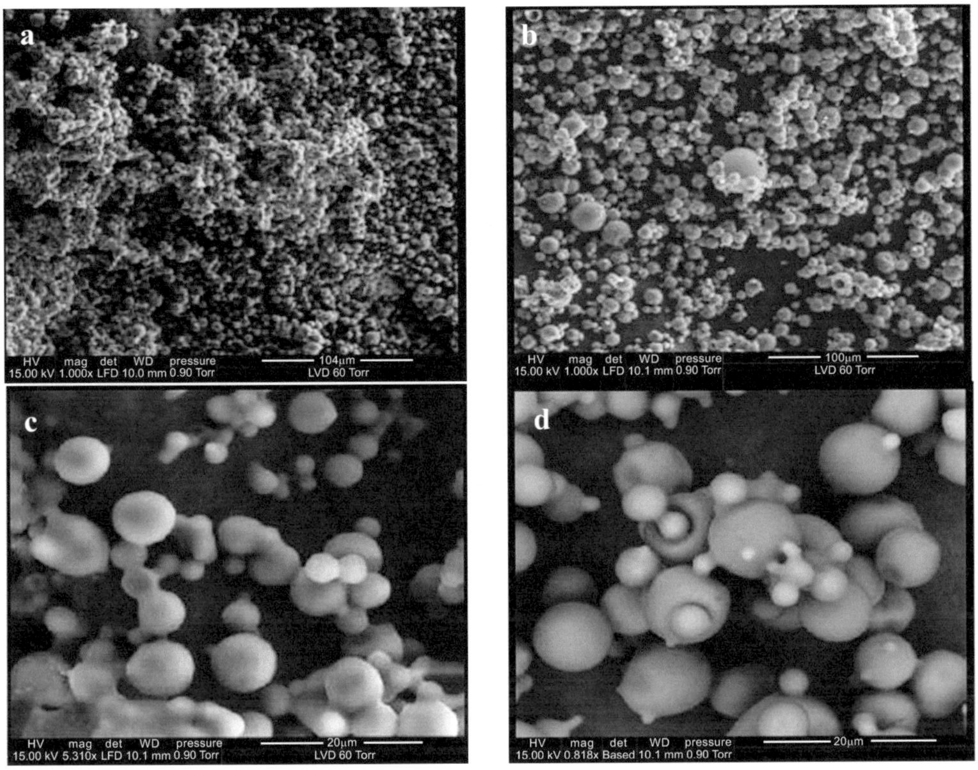

Fig. (8). Morphology of ascorbic acid microencapsulated with: **a** & **c**) Plantain maltodextrin (*Musa AAB*) (100 µm), and **b** & **d**) Green "topocho" maltodextrin (*Musa ABB*) (20 µm) [70].

Chitosan

Chitosan can be isolated from crustacean shells, insect cuticles and fungal membranes. The properties of chitosan depend on its source. Chitosan is a linear polysaccharide composed of randomly distributed chains of β-(1-4) D-glucosamine (deacetylated units) and N-acetyl-D-glucosamine (acetylated units). Also present is chitin, a monomer of chitosan, which is composed of units of N-acetylglucosamine (specifically, N-acetyl-D-glucos-2-amine), bonded by a β-(1-4) linkage [43].

Krasaekoopt *et al.* [71] showed that the survival of probiotic cultures (*Lactobacillus acidophilus* 547) in yoghurt was greater by approximately 1 log CFU when microencapsulated with chitosan compared to free (unencapsulated) cells.

Chitosan gel microcapsules can be ionically cross-linked with polyphosphates and alginate in order to improve their stability [5].

Microencapsulation with celluloses

Cellulose Acetate Phthalate

Cellulose acetate phthalate (CAP) is a thermoplastic material obtained through homogeneous and heterogeneous acetylation. This compound is derived from cellulose, which is a polydisperse linear polymer of vegetal origin composed of β-(1-4) D-glucose residues [43].

CAP is used widely as an enteric coating material for the controlled release of substances essential for the gastrointestinal tract. This is due to the fact that it is insoluble in an acidic medium (pH of 5 or less), but is soluble at a pH of 6 or more. CAP is physiologically inert when it is administered *in vivo* [72].

Various techniques have been employed for CAP-based microencapsulation. Rao *et al.* [73] reported that the microencapsulation of *Bifidobacterium pseudolongum* with CAP using the emulsion technique increased the survival of organisms (10^9 CFU mL^{-1}) when exposed to gastric fluids for a period of one hour. Cultures of *Bifidobacterium lactis* and *Lactobacillus acidophilus* microencapsulated with CAP using the spray drying method also survived well in high concentrations of salts and acid bile [72].

Microencapsulation with Proteins

The protein most commonly used for the encapsulation of food ingredients is gelatin. Its value as an encapsulating material is related to its low cost and low toxicity. Moreover, it has good film-forming potential due to its physico-chemical properties [9, 54].

Hyndman *et al.* [74] showed promising results for *Lactobacillus lactis* microencapsulated in 24% gelatin exposed to low pH values. Similarly, Guérin *et al.* [75] microencapsulated *Bifidobacterium bifidum* with alginate, pectin and whey protein, finding that the number of encapsulated organisms remained the same after 1 h at pH 2.5. Additionally, it has been reported that whey protein isolates formed extremely effective barriers that prevented the oxidation of microencapsulated orange oil [12].

Alfaro [76] reported a high retention capacity of d-limonene (approximately 100%) encapsulated with milk protein isolates under two emulsion systems homogenized by high pressure and ultrasound. This was verified by the fact that the concentrations of the degradation compounds of d-limonene (carvon and terpineol) were constant throughout the test period. Microencapsulation thus stabilized the d-limonene, as it would otherwise have degraded. Possibly for this reason the d-limonene was localized predominantly on the surface of the products rather than inside them.

Other proteins used as encapsulants are caseinates, milk whey and zein. These are highly complex biopolymers which show great diversity due to their chemistry. The spatial conformation of the proteins of all these polymers is strongly affected by the pH and ionic strength of the medium used. Thus, pH changes during the encapsulation process by coacervation (phase separation by chemical methods) produce size variations in the microparticles [9] as the expansion or contraction of the particles depends on intramolecular forces, which in turn affect the protein properties. An example of this is the decrease in the solubility of the proteins at the isoelectric point, due to the fact that at this point the loads of the proteins are neutral, thus reducing their solubility in water.

In the same way, the use of spray-dried whey proteins as a coating material influences the morphology of the microcapsules making them more spherical and with a smoother surface than polysaccharides. Proteins can thus be used to modify wall properties and particle size, either alone or in combination with carbohydrates [40]. Van der Waals forces are the principal interactions that occur between proteins and polysaccharides (amphiphilic macromolecules or biopolymers), and these depend on the pH, ionic strength, conformation, charge density and concentration of the biopolymers. These factors will determine whether complexes are formed, or if the two biopolymers are immiscibile (thermodynamically incompatibile) [77]. In many cases phase separation is observed depending on the affinity between the different polymers present and the solvent [78]. Two types of phase separation may occur when proteins and polysaccharides are mixed in an aqueous medium. The first type of phase separation is the result of strong protein-protein and polysaccharide-polysaccharide interactions, compared to the polymer-solvent interactions. The separation and segregation of the phases leads to a polysaccharide rich upper phase and a protein rich lower phase. This happens mainly if the polysaccharide-protein interactions are attractive (known as the Flory-Huggins interaction). The other type of phase separation is a product of repulsive forces between biopolymers. This may occur in the presence of a pH higher than the isoelectric point (IEP) of the protein. This is because in these conditions, both the protein and the polysaccharide could have the same (negative) charge. Reducing the pH of the mixture to below the isoelectric point of the protein, will favor interactions between the two now oppositely charged biopolymers, thus allowing the formation of a water soluble protein-polysaccharide complex.

Theoretically, there are limits to protein-polysaccharide interactions in an aqueous medium (if the protein has polymeric characteristics). The formation of the complex between the two charged biopolymers is a reversible process that depends on the pH and ionic strength of the solution that contains them. This is because the proteins could be dissociated to high values of ionic strength (0.2-0.3), or above the isoelectric point at which they form complexes with positively charged polysaccharides. Similarly, electrostatic interactions between anionic polysaccharides and positively charged proteins must occur at a pH below the isoelectric point of the proteins, otherwise the two biopolymers will be incompatible [77].

Finally, the velocity at which gravitational separation of the microcapsules occurs decreases with increased encapsulant concentration. This is because the movement of a droplet or particle is hampered by the presence of other drops. Thus, at high encapsulant concentrations gravitational separation diminishes [79].

CONCLUDING REMARKS

The term microencapsulation covers a wide variety of techniques and processes. These techniques result in microcapsules that range in size from fractions of microns to several millimeters. Microcapsules made from biopolymers offer the food industry an efficient way to protect sensitive food components, ensure against nutritional loss, utilize otherwise sensitive ingredients, incorporate unusual or time release mechanisms into the formulation, mask or preserve flavors and aromas, and transform liquids into easily handled solid ingredients. Various techniques have been employed to create microcapsules. These include spray drying, spray chilling or spray cooling, extrusion coating, fluidized-bed coating, liposome entrapment, coacervation, inclusion complexation, centrifugal extrusion, and rotational suspension separation. The selection of the appropriate microencapsulation technique and encapsulating material remains, however, a challenge since it depends on multiple factors. Thus, despite the wide range of encapsulated products that have been successfully commercialized in the pharmaceutical and cosmetic industries, microencapsulation has found a much smaller market in the food industry. Nevertheless, this technology will surely become a conventional tool in the food technologist's repertoire, since it provides a truly effective means to increase production, develop food fortification processes, incorporate neutraceuticals, improve organoleptic properties, or create novelty food products.

CONFLICT OF INTEREST

The author (editor) declares no conflict of interest, financial or otherwise.

ACKNOWLEDGEMENT

The authors would like to thank Dr. Mirian Carmona-Rodríguez.

REFERENCES

[1] B.F. Gibbs, S. Kermasha, I. Alli, and C.N. Mulligan, "Encapsulation in the food industry: a review", *Int. J. Food Sci. Nutr.*, vol. 50, no. 3, pp. 213-224, 1999. [http://dx.doi.org/10.1080/096374899101256] [PMID: 10627837]

[2] P. Sanguansri, and M.A. Augustin, "Nanoscale materials development- a food industry perspective", *Trends Food Sci. Technol.*, vol. 17, no. 10, pp. 547-556, 2006. [http://dx.doi.org/10.1016/j.tifs.2006.04.010]

[3] J. Ubbink, and J. Krüger, "Physical approaches for the delivery of active ingredients in foods", *Trends Food Sci. Technol.*, vol. 17, no. 5, pp. 244-254, 2006. [http://dx.doi.org/10.1016/j.tifs.2006.01.007]

[4] R.A. Jain, "The manufacturing techniques of various drug loaded biodegradable poly(lactide-c--glycolide) (PLGA) devices", *Biomaterials,* vol. 21, no. 23, pp. 2475-2490, 2000. [http://dx.doi.org/10.1016/S0142-9612(00)00115-0] [PMID: 11055295]

[5] C. Ragagnin, J. Smaniotto, A.J. Chicoskil, L. Queiroz, E. Jacob-Lopes, L. Martins, and N. Nascimento, "Microencapsulação de probióticos: avanços e perspectivas Ciência Rural", *Santa Maria. Online (Bergh.),* 2013.

[6] S.A. Hogan, B.F. McNamee, E.D. O'Riordan, and M. O'Sullivan, "Microencapsulating properties of whey protein concentrate", *J. Food Sci.*, vol. 66, no. 59, pp. 675-680, 2001. [http://dx.doi.org/10.1111/j.1365-2621.2001.tb04620.x]

[7] Vila Jato, "Sistemas dispersos heterogéneos. Suspensiones", In: *em Tecnología Farmacéutica: Formas Farmacéuticas,* vol. 1. Madrid, síntesis, 1997, pp. 299-316.

[8] M.E. Palomo, M.P. Ballesteros, and P. Frutos, "Microencapsulación mediante procesos físicos-químicos", In: *Tecnología Farmacéutica. Procedimientos de microencapsulación,* 1997, pp. 59-65.

[9] R. Pedroza-Islas, L. Cruz, D. Ricque-Marie, M. Tapia, M. Gaxiola, and N. Simoes, "Alimentos microencapsulados: particularidades de los procesos para la microencapsulación de alimentos para larvas de especies acuícolas", *Memorias del VI Simposio Internacional de Nutrición Acuícola,* 2002 Cancún, México

[10] G.A. Reineccius, *Carbohydrates for flavor encapsulation.* vol. 149. Food Technol: Chicago, 1991, pp. 144-146.

[11] C.S. Brazel, "Microencapsulation: offering solutions for the food industry", *Cereal Foods World,* vol. 44, no. 6, pp. 388-393, 1999.

[12] Y.D. Kim, C.V. Morr, and T.W. Schenz, "Microencapsulation properties of gum Arabic and several food proteins: liquid orange oil emulsion particles", *J. Agric. Food Chem.*, vol. 44, no. 5, pp. 1308-1313, 1996. [http://dx.doi.org/10.1021/jf950391e]

[13] T-Y. Sheu, and M. Rosenberg, "Microencapsulation by spray drying ethyl caprylate in whey protein and carbohydrate wall systems", *J. Food Sci.,* vol. 60, no. 1, pp. 98-103, 1995. [http://dx.doi.org/10.1111/j.1365-2621.1995.tb05615.x]

[14] I. Goubet, J-L. Le Quere, and A.J. Voilley, "Retention of aroma compounds by carbohydrates: influence of their physicochemical characteristics and of their physical state. A Review", *J. Agric. Food Chem.*, vol. 46, no. 5, pp. 1981-1990, 1998. [http://dx.doi.org/10.1021/jf970709y]

[15] K.S. Suslick, *Sonoluminescence and sonochemistry, Encyclopedia of physical science and technology.*

3rd ed Academic Press, Inc.: San Diego, USA, 2001.

[16] S. Avivi Levi, and A. Gedanken, "Are sonochemically prepared α-amylase protein microspheres biologically active?", *Ultrason. Sonochem.*, vol. 14, no. 1, pp. 1-5, 2007. [http://dx.doi.org/10.1016/j.ultsonch.2005.11.003] [PMID: 16403481]

[17] A. Kulmyrzaev, M.P. Sivestre, and D.J. McClements, "Rheology and stability of whey protein stabilized emulsions with high CaCl2 concentrations", *Food Res. Int.*, vol. 33, no. 1, pp. 21-25, 2000. [http://dx.doi.org/10.1016/S0963-9969(00)00019-3]

[18] R.G. Bownman, C.J. Rueb, and J.M. Finney, *"Encapsulation of oxidatively unstable compounds"*, US 2011/0020519 A1

[19] K.S. Suslick, "The chemistry of ultrasound", In: *The yearbook of science and the future.* Encyclopedia Britannica: Chicago, 1994, pp. 138-155.

[20] V. Kaushik, and Y.H. Roos, "“Limonene encapsulation in freeze-drying of gum Arabic–sucrose–gelatin systems”, LWT -", *Food Sci. Technol. (Campinas.),* vol. 40, no. 8, pp. 1381-1391, 2007.

[21] T-Y. Sheu, and M. Rosenberg, "Microstructure of microcapsules consisting of whey proteins and carbohydrates", *J. Food Sci.,* vol. 63, no. 3, pp. 491-494, 1998. [http://dx.doi.org/10.1111/j.1365-2621.1998.tb15770.x]

[22] D.K. Young, Ch.V. Morr, and T.W. Schenz, "Microencapsulation properties of gum arabic and several food proteins: Liquid orange oil emulsion particles", *J. Agric. Food Chem.,* vol. 44, no. 5, pp. 1308-1313, 1996. [http://dx.doi.org/10.1021/jf950391e]

[23] M.J. Martín, M.E. Morales, P. Gálvez, B. Clares, and M.A. Ruiz, "Desarrollo de una técnica para la microencapsulación de probióticos", *ARS Pharmaceutica,* vol. 51, no. 3, pp. 479-484, 2010.

[24] A. Matos, S. Quispe, and K. Mas, "Determinación de la eficiencia de microencapsulación de aceite esencial de orégano (*Origanum vulgare*) en β-Ciclodextrina", In: *I Congreso de Ciencia y Tecnología, Universidad Peruana Unión – UpeU,* Lima, Perú, 2011.

[25] M. Palma, *Extracción de capsantina a partir de hojuelas de pimentón (Capsicum annuum L.) y estabilización por microencapsulación.* Trabajo de Grado de Maestría, Facultad de Ciencias Químicas y Farmacéuticas, Universidad de Chile: Santiago, Chile, 2013.

[26] G. Nelson, "Application of microencapsulation in textiles", *Int. J. Pharm.,* vol. 242, no. 1-2, pp. 55-62, 2002. [http://dx.doi.org/10.1016/S0378-5173(02)00141-2] [PMID: 12176225]

[27] A. Romero, and S. Flores, "Reuso de relaves mineros como insumo para la elaboración de agregados de construcción para fabricar ladrillos y baldosas", *Industrial Data,* vol. 13, no. 2, pp. 75-82, 2014.

[28] H. Yoshizawa, *Trends in microencapsulation research.* vol. 22. Kona, 2004, pp. 23-31.

[29] D.A. Vignali, "Multiplexed particle-based flow cytometric assays", *J. Immunol. Methods,* vol. 243, no. 1-2, pp. 243-255, 2000. [http://dx.doi.org/10.1016/S0022-1759(00)00238-6] [PMID: 10986418]

[30] D. Defain, M. Murano, M. Stratico, and A. Lagomarsino, *Influencia de la temperatura en la microencapsulación de un oligopéptido altamente hidrosoluble,* 4º Jornadas de Desarrollo e Innovación, 2002.

[31] A. Yugcha, J. Kure, and P. Castillo, *Estudio del proceso de secado del látex de papaya (Carica papaya L.) deshidratado por aspersión.* Facultad de Ingeniería Mecánica y Ciencias de la Producción, Escuela Superior Politécnica del Litoral: Guayaquil, Ecuador, 2003.

[32] M. Villa, *"Sobrevivencia de Lactobacillus Acidophilus microencapsulado en función de la combinación de prebióticos y una cubierta entérica. Instituto Politécnico Nacional (IPN)", Centro de Investigaciones en Ciencia Aplicada y Tecnología (CICATA).* Trabajo Especial de Maestría: Distrito

Federal, México, 2009.

[33] S. Rojas, G.I. Osorio, and T.G. Gallardo, "Microencapsulación de aceite esencial de naranja por coacervación compleja", In: *XVIII Congreso Nacional de Ingeniería Bioquímica, VII Congreso Internacional de Ingeniería Bioquímica, X Jornadas Científicas de Biomedicina y Biotecnología Molecular, Ixtapa-Zihuatanejo Guerrero*, México, 2012.

[34] C.S. Quinde, N.V. Sánchez, and P. Castillo, *Extracción, purificación parcial y secado de la enzima bromelina obtenida a partir del corazón de la piña (Ananas comosus).* Artículo Tesis Grado, Escuela Superior Politécnica del Litoral, Facultad de Ingeniería Mecánica y Ciencias de la Producción: Guayaquil, Ecuador, 2013.

[35] V. Santacruz-Vázquez, C. Santacruz-Vázquez, and J.O. Laguna-Cortés, ""Diseño de un jugo mínimamente procesado adicionado con micro y nanocápsulas de áido fálico y su uso como vehículo para la Ingesta en ratas wistar", Rev. Mex. Ing", *Quím.*, vol. 12, no. 2, pp. 177-191, 2013.

[36] L. Vidal, M. Avello, C. Loyola, J. Campos, P. Aqueveque, S.R. Dungan, M. Galotto, and A. Guarda, "Microencapsulation of maqui (Aristotelia chilensis [Molina] Stuntz) leaf extracts to preserve and control antioxidant properties", *Chil. J. Agric. Res.*, vol. 73, no. 1, pp. 17-23, 2013. [http://dx.doi.org/10.4067/S0718-58392013000100003]

[37] C.L. Cooney, A.E. Humphrey, J. Klein, and K.D. , "Vorlop, Immobilization techniques: cells", *Comprehensive biotechnology.*, pp. 542-550, 1985.

[38] P. Audet, C. Paquin, and C. Lacroix, "Batch fermentations with a mixed culture of lactic acid bacteria immobilized separately in κ-carrageenan locust bean gum gel beads", *Appl. Microbiol. Biotechnol.*, vol. 32, no. 6, pp. 662-668, 1990. [http://dx.doi.org/10.1007/BF00164736]

[39] P. Audet, C. Paquin, and C. Lacroix, "Effect of medium and temperature of storage on viability of lactic acid bacteria immobilized in κ-carrageenan-locust bean gum gel beads", *Biotechnol. Tech.*, vol. 5, no. 4, pp. 307-312, 1991. [http://dx.doi.org/10.1007/BF02438669]

[40] M.J. Martín, M.E. Morales, V. Gallardo, and M.A. Ruiz, "Técnicas de microencapsulación: una propuesta para microencapsular probióticos", *ARS Pharmaceutica,* vol. 50, no. 1, pp. 43-50, 2009.

[41] D. Martínez, *"Efectos de la administración del probiótico Shewanella pdp 11 en juveniles de lenguado senegalés (Solea senegalensis, KAUP 1858) y desarrollo de un método para su microencapsulación", Tesis de Ingeniero Agrónomo, Escuela Superior de Ingeniería.* España: Universidad de Almería, Almería, 2011.

[42] A.K. Anal, and W.F. Stevens, "Chitosan-alginate multilayer beads for controlled release of ampicillin", *Int. J. Pharm.,* vol. 290, no. 1-2, pp. 45-54, 2005. [http://dx.doi.org/10.1016/j.ijpharm.2004.11.015] [PMID: 15664129]

[43] A.K. Anal, and K. Singh, "Recents advances in microencapsulation of probiotics for industrial applications and targeted delivery", *Trends Food Sci. Technol.,* vol. 18, no. 5, pp. 240-251, 2007. [http://dx.doi.org/10.1016/j.tifs.2007.01.004]

[44] A. Sohail, M.S. Turner, A. Coombes, T. Bostrom, and B. Bhandari, "Survivability of probiotics encapsulated in alginate gel microbeads using a novel impinging aerosols method", *Int. J. Food Microbiol.,* vol. 145, no. 1, pp. 162-168, 2011. [http://dx.doi.org/10.1016/j.ijfoodmicro.2010.12.007] [PMID: 21276627]

[45] P. Rosas-Ledesma, J.M. León-Rubio, F.J. Alarcón, M.A. Moriñigo, and M.C. Balebona, "Calcium alginate capsules for oral administration of fish probiotic bacteria: assessment of optimal conditions for encapsulation", *Aquacult. Res.,* vol. 43, no. 1, pp. 1-11, 2011.

[46] M.A. Moriñigo, V. Sánchez, and T.F. Martínez, "Encapsulation of a bacterial fish probiotic in alginate beads: protective effect under *in vitro* simulations of fish gastric conditions", *Abstracts of the International Conference on Fish Diseases and Fish Immunology,* , 2008

[47] L.T. Hansen, P.M. Allan-Wojtas, Y.L. Jin, and A.T. Paulson, "Survival of Ca-alginate microencapsulated Bifidobacterium spp. in milk and simulated gastrointestinal conditions", *Food Microbiol.,* vol. 19, no. 1, pp. 35-45, 2002. [http://dx.doi.org/10.1006/fmic.2001.0452]

[48] C. Saénz, S. Tapia, J. Chávez, and P. Robert, "Microencapsulation by spray drying of bioactive compounds from cactus pear (Opuntia ficus-indica)", *Food Chem.,* vol. 114, no. 2, pp. 616-622, 2009. [http://dx.doi.org/10.1016/j.foodchem.2008.09.095]

[49] D.L. Kaplan, *Biopolymers from renewable resources.* Springer: Berlín, 1998. [http://dx.doi.org/10.1007/978-3-662-03680-8]

[50] P. Trubiano, and N. Lacourse, "Emulsionstabilizing starches", *ACS Symposium Series,* vol. 370, , 1998pp. 45-54 [http://dx.doi.org/10.1021/bk-1988-0370.ch006]

[51] A. Soottitantawat, K. Takayama, K. Okamura, D. Muranaka, H. Yoshii, T. Furuta, M. Ohkawara, and P. Linko, "Microencapsulation of lmenthol by spray drying and its release characteristics", *Innov. Food Sci. Emerg.,* vol. 6, no. 2, pp. 163-170, 2005. [http://dx.doi.org/10.1016/j.ifset.2004.11.007]

[52] J. Shaikh, R. Bhosale, and R. Singhal, "Microencapsulation of black pepper oleoresin", *Food Chem.,* vol. 94, no. 1, pp. 105-110, 2006. [http://dx.doi.org/10.1016/j.foodchem.2004.10.056]

[53] D. Kanakdande, R. Bhosale, and R.S. Singhal, "Stability of cumin oleoresin microencapsulated in dierent combination of gum arabic, maltodextrina and modified starch", *Carbohydr. Polym.,* vol. 67, no. 4, pp. 536-541, 2007. [http://dx.doi.org/10.1016/j.carbpol.2006.06.023]

[54] F. Shahidi, and X-Q. Han, "Encapsulation of food ingredients", *Crit. Rev. Food Sci. Nutr.,* vol. 33, no. 6, pp. 501-547, 1993. [http://dx.doi.org/10.1080/10408399309527645] [PMID: 8216812]

[55] L. Marchal, H. Beeftink, and J. Tramper, "Towards a rational design of commercial maltodextrins", *Trends Food Sci. Technol.,* vol. 10, no. 11, pp. 345-355, 1999. [http://dx.doi.org/10.1016/S0924-2244(00)00018-2]

[56] P. Finotelli, and M. Rocha-Leao, "Microencapsulation of ascorbic acid in maltodextrin and capsul using spray drying", *2nd Mercosur Congress on Chemical Engineering, 4th Mercosur Congress on Process Systems Engineering,* 2005pp. 1-11

[57] W. Kolanowski, G. Laufenberg, and B. Kunz, "Fish oil stabilisation by microencapsulation with modified cellulose", *Int. J. Food Sci. Nutr.,* vol. 55, no. 4, pp. 333-343, 2004. [http://dx.doi.org/10.1080/09637480410001725157] [PMID: 15369987]

[58] R. Jonsdottir, M. Bragadottir, and G. Arnarson, "Oxidatively derived volatile compounds in microencapsulated fish oil monitored by solid-phase microextraction (SPME)", *J. Food Sci.,* vol. 70, no. 7, pp. 433-440, 2005. [http://dx.doi.org/10.1111/j.1365-2621.2005.tb11465.x]

[59] R. Crittenden, A. Laitila, P. Forssell, J. Mättö, M. Saarela, T. Mattila-Sandholm, and P. Myllärinen, "Adhesion of bifidobacteria to granular starch and its implications in probiotic technologies", *Appl. Environ. Microbiol.,* vol. 67, no. 8, pp. 3469-3475, 2001. [http://dx.doi.org/10.1128/AEM.67.8.3469-3475.2001] [PMID: 11472921]

[60] B.Z. Li, L.J. Wang, D. Li, B. Bhandari, S.J. Li, Y. Lan, X.D. Chen, and Z.H. Mao, "Fabrication of starch-based microparticles by an emulsification-crosslinking method", *J. Food Eng.,* vol. 92, no. 3, pp. 250-254, 2009. [http://dx.doi.org/10.1016/j.jfoodeng.2008.08.011]

[61] C. Iyer, and K. Kailasapathy, "Effect of co-encapsulation of probiotics with prebiotics on increasing

the viability of encapsulated bacteria under in vitro acidic and bile salt conditions and in yogurt", *J. Food Sci.,* vol. 70, no. 1, pp. 18-23, 2005.
[http://dx.doi.org/10.1111/j.1365-2621.2005.tb09041.x]

[62] K. Sultana, G. Godward, N. Reynolds, R. Arumugaswamy, P. Peiris, and K. Kailasapathy, "Encapsulation of probiotic bacteria with alginate-starch and evaluation of survival in simulated gastrointestinal conditions and in yoghurt", *Int. J. Food Microbiol.,* vol. 62, no. 1-2, pp. 47-55, 2000.
[http://dx.doi.org/10.1016/S0168-1605(00)00380-9] [PMID: 11139021]

[63] M. Calvo, *Oligosacaridos,* 2010. Available at: http://milksci.unizar.es/bioquimica/temas/azucares/disacaridos.html

[64] B. Wong, "Micorencapsulation of amino acid for prawn feed additives", In: *Department of chemi-cal Engineering the University of Queensland,* 1998, pp. 3-5.

[65] B. Brime, "Desarrollo y caracterización farmacotécnica y farmacológica de nuevas formulaciones parentales de anfotericina B", In: *España: Universidad Complutense de Madrid,* 2002, pp. 6-15.

[66] E. Leyva, E. Moctezuma, R. Leyva, and S. Oros, "“Estudios de los complejos de inclusión de ácido nalidíxico y ácido oxolínico con ciclodextrinas”, México", *Sociedad Química de México,* vol. 48, no. 3, pp. 189-195, 2004.

[67] J. Moreno, "Nuevos diseños galénicos de comprimidos de indometacina y tetraciclina con agentes hidrotrópicos y de inclusión para mejorar la velocidad de disolución y su estabilidad", In: *España: Universidad Complutense de Madrid,* 2004, pp. 11-19.

[68] L. Szente, and J. Szejtli, "Cyclodextrins as food ingredients", *Trends Food Sci. Technol.,* vol. 15, no. 3-4, pp. 137-142, 2003.
[http://dx.doi.org/10.1016/j.tifs.2003.09.019]

[69] J. Ayala, M. Villegas, and G. González, *¿La adición de compuestos antimicrobianos naturales afecta la calidad sensorial de los frutos frescos cortados?,* 2007. Available at: www.alfa-editores.com/web

[70] R. Guzmán, *Estudio integral de dos variedades de Musas (Musa Spp) I. Elaboración de productos a base de harinas II. Producción de maltodextrinas y uso en microencapsulación de antioxidantes.* Tesis Doctoral, Instituto de Ciencias y Tecnología de Alimentos, Facultad de Ciencias, Universidad Central de Venezuela: Caracas, Venezuela, 2011.

[71] W. Krasaekoopt, B. Bhandari, and H. Deeth, "The influence of coating materials on some properties of alginate beads and survivability of microencapsulated probiotic bacteria", *Int. Dairy J.,* vol. 14, no. 8, pp. 737-743, 2004.
[http://dx.doi.org/10.1016/j.idairyj.2004.01.004]

[72] C.S. Fávaro-Trindade, S.C. de Pinho, and G.A. Rocha, "Revisão: microencapsulação de ingredientes alimentícios", *Braz. J. Food. Technol.,* vol. 11, no. 2, pp. 103-112, 2008.

[73] A.V. Rao, N. Shiwnarain, and I. Maharaj, "Survival of microencapsulated Bifidobacterium pseudolongum in simulated gastric and intestinal juices", *Can. Inst. F. Sci. Tec. J.,* vol. 22, no. 4, pp. 345-349, 1989.
[http://dx.doi.org/10.1016/S0315-5463(89)70426-0]

[74] C.L. Hyndman, A.F. Groboillot, D. Poncelet, C.P. Champagne, and R.J. Neufeld, "Microencapsulation of Lactococcus lactis within cross-linked gelatin membranes", *J. Chem. Technol. Biotechnol.,* vol. 56, no. 3, pp. 259-263, 1993.
[http://dx.doi.org/10.1002/jctb.280560307]

[75] D. Guérin, J.C. Vuillemard, and M. Subirade, "Protection of bifidobacteria encapsulated in polysaccharide-protein gel beads against gastric juice and bile", *J. Food Prot.,* vol. 66, no. 11, pp. 2076-2084, 2003.
[http://dx.doi.org/10.4315/0362-028X-66.11.2076] [PMID: 14627286]

[76] M. Alfaro, *Desarrollo de metodologías de encapsulación utilizando aislado proteico de suero lácteo modificado con azúcares para la protección de ingredientes activos en alimentos.* Tesis Doctoral,

Instituto de Ciencias y Tecnología de Alimentos, Facultad de Ciencias, Universidad Central de Venezuela: Caracas, Venezuela, 2012.

[77] A. Benichou, A. Aserin, and N. Garti, "Protein-polysaccharide interactions for stabilization of food emulsions", *J. Dispers. Sci. Technol.,* vol. 23, no. 1-3, pp. 93-123, 2002. [http://dx.doi.org/10.1080/01932690208984192]

[78] J-L. Doublier, C. Garnier, D. Renard, and C. Sanchez, "Protein-polysaccharide interactions", *Curr. Opin. Colloid In.,* vol. 5, no. 3-4, pp. 202-214, 2000. [http://dx.doi.org/10.1016/S1359-0294(00)00054-6]

[79] D.J. McClements, E.A. Decker, and J. Weiss, "Emulsion-based delivery systems for lipophilic bioactive components", *J. Food Sci.,* vol. 72, no. 8, pp. R109-R124, 2007. [http://dx.doi.org/10.1111/j.1750-3841.2007.00507.x] [PMID: 17995616]

CHAPTER 7

Gels in Biomedical Applications: An Overview on Wound Healing and Tissue Engineering

M. Fernanda Horst* and **Mariela Agotegaray**

South Institute of Chemistry (INQUISUR), National Council of Scientific and Technical Research (CONICET), National University of South (UNS), Bahía Blanca, Buenos Aires, Argentina

Abstract: Biopolymer hydrogels present diverse applications in medicine due to their biocompatibility, biodegradability and low immunogenicity. The specific features related to swelling, holding a large amount of water while maintaining the structure, in addition to the ease of taking different shapes, make them the selected biomaterials as component for diverse bio-applications: contact lenses, injectable or implantable drug delivery devices and as platforms intended for wound healing and tissue engineering. In general, the restriction associated to the use of hydrogels in biomedicine lies in the poor mechanical properties associated to the natural biopolymers. This problem can be solved by the use of other materials during the synthesis procedure. In this chapter, general synthesis methodologies and latest innovations in terms of gelatin, collagen and hyaluronic acid hydrogels for wound care and tissue engineering are reviewed. The selection of the biopolymers is based on their suitable features for biomedical applications. The focusing of specific clinical challenges for wound healing and tissue engineering can prove to be beneficial for rapid development in science and marketing. This implies considering the current increasing market associated to hydrogel employment for wound care and treatment as well as the requirement to develop concise clinical hydrogels implementation on the replacement of diverse tissue and organs.

Keywords: Biopolymers, Collagen, Gelatin, Gels, Hyaluronic acid, Synthesis and characterization, Tissue Engineering, Wound healing.

INTRODUCTION

Natural polymers have long been studied and widely used in the last decades in many applications issues such as food and pharmaccutics industries as well as in medicine field. In the recent years, they have become promissory materials for

* **Corresponding author M. Fernanda Horst:** South Institute of Chemistry, National University of South (INQUISUR-CONICET-UNS), Bahía Blanca, Buenos Aires, Argentina; Tel: (54) 2914595101; Int: 3556; Fax: (54) 2914595460; E-mail: mfhorst@uns.edu.ar

biomedical applications owing to their biodegradability, biocompatibility, non-toxicity and specific therapeutic activities [1]. Biopolymers contain different functional groups, such as hydroxyl, amino, carboxylic acid, aldehydes, and others which make them ideal for conjugation. The study of biopolymers-based hydrogels and the possibility of their employment in living tissues begun analyzed in the early 1960s by Wichterle and Lim who proposed the design of new biomaterials for ophthalmology [2].

In general terms, gels consist of a solid three-dimensional network that extends the volume of a liquid medium and entangles through surface tension effects. The gels could contain crystallites or other junctions that remain intact within the extending fluid. For all practical purposes, any fluid can be used as an extender including water, oil and air. In the case when the extender is water the gels are usually known as *hydrogels* and when the extender is air, they are named as *aerogels*.

At that point, hydrogels consist of three-dimensional, hydrophilic, polymeric networks with soft and rubbery consistence being thus similar to living tissues. They are capable to retain large amounts of water or biological fluids, so they can absorb from 20% up to thousands of times their dry weight in water. They may be chemically stable or could degrade and eventually disintegrate and dissolve [3, 4].

Softness and rapid diffusion ability of molecules make them useful for a wide variety of bio-applications (Fig. **1**). Soft contact lenses made of hydrogels became commercially successful, since these polymeric systems allow controlling all necessary features through the molecular design. Gels composed by diverse polymers such as poly(2-hydroxyethyl methacrylate) and its copolymers with methacrylic acid; 1-vinyl-2-pyrrolidone and its copolymers with 2-hydroxyethyl methacrylate and alkyl methacrylates; and poly(vinyl alcohol) and its derivatives are considered unique materials for the development of contact lenses [5].

Hydrogels are extensively employed in the development of drug delivery systems (DDS) due to their common properties related to swelling and holding a large amount of water while maintaining the structure. Moreover, they present a controllable diffusion rate through the induction of changes in the crosslink density [6]. These features make them suitable for drug protection from hostile environments in the body, such as the presence of enzymes and low pH in stomach, and also to control drug release [7]. Hydrogels present different devices forms as DDS: injectable or implantable depots and nanogels. The latest trend is the development of smart-hydrogels: environmentally sensitive gels which respond to pH or temperature changes in the body and self-regulated DDS which respond to specific stimuli, for instance, self-regulated insulin delivery hydrogels sense glucose concentration to control swelling to release the hormone [8].

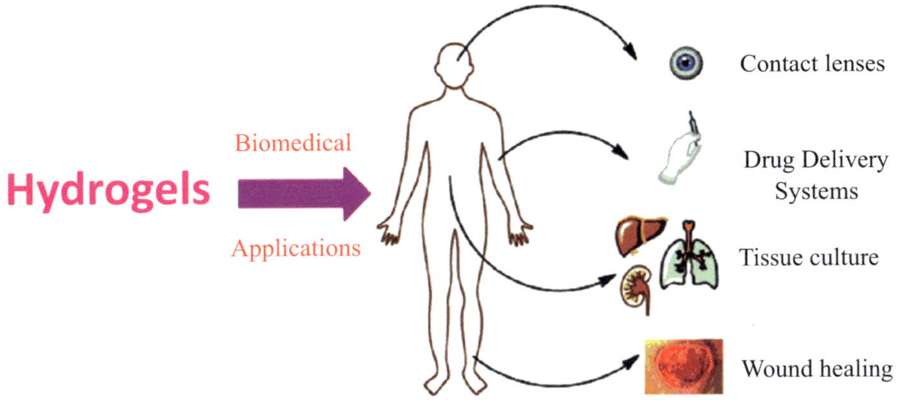

Fig. (1). General biomedical applications of hydrogels.

There are two areas of interest in biomedicine where hydrogels application is of greatest importance: wound healing and tissue engineering. The skin lesions induce a cascade of processes leading to re-epithelialization and re-establishment of skin function. If this process misleads, it becomes a stagnation of the repair process causing a chronic wound. Therefore, hydrogels arise as a malleable alternative for wound healing. In tissue engineering insights, hydrogels use attempts to mimic extracellular matrix and lead to the replacements of organs or implantations.

Wide and diverse biopolymers have been used to fabricate hydrogels for biomedical applications. This chapter focuses on the latest trends and innovations of collagen, gelatin and hyaluronic acid hydrogels for wound healing and tissue engineering.

HYDROGELS: SYNTHESIS AND PHYSICOCHEMICAL PROPERTIES

Collagen, gelatin and hyaluronic acid among others, are protein and polysaccharides widely used and investigated to obtain hydrogels for wound healing and tissue engineering applications. In this section, we present a brief description of the protein and polysaccharides nature, and chemical structure and describe the synthesis methodologies and their physicochemical characteristics taking into account the biomedical applications.

Chemical Structures of Collagen, Gelatin and Hyaluronic Acid

Collagen is the most abundant protein constituting more than one-third of weight of total body protein present in mammals. This protein has a complex hierarchical conformation divided in four structures: i) the primary structure constituted by amino acids triplet, ii) the secondary structure is α-helix, iii) tertiary structure, a triple helix and iv) the last structure consists of fibrils [9]. A scheme of its chemical structure is depicted in Fig. (**2**). This natural substance can be found in all the connective tissues in human body: bones, skin, *etc.* and it is crucial in providing structure for all of the internal organs. Collagen can be processed in different physical forms due to its technological application. Regarding biomedical applications, collagen hydrogels have been widely used for wound healing [10] and tissue engineering [11]. In terms of tissues repair and regeneration, collagen has been widely used for bone and teeth reconstruction [12].

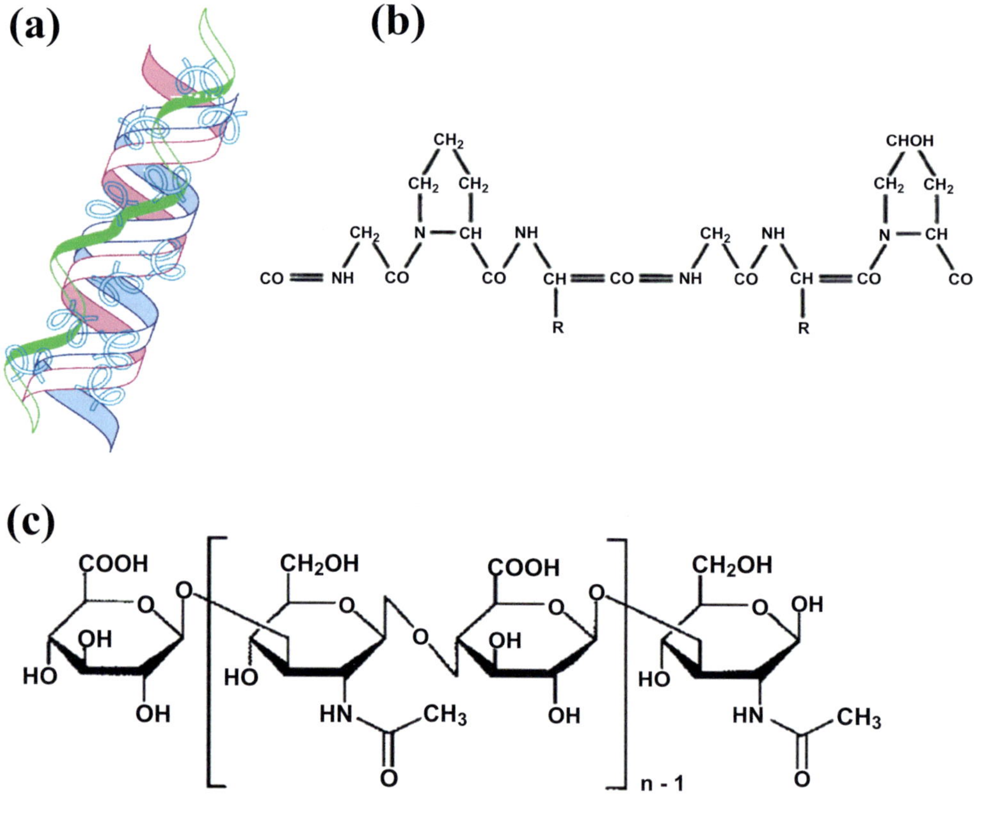

Fig. (2). Scheme of chemical structure of (a) collagen, (b) gelatin and (c) hyaluronic acid.

Gelatin (GE) derived from collagen, and its composition is a mixture of peptides and proteins. Chemical structure is displayed in Fig. (**2**). As a protein based biomaterial, gelatin is biodegradable and has excellent biocompatibility.

Therefore, it has received great attention for biomedical applications [13, 14]. In addition, gelatin can be crosslinked or modified with the inclusions of other materials to significantly alter its mechanical and biomedical properties.

Hyaluronic acid (HA) also called hyaluronan or hyaluronate, is an anionic nonsulfated glycosaminoglycan distributed widely throughout connective, epithelial and neural tissues. The chemical structure of hyaluronic acid is based on a long chain composed of disaccharide repeating units containing D-glucuronic acid and N-acetyl-D-glucosamine. In Fig. (**2**), HA chemical structure can be seen. The average molecular weight could reach 8.106Da, 107Da or 108Da, due to this high molecular weight and strong intermolecular forces. Hyaluronic acid aqueous solutions are highly viscous and can retain high amount of water [15]. Owing to unique HA properties, it has become important for many medical and esthetical applications [16].

Hydrogels: Classification and Synthesis Methodology

Regarding the chapter scope, in this section, we describe hydrogel characteristics as function of their biomedical applications and specifically the synthesis procedure to obtain hydrogels for wound healing and tissue engineering.

In general terms, hydrogels can be classified based on:

I. *Crosslinking.* They are divided in two groups: physical and chemical hydrogels. This classification is based on procedure employed for network synthesis, the crosslinking mechanism. In physical hydrogels, the nature of crosslinking comes from hydrophobic association, chain aggregation, crystallization, chain complexation or hydrogen bonding. Due to the weak nature of the forces between different networks, these hydrogels are also known as reversible and they could disintegrate and dissolve in water. On the other hand, in chemically crosslinked hydrogels, strong covalent bonding exists between the different hydrogels polymeric networks. They are also called as permanent hydrogels and present suitable resistance properties, however degradation can be slower. Fig. (**3**) displays a scheme of physical and chemical crosslinked hydrogels. The applications of physical hydrogels are limited due to lack of thermal, pH, or salt stability, since they are formed through hydrogen bonding or intermolecular electrostatic interactions. Therefore, chemically crosslinked hydrogels have greatly broadened the scope of polysaccharides hydrogels and their applications in huge and diverse fields, but more precisely in wound healing and tissue engineering [17].

II. *Physical appearance.* The hydrogels can be obtained as matrix, films or microspheres depending on synthesis procedure. Moreover, the physical appearance of the hydrogels is directly related to the specific biomedical application. Regarding wound healing applications, films hydrogels may be suitable for the treatment of these diseases. With regard to tissue engineering, injectable hydrogels and/or microspheres are better as junctions in the body, recovering material of tissue injured.

III. *Network electrical charge.* Hydrogels may be categorized on the basis of absence or presence of electric charge located in crosslinked chains. The groups are: nonionic (neutral), ionic (cationic or anionic), amphoteric electrolyte containing acidic and basic groups, and zwitterionic with both anionic and cationic groups in each structural repeating unit.

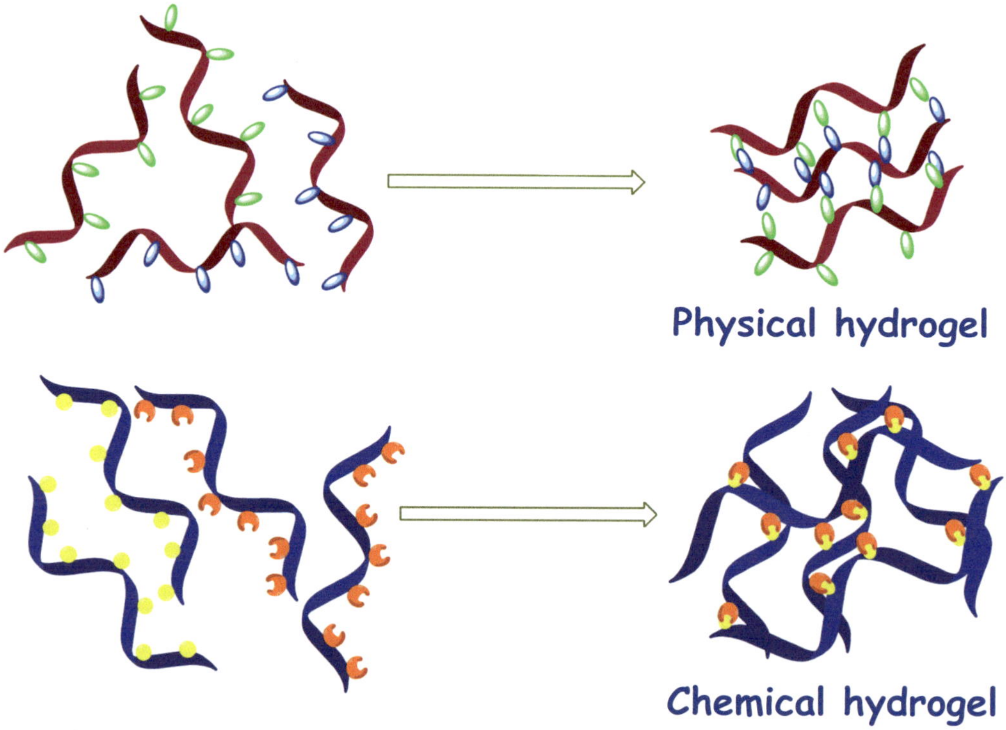

Fig. (3). Scheme of chemical and physical crosslinked hydrogel.

Considering wound healing, hydrogels should act as a moisture dressing material with the ability to absorb and retain exudates along with foreign bodies, such as bacteria within its network structure. So it is important to design optimum hydrogels in terms of water absorption capability, biocompatibility for tissues, and biodegradability at ideal rate corresponding to the rate of new tissue formation, nontoxicity and non-immunogenicity. Besides, hydrogel should have optimal mechanical properties and adequate porosity and morphology for transporting cells, gaseous metabolites, nutrients and molecules both within the biomaterial and between the hydrogel and local environment [18].

The importance of collagen in wound healing has been appreciated for a long time for the simple reason that the final result of most repairs in body is the formation of scar tissue composed of collagenous fibres. However, wound healing and regeneration are principally, developing processes involving a variety of phenomena such as cell proliferation, cell migration, cell differentiation and interactions between the different tissue components where collagen may affect healing not only at its final stage, but also in the very early stages of healing. The hydrophilic nature of collagen due to its molecular structure characterized by high content of diamine dicarboxylic amino acids and carbohydrate moieties provides a surface geometry suitable for cell adhesion. In this sense, collagen hydrogels appear as promising biomaterials for skin repair. Unfortunately, their extensive contraction and their poor mechanical properties constitute major disadvantages toward their utilization as wound healing management. So, many studies have analyzed the microstructural features and bulk stiffness of collagen hydrogels [19 - 22]. However, these studies have employed diverse collagen hydrogel composition, synthesis temperature and pH, which alter physiological properties. These modifications hinder the hydrogels application for wound healing, as well as the analysis of the impact of these parameters on the cell response.

Gelatin is considered to form hydrogels for wound healing taking into account the fact that its protein nature turns it as hemostatic agent. At 40°C, approximately gelatin melts and changes into gels when the solution is cooled at room temperature. The sol-gel transformation is due to a conformational disorder-order transition of the gelatin chains which form thermo-reversible networks by associating helices in junction zones stabilized by hydrogen bonds [23]. Nevertheless, gelatin hydrogel presents two important disadvantages such as: 1- it dissolves in water at physiological temperature and 2- its poor mechanical properties. To overcome these features, crosslinking treatment is an alternative to maintain the demanded scaffolds shape. Thus, physical and chemical crosslinking techniques have been developed. Physical methods include irradiation with UV light, visible light, electron beam or γ rays, as well as by thermal treatment, by the formation polyion complexes with other polymers or by gelling biopolymers [24]. Moreover, gelatin readily undergoes chemical crosslinking due to the large number of functional side groups. This is usually performed using bifunctional reagents such as glutaraldehyde and diisocyanates, as well as carboiimides, polyepoxy compounds and acyl azide methods [25]. As glutaralgehyde may lead to apparent cytotoxicity, several efforts have been developed to substitute glutaraldehyde with natural and low-toxic crosslinking reagents such as genipin. This naturally occurring crosslinking agent can be isolated from Gardenia fruits. It was studied that gelatin hydrogels crosslinked with genipin demonstrated good stability behavior and mechanical, thermal and swelling properties close to those obtained for glutaraldehyde gelatin crosslinked hydrogel, which is by far less cytotoxic [26 - 28].

Finally, last year's hyaluronic acid hydrogels have been in the focus for skin diseases treatment [29, 30]. The degradation and synthesis of hyaluronic acid are essential part for extracellular matrix maintenance and may be critical for appropriate wound healing. Nevertheless, the poor mechanical properties of hyaluronan and its rapid degradation *in vivo* hinder its use for wound healing. In order to obtain better mechanically resistant materials, chemical crosslinking strategies can be employed to engineer insoluble hyaluronic acid hydrogel with specified mechanical properties and resorption rates suitable for the proposed clinical use. Because of high functionality and charge density of hyaluronic acid, it can be cross-linked by a variety of physical and chemical methods. Many derivatives of hyaluronic acid have been prepared by modifying its functional group, which improves its properties and potentiates its application as hydrogel for wound healing treatment [31 - 33].

Regarding tissue engineering, the principle is to provide a functional biological tissue equivalent to replace tissue lost by disease, congenital abnormalities or traumatic events. In this sense, hyaluronic acid, gelatin and collagen hydrogels scaffolds have been studied in tissue engineering. Cells can be encapsulated during the hydrogel formation process and hydrogels may be formed *in situ* within a defect site in the body, facilitating application to tissue engineering strategies. Despite the advantages of hydrogels, the widespread use of them as an ideal cell culture system in tissue engineering is limited due to their poor mechanical properties, and weak and brittle nature upon dehydration [34]. Physical properties such as hydrogels porosity can be modulated by altering the crosslinking density which in effect changes the volume water absorbed in the hydrogel and the mechanical properties. These physical properties are important for tissue engineering applications as water content influences the viability of cells encapsulated in the hydrogel and the rate of tissue development. To overcome these limitations, biopolymer hydrogels can be formed by covalent crosslinking, chemical conjugation, esterification and polymerization. In addition, they have been combined with other polysaccharides or proteins to form an interpenetrating network or composite hydrogels [35 - 37]. Furthermore, other simple and promising way to enhance not only bioactivity but also mechanical properties of hydrogels is the incorporation of inorganic nanoparticles, such as hydroxyapatite, bioactive glass or carbon nanotubes into hydrogel matrix [12, 38, 39].

Summarizing in Table **1** describes some works based on polysaccharides and proteins hydrogels for tissue engineering applications.

Table 1. Natural polymer Hydrogels for tissue engineering.

Natural polymer	Application	Ref.
Hyaluronic acid/gelatin	Cartilage tissue engineering	[40]
hyaluronic acid/gellan gum/calcium chloride	Osteochondral defects reparation	[41]
Alginate/collagen	Bone tissue regeneration	[42]
Collagen/silica nanoparticles	Scaffold for bone tissue engineering	[43]
Collagen/carbonated/hydroxyapatite	Bone tissue regeneration	[44]
Collagen/nanosilicate	Bone tissue engineering	[45]

PROPERTIES AND CHARACTERIZATION OF HYDROGELS

This section presents most required characteristics of hydrogel for biomedical applications with emphasis on wound healing and tissue engineering. In order to design hydrogel with desired performance and structure, characterization of hydrogel network parameters is of great significance. The common techniques employed to provide the insight into internal structure of hydrogels are infrared spectroscopy, X-ray diffraction analysis (DRX), atomic force microscopy, and electron microscopy.

Infrared spectroscopy (IR) is one of the most widely used techniques for identifying the chemical structure of natural polymers to measure vibrational energy transitions, yielding information about the types of chemical bonds, the atoms involved and local chemical environment present within a material and on the surface [46].

Some natural polymers have mainly amorphous structure, but some of the hydrogels when prepared by freeze-thawing method present crystalline parts in their structure known as semi-crystalline. This structure can be identified by DRX. Hydrogels with different "degree of crystallinity" display difference in mechanical and chemical properties, which are often represented as crystallinity percentage or crystalline/amorphous ratio.

Atomic force microscopy (AFM) is a useful technique to monitor the surface of materials. This methodology opened whole new horizons for biomaterials characterization as it is able to capture images of non-conducting materials. The main advantage of this method is the non-destructive nature [47].

Another useful technique is differential scanning calorimetry (DSC). It is a sensitive method to measure conformational transitions of polymers as a function of temperature through the changes in heat capacity.

It serves as a tool for studying glass transition temperature, crystal structure and crystal transition temperature. Generally, this method can be used to investigate crystalline nature of hydrogels, especially in hydrogels prepared by freezing-thawing processes. Also, DSC methodology can be applied to determine crystallinity degree and crystal size distributions of samples in the initial state (before swelling) and at different times during swelling.

Optical methods have been applied to determine hydrogels microstructure, yielding a three-dimensional structure image. In this sense, electron microscopies, particularly scanning electron microscopy (SEM) and transmission electron microscopy (TEM), are two methods often used for gel dispersions analysis. From micrograph images, it should be possible to evaluate size, shape and size distribution of the particles. The main feature of electron microscopy is, its ability to directly observe inter-particle bridge formation, and in some cases anomalous particle formation and presence of smaller particles fraction resulting from secondary nucleation [48]. Nevertheless, this methodology cannot be used in aqueous phase. Thus the material should be dry, and therefore it can be observed in the collapsed state only. Besides, these methods are not suitable for accurate measurement of swollen diameters. Also, swollen gel may undergo aggregation under conditions of reduced pressure and electron beam irradiation. It has been proved that SEM can be applied to the study of network morphology.

One of the most important characterizations of hydrogels is the water content. Hoffman [49] well described hydrogels characteristics for biomedical applications. The role of water in a hydrogel can determine the overall permeation of nutrients into and cellular products out of the hydrogel. When a dry hydrogel begins to absorb water, the first water molecules entering the matrix will hydrate the hydrophilic groups (primary bound water). As the polar groups are hydrated, the network swells, and exposes hydrophobic groups, which also interact with water molecules, leading to hydrophobic-bound water (secondary bound water). Primary and secondary bound water are often combined and called the 'total bound water'. After this, the network will absorb additional water due to osmotic driving force of network chains towards infinite dilution. Thus, the hydrogel will reach an equilibrium swelling level. The additional swelling water which is absorbed after ionic, polar and hydrophobic groups becomes saturated with bound water is named 'free water' or 'bulk water', and it is assumed to fill the space between the network chains and/or the center of large pores, macro-pores or voids. As the network swells, if network chains or crosslinks are degradable, the gel will begin to disintegrate, at a rate depending on its composition. It should be noted that a hydrogel used as a tissue engineering matrix may never be dried, but the total water in the hydrogel is still comprised of 'bound' and 'free' water. Therefore it is important to design the hydrogel with required features for wound healing or tissue engineering applications.

The water absorbed in a hydrogel will determine the mobility and absorption of entities within the hydrogel. This feature is directly related with the presence of pores in hydrogel structure. Pores may be formed during the hydrogel synthesis. The average pore size, the pore size distribution, and the pore interconnections are important factors of a hydrogel matrix that usually called the parameter 'tortuosity'. This parameter is directly related to the release of animal or human cells encapsulated to the water content. Considering wound healing, hydrogels can clean wounds by removing necrotic and infected tissue from the wound bed. And due to the high water content and hydrogel consistency, it also feels cooling and soothing on the wounded area.

So hydrogels act as a wet wound dressing material and have the ability to absorb and retain the wound exudates along with the foreign bodies, such as bacteria, within its network structure. In addition, hydrogels could be employed to deliver topical wound drugs (*e.g.*, metronidazole and silver sulfadiazine). Regarding tissue engineering, hydrogels should contain pores large enough to accommodate living cells, or they may be designed to dissolve or degrade away, releasing growth factors and creating pores into which living cells may penetrate and proliferate. The discussion of this property is not concerned to this part of the book. This issue will be addressed in the following chapters.

HYDROGELS FOR WOUND HEALING

Skin injuries and wounds trigger a physiological cascade of processes aimed to re-epithelialization of the damaged area to re-establish the functionality [50]. Injury is a general term referring to harm in the body caused by accidents, falls, hits, *etc*. Wounds result from physical (surgery, sutures and stitches), thermal damage or the presence of an underlying medical or physiological condition.

The physiological wound healing process is described in a sequence of four events:

1. In the *Homeostasis Phase* the initial reparative coagulum is formed. Vasoconstriction occurs and platelets adhere to the damaged endothelium to promote thrombocyte clumping. Then, vasodilatation comes so to allow influx of more thrombocytes and other cells such as white blood cells. These

components release numerous factors which promote collagen degradation, collagen-genesis and angiogenesis to originate re-epithelialization. This process may continue for a few days.

2. The early component of the *Inflammatory Phase* occurs concomitantly with the homeostasis phase, where polymorphonuclear leukocytes (PMNs) arrive to the area. The later component is governed by monocytes/macrophages, which are responsible for secreting many factors influencing wound healing.
3. *Granulation/Proliferation Phase* is divided into different sub-phases which are fibroplasias, matrix depositions, angiogenesis and re-epithelialization. All these stages constitute an overall and ongoing process.
4. The last phase, named as Remodeling/Maturation undergoes just after the third week and it is characterized by collagen deposition and contraction of the wound due to proliferation of myofibroblasts. These phases usually happen after surgery. In general, the scar presents the 80% of the tensile strength in comparison to the healthy skin [51].

Chronic wounds ensue when the coordinated physiological mechanisms of the repair process fails, inducing a prolonged healing time. Many factors can be associated to failure of wound healing: underlying vascular disorders such as venous insufficiency, arterial occlusive disease and diabetic angiopathy and neuropathy, infections with foreign microorganisms or excessive production of exudates. The term "wound exudates" refers to the liquid produced once haemostasis has been achieved. Essentially, it is composed by blood components without red cells and platelets [52]. Its function consists of irrigating and keeping the wound moist. Among different types of chronic ulcers known, the most common are leg ulcers, pressure ulcers and diabetic foot ulcers [53].

Traditional wound healing agents consist of topical pharmaceutical formulations and dressings including cotton wool, natural or synthetic bandages. These materials are being replaced for the treatment of chronic wounds. Topical liquid or semi-solid formulations do not remain on the wound for the necessary time to ensure pharmacological effects. While dry dressings do not provide a moist wound environment. Different modern dressings have been developed to improve traditional wound healing treatments. The main feature is to create a moist environment around the wound to ease the healing process. Hydrocolloids and hydrogels represent the most studied systems. They generally occur in the form of gels, thin films and foam sheets.

Hydrogels exist either as amorphous gels, sheets and films. Amorphous gels contain a significant water content of about 70-90%, this is an advantage because they could be applied in wounds with low exudates production. They need a secondary coating such gauze which should be frequently changed. The goal is to employ hydrogels as sheets or films which are malleable so they can be cut to fit in the wound area [54].

Other important benefit of these materials is that they are usable in dry necrotic tissues keeping moisture and enhancing autolytic debridement. This process is essential in removal necrotic tissue and the cleansing of foreign materials to prevent from infection and to induce healing [55].

The properties of an ideal wound dressing are described as follows (Fig. **4**):

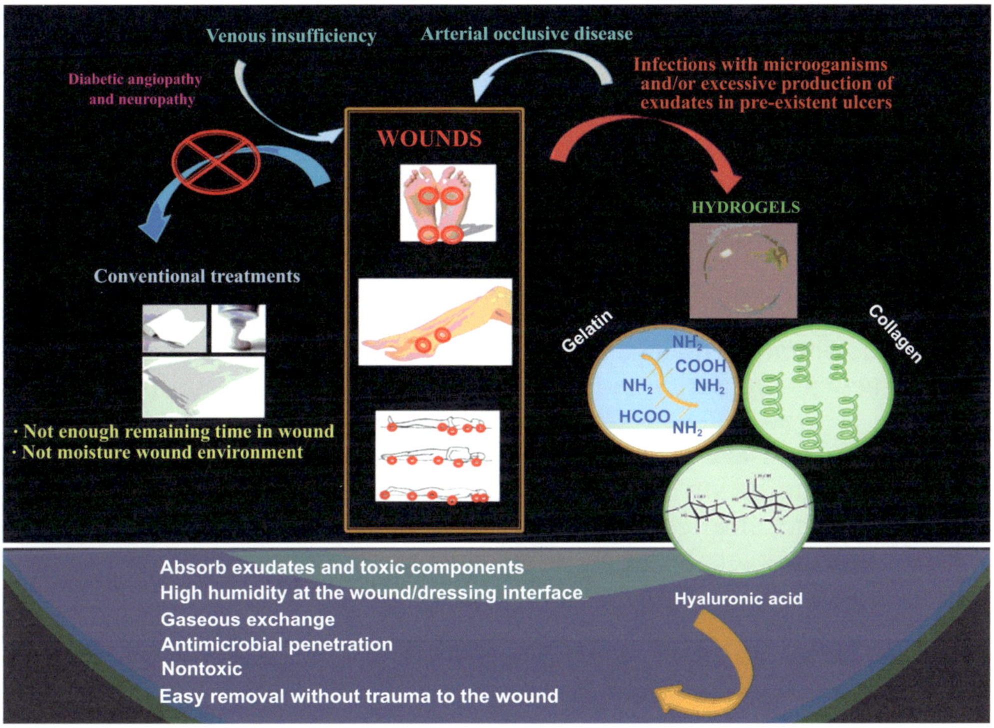

Fig. (4). General features of treatments employing hydrogels against conventional therapeutic agents in wound healing.

- Absorb exudates and toxic components from the wound's surface.
- Maintain a high humidity at the wound/dressing interface.
- Allow gaseous exchange.
- Provide thermal insulation.
- Protect the wound from bacterial penetration.
- Be nontoxic.
- Be able to remove easily without trauma to the wound.

In order to achieve hydrogels with these features not only the structure is important, but also all the components which can be added to the formulation aiming to increase the therapeutic performance. By this way, last trend in hydrogels development lies in the incorporation of diverse agents such as antimicrobials, anti-inflammatory agents and analgesics intending to healing the wound and diminishing other diseases associated to the pathology such as pain and infection.

Collagen

Collagen-based hydrogels are considered effective for chronic wounds treatment. Among the twenty five different types of collagen that have been identified [56], fibrillar collagen, which is formed by large sections of homologous sequences of polypeptides, constitutes the most commonly used form in biomaterials for wound healing. Common sources of collagen intended for biomedicine come from bovine skin and tendons, porcine skin and rat tail [57]. However, probable allergic reactions and pathogen transmission derived from animal sources suggest the use of alternatives such as collagen production by heterologous expression in different types of cells (mammalian, insects, yeast, and bacteria) [58].

Collagen-based wound dressings present simple structure which makes this biopolymer practical and easy remodeling. In addition, the large uniform surface area can serve as drug delivery system. In this sense, incorporation of diverse drugs is feasible, from antibiotics to analgesics, to achieve an integral healing treatment. The combination of collagen with synthetic polymers imparts to hydrogels higher mechanical strength, maintaining biocompatibility. The incorporation of biomolecules such as growth factors with controlled release claims as an improvement of conventional treatments [59].

Gentamicin antibiotic was encapsulated in plain silica nanoparticles which were then immobilized inside a concentrated collagen type I hydrogel. The hydrogel presented antimicrobial efficiency against *Pseudomonas aeruginosa* and *Staphylococcus aureus*, offering a prolonged antibacterial activity up to 7 days. In addition, silica incorporation impacted on the decrease of cytotoxicicty of fibroblasts but did not exert modifications on the fibrillar gel structure nor on the rheological behavior [60].

The incorporation of nano graphene oxide sheets was achieved into collagen hydrogel for the control release of growth factors such as Fibroblast growthfactor-2 (FGF-2). The resultant device would offer a new approach to stimulate fibroblasts growth in the wound area [61].

Functionalization of type I collagen with 4-vinylbenzyl chloride, glycidyl methacrylate and methacrylic anhydride was assessed to obtain biocompatible collagen hydrogels with improved swelling and mechanical properties for the treatment of chronic wounds [62].

The latest trends in research on collagen hydrogels indicate that the main goal is to obtain tunable and ductile collagen gels for the incorporation of agents such as growth factors and antimicrobials to provide comprehensive treatment for chronic wounds.

Gelatin

Gelatin is widely used in hydrogels composition because of its biocompatibility, biodegradability and non-immunogenicity as well as by presenting a great capacity for chemical modification at the level of amino acids [63]. It is available from nature, inexpensive and facile processing into different forms and shapes turn it suitable for its usage in biomedical hydrogels [64]. The only limiting factor in the use of gelatin hydrogels lies in its poor mechanical properties [65]. However this can be overcome by the combination with other materials. Hence this material provides many possibilities for application in wounds treatment. A gel composed by chitosan, honey and gelatin presented not only high antibacterial performance to *Staphilococcus aureus* and *Esccherichia coli*, but also induced a significant effect on wound contraction after 12 days of treatment in a rabbit model. The hydrogel was non-toxic and non-irritant. A comparative study was developed in comparison to a commercialized formulation, demonstrating an improvement in therapeutic properties [66]. Gelatin films containing glycerol as plasticizer agent and glutaraldehyde as cross-linker were obtained aiming to improve mechanical properties of gelatin. Thymol incorporation to gelatin hydrogel structure imparted appropriate mechanical, physical and water activity properties. In addition, the hydrogel exhibited antibacterial activity against *Staphilococcus aureus, Bacillus subtilis, Escherichia coli and Pseudomonas aeruginosa* [67]. Hydrogel consisting in a bi-layer dressing of gelatin nanofibers mat loaded with epigallocatechin gallate/polyvinyl alcohol improved healing rate probably due to an increase in DNA and protein contents. The *in vivo* study on Wistar rats revealed that topical application of the dressing promoted a faster wound contraction. The stimulation of angiogenesis, re-epithelialization and diminution in inflammatory response revealed the hydrogel as an ideal device for uninterrupted wound healing [68].

A gelatin hydrogel was developed with a platelet-rich plasma (PRP) impregnation. PRP consists in plasma with a high content of platelets. It can release numerous growth factors through activation, stimulating the healing process. It was demonstrated that this novel system induces angiogenesis and accelerates the healing process in comparison to a single application of PRP in mice [69].

Recently it was achieved the combination of synthetic polymers and biopolymers to obtain optimum mechanic and physical properties concomitantly with biocompatibility. The resultant device was a hydrogel patch which combines poly(acrylamide) and gelatin with poly(caprolactone) diacrylate as crosslinker agent. The device showed proper features in terms of wound contraction, inflammatory response, and improvement in collagen biosynthesis. It may be successfully employed for wound care and healing [70].

Hyaluronic Acid

Hyaluronic acid, also called hyaluronan is a wide spread biomolecule in different tissues of human body including synovial fluids in joints, cartilage, eye tissues, skin and extracellular matrix [71]. Immune-neutrality and tunable features make this biopolymer of choice for clinical use in wound healing. Different studies on Hyaluronic acid hydrogels agreed that this biopolymer facilitates re-epithelialization, leads to the formation of elastic new tissue and induces angiogenesis [72, 73].

There are different forms of hyaluronic acid hydrogels which can be used for healing wounds: from macroscopic structures as bulk gels to Hyaluronic acid hydrogel particles as microgels and nanogels [74]. The great versatility of this naturally occurring biopolymer gives rise to diverse macro, micro and nano structures capable to repair and regenerate skin.

Combination of Hyaluronic acid and chitosan deals hydrogels with high water holding capacity and antimicrobial activity. These gels do not exert inflammatory and toxicological reactions in animals, so they constitute a promising alternative in wound treatments [75].

Integration of Hyaluronic acid in a physically cross-linked alginate hydrogel resulted in a biofunctional platform which significantly promoted wound closure on a rat model after 5 days of treatment. This system was considered possible to translate in a clinical setting [76].

Hyaluronic acid composed hydrogels facilitates topical drug absorption and moisturizes the wound area. These features turn the biopolymeric structure as one of the preferable components in gels for healing wounds.

The study performed to evaluate Hyaluronic acid hydrogels as delivery system for biomacromolecules as bovine serum albumin, revealed that the biopolymeric active agent acted as an enhancer for delivery of bioactive compounds by the combination of co-transport, modification of stratum corneum of skin and moisturizing [77].

Considering that porous hydrogels promotes angiogenesis even in the absence of pro-angiogenic factors, a porous Hyaluronic acid hydrogel was designed and added with proangiogenic plasmids to mediate gene delivery in a diabetic wound healing murine model. Faster wound closure was achieved and granulation tissue formation was promoted. This research showed promising insights of hyaluronic hydrogels uses for local gene delivery in wound healing [78].

Future trends in the development of Hyaluronic acid based hydrogels lie in including bioactive agents such as anti-inflammatory, analgesic and protease-modulating agents considering the functional groups available from the biopolymer to interact and constitute a complete network for an integral treatment of chronic wounds.

HYDROGELS FOR TISSUE ENGINEERING

Tissue engineering is a discipline devoted to the harvest of tissues or organs from different sources. Traditional techniques involve three categories as either from self-tissues, called autologous; tissues from other human donor, named as allogenic and tissues corresponding to another species, known as xenogenic [79].

The limitations associated with this type of practice are morbidity of receptor and donor patients, the risk of transmitting harmful diseases and triggering immune responses.

New trends in the issue of tissue engineering lie in the development of replacement materials to repair damaged tissue or regenerate defective organs based on the implantation of three-dimensional scaffolds. The idea is to obtain templates for cellular activities to repair, rebuild and regenerate damaged or lost tissues. These materials should possess specific characteristics to afford the desired applications such as:

- Bioactivity to promote cell attachment, growth proliferation and differentiation.
- Biocompatibility.
- Three-dimensional architecture.
- Porous structure and interconnected network to induce flow transport of metabolic moieties.
- Biodegradability without cytotoxic effects.
- Optimum mechanical properties for the implantation site and ease to form diverse sizes and shapes.

Hydrogels are materials that meet these requirements to be used in the field of tissue engineering. The main goal is to achieve a three-dimensional polymeric network to make the gel similar to biological tissues attempting to mimic the extracellular matrix. The organization of different cell types precisely in space as organ functional subunits is possible with the ability of hydrogels to structure on desired scales. This feature is addressable with particular emphasis on the degree and type of crosslinking during the preparation. In addition the hydrogel should also play a principal role in the modulation of cell-signaling events involved in cell proliferation and migration. Growth factors can be incorporated to meet this challenge and the inclusion of angiogenic growth factors is a key issue in the design of scaffolds in order to promote and ease revascularization.

Among the two types of hydrogels that can be distinguished in terms of the interactions involved in the building up of the networks, that is physical and chemical hydrogels, chemical ones present high reproducible mechanical properties and enable the modulation of the crosslinking degree in order to obtain specific features with application in tissue engineering [80].

Among the different polymers that may be employed in hydrogels for tissue and organs harvesting, gelatin, collagen and hyaluronic acid present several positive features which enable them to compose appropriate networks to enhance the biomedical applications of these devices.

Collagen

Despite collagen immunogenicity, it is an essential biopolymer for designing hydrogels for tissue engineering applications. The limitation of collagen-based hydrogels in terms of strength has been solved by reinforcement with solid components to improve culture performance.

Recent advances in collagen-based biomaterials tend to the design of devices owing to mimic the extracellular matrix of many organs. Thus, collagen hydrogels find applications in many body sites.

A microscale collagen structure has been designed mimicking human intestinal villi by the combination of laser ablation and molding technique [81]. A colon cell line was employed to evaluate cell growth on the surface of hydrogel, and successful results were found considering that the scaffold correctly mimics intestinal villi epithelium.

Methacrylated glycol chitosan and semi-interpenetrating collagen have been employed to develop composite hydrogels useful in promoting bone regeneration. The incorporation of collagen induced cellular attachment, proliferation and osteogenic differentiation of mouse bone marrow stromal cells [82].

Aiming to going on in the improvement of collagen fibers as hydrogels considering its capacity to promote cell adhesion to detriment of its low stiffness, strength and cell mediated contraction, a collagen hydrogel was infused with poly(ethylene glycol) diacrylate and crosslinked by the exposure to UV light [83]. The device resulted in a good support in the initial stage of smooth muscle cells being a potential hydrogel for vascular tissue engineering.

The design of osteochondral scaffolds is a challenge issue in biomedicine with the main goal focused on the obtaining of tissue-specific architecture for osteochondral repair in short times. Layered cell-free scaffolds represent a new option for the treatment of osteochondral injuries. In a recent clinical study it was evaluated the osteochondral repair in patients treated with a biomimetic scaffold composed by type I collagen and hydroxyapatite [84]. The study performed in knee and talus joints revealed an incomplete cartilage and subchondral bone repair after 2.5 years follow-up. Authors advised about biomedical concerns of this treatment. This study reflects that although there are many *in vitro* researches on the development of potential collagen-based biomaterials for osteochondral repair, clinical studies are further needed to evaluate the real biomedical potential at least in this particular issue.

Collagen-based scaffolds were improved to be potentially applied in meniscus tissue engineering for the treatment of meniscus tear which is a common knee injury. Photo-crosslinking with the vitamin riboflavin and UV exposure rendered a device with better mechanical properties than collagen itself and delayed degradation of the biopolymer. Encapsulation of fibrochondrocytes led to a reduced scaffold contraction and enhanced the expression of type II collagen [85].

New trend in the development of collagen hydrogels for tissue engineering consists in improvement of biopolymeric features but further research is necessary in clinical stages to ensure the use of collagen hydrogels in humans.

Gelatin

The breakdown of the collagen triple helix by hydrolysis leads to single-strand molecules of gelatin which retains signals like the Arg–Gly–Asp (RGD) sequence. This amino acid series would be important for the promotion of cell adhesion, migration, differentiation and also proliferation [86]. Thus, gelatin is a promising component of hydrogels structure as scaffolds which enable the employment in a wide variety of organs and tissues.

A gelatin/alginate-containing device composed by 80 μm fibers was developed to include rat liver and feeding cells aiming to contribute to liver regeneration [87]. Successful results were obtained by the observation of cell migration into the hydrogel structure and self-organization. Micro-organoids were formed as reminiscence of the hepatic cord of liver lobule.

Gelatin-based hydrogels are promising devices with application in ocular tissue engineering. Extensive research has been developed on crosslinked gelatin biomaterials as bio-adhesives to stabilize retinal tissue; carriers for corneal endothelial cells and scaffolds to repair and regenerate corneal stroma [88].

Bone tissue engineering is also a scope for gelatin hydrogels where the biomaterials should present good affinity for bone tissue cells, bioactive properties and enough mechanical strength to serve as support. A three dimensional bi-layer scaffold composed by gelatin modified with calcium phosphate and polycaprolactone was developed to bind the bone. The new biomaterial exhibited good mechanical properties and accelerated the nucleation and growth of apatite crystals on the surface. The performance of the scaffold was analyzed in presence of human osteoblasts *in vitro*, revealing as a good skeleton for bone cell spreading and mineralization [89].

The incorporation of gelatin into the synthesis procedure to obtain tubular composite scaffolds made of tecophilic to serve as simulating native blood vessels, lead to improved adhesion of vascular smooth muscle cells. Thus, this blending allows a blood vessel platform with non-thrombogenic properties and potential feasibility toward functioning as a vascular graft [90].

Engineering of fatty tissue is important for the treatment of congenital deformities, tumor resections and burns. Methacrylated gelatin was obtained by a photo-crosslinkable technique as matrix for mature adipocytes. Cultures of adipocytes for 14 days within de device lead to an increased viability of the cells and to the expression of markers which demonstrates biocompatibility and tunable properties of gelatin scaffold [91].

These new trends about modified gelatin as scaffolds demonstrate that this biopolymer is useful for several applications in diverse tissues and organs due to its biocompatibility and to the possibility to give a suitable matrix for growth of diverse cells types.

Hyaluronic Acid

The predominant features that turn hyaluronic acid the chosen component when developing biomaterials for tissue engineering are biodegradability, biocompatibility and bioresorbability, in addition to the role that it plays in lubrication, cell differentiation and cell growth in connective tissues. There are many applications related to this biopolymer in tissue engineering, among which may be mentioned hydrogels for bone and cartilage tissue repair and regeneration; nerve and brain tissue repair; scaffolds for bioactive molecules and cell delivery; soft tissue repair and smooth muscle engineering. Collins *et al.* have described in their review the innovations in the field of hyaluronic acid for tissue engineering achieved until 2012 [92]. New trends in the development of hyaluronic acid-based hydrogels for tissue engineering consist in, not only the design of proper materials but also in the incorporation of diverse components which could improve tissue regeneration. Gels composed of this biopolymer are applicable to diverse tissues and organs.

The treatment of atherosclerosis is a constant challenge in biomedicine, because the implantation of cardiovascular stents carries the risk of breakage and hyperplasia. The middle layer of the vascular is composed by smooth muscle cells which are responsible for the maintenance of the vascular wall. Synthetic and contractile types of smooth muscle cells are the phenotypes existing of these cells, being the transition from contractile to synthetic type an early event in the development of atherosclerosis. Commonly stent used in clinical treatment of atherosclerosis are cytotoxic for these cells and stop proliferation instead of promoting smooth muscle cells to contractile type phenotype proliferation. The co-immobilization of collagen and hyaluronic acid on titanium stent material promoted the expression of contractile smooth muscle cell phenotype improving the functionality and diminished the side effects associated to commonly used stents [93].

Regarding to cartilage engineering, polycaprolactone scaffolds modified with cross-linked hyaluronic acid promoted cells arrival from bone marrow in a chondral defect studied in rabbits. The presence of hyaluronic acid improved the scaffold performance by promoting a favorable repair response without biomaterial pre-seeding [94].

Skin tissue engineering is a challenging area because of difficulties involved in the combination of specific properties of dermis and epidermis in a single scaffold. A multilayer device composed of hyaluronic acid and poly-L-lysine, mimicking epidermis, on a porous hyaluronic acid scaffold, mimicking dermis, was designed. The structure permitted the growth of keratinocytes and fibroblasts in an ordered structure revealing a great potential application in the treatment of full-thickness skin defects [95].

Adipose tissue engineering not only needs platforms for adipocytes growth but also requires controlled and sustained release of adipogenic factors to induce adipogenesis. A conjugate hydrogel composed by hyaluronic acid derivatives was developed for the release of dexamethasone. The hybrid material led to proliferation of human adipose-derived stem cells with non-toxicity providing a new platform for both structural and release issues [96].

Corneal endothelial regenerative medicine finds promising insights in hyaluronic acid hydrogels considering that this biopolymer is a natural occurring component of eyes. Recently and by first time a hyaluronic acid hydrogel was obtained by crosslinking with carboiimide for corneal tissue engineering. The device demonstrated great biological stability, capacity to regenerate corneal endothelium and restore corneal transparency after 4 weeks post implantation in a rabbit model with corneal endothelial dysfunction [97].

Regenerating periodontium is a current challenge for the treatment of periodontitis. Hyaluronic acid becomes once again to give a possible alternative as scaffold in combination with chitosan to create a hybrid hydrogel with porous structure suitable for odontological applications [98].

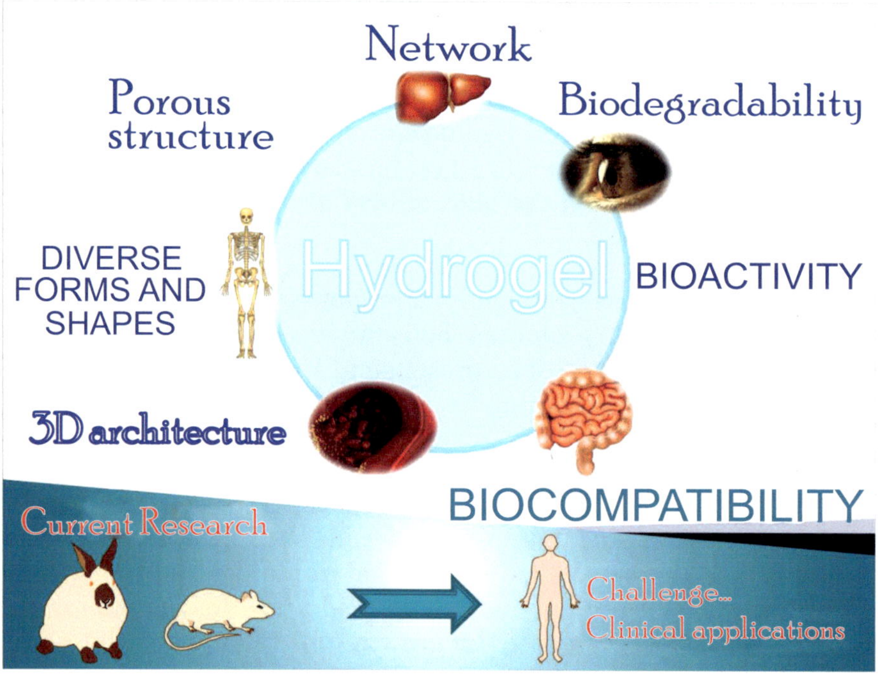

Fig. (5). Hydrogels features for tissue engineering and main tissues and organs where the devices can be employed for regeneration. Current research is being developed in mice and rabbits but studies and applications in human beings is still a challenge to achieve, approachable considering the boom of research aimed to improve the life quality of patients around the world.

Commercially Available Hydrogels for Wound Healing and Tissue Engineering

There are some commercially available hydrogel for wound healing based on biopolymers here described, such as Kikgel® and Aqua-gel® which are composed of gamma radiation-crosslinked natural gelatin or agar. Biostep® consists in a collagen matrix dressing with antibacterial activity imparted by silver.

This hydrogel has a 6-day wear time and provides a prolonged control of moist wound environment. Corgel® BioHyodrogel is a patented hyaluronan hydrogel based on di-hydroxyphenyl linkages of tyramine substituted sodium hyaluronate. It is currently available for research use only but it presents potential applications in wound healing and tissue engineering also, as orthopedics, wound care, vascular surgery and ophthalmology among others.

Regarding to tissue engineering, the commercial availability of hydrogels is limited due to the need of more research in terms of ensure long term safe use in human. Fig. (**5**) displayed an overview of the features that will be overcome and the advantages for their application in the future.

CONCLUDING REMARKS AND FUTURE INSIGHTS

Research related to the application of biopolymeric hydrogels in wound healing and tissue engineering is currently arising. Biocompatibility and versatility methodology of synthesis to obtain bio-devices adjustable with desired properties turn them into current biomaterials of election for wound care and tissue or organ replacement. From a commercial point of view it is expected a great market growth of biopolymer hydrogels in the next six years for wounds treatments.

According to bibliography here reviewed, the most important feature to resolve during the synthesis is the improvement of mechanical properties. In addition, more clinical trials in humans are needed to increase commercially available hydrogels for tissue engineering.

CONFLICT OF INTEREST

The author (editor) declares no conflict of interest, financial or otherwise.

ACKNOWLEDGEMENTS

Declared None.

REFERENCES

[1] J. Cabral, and S.C. Moratti, "Hydrogels for biomedical applications", *Future Med. Chem.* , vol. 3, no.

15, pp. 1877-1888, 2011.
[http://dx.doi.org/10.4155/fmc.11.134] [PMID: 22023032]

[2] O. Wichterle, and D. Lim, "Hydrophilic gels in biologic use", *Nature,* , vol. 185, pp. 185-117, 1960.

[3] J.Y. Lai, "Biocompatibility of chemically cross-linked gelatin hydrogels for ophthalmic use", *J. Mater. Sci. Mater. Med.* , vol. 21, no. 6, pp. 1899-1911, 2010.
[http://dx.doi.org/10.1007/s10856-010-4035-3] [PMID: 20238149]

[4] A.N. Jatariu, A.N. Cadinoiu, M. Popa, S. Curteanu, and C.A. Peptu, "Covalent and ionic co-cros--linking-An original way to prepare chitosan–gelatin hydrogels for biomedical applications", *J. Biomed. Mater. Res. Part A,* , vol. 98A, pp. 342-350, 2011.

[5] J. Michalek, R. Hobzova, and M. Pradny, *Biomedical Applications of Hydrogels Handbook,* Springer Science + Business Media, 2010, pp. 303-315.

[6] S. Koutsopoulos, L.D. Unsworth, Y. Nagai, and S. Zhang, "Controlled release of functional proteins through designer self-assembling peptide nanofiber hydrogel scaffold", *Proc. Natl. Acad. Sci. USA,* , vol. 106, no. 12, pp. 4623-4628, 2009.
[http://dx.doi.org/10.1073/pnas.0807506106] [PMID: 19273853]

[7] Y. Qiu, and K. Park, "Environment-sensitive hydrogels for drug delivery", *Adv. Drug Deliv. Rev.* , vol. 64, pp. 49-60, 2012.
[http://dx.doi.org/10.1016/j.addr.2012.09.024] [PMID: 11744175]

[8] S.C. Lee, I.K. Kwon, and K. Park, "Hydrogels for delivery of bioactive agents: a historical perspective", *Adv. Drug Deliv. Rev.* , vol. 65, no. 1, pp. 17-20, 2013.
[http://dx.doi.org/10.1016/j.addr.2012.07.015] [PMID: 22906864]

[9] A.M. Ferreira, P. Gentile, V. Chiono, and G. Ciardelli, "Collagen for bone tissue regeneration", *Acta Biomater.* , vol. 8, no. 9, pp. 3191-3200, 2012.
[http://dx.doi.org/10.1016/j.actbio.2012.06.014] [PMID: 22705634]

[10] H. Engel, S.W. Kao, J. Larson, S. Uriel, B. Jiang, E.M. Brey, and M.H. Cheng, "Investigation of Dermis-derived hydrogels for wound healing applications", *Biomed. J.* , vol. 38, no. 1, pp. 58-64, 2015.
[http://dx.doi.org/10.4103/2319-4170.132899] [PMID: 25179708]

[11] P.A. Parmar, L.W. Chow, J.P. St-Pierre, C.M. Horejs, Y.Y. Peng, J.A. Werkmeister, J.A. Ramshaw, and M.M. Stevens, "Collagen-mimetic peptide-modifiable hydrogels for articular cartilage regeneration", *Biomaterials,* , vol. 54, pp. 213-225, 2015.
[http://dx.doi.org/10.1016/j.biomaterials.2015.02.079] [PMID: 25907054]

[12] J.H. Lee, A. El-Fiqi, C.M. Han, and H.W. Kim, "Physically-strengthened collagen bioactive nanocomposite gels for bone: A feasibility study", *Tissue Eng. Regen. Med.* , vol. 12, pp. 1-8, 2015.
[http://dx.doi.org/10.1007/s13770-015-0102-7]

[13] X. Sun, X. Zhao, L. Zhao, and Q. Li, "M. DÓrtenzio, B. Nguyen, X. Xu, Y. Wen. "Development of a hybrid gelatin hydrogel plataform for tissue engineering and protein delivery applications", *J. Mater. Chem. B Mater. Biol. Med.* , vol. 3, pp. 6368-6376, 2015.
[http://dx.doi.org/10.1039/C5TB00645G]

[14] J.S. Varghese, N. Chellappa, and N.N. Fathima, "Gelatin-carrageenan hydrogels: role of pore size distribution on drug delivery process", *Colloids Surf. B Biointerfaces,* , vol. 113, pp. 346-351, 2014.
[http://dx.doi.org/10.1016/j.colsurfb.2013.08.049] [PMID: 24126319]

[15] J.R. Fraser, T.C. Laurent, and U.B. Laurent, "Hyaluronan: its nature, distribution, functions and turnover", *J. Intern. Med.* , vol. 242, no. 1, pp. 27-33, 1997.
[http://dx.doi.org/10.1046/j.1365-2796.1997.00170.x] [PMID: 9260563]

[16] C.E. Schanté, G. Zuber, C. Herlin, and T.F. Vandamme, "Chemical modifications of hyaluronic acid for the synthesis of derivatives for a broad range of biomedical applications", *Carbohydr. Polym.* , vol. 85, pp. 469-489, 2011.

[http://dx.doi.org/10.1016/j.carbpol.2011.03.019]

[17] Y. Tao, R. Zhang, W. Yang, H. Liu, H. Yang, and Q. Zhao, "Carboxymethylated hyperbranched polysaccharide: Synthesis, solution properties, and fabrication of hydrogel", *Carbohydr. Polym.* , vol. 128, pp. 179-187, 2015.
[http://dx.doi.org/10.1016/j.carbpol.2015.04.012] [PMID: 26005154]

[18] P. Agrawal, S. Soni, G. Mittal, and A. Bhatnagar, "Role of polymeric biomaterials as wound healing agents", *Int. J. Low. Extrem. Wounds,* , vol. 13, no. 3, pp. 180-190, 2014.
[http://dx.doi.org/10.1177/1534734614544523] [PMID: 25056991]

[19] V.K. Lai, C.R. Frey, A.M. Kerandi, S.P. Lake, R.T. Tranquillo, and V.H. Barocas, "Microstructural and mechanical differences between digested collagen-fibrin co-gels and pure collagen and fibrin gels", *Acta Biomater.* , vol. 8, no. 11, pp. 4031-4042, 2012.
[http://dx.doi.org/10.1016/j.actbio.2012.07.010] [PMID: 22828381]

[20] S. Münster, L.M. Jawerth, B.A. Leslie, J.I. Weitz, B. Fabry, and D.A. Weitz, "Strain history dependence of the nonlinear stress response of fibrin and collagen networks", *Proc. Natl. Acad. Sci. USA,* , vol. 110, no. 30, pp. 12197-12202, 2013.
[http://dx.doi.org/10.1073/pnas.1222787110] [PMID: 23754380]

[21] O. Moreno-Arotzena, J.G. Meier, C. Del Amo, and J.M. García-Aznar, "Characterization of Fibrin and Collagen Gels for Engineering Wound Healing Models", *Materials (Basel),* , vol. 8, no. 4, pp. 1636-1651, 2015.
[http://dx.doi.org/10.3390/ma8041636] [PMID: 26290683]

[22] C. Helary, M. Zarka, and M.M. Giraud-Guille, "Fibroblasts within concentrated collagen hydrogels favour chronic skin wound healing", *J. Tissue Eng. Regen. Med.* , vol. 6, no. 3, pp. 225-237, 2012.
[http://dx.doi.org/10.1002/term.420] [PMID: 22362469]

[23] S.B. Ross-Murphy, "Structure and rheology of gelatin gels: recent progress", *Polymer (Guildf.),* , vol. 33, pp. 2622-2627, 1992.
[http://dx.doi.org/10.1016/0032-3861(92)91146-S]

[24] C. Yang, L. Xu, Y. Zhou, X. Zhang, X. Huang, M. Wang, Y. Han, M. Zhai, S. Wei, and J. Li, "A green fabrication approach of gelatin/CM-chitosan hybrid hydrogel for wound healing", *Carbohydr. Polym.* , vol. 82, pp. 1297-1305, 2010.
[http://dx.doi.org/10.1016/j.carbpol.2010.07.013]

[25] Y. Tao, R. Zhang, W. Yang, H. Liu, H. Yang, and Q. Zhao, "Carboxymethylated hyperbranched polysaccharide: Synthesis, solution properties, and fabrication of hydrogel", *Carbohydr. Polym.* , vol. 128, pp. 179-187, 2015.
[http://dx.doi.org/10.1016/j.carbpol.2015.04.012] [PMID: 26005154]

[26] S. Amadori, P. Torricelli, K. Rubini, M. Fini, S. Panzavolta, and A. Bigi, "Effect of sterilization and crosslinking on gelatin films", *J. Mater. Sci. Mater. Med.* , vol. 26, no. 2, p. 69, 2015.
[http://dx.doi.org/10.1007/s10856-015-5396-4] [PMID: 25631265]

[27] T.T. Lau, C. Wang, and D.A. Wang, "Cell delivery with genipin crosslinked gelatin microspheres in hydrogel/microcarrier composite", *Compos. Sci. Technol.* , vol. 70, pp. 1909-1914, 2010.
[http://dx.doi.org/10.1016/j.compscitech.2010.05.015]

[28] A. Bigi, G. Cojazzi, S. Panzavolta, N. Roveri, and K. Rubini, "Stabilization of gelatin films by crosslinking with genipin", *Biomaterials,* , vol. 23, no. 24, pp. 4827-4832, 2002.
[http://dx.doi.org/10.1016/S0142-9612(02)00235-1] [PMID: 12361622]

[29] D. Seol, M.J. Magnetta, P.S. Ramakrishnan, G.L. Kurriger, H. Choe, K. Jang, J.A. Martin, and T. Lim, "Biocompatibility and preclinical feasibility tests of a temperature-sensitive hydrogel for the purpose of surgical wound paincontrol and cartilage repair", *J. Biomed. Mater. Res. Part B,* , vol. 101B, pp. 1508-1515, 2013.
[http://dx.doi.org/10.1002/jbm.b.32981]

[30] M.E. Smithmyer, L.A. Sawicki, and A.M. Kloxin, "Hydrogel scaffolds as in vitro models to study fibroblast activation in wound healing and disease", *Biomater. Sci.* , vol. 2, no. 5, pp. 634-650, 2014. [http://dx.doi.org/10.1039/C3BM60319A] [PMID: 25379176]

[31] G. Pitarresi, F.S. Palumbo, R. Calabrese, E.F. Craparo, and G. Giammona, "Crosslinked hyaluronan with a protein-like polymer: novel bioresorbable films for biomedical applications", *J. Biomed. Mater. Res. A*, , vol. 84, no. 2, pp. 413-424, 2008. [http://dx.doi.org/10.1002/jbm.a.31316] [PMID: 17618478]

[32] M. Witting, A. Boreham, R. Brodwolf, K. Vávrová, U. Alexiev, W. Friess, and S. Hedtrich, "Interactions of hyaluronic Acid with the skin and implications for the dermal delivery of biomacromolecules", *Mol. Pharm.* , vol. 12, no. 5, pp. 1391-1401, 2015. [http://dx.doi.org/10.1021/mp500676e] [PMID: 25871518]

[33] R. Buffa, P. Šedová, I. Basarabová, M. Moravcová, L. Wolfová, T. Bobula, and V. Velebný, "α,β-Unsaturated aldehyde of hyaluronanSynthesis, analysis and applications", *Carbohydr. Polym.* , vol. 134, pp. 293-299, 2015. [http://dx.doi.org/10.1016/j.carbpol.2015.07.084] [PMID: 26428127]

[34] T. Billiet, M. Vandenhaute, J. Schelfhout, S. Van Vlierberghe, and P. Dubruel, "A review of trends and limitations in hydrogel-rapid prototyping for tissue engineering", *Biomaterials*, , vol. 33, no. 26, pp. 6020-6041, 2012. [http://dx.doi.org/10.1016/j.biomaterials.2012.04.050] [PMID: 22681979]

[35] N. Davidenko, J.J. Campbell, E.S. Thian, C.J. Watson, and R.E. Cameron, "Collagen-hyaluronic acid scaffolds for adipose tissue engineering", *Acta Biomater.* , vol. 6, no. 10, pp. 3957-3968, 2010. [http://dx.doi.org/10.1016/j.actbio.2010.05.005] [PMID: 20466086]

[36] B. Dhandayuthapani, U.M. Krishnan, and S. Sethuraman, "Fabrication and characterization of chitosangelatin blend nanofiber for skin tissue engineering", *J. Biomed. Mater. Res. B*, , vol. 94, no. 1, pp. 264-272.

[37] P.A. Levett, D.W. Hutmacher, J. Malda, and T.J. Klein, "Hyaluronic acid enhances the mechanical properties of tissue-engineered cartilage constructs", *PLoS One*, , vol. 9, no. 12, p. e113216, 2014. [http://dx.doi.org/10.1371/journal.pone.0113216] [PMID: 25438040]

[38] J.H. Lee, J.Y. Lee, S.H. Yang, E.J. Lee, and H.W. Kim, "Carbon nanotube-collagen three-dimensional culture of mesenchymal stem cells promotes expression of neural phenotypes and secretion of neurotrophic factors", *Acta Biomater.* , vol. 10, no. 10, pp. 4425-4436, 2014. [http://dx.doi.org/10.1016/j.actbio.2014.06.023] [PMID: 24954912]

[39] A. El-Fiqi, J.H. Lee, E.J. Lee, and H.W. Kim, "Collagen hydrogels incorporated with surface-aminated mesoporous nanobioactive glass: Improvement of physicochemical stability and mechanical properties is effective for hard tissue engineering", *Acta Biomater.* , vol. 9, no. 12, pp. 9508-9521, 2013. [http://dx.doi.org/10.1016/j.actbio.2013.07.036] [PMID: 23928332]

[40] P.A. Levett, F.P.W. Melchels, K. Schrobback, D.W. Hutmacher, J. Malda, and T.J. Klein, "A biomimetic extracellular matrix for cartilage tissue engineering centered on photocurable gelatin, hyaluronic acid and chondroitin sulfate", *Acta Biomaterialia*, , vol. 10, no. 10, pp. 214-223, 2014. [http://dx.doi.org/10.1016/j.actbio.2013.10.005]

[41] D. Bellini, C. Cencetti, J. Meraner, D. Stoppoloni, A.S. D'Abusco, and P. Matricardi, "An in situ gelling system for bone regeneration of osteochondral defects", *Eur. Polym. J.* , vol. 72, pp. 645-650, 2015. [http://dx.doi.org/10.1016/j.eurpolymj.2015.02.043]

[42] S.T. Bendtsen, and M. Wei, "Synthesis and characterization of a novel injectable alginate–collagen–hydroxyapatite hydrogel for bone tissue regeneration", *J. Mater. Chem. B Mater. Biol. Med.* , vol. 3, pp. 3081-3090, 2015. [http://dx.doi.org/10.1039/C5TB00072F]

[43] J. Lewandowska-Łańcucka, S. Fiejdasz, Ł. Rodzik, M. Kozieł, and M. Nowakowska, "Bioactive hydrogel-nanosilica hybrid materials: a potential injectable scaffold for bone tissue engineering", *Biomed. Mater.* , vol. 10, no. 1, p. 015020, 2015. [http://dx.doi.org/10.1088/1748-6041/10/1/015020] [PMID: 25668107]

[44] Q. Wei, J. Lu, Q. Wang, H. Fan, and X. Zhang, "Novel synthesis strategy for composite hydrogel of collagen/hydroxyapatite-microsphere originating from conversion of $CaCO_3$ templates", *Nanotechnology,* , vol. 26, no. 11, pp. 115605-115613, 2015. [http://dx.doi.org/10.1088/0957-4484/26/11/115605] [PMID: 25719911]

[45] J.R. Xavier, T. Thakur, P. Desai, M.K. Jaiswal, N. Sears, E. Cosgriff-Hernandez, R. Kaunas, and A.K. Gaharwar, "Bioactive nanoengineered hydrogels for bone tissue engineering: a growth-factor-free approach", *ACS Nano,* , vol. 9, no. 3, pp. 3109-3118, 2015. [http://dx.doi.org/10.1021/nn507488s] [PMID: 25674809]

[46] S. Wakhet, V.K. Singh, S. Sahoo, S.S. Sagiri, S. Kulanthaivel, M.K. Bhattacharya, N. Kumar, I. Banerjee, and K. Pal, "Characterization of gelatin–agar based phase separated hydrogel, emulgel and bigel: a comparative study", *J. Mater-Sci. M.* , vol. 26, p. 118, 2015. [http://dx.doi.org/10.1007/s10856-015-5434-2]

[47] K.A. Smeds, A. Pfister-Serres, D. Miki, K. Dastgheib, M. Inoue, D.L. Hatchell, and M.W. Grinstaff, "Photocrosslinkable polysaccharides for in situ hydrogel formation", *J. Biomed. Mater. Res.* , vol. 54, no. 1, pp. 115-121, 2001. [http://dx.doi.org/10.1002/1097-4636(200101)54:1<115::AID-JBM14>3.0.CO;2-Q] [PMID: 11077410]

[48] M. Guvendiren, H.D. Lu, and J.A. Burdick, "Shear-thinning hydrogels for biomedical applications", *Soft Matter,* , vol. 8, pp. 260-272, 2012. [http://dx.doi.org/10.1039/C1SM06513K]

[49] A.S. Hoffman, "Hydrogels for biomedical applications", *Adv. Drug Deliv. Rev.* , vol. 64, pp. 18-23, 2012. [http://dx.doi.org/10.1016/j.addr.2012.09.010] [PMID: 11755703]

[50] P. Martin, "Wound healingaiming for perfect skin regeneration", *Science,* , vol. 276, no. 5309, pp. 75-81, 1997. [http://dx.doi.org/10.1126/science.276.5309.75] [PMID: 9082989]

[51] C.Y. Cho, and J.S. Lo, "Dressing the part", *Dermatol. Clin.* , vol. 16, no. 1, pp. 25-47, 1998. [http://dx.doi.org/10.1016/S0733-8635(05)70485-X] [PMID: 9460576]

[52] S. Thomas, "Assessment and management of wound exudate", *J. Wound Care,* , vol. 6, no. 7, pp. 327-330, 1997. [PMID: 9325827]

[53] C. Helary, M. Zarka, and M.M. Giraud-Guille, "Fibroblasts within concentrated collagen hydrogels favour chronic skin wound healing", *J. Tissue Eng. Regen. Med.* , vol. 6, no. 3, pp. 225-237, 2012. [http://dx.doi.org/10.1002/term.420] [PMID: 22362469]

[54] J.S. Boateng, K.H. Matthews, H.N. Stevens, and G.M. Eccleston, "Wound healing dressings and drug delivery systems: a review", *J. Pharm. Sci.* , vol. 97, no. 8, pp. 2892-2923, 2008. [http://dx.doi.org/10.1002/jps.21210] [PMID: 17963217]

[55] A.F. Falabella, "Debridement and wound bed preparation", *Dermatol. Ther. (Heidelb.),* , vol. 19, no. 6, pp. 317-325, 2006. [http://dx.doi.org/10.1111/j.1529-8019.2006.00090.x] [PMID: 17199674]

[56] D.J. Prockop, and K.I. Kivirikko, "Collagens: molecular biology, diseases, and potentials for therapy", *Annu. Rev. Biochem.* , vol. 64, pp. 403-434, 1995. [http://dx.doi.org/10.1146/annurev.bi.64.070195.002155] [PMID: 7574488]

[57] S.F. Badylak, "Xenogeneic extracellular matrix as a scaffold for tissue reconstruction", *Transpl.*

Immunol. , vol. 12, no. 3-4, pp. 367-377, 2004.
[http://dx.doi.org/10.1016/j.trim.2003.12.016] [PMID: 15157928]

[58] D. Olsen, C. Yang, M. Bodo, R. Chang, S. Leigh, J. Baez, D. Carmichael, M. Perälä, E.R. Hämäläinen, M. Jarvinen, and J. Polarek, "Recombinant collagen and gelatin for drug delivery", *Adv. Drug Deliv. Rev.* , vol. 55, no. 12, pp. 1547-1567, 2003.
[http://dx.doi.org/10.1016/j.addr.2003.08.008] [PMID: 14623401]

[59] S. Chattopadhyay, and R.T. Raines, "Review collagen-based biomaterials for wound healing", *Biopolymers,* , vol. 101, no. 8, pp. 821-833, 2014.
[http://dx.doi.org/10.1002/bip.22486] [PMID: 24633807]

[60] G.S. Alvarez, C. Hélary, A.M. Mebert, X. Wang, and T. Coradin, "Antibiotic-loaded silica nanoparticle collagen composite hydrogels with prolonged antimicrobial activity for wound infection prevention", *J. Mater. Chem. B Mater. Biol. Med.* , vol. 2, no. 29, pp. 4660-4670, 2014.
[http://dx.doi.org/10.1039/c4tb00327f]

[61] M. Choi, J.H. Chung, Y. Cho, B.Y. Hong, and J. Hong, "Nano-film modification of collagen hydrogels for controlled growth factor release", *Chem. Eng. Sci.* , vol. 137, pp. 626-630, 2015.
[http://dx.doi.org/10.1016/j.ces.2015.07.011]

[62] G. Tronci, C.A. Grant, N.H. Thomson, S.J. Russell, and D.J. Wood, "Multi-scale mechanical characterization of highly swollen photo-activated collagen hydrogels", *J. R. Soc. Interface,* , vol. 12, no. 102, p. 20141079, 2015.
[http://dx.doi.org/10.1098/rsif.2014.1079] [PMID: 25411409]

[63] S.S. Silva, J.F. Mano, and R.L. Reis, "Potential applications of natural origin polymer-based systems in soft tissue regeneration", *Crit. Rev. Biotechnol.* , vol. 30, no. 3, pp. 200-221, 2010.
[http://dx.doi.org/10.3109/07388551.2010.505561] [PMID: 20735324]

[64] S. Huang, and X. Fu, "Naturally derived materials-based cell and drug delivery systems in skin regeneration", *J. Control. Release,* , vol. 142, no. 2, pp. 149-159, 2010.
[http://dx.doi.org/10.1016/j.jconrel.2009.10.018] [PMID: 19850093]

[65] Q. Xing, K. Yates, C. Vogt, Z. Qian, M.C. Frost, and F. Zhao, "Increasing mechanical strength of gelatin hydrogels by divalent metal ion removal", *Sci. Rep.* , vol. 4, pp. 4706-4716, 2014.
[http://dx.doi.org/10.1038/srep04706] [PMID: 24736500]

[66] T. Wang, X.K. Zhu, X.T. Xue, and D.Y. Wu, "Hydrogel sheets of chitosan, honey and gelatin as burn wound dressings", *Carbohydr. Polym.* , vol. 88, pp. 75-83, 2012.
[http://dx.doi.org/10.1016/j.carbpol.2011.11.069]

[67] G. Kavoosi, S.M. Dadfar, and A.M. Purfard, "Mechanical, physical, antioxidant, and antimicrobial properties of gelatin films incorporated with thymol for potential use as nano wound dressing", *J. Food Sci.* , vol. 78, no. 2, pp. E244-E250, 2013.
[http://dx.doi.org/10.1111/1750-3841.12015] [PMID: 23317304]

[68] M. Jaiswal, A. Gupta, A.K. Agrawal, M. Jassal, A.K. Dinda, and V. Koul, "Bi-layer composite dressing of gelatin nanofibrous mat and poly vinyl alcohol hydrogel for drug delivery and wound healing application: in-vitro and in-vivo studies", *J. Biomed. Nanotechnol.* , vol. 9, no. 9, pp. 1495-1508, 2013.
[http://dx.doi.org/10.1166/jbn.2013.1643] [PMID: 23980498]

[69] P.V. Notodihardjo, N. Morimoto, N. Kakudo, M. Matsui, M. Sakamoto, P.H. Liem, K. Suzuki, Y. Tabata, and K. Kusumoto, "Gelatin hydrogel impregnated with platelet-rich plasma releasate promotes angiogenesis and wound healing in murine model", *J. Artif. Organs,* , vol. 18, no. 1, pp. 64-71, 2015.
[http://dx.doi.org/10.1007/s10047-014-0795-8] [PMID: 25326193]

[70] M. Jaiswal, A. Gupta, A.K. Dinda, and V. Koul, "An investigation study of gelatin release from semi-interpenetrating polymeric network hydrogel patch for excision wound healing on Wistar rat model", *J. Appl. Polym. Sci.* , vol. 132, no. 25, p. 42120, 2015.
[http://dx.doi.org/10.1002/app.42120]

[71] M.G. Neuman, R.M. Nanau, L. Oruña-Sanchez, and G. Coto, "Hyaluronic acid and wound healing", *J. Pharm. Pharm. Sci.* , vol. 18, no. 1, pp. 53-60, 2015. [http://dx.doi.org/10.18433/J3K89D] [PMID: 25877441]

[72] J.Y. Zhao, J.K. Chai, H.F. Song, J. Zhang, M.H. Xu, and Y.D. Liang, "Influence of hyaluronic acid on wound healing using composite porcine acellular dermal matrix grafts and autologous skin in rabbits", *Int. Wound J.* , vol. 10, no. 5, pp. 562-572, 2013. [http://dx.doi.org/10.1111/j.1742-481X.2012.01023.x] [PMID: 22682212]

[73] N. Shimizu, D. Ishida, A. Yamamoto, M. Kuroyanagi, and Y. Kuroyanagi, "Development of a functional wound dressing composed of hyaluronic acid spongy sheet containing bioactive components: evaluation of wound healing potential in animal tests", *J. Biomater. Sci. Polym. Ed.* , vol. 25, no. 12, pp. 1278-1291, 2014. [http://dx.doi.org/10.1080/09205063.2014.929427] [PMID: 24959914]

[74] X. Xu, A.K. Jha, D.A. Harrington, M.C. Farach-Carson, and X. Jia, "Hyaluronic Acid-Based Hydrogels: from a Natural Polysaccharide to Complex Networks", *Soft Matter,* , vol. 8, no. 12, pp. 3280-3294, 2012. [http://dx.doi.org/10.1039/c2sm06463d] [PMID: 22419946]

[75] C. Vasile, D. Pieptu, R.P. Dumitriu, A. Pânzariu, and L. Profire, "Chitosan/hyaluronic acid polyelectrolyte complex hydrogels in the management of burn wounds", *Rev. Med. Chir. Soc. Med. Nat. Iasi,* , vol. 117, no. 2, pp. 565-571, 2013. [PMID: 24340547]

[76] O. Catanzano, V. D'Esposito, S. Acierno, M.R. Ambrosio, C. De Caro, C. Avagliano, P. Russo, R. Russo, A. Miro, F. Ungaro, A. Calignano, P. Formisano, and F. Quaglia, "Alginate–hyaluronan composite hydrogels accelerate wound healing process", *Carbohyd. Polym.* , vol. 131, pp. 407-414, 2015. [http://dx.doi.org/10.1016/j.carbpol.2015.05.081]

[77] M. Witting, A. Boreham, R. Brodwolf, K. Vávrová, U. Alexiev, W. Friess, and S. Hedtrich, "Interactions of hyaluronic Acid with the skin and implications for the dermal delivery of biomacromolecules", *Mol. Pharm.* , vol. 12, no. 5, pp. 1391-1401, 2015. [http://dx.doi.org/10.1021/mp500676e] [PMID: 25871518]

[78] T. Tokatlian, C. Cam, and T. Segura, "Porous hyaluronic acid hydrogels for localized nonviral DNA delivery in a diabetic wound healing model", *Adv. Healthc. Mater.* , vol. 4, no. 7, pp. 1084-1091, 2015. [http://dx.doi.org/10.1002/adhm.201400783] [PMID: 25694196]

[79] T. Ahsan, R. Bellamkonda, and R.M. Nerem, *Translational approaches in tissue engineering and regenerative medicine.* Artech House, 2008.

[80] F. Camponeschi, A. Atrei, G. Rocchigiani, L. Mencuccini, M. Uva, and R. Barbucci, "New Formulations of Polysaccharide-Based Hydrogels for Drug Release and Tissue Engineering", *Gels,* , vol. 1, pp. 3-23, 2015. [http://dx.doi.org/10.3390/gels1010003]

[81] J.H. Sung, J. Yu, D. Luo, M.L. Shuler, and J.C. March, "Microscale 3-D hydrogel scaffold for biomimetic gastrointestinal (GI) tract model", *Lab Chip,* , vol. 11, no. 3, pp. 389-392, 2011. [http://dx.doi.org/10.1039/C0LC00273A] [PMID: 21157619]

[82] C. Arakawa, S. Tan, S. Kim, B. Wu, and M. Lee, "Photopolymerizable chitosan–collagen hydrogels for bone tissue engineering", *J. Tissue Eng. Regen. Med.* , 2014. [http://dx.doi.org/10.1002/term.1896] [PMID: 24771649]

[83] D.J. Muñoz-Pinto, A.C. Jimenez-Vergara, T.P. Gharat, and M.S. Hahn, "Characterization of sequential collagen-poly(ethylene glycol) diacrylate interpenetrating networks and initial assessment of their potential for vascular tissue engineering", *Biomaterials,* , vol. 40, pp. 32-42, 2015. [http://dx.doi.org/10.1016/j.biomaterials.2014.10.051] [PMID: 25433604]

[84] B.B. Christensen, C.B. Foldager, J. Jensen, N.C. Jensen, and M. Lind, "Poor osteochondral repair by a biomimetic collagen scaffold: 1- to 3-year clinical and radiological follow-up", *Knee Surg. Sports Traumatol. Arthrosc.* , 2015. [http://dx.doi.org/10.1007/s00167-015-3538-3] [PMID: 25691368]

[85] J. Heo, R.H. Koh, W. Shim, H.D. Kim, H.G. Yim, and N.S. Hwang, "Riboflavin-induced photo-crosslinking of collagen hydrogel and its application in meniscus tissue engineering", *Drug Deliv. Transl. Res.* , vol. 6, no. 2, pp. 148-158, 2016. [http://dx.doi.org/10.1007/s13346-015-0224-4] [PMID: 25809935]

[86] J. Zhu, and R.E. Marchant, "Design properties of hydrogel tissue-engineering scaffolds", *Expert Rev. Med. Devices,* , vol. 8, no. 5, pp. 607-626, 2011. [http://dx.doi.org/10.1586/erd.11.27] [PMID: 22026626]

[87] M. Yamada, R. Utoh, K. Ohashi, K. Tatsumi, M. Yamato, T. Okano, and M. Seki, "Controlled formation of heterotypic hepatic micro-organoids in anisotropic hydrogel microfibers for long-term preservation of liver-specific functions", *Biomaterials,* , vol. 33, no. 33, pp. 8304-8315, 2012. [http://dx.doi.org/10.1016/j.biomaterials.2012.07.068] [PMID: 22906609]

[88] J.B. Rose, S. Pacelli, A.J. El Haj, H.S. Dua, A. Hopkinson, L.J. White, and F. Rose, "Gelatin-Based Materials in Ocular Tissue Engineering", *Materials (Basel),* , vol. 7, pp. 3106-3135, 2014. [http://dx.doi.org/10.3390/ma7043106]

[89] I. Rajzer, E. Menaszek, R. Kwiatkowski, J.A. Planell, and O. Castano, "Electrospun gelatin/poly(ε-caprolactone) fibrous scaffold modified with calcium phosphate for bone tissue engineering", *Mater Sci Eng C Mater Biol Appl,* , vol. 44, pp. 183-190, 2014. [http://dx.doi.org/10.1016/j.msec.2014.08.017] [PMID: 25280695]

[90] E. Vatankhah, M.P. Prabhakaran, D. Semnani, S. Razavi, M. Morshed, and S. Ramakrishna, "Electrospun tecophilic/gelatin nanofibers with potential for small diameter blood vessel tissue engineering", *Biopolymers,* , vol. 101, no. 12, pp. 1165-1180, 2014. [http://dx.doi.org/10.1002/bip.22524] [PMID: 25042000]

[91] B. Huber, K. Borchers, G.E. Tovar, and P.J. Kluger, "Methacrylated gelatin and mature adipocytes are promising components for adipose tissue engineering", *J. Biomater. Appl.* , vol. •••, pp. 1-12, 2015. [PMID: 26017717]

[92] M.N. Collins, and C. Birkinshaw, "Hyaluronic acid based scaffolds for tissue engineeringa review", *Carbohydr. Polym.* , vol. 92, no. 2, pp. 1262-1279, 2013. [http://dx.doi.org/10.1016/j.carbpol.2012.10.028] [PMID: 23399155]

[93] J. Li, K. Zhang, H. Chen, T. Liu, P. Yang, Y. Zhao, and N. Huang, "A novel coating of type IV collagen and hyaluronic acid on stent material-titanium for promoting smooth muscle cell contractile phenotype", *Mater Sci Eng C Mater Biol Appl,* , vol. 38, pp. 235-243, 2014. [http://dx.doi.org/10.1016/j.msec.2014.02.008] [PMID: 24656374]

[94] M. Lebourg, S. Martínez-Díaz, N. García-Giralt, R. Torres-Claramunt, J.A. Gómez-Tejedor, J.L. Ribelles, G. Vila-Canet, and J.C. Monllau, "Cell-free cartilage engineering approach using hyaluronic acid-polycaprolactone scaffolds: a study in vivo", *J. Biomater. Appl.* , vol. 28, no. 9, pp. 1304-1315, 2014. [http://dx.doi.org/10.1177/0885328213507298] [PMID: 24108064]

[95] I.P. Monteiro, A. Shukla, A.P. Marques, R.L. Reis, and P.T. Hammond, "Spray-assisted layer-by-layer assembly on hyaluronic acid scaffolds for skin tissue engineering", *J. Biomed. Mater. Res. A,* , vol. 103, no. 1, pp. 330-340, 2015. [http://dx.doi.org/10.1002/jbm.a.35178] [PMID: 24659574]

[96] M. Fan, Y. Ma, Z. Zhang, J. Mao, H. Tan, and X. Hu, "Biodegradable hyaluronic acid hydrogels to control release of dexamethasone through aqueous Diels–Alder chemistry for adipose tissue engineering", *Mater. Sci. Eng. C,* , vol. 56, pp. 311-317, 2015. [http://dx.doi.org/10.1016/j.msec.2015.04.004]

[97] J.Y. Lai, H.Y. Cheng, and D.H. Ma, "Investigation of overrun-processed porous hyaluronic acid carriers in corneal endothelial tissue engineering", *PLoS One, ,* vol. 10, no. 8, p. e0136067, 2015. [http://dx.doi.org/10.1371/journal.pone.0136067] [PMID: 26296087]

[98] D.G. Miranda, S.M. Malmonge, D.M. Campos, N.G. Attik, B. Grosgogeat, and K. Gritsch, "A chitosan-hyaluronic acid hydrogel scaffold for periodontal tissue engineering", *J. Biomed. Mater. Res. B, ,* 2015.

CHAPTER 8

Approaches for Improving the Mechanical Properties of Collagen Biomaterials

María Victoria Tuttolomondo[1,2], Maria Lucia Foglia[1,2], Juan Manuel Galdopórpora[2] and Martin Federico Desimone[1,2,*]

[1] *IQUIMEFA-CONICET, Universidad de Buenos Aires, Facultad de Farmacia y Bioquímica, Cátedra de Química Analítica Instrumental. (1113) Junin 956 Piso 3. Ciudad de Buenos Aires, Argentina*

[2] *Universidad de Buenos Aires, Facultad de Farmacia y Bioquímica, Cátedra de Química Analítica Instrumental. (1113) Junin 956 Piso 3. Ciudad de Buenos Aires. Argentina*

Abstract: Collagen is a natural constituent of the extracellular matrix of tissues. It is well known that, when prepared without additives, diluted collagen hydrogels present poor mechanical properties limiting their use in tissue engineering. Hence, collagen's structural properties need to be enhanced. For this purpose, this chapter presents an overview of the recent advances and alternatives developed to improve the mechanical properties of collagen biomaterials.

Keywords: Biomaterial, Collagen, Crosslink, Fibrils, Hydrogel, Mechanical properties, Thermal stability, Tissue engineering, Wound healing.

INTRODUCTION

Within the biomaterials field, efforts are made in order to mimic the extracellular environment, taking into account its natural biocompatibility and mechanical properties. The collaboration between different disciplines is therefore sought to reproduce the complexity of the affected tissue. Among the different types of tissues found within the body, the connective tissue provides support and connects different tissues or organs. For its supportive function, the mechanical behavior is primarily determined by the composition and organization of the collagen present in it, a reason for which the characterization of fibril morphology and organization is so important in the design of new materials [1]. Herein, we present an overview of those recent approaches made to improve the mechanical properties of collagen involving the use of higher collagen concentrations as well as modifications in

* **Corresponding autor Martín F. Desimone**: IQUIMEFA-CONICET. Universidad de Buenos Aires, Facultad de Farmacia y Bioquímica, Cátedra de Química Analítica Instrumental. (1113) Junin 956 Piso 3. Ciudad de Buenos Aires, Argentina; Tel/Fax: +54-11-49648254; E-mail: desimone@ffyb.uba.ar

Martin Masuelli & Denis Renard (Eds.)

pH, temperature or ionic strength along with cross linking methods, all of which result in a significant enhancement of the thermal stability and rheological properties of the collagen network.

Biological Importance of Collagen and its Structure

Within the extracellular matrix, collagen molecules assemble into 25 to 500 nm fibrils, which further aggregate giving rise to fibers, the latter with a width up to hundreds of micrometers. Fibril width within a given tissue is associated with the capacity of it to withstand plastic deformation and high mechanical forces. While small fibrils allow the tissue to support plastic deformation, thicker ones are related to its ability to tolerate high mechanical loads. Additionally, glycosaminoglycans, hyaluronan and elastin are present in connective tissues in a smaller proportion when compared to collagen, though they have been found to contribute to the tissue′s overall mechanical stability. Oxlund *et al.*, investigated the roles of hyaluronic acid, collagen and elastin in the mechanical properties of connective tissues. By means of enzymatic degradation, they were able to measure the response of the digested specimens after a load was applied. As a result, stress-strain curves showed that collagen is the component which is mainly responsible for the tensile strength of skin and aorta [2]. In a later work, the sole contribution of elastin was assessed which suggested that elastin fibers might be responsible for the recoiling mechanism after a stress or deformation is applied [3].

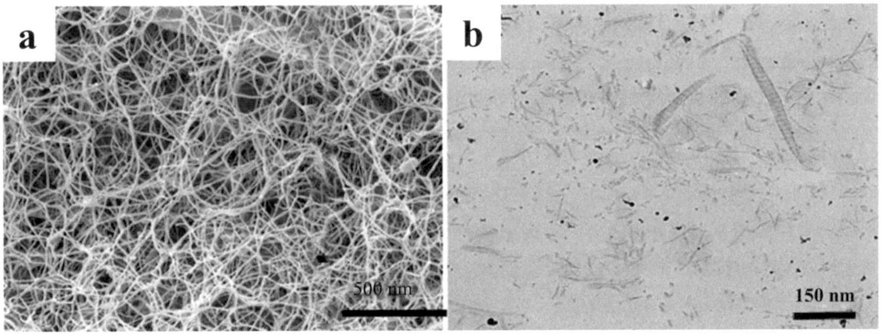

Fig. (1). a) representative SEM image of a collagen hydrogel network. **B)** TEM image where the typical cross striations are seen in the collagen fibre.

Type I collagen is present in connective tissues in concentrations ranging from 30 to 40 mg.mL^{-1}. In the presence of acid, the net charge of the polypeptide chain is altered and, as a result, hydrogen bonds that hold tropocollagen molecules together are disrupted leading to their release from the fibrils. After its solubilisation in diluted acid medium, collagen suspensions are obtained in concentrations which lay far behind from the ones normally found in living tissues. Such collagen suspensions give rise, upon neutralization, to matrices or gels composed of around 60 to 100 nm fibrils with the characteristic axial D periodic spacing. The gel consists of a fibrous network (with fibers ca. 60 nm in diameter) exhibiting a high porosity level, with an average pore size larger than 200 nm, characteristic of these hydrogel structures. Moreover, the characteristic transversal cross-striations with a periodicity of 67 nm were clearly evidenced on TEM images (Fig. **1**). However, one of the main limitations of using these types of collagen gels as biomedical materials is related to their poor mechanical properties [4].

Properties of *in vitro* Reconstituted Collagen Matrices

in vitro, the mechanical properties of reconstituted collagen matrices have been extensively studied by evaluating the individual contribution of changes in the pH, temperature, protein concentration and ionic strength of different preparations [5 - 7]. In relation to this, Achilli *et al.*, prepared collagen gels under different conditions (pH, temperature and ionic strength) and observed that at pH 10, fibril packing was increased along with the mechanical resistance of the obtained gel, even when the fibril diameter decreased at a higher pH. Higher pH involves higher number of electrostatic bonds between fibrils, hence resulting in a gel with better mechanical properties [8]. On the other hand, differences in collagen concentration led to variations in fibril density but not in fibril diameter, which appeared constant around *ca.* 60 nm [7]. Even though fibril diameter remains unchanged, a denser collagen network is achieved as its concentration rises, leading to gels with improved mechanical properties.

Through the commonly used extraction methods, collagen concentrations around 2 $mg.mL^{-1}$ are obtained, which after proper solvent evaporation give rise to collagen solutions up to 40 $mg.mL^{-1}$. Recently, Hélary *et al.*, managed to control the self-assembly conditions and in this way, increased the collagen concentration up to 5 mg mL^{-1}. The higher collagen concentration resulted in lower cell-mediated contraction, increased cell proliferation, and superior *in vivo* integration [9, 10]. The so obtained matrices proved useful for the cell culture of human dermal fibroblasts as their secreted metalloproteinases were able to remodel collagen, stimulating cell migration within the scaffolds [11]. Manipulation of dense acid collagen solutions remains a challenge but neutralization is possible after their exposure to ammonia vapors [12]. Dense hydrogels with concentrations higher than 100 mg mL^{-1} were also recently obtained by injection, dialysis and combination of the two methods [13, 14].

Similarly, changes in temperature lead to changes in the mechanical strength. As a matter of fact, fibrils formed at low temperatures (*i.e.*, 20°C) have larger diameters than those formed at high temperatures (*i.e.*, 37°C) [15]. Interestingly, the mechanical resistance of the obtained hydrogels increases with time and remains constant after 10 hours, even when turbidimetry results indicate that fibril assembly is completed in 30 to 120 minutes [16].

The complexity of the connective tissue *in vivo* is such that it is difficult to evaluate the individual contribution of each component to the observed mechanical properties. It is not surprising that so many works have been published with contradicting results, for example in relation to the pH´s influence over fibril diameter or mechanical resistance of the obtained diluted hydrogels. On the other hand, even when high concentrations of collagen are employed, assembly of individual fibrils into fibers is restricted. In relation to this, Gentleman *et al.*, evaluated the performance of extruded collagen fibers when compared to the native ones obtained from rat tail tendons. Extruded fibers exhibited mechanical properties which were significantly affected by their diameter. Even when the tangent moduli and peak stress of the extruded fibers were close to the ones reported for human ligaments, the scaffolds obtained *in vitro* did not exhibit the same viscoelastic creep properties. Interestingly, fibroblast cultures for over 25 days increased the resistance of the materials, probably due to tissue remodeling.

Overall, the *in vitro* collagen assembly process originates fibrils, whose diameter, length and ultimate tensile strength can be manipulated from the working conditions previously chosen. However, given the extraction procedure, collagen´s spontaneous assembly generates gels which in many cases require further crosslinking. The latter is commonly achieved through physical, enzymatic or chemical methods.

Effects of Metalloproteinases on Collagen´s Remodeling

Collagen remodeling *in vivo* involves the degradation, deposition and reinforcement of fibrils within a given tissue. Matrix metalloproteinases (MMP), among which we find collagenase or MMP-1, were initially thought as the enzymes responsible for the degradation of structural components of the extracellular matrix. Indeed, MMP-1 is involved in the degradation of native collagen fibrils, a task achieved through the cleavage of the peptide bond between Gly775–Ile776 or Gly775–Lys776 in native type I, II or III collagen molecules. The proteolysis, however, may lead to the regulation of tissue architecture, to the stimulation of cell migration and even play major roles in inflammation and disease [17]. In this sense, the up regulation of the expression of some metalloproteinases may lead to impaired tissue remodeling, which may be related to the decreased mechanical properties observed in some pathologies such as osteoarthritis and ascending aorta aneurysm, as well as in the increased arterial stiffness associated with both aging and hypertension [18 - 20]. The opposite process, namely the deposition of collagen, occurs naturally during the healing process, with the extent of collagen alignment determining the severity of scar tissue formation. Indeed, the most significant difference between normal tissue and scar tissue is the orientation of the fibrous matrix, leading, as a consequence, to a denser, less elastic connective tissue [21].

Effects of Architecture in Dense Collagen

The orientation of the fibers provides strength and resistance and acts as a guide for the cells to attach and grow in an oriented pattern. Such orientations, as the one observed in bone, are essential for the complete understanding of the resulting tissue´s mechanical properties. Giraud-Guille and collaborators studied the organization of decalcified human compact bone sections and compared the obtained results to the organization of highly concentrated collagen solutions. The data obtained supported the theory involving liquid crystalline phases in the morphogenesis of dense collagen matrices. They have shown that at concentrations of 50 $mg.mL^{-1}$ or higher, acid soluble collagen undergoes liquid crystalline ordering under acidic conditions. Furthermore, they have demonstrated that procollagen, the soluble precursor of collagen, undergoes liquid crystalline ordering as well, suggesting the higher order architecture in dense collagen observed *in vivo* may be determined prior to fibril assembly [22].

The Contribution of Cells

The contribution of cells to the collagen patterns observed *in vivo* should be considered as well, as cells are actively involved in matrix remodeling *via* the secretion of metalloproteinases and crosslinking enzymes, as well as the deposition of collagen itself and contraction of the fibers already present in the extracellular matrix.

Integrins present in the cell membrane act as mechanotransducers, which give information to the cell about the mechanical properties, the dimensionality and other physical and biochemical characteristics of the surrounding media [23]. The exact mechanisms through which the cells remodel the collagen architecture in order to adapt to functional demands remain unknown. However, it is generally accepted that collagen synthesis, accumulation and organization are driven by mechanical stimuli, leading to modifications in the mechanical properties of the tissue involved [24]. To study the contribution of the different factors involved in tissue remodeling, fibroblast orientation during wound healing following a myocardial infarction was determined probabilistically, for which four guidance cues were assayed. Namely, the mechanical cues defined by strain anisotropy, structural cues defined by existing fiber alignment, chemical cues defined by chemokine gradients and lastly persistence cues defined by a pre-existing fibroblast orientation. As a result, when compared to experimental data, it was observed that the healing process and the resulting alignment was consistent with the model guided by mechanical cues, suggesting that mechanical forces play a pivotal role during wound healing, affecting fibroblast orientation and collagen organization [25].

Collagen Cross-linking Methods: Enzymatic

In addition, no review focusing on the mechanical properties of collagen would be complete without mentioning the enzyme responsible for its crosslinking *in vivo*: the copper dependent lysyl oxidase. Shortly after fibril assembly, crosslinking is initiated by the oxidative deamination of specific lysine or hydroxylysine residues, resulting in an increase in tensile strength due to the formation of intermolecular crosslinks [26]. The crosslinking mechanism involves aldehyde formation from specific telopeptide lysine or hydroxylysine residues. Immature ketoimine formation occurs through the condensation of the newly formed telopeptide aldehydes with helical lysine, hydroxylysine or hystidine residues.

Maturation to the pyridinoline and pyrrole crosslinks occurs by further reaction of the ketoimines with telopeptide aldehydes [27]. Increased enzymatic and non-enzymatic collagen crosslinking (the latter due to glycation of certain residues) reduces collagen digestibility *in vitro*, though it has been shown that not only the amount of crosslinks but also the type of crosslinks might affect collagen´s degradation, as it has been noted that the hydroxyallysine route is enhanced during irreversible fibrosis, leading to a decrease in the capacity of matrix metalloproteinase-1 (MMP-1) to degrade collagen [28].

Collagen Cross-linking Methods: Synthetic Methods

Similarly, synthetic cross-linking of collagen, which will be thoroughly reviewed within this chapter, reduces its degradation, enabling the use of less concentrated, hence less dense collagen matrices, while providing at the same time strength and stability. Consequently, different methods have been developed to enhance the mechanical properties of diluted collagen gels, for example, by means of creating co-polymers of collagen and chitosan [29], through the design of interpenetrating hyaluronic acid-collagen [30] or collagen and fibrin networks [31], through the cross-linking of gelatin gels [32] and non-enzymatic glycation [16].

Within the vast majority of collagen-based biomaterials, native or denatured collagen is cross-linked in order to improve their mechanical properties and to decrease the susceptibility towards enzymatic degradation. Numerous physical and chemical methods have been reported for collagen cross-linking, but the chemical ones are often preferred because they allow a better control over the extent of cross-linking.

There are several reviews dealing with the uses of collagen for tissue engineering, drug delivery and the developments of new biomaterials in general [33 - 37]. When dealing with collagen gels and hydrogels, several cross linking methods were developed in order to increase stability, duration and strength of gels as can be seen in Table **1**. In an extensive review, Hennink and Nostruem cover a wide selection of hydrogel cross linking methods [38, 39].

Regarding chemical cross-linking strategies, Bertrand *et al.*, compared two cross-linking methods for collagen gels to be implanted in rat transected spinal cords [40]. Their results showed co-precipitation with chondroitin 6-sulphate and carbodiimide cross-linkage improving the mechanical properties of the matrix, modifying the normal scarring process, and favoring axonal regeneration.

Al-Ammar *et al.*, obtained an increase in bond strength by means of cross linking collagen with glutaraldehyde and grape seed extract within the dentin matrix [41], a result that shows that the application of selective collagen cross-linkers during adhesive restorative procedures improves dentin bond strength properties.

On the other hand, Sundararaghavan *et al.* [45]; Sung *et al.* [46], and Chang *et al.* [48], used genipin as a covalent collagen crosslinker. Their results show that genipin-induced cross-linking improves the mechanical properties of tissue constructs, but cell viability has to be carefully taken into account, even when the antigenicity of the genipin is low compared to other reagents. These last authors found out that before implantation, the denaturalization temperatures of acellular glutaraldehyde fixed gels *vs* genipin fixed ones was higher, but no difference was observed in those gels containing cells. They also found out that tensile strength values of all test samples were similar before implantation. These values in all test samples declined significantly with the increase in the duration of implantation prior to 4-week postoperatively. However, with the exception of the glutaraldehyde- fixed acellular tissue, the tensile-strength values of all other test samples increased steadily afterwards.

Collagen treated with high concentrations of glutaraldehyde gives materials with high mechanical properties, good resistance to proteolytic enzymes, and low adhesion properties towards cells. However, after implantation of such collagen-based biomaterials, polymeric chains of glutaraldehyde are slowly hydrolyzed into cytotoxic monomers, and calcification occurs. When low concentrations of glutaraldehyde are used, cross-linked collagen exhibits a high degree of biocompatibility, and can be used for example, for soft tissue augmentation in plastic surgery and urology [53].

Table 1. Overview of collagen cross-linking methods.

Crosslinktype	Advantages	Uses	References
Carbodiimide	Improved mechanical properties	Transected spinal cord implant.	[40]
Grape seed extract	Improved dentin bond strength	Dental reconstitution	[41]
Glutaraldehyde	Improved dentin bond strength	Dental reconstitution	[41]
	Improved mechanical properties	Basic study on cellular and acellular collagen	[42]
	Increased stiffness	Basic study on cellular and acellular collagen	[43]
	Improved mechanical properties, increased stiffness, drug liberation	Subdermal implants	[44]
Genipin	No difference with control	Dental reconstitution	[41, 45]
	Improved mechanical properties of tissues	Collagen implants	[46]
	Changes in denaturation temperature		[47]
	Tensile strength value increased		[48]
Dimethyl 3,3'-dithiobispropionimidate	Increased tensile strength, less cytotoxic than glutaraldehyde	Bioprosthetic value and tendon replacement	[42]
Dyes	Increased denaturation temperature, improved mechanical properties	Prosthetics	[49]
	Higher thermal relaxation	Prosthetic devices	[50]
Nonenzymatic glycation	Higher tensile stress, higher elasticity modulus, higher energy absorbed and tougher than control	Implants	[51]
Polymeric polyaldehydes	Increase in denaturalization temperature, reduced swelling, less citotoxicity *vs* glutaraldehyde	Implants	[52]

Indeed, when taking into account glutaraldehyde as cross-linking agent, several authors have reported an increase in the mechanical properties and an improvement in the assays compared to control groups. Glutaraldehyde cross-linking involves lysines and hydroxylysines of collagen. As stated before, the use of glutaraldehyde in dentine has a favorable effect in the dentin bond strength properties [41].

Chandran *et al.*, studied glutaraldehyde crosslinks at the microstructural levels.

Their results show that collagen cross-linkage was increased by either increasing the glutaraldehyde concentration or the reaction time. Increasing cross-linking resulted in decreased strain response and decreased coupling between the fiber and cross-fiber direction. The study concludes that the mechanical changes in cross-linked collagen gels are caused by the microstructural mechanism of increased resistance to fibril rotation. Furthermore, Raub *et al.*, worked on the microstructure and bulk mechanical properties relationship by studying the Young modulus from cellular and acellular collagen [43]. They observed that the processes of matrix synthesis, degradation and remodeling may explain the generally reduced Young modulus of cellularized gels *versus* the acellular ones, where the indentation tests showed agreement with a semiflexible network. In a work by Haugh *et al.*, in 2011, the conclusion reached after performing crosslink studies with both glutaraldehyde and 1-ethyl-3-3-dimethyl aminopropylcarbodiimide was that crosslinking can be used to produce collagen-glycosaminoglycan scaffolds with a range of compressive moduli, and that increased stiffness enhances cellular activity within the scaffolds. When cells were seeded onto the scaffolds, the stiffest scaffolds also showed increased cell numbers and enhanced cellular distributions when compared to the other groups studied [54].

Foglia *et al.,* on the other hand, managed to obtain hybrid collagen-silica matrices by covalently linking an aminosilane to glutaraldehyde fixed collagen hydrogels. This silica grafted collagen gels conserved the fibrillar network of pure collagen hydrogels and presented enhanced mechanical properties (Fig. **2**) [44]. When considering silica reinforced collagen hydrogels, these have also been obtained by mixing either silica precursors like sodium silicate or silica nanoparticles with an acid solubilised collagen type I mixture. In the first case, the materials were obtained by *in situ* polymerization of silicates, resulting in a continuous coating of collagen by silica. As a result, a significant increase in collagen thermal stability was observed. Alternatively, the peak temperature of collagen denaturation for pure collagen did not change significantly when silica nanoparticles, irrespective of their size and concentration, were present in the system, while both storage and loss moduli, G' and G", increased (Fig. **3**) [9, 55 - 57].

Charulatha and Rajaram studied the difference between crosslinking treatments on collagen membranes and concluded that glutaraldehyde had the highest efficiency and the highest value of tensile strength [42]. On the other hand, dimethyl 3,3'-dithiobispropionimidate (DTBP) crosslinked gels had physical properties comparable to those crosslinked by glutaraldehyde. However, DTBP treated matrices are much more biocompatible than glutaraldehyde treated matrices. Hence, it was suggested that DTBP may be a better reagent to stabilize collagen matrices for bioprosthetic valve and tendon replacements.

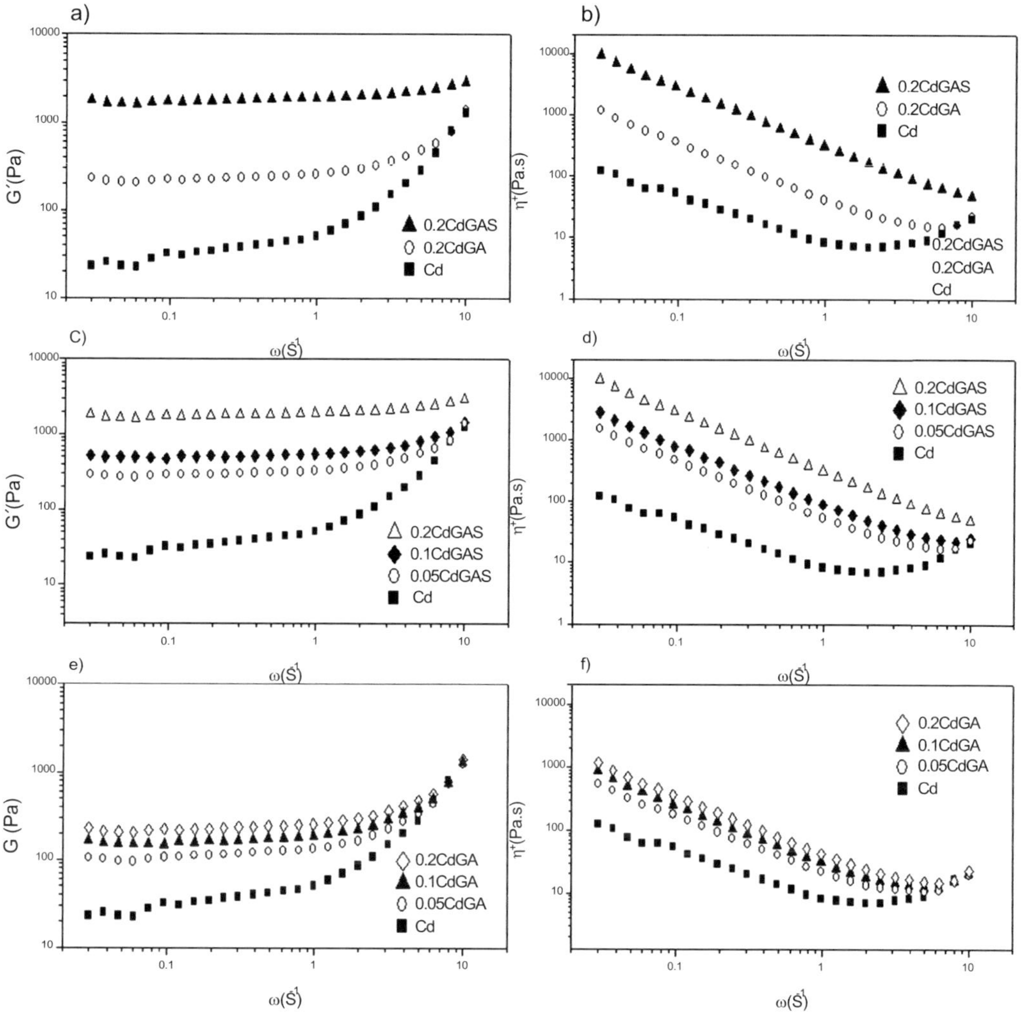

Fig. (2). Frequency dependence of the elastic modulus (G', in Pa) and the complex viscosity (η, in Pas) of all the materials under study at 22°C. (**a**) and (**b**) show the influence of glutaraldehyde fixation and posterior silica grafting for the highest glutaraldehyde concentration (0.2 M), (**c** and **d**) the effect of silica content and (**e** and **f**) the effect of glutaraldehyde fixation on the rheological properties assayed in comparison with pure collagen scaffolds [44]. -Reproduced by permission of The Royal Society of Chemistry.

Alternatively, Tuttolomondo *et al.*, managed to crosslink collagen fibers by using a reactive azo dye. Their results show a significant enhancement of the thermal stability and rheological properties of the protein network [49] (Fig. **4**).

Adams *et al.*, crosslinked collagen fibers by means of dye mediated photooxidation. Their results show that the histidine is the aminoacid involved in the reaction. Thermal relaxation analysis indicated a significantly higher thermal stability for photooxidized porcine cusps *versus* untreated controls, with mean relaxation times of 14000 ± 4650 and 22900 ± 2480 s respectively. In summary, porcine aortic valve tissues treated by dye-mediated photooxidation contain new chemical species and exhibit properties consistent with intermolecular crosslinks formation, which explain the increased biostability of this material and its potential use in bioprosthetic devices [50].

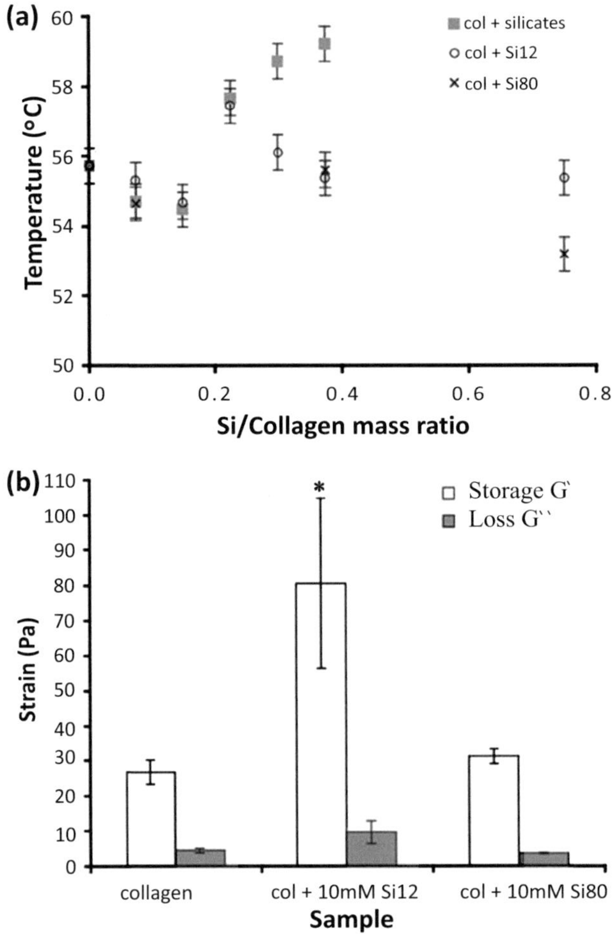

Fig. (3). Effect of silica addition on (**a**) the denaturation temperature of collagen and (**b**) the storage (G`) and loss (G`` moduli of selected materials, All tests were performed at 37°C [9]. - Reproduced by permission of The Royal Society of Chemistry.

Additionally, connective tissues are known to undergo non enzymatic glycation. Reddy *et al.*, studied the extent of this effect in the rabbit Achilles tendon. Their results showed that after 4 weeks of incubation the differences between treatments were significant. Glycated tendons withstood more load and tensile stress, attained significantly higher modulus of elasticity, absorbed more energy, and became tougher than controls. The maximum increases in the biomechanical measurements as a result of glycation were 29% for maximum load, 125% for stress, 19% for strain, 106% for Young's modulus of elasticity, 14% for energy to yield, and 57% for toughness. Biochemical analysis showed a 61% reduction in the extractability of neutral salt-soluble collagen, a 48% decrease in acid soluble collagen, and a 29% decline in pepsin-soluble collagen in glycated tendons. In contrast, there was a 28% increase in the amount of insoluble collagen and significantly higher amounts of pentosidine links in glycated tendons [51]. Even when the study was performed with the aim to show the effect of diabetes in tendons, crosslinking by glycation is an alternative way to change collagen mechanical properties.

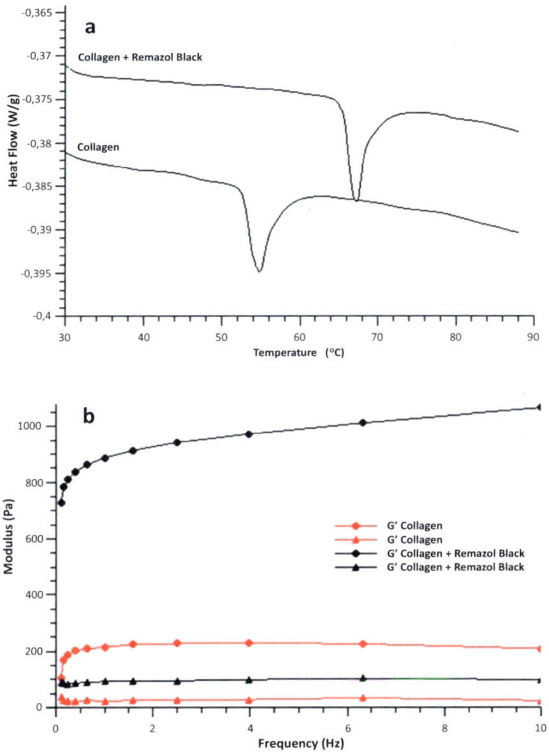

Fig. (4). a) difference in the decomposition temperature between collagen and crosslinked collagen with Remazol black. **B**) Difference in G" between .collagen and crosslinked collagen with Remazol black. The rheological behavior of the samples was investigated under sinusoidal deformations with a frequency of oscillations ranging from 0.1 to 100 Hz at set strain (0.1%) and under a normal force of 0.04 N [49]. Reproduced by permission of The Royal Society of Chemistry.

Rousseau *et al.*, concerned with the byproducts of glutaraldehyde degradation and toxic products liberation, presented another way of crosslinking collagen fibers by using polymeric polyaldehydes. These modified polysaccharides with free -CHO groups included oxidized regenerated cellulose and carboxymethylcellulose, oxidized starch, dextran dialdehyde and other polysaccharides. Their experiments showed an increase in the denaturation temperature *vs* pure collagen, reduced swelling and decreased cytotoxicity [52].

CONCLUDING REMARKS

Different extraction and reconstitution conditions influence the mechanical properties, thermal stability, enzymatic digestion and biocompatibility of collagen based biomaterials. These developments involve the utilization of higher collagen concentrations as well as modifications in pH, temperature or ionic strength. In addition, various cross linking methods have been developed to improve collagen´s properties. All together, these advances result in a significant enhancement of the thermal stability and rheological properties of the protein network.

CONFLICT OF INTEREST

The author (editor) declares no conflict of interest, financial or otherwise.

ACKNOWLEDGEMENT

The authors would like to acknowledge the support of grants from the University of Buenos Aires UBACYT 20020110100081, from CONICET PIP 11220120100657CO and from Agencia Nacional de Promoción Científica y Técnológica PICT 2012-1441.

REFERENCES

[1] P.P. Provenzano, and R. Vanderby Jr, "Collagen fibril morphology and organization: implications for force transmission in ligament and tendon", *Matrix Biol.,* vol. 25, no. 2, pp. 71-84, 2006. [http://dx.doi.org/10.1016/j.matbio.2005.09.005] [PMID: 16271455]

[2] H. Oxlund, and T.T. Andreassen, "The roles of hyaluronic acid, collagen and elastin in the mechanical properties of connective tissues", *J. Anat.,* vol. 131, no. Pt 4, pp. 611-620, 1980. [PMID: 7216901]

[3] H. Oxlund, J. Manschot, and A. Viidik, "The role of elastin in the mechanical properties of skin", *J. Biomech.,* vol. 21, no. 3, pp. 213-218, 1988. [http://dx.doi.org/10.1016/0021-9290(88)90172-8] [PMID: 3379082]

[4] E. Bell, B. Ivarsson, and C. Merrill, "Production of a tissue-like structure by contraction of collagen lattices by human fibroblasts of different proliferative potential in vitro", *Proc. Natl. Acad. Sci. USA,* vol. 76, no. 3, pp. 1274-1278, 1979. [http://dx.doi.org/10.1073/pnas.76.3.1274] [PMID: 286310]

[5] M.K. Keech, R. Reed, and M.J. Wood, "Further observations on the transformation of collagen fibrils

into elastin: an electron-microscopic study", *J. Pathol. Bacteriol.,* vol. 71, no. 2, pp. 477-493, 1956. [http://dx.doi.org/10.1002/path.1700710221] [PMID: 13398892]

[6] G.C. Wood, and M.K. Keech, "The formation of fibrils from collagen solutions. 1. The effect of experimental conditions: kinetic and electron-microscope studies", *Biochem. J.,* vol. 75, pp. 588-598, 1960. [http://dx.doi.org/10.1042/bj0750588] [PMID: 13845809]

[7] B.A. Roeder, K. Kokini, J.E. Sturgis, J.P. Robinson, and S.L. Voytik-Harbin, "Tensile mechanical properties of three-dimensional type I collagen extracellular matrices with varied microstructure", *J. Biomech. Eng.,* vol. 124, no. 2, pp. 214-222, 2002. [http://dx.doi.org/10.1115/1.1449904] [PMID: 12002131]

[8] M. Achilli, and D. Mantovani, "Tailoring mechanical properties of collagen-based scaffolds for vascular tissue engineering: the effects of pH, temperature and ionic strength on gelation", *Polymers (Basel),* vol. 4, no. 2, p. 664, 2010. [http://dx.doi.org/10.3390/polym2040664]

[9] M.F. Desimone, C. Hélary, I.B. Rietveld, I. Bataille, G. Mosser, M.M. Giraud-Guille, J. Livage, and T. Coradin, "Silica-collagen bionanocomposites as three-dimensional scaffolds for fibroblast immobilization", *Acta Biomater.,* vol. 6, no. 10, pp. 3998-4004, 2010. [http://dx.doi.org/10.1016/j.actbio.2010.05.014] [PMID: 20493975]

[10] C. Helary, A. Abed, G. Mosser, L. Louedec, A. Meddahi-Pellé, and M.M. Giraud-Guille, "Synthesis and in vivo integration of improved concentrated collagen hydrogels", *J. Tissue Eng. Regen. Med.,* vol. 5, no. 3, pp. 248-252, 2011. [http://dx.doi.org/10.1002/term.326] [PMID: 20665712]

[11] C. Helary, A. Foucault-Bertaud, G. Godeau, B. Coulomb, and M.M. Guille, "Fibroblast populated dense collagen matrices: cell migration, cell density and metalloproteinases expression", *Biomaterials,* vol. 26, no. 13, pp. 1533-1543, 2005. [http://dx.doi.org/10.1016/j.biomaterials.2004.05.016] [PMID: 15522755]

[12] M.M. Giraud Guille, C. Helary, S. Vigier, and N. Nassif, "Dense fibrillar collagen matrices for tissue repair", *Soft Matter,* vol. 20, no. 6, pp. 4963-4967, 2010. [http://dx.doi.org/10.1039/c0sm00260g]

[13] Y. Wang, J. Silvent, M. Robin, F. Babonneau, A. Meddahi-Pelle, and N. Nassif, "Controlled collagen assembly to build dense tissue-like materials for tissue engineering", *Soft Matter,* vol. 20, no. 7, pp. 9659-9664, 2011. [http://dx.doi.org/10.1039/c1sm05868a]

[14] S. Vigier, C. Catania, B. Baroukh, J.L. Saffar, M.M. Giraud-Guille, and M.L. Colombier, "Dense fibrillar collagen matrices sustain osteoblast phenotype in vitro and promote bone formation in rat calvaria defect", *Tissue Eng. Part A,* vol. 17, no. 7-8, pp. 889-898, 2011. [http://dx.doi.org/10.1089/ten.tea.2010.0336] [PMID: 21034313]

[15] P. Fratzl, "Collagen: Structure and mechanics, an introduction", In: *Collagen: Structure and Mechanics.* Springer US: New York City, NY, 2008, pp. 1-13. [http://dx.doi.org/10.1007/978-0-387-73906-9_1]

[16] B.N. Mason, and C.A. Reinhart-King, "Controlling the mechanical properties of three-dimensional matrices *via* non-enzymatic collagen glycation", *Organogenesis,* vol. 9, no. 2, pp. 70-75, 2013. [http://dx.doi.org/10.4161/org.24942] [PMID: 23811696]

[17] A. Page-McCaw, A.J. Ewald, and Z. Werb, "Matrix metalloproteinases and the regulation of tissue remodelling", *Nat. Rev. Mol. Cell Biol.,* vol. 8, no. 3, pp. 221-233, 2007. [http://dx.doi.org/10.1038/nrm2125] [PMID: 17318226]

[18] K. Sakao, K.A. Takahashi, O. Mazda, Y. Arai, H. Tonomura, A. Inoue, M. Saito, M. Fujioka, H. Takamiya, J. Imanishi, and T. Kubo, "Enhanced expression of interleukin-6, matrix metalloproteinase-13, and receptor activator of NF-kappaB ligand in cells derived from osteoarthritic subchondral bone",

J. Orthop. Sci., vol. 13, no. 3, pp. 202-210, 2008.
[http://dx.doi.org/10.1007/s00776-008-1227-5] [PMID: 18528653]

[19] O. Irtyuga, N. Pimenova, I. Voronkina, V. Uspensky, E. Moguchaya, and M. Boyarinova, "Role of extracellular matrix proteins, matrix metalloproteinase activity and arterial stiffness in ascending aorta dilatation", *Eur. Heart J.,* vol. 34, suppl. Suppl. 1, p. 861, 2013.
[http://dx.doi.org/10.1093/eurheartj/eht310.P4763]

[20] M. McNulty, A. Mahmud, P. Spiers, and J. Feely, "Collagen type-I degradation is related to arterial stiffness in hypertensive and normotensive subjects", *J. Hum. Hypertens.,* vol. 20, no. 11, pp. 867-873, 2006.
[http://dx.doi.org/10.1038/sj.jhh.1002015] [PMID: 16598292]

[21] S. McDougall, J. Dallon, J. Sherratt, and P. Maini, "Fibroblast migration and collagen deposition during dermal wound healing: mathematical modelling and clinical implications", *Philos Trans A Math Phys Eng Sci.,* vol. 364, no. 1843, pp. 1385-1405, 2006.

[22] M.M. Giraud-Guille, E. Belamie, G. Mosser, C. Helary, F. Gobeaux, and S. Vigier, "Liquid crystalline properties of type I collagen: Perspectives in tissue morphogenesis", *C. R. Chim.,* vol. 3, no. 11, pp. 245-252, 2008.
[http://dx.doi.org/10.1016/j.crci.2007.05.005]

[23] M. Larsen, V.V. Artym, J.A. Green, and K.M. Yamada, "The matrix reorganized: extracellular matrix remodeling and integrin signaling", *Curr. Opin. Cell Biol.,* vol. 18, no. 5, pp. 463-471, 2006.
[http://dx.doi.org/10.1016/j.ceb.2006.08.009] [PMID: 16919434]

[24] F. Baaijens, C. Bouten, and N. Driessen, "Modeling collagen remodeling", *J. Biomech.,* vol. 43, no. 1, pp. 166-175, 2010.
[http://dx.doi.org/10.1016/j.jbiomech.2009.09.022] [PMID: 19818962]

[25] A.D. Rouillard, and J.W. Holmes, "Mechanical regulation of fibroblast migration and collagen remodelling in healing myocardial infarcts", *J. Physiol.,* vol. 590, no. 18, pp. 4585-4602, 2012.
[http://dx.doi.org/10.1113/jphysiol.2012.229484] [PMID: 22495588]

[26] R.J. McCormick, "Extracellular modifications to muscle collagen: implications for meat quality", *Poult. Sci.,* vol. 78, no. 5, pp. 785-791, 1999.
[http://dx.doi.org/10.1093/ps/78.5.785] [PMID: 10228977]

[27] L. Knott, and A.J. Bailey, "Collagen cross-links in mineralizing tissues: a review of their chemistry, function, and clinical relevance", *Bone,* vol. 22, no. 3, pp. 181-187, 1998.
[http://dx.doi.org/10.1016/S8756-3282(97)00279-2] [PMID: 9514209]

[28] A.J. Van Der Slot-Verhoeven, E.A. Van Dura, J. Attema, B. Blauw, J. DeGroot, and T.W. Huizinga, "The type of collagen cross-link determines the reversibility of experimental skin fibrosis", *acta, Mol. basis dis.,* vol. 1, no. 1740, pp. 60-67, 2005.

[29] L.A. Reis, L.L. Chiu, Y. Liang, K. Hyunh, A. Momen, and M. Radisic, "A peptide-modified chitosan-collagen hydrogel for cardiac cell culture and delivery", *Acta Biomater.,* vol. 8, no. 3, pp. 1022-1036, 2012.
[http://dx.doi.org/10.1016/j.actbio.2011.11.030] [PMID: 22155066]

[30] S. Suri, and C.E. Schmidt, "Photopatterned collagen-hyaluronic acid interpenetrating polymer network hydrogels", *Acta Biomater.,* vol. 5, no. 7, pp. 2385-2397, 2009.
[http://dx.doi.org/10.1016/j.actbio.2009.05.004] [PMID: 19446050]

[31] V.K. Lai, S.P. Lake, C.R. Frey, R.T. Tranquillo, and V.H. Barocas, "Mechanical behavior of collagen-fibrin co-gels reflects transition from series to parallel interactions with increasing collagen content", *J. Biomech. Eng.,* vol. 134, no. 1, p. 011004, 2012.
[http://dx.doi.org/10.1115/1.4005544] [PMID: 22482659]

[32] A. Bigi, G. Cojazzi, S. Panzavolta, K. Rubini, and N. Roveri, "Mechanical and thermal properties of gelatin films at different degrees of glutaraldehyde crosslinking", *Biomaterials,* vol. 22, no. 8, pp. 763-

768, 2001.
[http://dx.doi.org/10.1016/S0142-9612(00)00236-2] [PMID: 11246944]

[33] C. Helary, and M.F. Desimone, "Recent advances in biomaterials for tissue engineering and controlled drug delivery", *Curr. Pharm. Biotechnol.,* vol. 16, no. 7, pp. 635-645, 2015.
[http://dx.doi.org/10.2174/138920101607150427112208] [PMID: 25934973]

[34] C. Helary, and M.F. Desimone, "Editorial: pharmaceutical biotechnology for tissue repair", *Curr. Pharm. Biotechnol.,* vol. 16, no. 7, pp. 580-581, 2015.
[http://dx.doi.org/10.2174/138920101607150427111433] [PMID: 25934970]

[35] S. Heinemann, T. Coradin, and M.F. Desimone, "Bio-inspired silica-collagen materials: Applications and perspectives in the medical field", *Biomater. Sci.,* vol. 7, no. 1, pp. 688-702, 2013.
[http://dx.doi.org/10.1039/c3bm00014a]

[36] D.G. Wallace, and J. Rosenblatt, "Collagen gel systems for sustained delivery and tissue engineering", *Adv. Drug Deliv. Rev.,* vol. 55, no. 12, pp. 1631-1649, 2003.
[http://dx.doi.org/10.1016/j.addr.2003.08.004] [PMID: 14623405]

[37] Rm. Parenteau-Bareil, R. Gauvin, and Fo. Berthod, "Collagen-Based Biomaterials for Tissue Engineering Applications", *Material,* vol. 3, no. 3, pp. 1863-1887, 2012.
[http://dx.doi.org/10.3390/ma3031863]

[38] WE Hennink, and CF van Nostrum, "Novel crosslinking methods to design hydrogels", *Adv. Drug Deliv. Rev.,* vol. suppl., no. 64, pp. 223-236, 2012.
[http://dx.doi.org/10.1016/j.addr.2012.09.009]

[39] W.E. Hennink, and C.F. van Nostrum, "Novel crosslinking methods to design hydrogels", *Adv. Drug Deliv. Rev.,* vol. 54, no. 1, pp. 13-36, 2002.
[http://dx.doi.org/10.1016/S0169-409X(01)00240-X] [PMID: 11755704]

[40] R. Marchand, S. Woerly, L. Bertrand, and N. Valdes, "Evaluation of two cross-linked collagen gels implanted in the transected spinal cord", *Brain Res. Bull.,* vol. 30, no. 3-4, pp. 415-422, 1993.
[http://dx.doi.org/10.1016/0361-9230(93)90273-E] [PMID: 8457891]

[41] A. Al-Ammar, J.L. Drummond, and A.K. Bedran-Russo, "The use of collagen cross-linking agents to enhance dentin bond strength", *J. Biomed. Mater. Res. B Appl. Biomater.,* vol. 91, no. 1, pp. 419-424, 2009.
[http://dx.doi.org/10.1002/jbm.b.31417] [PMID: 19507140]

[42] V. Charulatha, and A. Rajaram, "Influence of different crosslinking treatments on the physical properties of collagen membranes", *Biomaterials,* vol. 24, no. 5, pp. 759-767, 2003.
[http://dx.doi.org/10.1016/S0142-9612(02)00412-X] [PMID: 12485794]

[43] CB Raub, AJ Putnam, BJ Tromberg, and SC George, "Predicting bulk mechanical properties of cellularized collagen gels using multiphoton microscopy", *Acta Biomater.,* vol. 12, no. 6, pp. 4657-4665, 2010.
[http://dx.doi.org/10.1016/j.actbio.2010.07.004]

[44] M.L. Foglia, D.E. Camporotondi, G.S. Alvarez, S. Heinemann, T. Hanke, C.J. Perez, L.E. Díaz, and M.F. Desimone, "A new method for the preparation of biocompatible silica coated-collagen hydrogels", *J. Mater. Chem. B Mater. Biol. Med.,* vol. 45, no. 1, pp. 6283-6290, 2013.
[http://dx.doi.org/10.1039/c3tb21067g]

[45] H.G. Sundararaghavan, G.A. Monteiro, N.A. Lapin, Y.J. Chabal, J.R. Miksan, and D.I. Shreiber, "Genipin-induced changes in collagen gels: correlation of mechanical properties to fluorescence", *J. Biomed. Mater. Res. A,* vol. 87, no. 2, pp. 308-320, 2008.
[http://dx.doi.org/10.1002/jbm.a.31715] [PMID: 18181104]

[46] H-W. Sung, Y. Chang, C-T. Chiu, C-N. Chen, and H-C. Liang, "Mechanical properties of a porcine aortic valve fixed with a naturally occurring crosslinking agent", *Biomaterials,* vol. 20, no. 19, pp. 1759-1772, 1999.

[http://dx.doi.org/10.1016/S0142-9612(99)00069-1] [PMID: 10509186]

[47] B. Xu, H. Li, and Y. Zhang, "Understanding the viscoelastic behavior of collagen matrices through relaxation time distribution spectrum", *Biomatter,* vol. 3, no. 3, p. e24651, 2013. [http://dx.doi.org/10.4161/biom.24651] [PMID: 23628869]

[48] Y. Chang, C-C. Tsai, H-C. Liang, and H-W. Sung, "in vivo evaluation of cellular and acellular bovine pericardia fixed with a naturally occurring crosslinking agent (genipin)", *Biomaterials,* vol. 23, no. 12, pp. 2447-2457, 2002. [http://dx.doi.org/10.1016/S0142-9612(01)00379-9] [PMID: 12033592]

[49] M.V. Tuttolomondo, J.M. Galdopórpora, L. Trichet, H. Voisin, T. Coradin, and M.F. Desimone, "Dye-collagen interactions. Mechanism, kinetic and thermodynamic analysis", *RSC Advances,* vol. 71, no. 5, pp. 57395-57405, 2015. [http://dx.doi.org/10.1039/C5RA08611F]

[50] A.K. Adams, E.A. Talman, L. Campbell, B.K. McIlroy, and M.A. Moore, "Crosslink formation in porcine valves stabilized by dye-mediated photooxidation", *J. Biomed. Mater. Res.,* vol. 57, no. 4, pp. 582-587, 2001. [http://dx.doi.org/10.1002/1097-4636(20011215)57:4<582::AID-JBM1205>3.0.CO;2-6] [PMID: 11553889]

[51] G.K. Reddy, L. Stehno-Bittel, and C.S. Enwemeka, "Glycation-induced matrix stability in the rabbit achilles tendon", *Arch. Biochem. Biophys.,* vol. 399, no. 2, pp. 174-180, 2002. [http://dx.doi.org/10.1006/abbi.2001.2747] [PMID: 11888203]

[52] C.F. Rousseau, and C.H. Gagnieu, "in vitro cytocompatibility of porcine type I atelocollagen crosslinked by oxidized glycogen", *Biomaterials,* vol. 23, no. 6, pp. 1503-1510, 2002. [http://dx.doi.org/10.1016/S0142-9612(01)00276-9] [PMID: 11829447]

[53] "Gagnieu CH. "in vitro cytocompatibility of porcine type I atelocollagen crosslinked by oxidized glycogen", *Biomaterials,* vol. 6, no. 23, pp. 1503-1510, 2002.

[54] M.G. Haugh, C.M. Murphy, R.C. McKiernan, C. Altenbuchner, and F.J. OBrien, "Crosslinking and mechanical properties significantly influence cell attachment, proliferation, and migration within collagen glycosaminoglycan scaffolds", *Tissue Eng. Part A,* vol. 17, no. 9-10, pp. 1201-1208, 2011. [http://dx.doi.org/10.1089/ten.tea.2010.0590] [PMID: 21155630]

[55] S. Quignard, G.J. Copello, C. Aime, I. Bataille, C. Helary, M.F. Desimone, and T. Coradin, "Influence of silicification on the structural and biological properties of buffer-mediated collagen hydrogels", *Adv. Eng. Mater.,* vol. 3, no. 14, pp. B51-B55, 2012. [http://dx.doi.org/10.1002/adem.201180063]

[56] M.F. Desimone, C. Hélary, S. Quignard, I.B. Rietveld, I. Bataille, G.J. Copello, G. Mosser, M.M. Giraud-Guille, J. Livage, A. Meddahi-Pellé, and T. Coradin, "in vitro studies and preliminary in vivo evaluation of silicified concentrated collagen hydrogels", *ACS Appl. Mater. Interfaces,* vol. 3, no. 10, pp. 3831-3838, 2011. [http://dx.doi.org/10.1021/am2009844] [PMID: 21910471]

[57] MF Desimone, C Helary, G Mosser, MM Giraud-Guille, J Livage, and T Coradin, "Fibroblast encapsulation in hybrid silica-collagen hydrogels", *J. Mater. Chem.,* vol. 4, no. 20, pp. -68, 2010. [http://dx.doi.org/10.1039/B921572G]

CHAPTER 9

Protein Plastic Foams

Chanelle Gavin, Mark C. Lay*, Casparus J.R. Verbeek* and **Anuradha Walallavita**

School of Engineering, Faculty of Science and Engineering, University of Waikato, Hamilton, New Zealand

Abstract: The development of protein-based biopolymers has been driven by the increasing global demand for polymer products, the need for sustainable practice within this industry and the availability of low cost by-products, such as high protein content meals. The continuing development of new and existing protein-based biopolymers will enable these materials to help supplement the increasing global demand for polymer products and to develop new markets with their niche applications. To date various compositions of protein-based biopolymers have been successfully used to produce injection moulded articles, films and foams. Biopolymers typically display poor foaming behavior and commonly produce foams with irregular morphology and high densities. Protein-based biopolymers are no exception, therefore it is important to fully understand how the foaming mechanisms of bubble nucleation, growth and stabilization are affected by the inherently different properties of these materials.

This chapter aims to review the production of stable protein-based foams for use in applications such as cushioning, insulation and packaging through a variety of methods. The review specifically focuses on the production of protein-based foams through thermosetting, the emerging role of proteins as a renewable substitute in polyurethane production and the application of thermoplastic foam technologies to protein-based thermoplastics, with an emphasis on batch and extrusion foaming methods. The similarities and differences between the production of traditional foams and those produced from proteins are highlighted here. Discussion of foam morphologies, properties and processing conditions is also included. Overall, this chapter intends to provide the reader with a greater understanding of the existing research and the current challenges associated with the production of protein-based thermoplastic and thermoset foams.

Keywords: Batch foaming, Extrusion foaming, Polyurethane foams, Protein biopolymers, Protein foams, Thermoplastic proteins, Thermoset foams.

* **Corresponding authors Mark C. Lay and Casparus J.R. Verbeek:** School of Engineering, Faculty of Science and Engineering, University of Waikato, Hamilton, New Zealand; Tel: +64 7 856 2889; E-mails: mark.lay@waikato.ac.nz, johan.verbeek@waikato.ac.nz

Martin Masuelli & Denis Renard (Eds.)

INTRODUCTION

This review examines the application of protein-based biopolymers to the polymer foam industry and discusses potential manufacturing methods for stable protein-based foams intended for use as packaging, insulation or cushioning. Until now, reviews regarding protein foaming have typically been limited to those relevant to the food industry with very little emphasis on their application to or use in thermoset or thermoplastic foams [1]. However with increasing interest in thermoplastic proteins for producing injection molded articles and films it is only logical that the industry also begins to assess the foaming ability of these materials on a wider scale.

The development of protein-based foams with similar or enhanced properties in comparison to their traditional polymer counterparts creates an entirely new market at the intersection of the polymer foam industry and the growing biopolymers sector. Recent reports regarding the polymer processing sector show that the global demand for polymer foam products is increasing, such that by 2019 it is predicted that the sector will generate 25 million tons of material. This represents an increase in production of nearly 32% on the 2013 figures, when the industry totaled 19 million tons (equivalent to $USD 87 billion) [2]. The majority of materials consumed by this sector are non-renewable and/or non-biodegradable including polyurethane, polystyrene and polyolefin materials such as polypropylene and polyethylene [2].

Traditional polymeric materials are being supplemented, and to some extent replaced, by biopolymers to address issues regarding the sustainability of this industry. The leading biopolymers are currently starch and PLA, however the use of protein-based biopolymers is also increasing, especially within the US [3]. Combined, this sector is expected to generate $USD 3.67 billion in revenue in 2018 [4, 5]. Unfortunately, the economics relating to the proportion of these materials which are foamed are as yet unpublished, hence the exact size and value of this market is unknown. Furthermore, a number of barriers exist for these and other biopolymer foams including consumer acceptance, poor thermal stability and the associated difficulty and cost of manufacture [6].

In general, the production of foamed polymeric products involves the introduction of a gaseous phase to the polymer system through chemical blowing agents (CBA) or physical blowing agents (PBA), the expansion or evolution of this gas due to changes in physical parameters such as pressure or temperature and the subsequent solidification of the resulting structure (usually by temperature changes or crosslinking) before the cells rupture or collapse due to condensation of the gaseous phase [7, 8]. Following stabilization the blowing agent is eventually replaced by air as the gaseous species exchange across the cellular matrix (Fig. **1**). This phenomenon has been extensively studied and reviewed with respect to polyurethane, polystyrene, polypropylene and polyethylene foams where the behavior of these materials can be predicted relatively well in relation to physical foaming behavior; however the foaming mechanisms of bubble nucleation, growth and stabilization are yet to be fully understood.

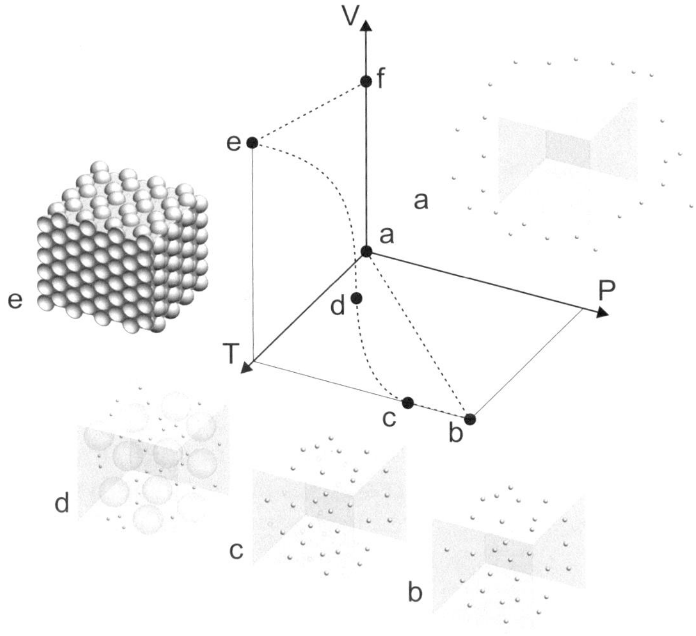

Fig. (1). Polymeric foaming process represented using a pressure, volume, and temperature diagram [7], with each stage shown by a letter and picture. The polymer and gas (**A**) is heated and pressurised to saturate the polymer (**B**), which is then depressurised so the gas changes phase and nucleates in the polymer (**C**), and as pressure drops, the remaining gas diffuses into the bubbles and the bubbles grow (**D**), until the final expanded foam is produced (**E**), and cooled (**F**).

Furthermore, the behavior of protein based biopolymers is still being characterized in relation to their structure, reactivity and processing ability. These are all related to their primary structure (amino acid sequence) and secondary structure (α-helices and β-sheets). Research is limited regarding the behavior of proteins within foaming systems. The properties of these materials, including poor rheological properties, low solubility of blowing agents and diffusion behaviour, severely hinder the foaming process. Consequently foamed biopolymers typically demonstrate non-uniform morphology and high density [6].

To date, efforts have been made to foam proteins through thermosetting methods based on existing knowledge of the crosslinking behavior of casein and formaldehyde (although the possibility of using glutaraldehyde has also been noted) [9]. Aldehydes can react with the amine group from lysine and amino acids with side chains including cysteine, tyrosine, histidine, tryptophan, and arginine (Fig. **2**). Cereal and animal-based proteins are typically high in these amino acids, which enables good crosslinking. In comparison, glutaraldehyde is more selective reacting with lysine, cysteine, histidine, and tyrosine residues only [10]. Reaction with either crosslinking agent reduces the proportion of hydrophilic amino acids in protein blends and consequently these materials display lower water absorption capacities, thereby improving the properties of the resulting foams.

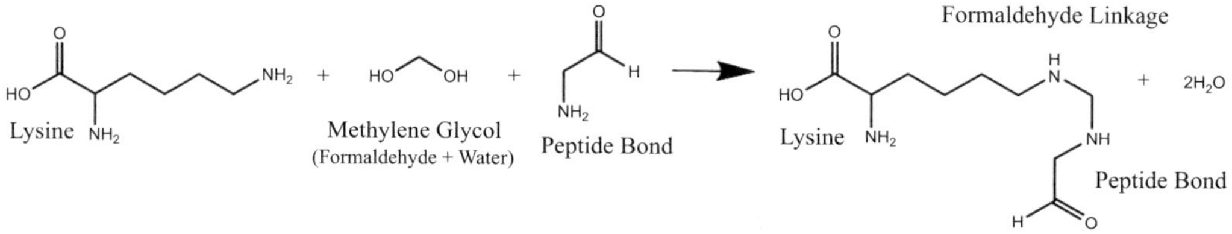

Fig. (2). Protein formaldehyde reaction chemistry [11]. Reproduced with permission, see note[1].

Proteins are also finding applications in polyurethane products in an effort to reduce production costs and increase the renewable fraction. Proteins are naturally reactive with these systems and the reactions are self-catalytic in water blown systems demonstrating how the unique chemistry of protein biopolymers, *i.e.* the presence of amide groups, can benefit this part of the market as well.

The increased interest in the production of value-added thermoplastic materials from waste animal and plant proteins has led to investigations regarding their extrudability, injection moulding capacity, film forming ability and recently their foaming ability. These protein based biopolymers have great potential to replace traditional thermoplastic materials in applications such as packaging as they can remain biodegradable after processing.

To enable thermoplastic processing, most proteins need to be blended with denaturants and plasticisers to disrupt protein bonding, lower the glass transition temperature (T_g) and enable chain mobility. A variety of proteins have been successfully utilized to produce thermoplastics although only a small proportion of these have been foamed *via* thermoplastic methods, which will be discussed in depth in this review.

This chapter will review the production and properties of thermoset protein foams, the role these protein biopolymers have in polyurethane systems, the production of thermoplastic proteins and thermoplastic foaming. The application of these materials to these techniques is discussed by drawing on existing knowledge of these systems and protein behavior in general. This review includes an overview of the efforts to foam both plant and animal proteins including: zein, gluten, collagen, myofibrillar and bloodmeal and how these materials differ from traditional thermoplastic materials with respect to their physiochemistry. The underlying principles governing the foaming phenomenon are also briefly discussed including bubble nucleation, growth and stabilization. Attention is given to the factors which govern the foaming ability of these materials such as crystallinity, glass transition temperature, molecular weight distribution and the melt behavior including shear viscosity, extensional rheology, melt strength, surface tension and strain hardening behavior.

PROTEINS AND THEIR SOURCES

Proteins are genetically encoded molecules found in living organisms where they perform vital structural or functional roles. However, these molecules find secondary applications within the plastics industry. Proteins which have been identified as possible plastic feedstocks include: animal proteins, milk (casein and whey), collagen, gelatin, keratin, egg white, myofibrillar and bloodmeal; and plant and cereal sources. While the structure of animal proteins is very diverse, as they carry out a wide range of biological functions, the proteins found in plants can be described as more or less analogous to each other with most acting as storage proteins. For example, zein is the major prolamine protein in corn/maize while kafirin is analogous, performing the same function in sorghum. Other plant proteins which have been recognised as potential feedstocks include wheat gluten, soy, sunflower, barley and peanut proteins.

Proteins are long chains of amino acids, a fraction of which are folded into structured regions containing α-helices, β-sheets and β-strands while the rest remains amorphous. Their structure is held together by hydrogen bonds, hydrophobic interactions, covalent bonds, ionic interactions and van der Waals forces. Some proteins can be readily incorporated into foams, for example native/soluble proteins can be mixed in solution and mechanically whipped or aerated to produce foams which can be set by a combination of cooking and drying. The available amide groups on native proteins and peptides can also enable them to react with isocyanates to produce polyurethane foams.

Other proteins can be manipulated to behave like thermoplastic protein by adding plasticisers and denaturants which reduce the existing bonding within and between polymer chains to reduce their glass transition temperature (dried proteins have glass transition temperatures >180°C). Sodium sulphite or sodium bisulphite break cysteine-cysteine covalent bonds between chains [12], while compounds like urea and sodium dodecyl sulphate disrupt hydrogen bonding and hydrophobic interactions respectively [13]. Plasticisers also disrupt hydrogen bonding and increase the free volume between chains allowing greater mobility [13].

Typical plasticisers include glycerol, triethylene glycol, water and other compounds which contain large proportions of hydroxyl groups [13]. This treatment is required to enable these materials to be extruded below their thermal degradation temperature. However these materials can still crosslink during processing leading to excessive pressure and torque in some cases [12]. Thermoplastic proteins have been produced from bloodmeal [14 - 18], gelatin [19, 20], keratin [21, 22], soy [23 - 26], kafirin [27], and zein [20, 27 - 30].

The feasibility of incorporating these molecules into protein-based thermoplastics is dependent upon a combination of the cost and continuity of supply of the proteins and their processability. These heteropolymers each display unique biochemistry which enables them to perform their primary role. Their structure and function of these molecules has been extensively reviewed elsewhere, although only briefly for application in bioplastics. Due to the large amount of available literature on these materials the structures, sources and function of these molecules will not be reviewed here; instead their application shall be the focus.

THERMOSET PROTEIN-BASED FOAMS

Protein foams (*e.g.* from whey or egg white) can be produced through thermosetting methods where the foamed structure is stabilized by crosslinking the protein chains by heating (*e.g.* meringue/pavlova) or chemical reagents (typically aldehydes).

In the food industry protein foaming is usually conducted by mechanical mixing (whipping) of a protein in a liquid phase to incorporate the air phase [31]. The resulting foam is dependent upon the physical-chemical properties of the solution such as viscosity, surface tension, and surface elasticity [32]. This is dependent on: (i) aggregation properties, protein-protein interactions, (ii) hydration properties, protein-water interactions and (iii) interfacial properties, such as surface tension [33].

Solidified whey protein foams have been produced from whey mixed with NaCl, Na_2CO_3 and water by high speed mechanical whipping and baking in an oven at 160°C for 40-50 minutes [33]. Whey foam blends with starch, gluten, and soy protein have also been produced. Such products are of interest to the food industry as whey has a high nutritional value and reasonable cost [34]. However whey protein is seldomly used alone in applications other than food because it is brittle and has limited mechanical strength. These properties can be improved by blending whey with cross linking agents, plasticizers or other polymers such as alginate [35].

The compressive and flexural strength of hot moulded starch-based foams have been improved by adding gluten and zein proteins [36] while adding palm oil provides water resistance. Foamed combinations of sunflower protein, cassava starch and cellulose fibres have also been produced with densities of 0.45-0.51 g/cm^3 [37].

Thermoset plastic foams have also been produced *via* chemical reaction through mechanical mixing of a solution of egg albumin, 37% formaldehyde (cross linking agent) and camphor (waxy plasticizer). Once mixed the material was microwaved for 4-10 minutes to cure and partially dry the resulting foam. The foams had densities between 0.18-0.39 g/cm^3, a porosity of 77-79% and thermal conductivities between 0.06-0.065 W/mK, with little dependence of the thermal conductivity on foam density [38]. Blends with higher water content resulted in a softer material which was more elastic. Increasing the formaldehyde content gave a more rigid foam and camphor acted as an external plasticiser which resulted in more elastic and softer foams. An optimal formulation was 4 g camphor, 30 g water, and 8 g formaldehyde to 15 g of albumin protein, and a curing time of four minutes. Adding glycerol enabled foams to retain their elasticity over the course of two months [38]. Foam flexibility was temperature dependent for glycerol containing foams but not for foams produced without this plasticizer [39].

PROTEINS INCORPORATED INTO POLYURETHANE FOAMS

Polyurethane foams require different processing technologies and have different cell morphologies compared to thermoplastic foams and other thermosets previously described. In general, polyurethane foams contain mainly polyurethane chains with some polyurea blocks and foaming is strongly linked to surface tension effects. Foam formation is governed by the reaction (Fig. **3**) between a diisocyanate and a diol to form a urethane linkage [40]. Carbon dioxide is evolved as a blowing agent through the reaction of water (1-4%) with the diisocyanate monomer. Amide groups on both terminals of the monomer also react with diisocyanate monomers to form urea linkages. Recently, research has been conducted to incorporate protein isolates, concentrates and meals into polyurethane foams due to the availability of reactive amino acid groups which can reduce production costs and give unique morphologies. To date, protein products have been successfully incorporated into both flexible [40, 41] and rigid polyurethane products [42, 43].

Flexible polyurethane foams have been produced with soy protein isolate (SPI), concentrate (SPC) or defatted soy flower (DFS) in concentrations between 0-30 wt% with varying purity (50-90%) [41]. These proteins were mixed with glycerol/propylene oxide polyether triol, a tertiary amine as catalyst, a surfactant, triethnolamine as a crosslinking agent and water (to evolve CO_2 as a blowing agent). Polymeric methylene-4,4'-diphenyldiisocyanate (pMDI) was used as the isocyanate component and the foam was allowed to rise for one hour at room temperature and cured at 25°C for one week.

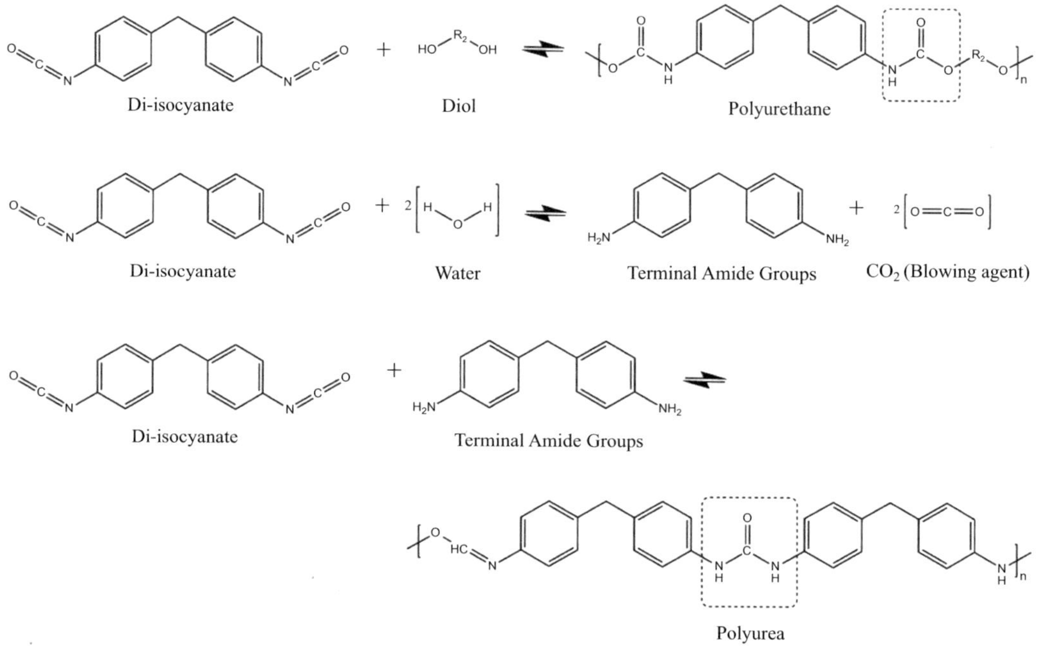

Fig. (3). Polyurethane reaction chemistry [8].

Foam density increased with higher contents of soy protein products. Foams containing DFS had a greater expansion ratio than those containing SPI or SPC due to the lower availability of active hydrogen atoms in DFS for reaction with the isocyanate. Foams containing up to 20% SPI demonstrated a higher compressive strength before reaching a plateau. This plateau was due to the protein reacting with all of the available isocyanate, resulting in a lower level of isocyanate available to evolve carbon dioxide and the excess protein weakened the foam structure. Furthermore, all foams containing soy protein products in this study showed greater resilience in comparison to polyurethane, suggesting that these may be suitable for packaging or cushioning applications [41].

A similar study produced polyurethane foams containing SPI, soy fiber and corn starch with toluene di-isocyanate (TDI) and demonstrated that using protein rather than carbohydrates (starch) produced foams with better properties [40]. At lower amounts of additives (10%), foams containing SPI had a greater resilience compared to the other two formulations at the same level. At 20 wt% all formulations displayed similar resilience while 30% resulted in higher resilience for the soy based foams (SPI and soy fibre) than corn starch.

SPI and soybean meal (SM) have also been used to produce rigid polyurethane foams [43]. This was achieved by blending polyol (PEG-400), catalyst (DABCO), foam stabiliser (L-6865), water, crosslinker (triethanolamine) and protein (SM or SPI in either an activated or non-activated state). Polymeric isocyanate (PAPI-200) was then added and the mixture stirred before curing at room temperature for 30 minutes and aged at 35°C for 24 hours [43]. Up to 30% activated SM gave rigid foams with better thermal and mechanical properties than foams with SPI. Activation of the soybean meal *via* alkali treatment and sulphur bonding reduction increased availability of reactive protein groups to react with isocyanate, increasing foam density. Fibres in SM gave a more compact structure causing higher foam densities (67.6 kg/m^3 at 30 wt% SM) compared to foams with SPI (65.3 kg/m^3) [43].

The compressive strength of foams increased with increasing SM and SPI and foams containing activated fractions demonstrated higher compressive strength (348 and 262 kPa respectively at 30 wt%) than those which were not activated (242 and 221 kPa respectively) [43]. The polysaccharide and fibre fractions in SM (absent in SPI) may have helped increase compressive strength. Activated SM and SPI gave foams with smaller cell sizes, resulting in lower thermal conductivities, 0.0265 W/mK for activated SM at 30 wt% compared to 0.0312 W/mK for inactivated SM. SM based foams had lower thermal conductivities than those with SPI [43].

Similar thermal conductivities were achieved by another study which used soy flours (SF). SF has a protein content of 50-53% compared to 46% for SM from the previous study. Polyurethane foams containing SF were prepared in a similar manner. Typical thermal conductivities for these foams with 30% SF with the same blowing agent concentration were between 0.0263-0.0266 W/mK [42]. Densities for 30% SF polyurethane foams were 43.3-38.0 kg/m^3 which is lower than those from SM. Compressive strengths were also reported and ranged from 247-265 kPa which is comparable to the non-activated (native) SM polyurethane foams previously described.

A study investigated the use of algal protein, a co-product of the biofuels industry, with polyurethane. The algal proteins were first extracted, hydrolysed and the soluble peptides and amino acids retained for freeze drying. The freeze dried material was then reacted with diamine and ethylene carbonate to produce the necessary polyol structure required for producing polyurethane. The algae polyol was then combined at a concentration of 5% with the other polyols, surfactant, water and isocyanate to produce the polyurethane foam. The algal polyols were high in hydroxyl content and very compatible with other polyols. The polymerisation process was self-catalytic and gave a rigid polyurethane foam which was less flammable than traditional polyurethane foams [44]. The resulting cellular structure was irregular open cell pores with a cell size of 0.6 ± 0.23 mm. Compared to the reference foam from commercial polyols, there was no change in core density (53 kg/m^3) but there was a decrease in resiliency, tensile strength, elongation and compressive strength when algal polyol was incorporated.

A review of the applications for a natural silk protein called sericin suggests that this protein may also behave favourably in polyurethane systems. A large number of Japanese patents are cited for the incorporation of sericin into polyurethane foams [45], however there is little available literature on these foaming systems.

THERMOPLASTIC FOAMING PRINCIPLES

The mechanisms which define foaming of thermoplastic polymers are bubble nucleation, growth and stabilisation and have been reviewed at length elsewhere for traditional systems. However, to briefly summarise, the creation of a foamed product is dependent upon the development of gas bubbles within a liquid/melt or softened phase. These bubbles must grow and stabilise to result in a permanent cellular product [46 - 48]. These processes are thermodynamically controlled and polymer must exhibit suitable rheology to successfully produce a cell structure [8].

Initial bubble formation is referred to as cell nucleation. From a thermodynamic perspective, the evolution of gas bubbles within a polymer melt requires instabilities, *i.e.* the presence of a driving force [8]. During bubble nucleation the decrease in the free energy of the system which occurs with the development of the gaseous phase (molecules have a lower Gibb's free energy when present as a distinct phase rather than as individual molecules) provides this driver [49]. Later, as the cells grow to reach a stable state, there is a corresponding increase in Gibb's free energy as new interfaces are formed [47]. Overall, the free energy of the system is said to increase during the foaming process.

Polymer melts can be described as either homogenous or heterogenous which affects the nucleation mechanism. Self-nucleation occurs when a homogenous melt (the primary phase) develops gas bubbles (a secondary phase) without external assistance. In reality, self-nucleation is very seldomly achieved as such a process would encounter a large activation energy barrier. However, should a number of dissolved gas molecules cluster together, it is theoretically possible [47, 48]. Polymer melts typically contain additives (usually talc 1-3%) or impurities resulting in an inhomogeneous system to which homogenous nucleation theory cannot be directly applied. The presence of these particles and cavities reduces the activation energy required to achieve a stable nucleus, therefore foaming more commonly occurs *via* this mechanism [47 - 50]. The efficiency of the nucleating agent depends on the type and shape of the nucleating particles as well as the interfacial tension between the polymer and nucleating agent [50].

Once nucleated, cells grow as gas continues to diffuse from the polymer into the cells. Growth ceases when the cells are stabilized (through cooling) or when they rupture (*i.e.* the polymer is unable to contain the gas). Bubble coalescence and rupture will continue to occur simultaneously with bubble growth. A permanent cellular product will only result if the cell structure can be stabilised. Typically, a foamed structure becomes stabilised through the cooling process which raises the viscosity of the polymer melt. Eventually, the expansion ceases and the cellular structure appears to stabilize when the polymer has cooled sufficiently to return to a solid state [51].

FACTORS WHICH AFFECT THERMOPLASTIC FOAMING

Excluding processing conditions, the foaming ability of biopolymers and proteins is similar to that of traditional polymers. Factors which affect foaming include: solubility and diffusivity of the gaseous phase, melt rheology, melt strength and strain hardening behaviour, all of which can be related to the crystallinity of the material. The relationship between these effects are well established with traditional polymers and can easily be manipulated. An improvement of foaming ability has been linked to the introduction of branching, or through increasing the molecular weight distribution in traditional systems [6]. This section aims to discuss the material properties which affect foaming ability in relation to foaming in general and to highlight the factors which are more important in batch and extrusion foaming respectively before these methods are discussed in depth.

Crystallinity and Thermal Transitions in Polymers and Proteins

Polymers are described as either amorphous, semi-crystalline or crystalline depending upon the degree of ordered structures they contain. For traditional polymers, crystallinity is defined by the proportion of ordered regions called lamellae made by regular folding of the polymer chain. The subsequent production of spherulites from these lamellae results in larger crystalline structures which can reach up to several millimetres in size [15, 52]. In protein thermoplastics, the semi-crystalline nature of these materials is determined by the secondary structure of the proteins and the relative proportion of α-helices and β-sheets in comparison to less structured regions such as random coils. The crystalline regions within protein polymers are smaller than those observed in traditional polymers, approximately 10 nm in size, but have been shown to dictate the processing temperature of protein thermoplastics and the ability of these materials to form stable products [15, 53].

During processing, ordered regions are disrupted through the addition of heat which enables the production of a melt. At the melting temperature (T_m) complete disruption of these ordered structures occurs in the case of traditional polymers. The semi-crystalline nature and the transitional behaviour of protein-based polymers has been extensively reviewed in comparison to the responses demonstrated by traditional polymers [15].

Dry proteins typically display glass transition temperatures which are close to their degradation temperature. This is overcome when denaturants and plasticisers are added to lower the T_g. A denaturing temperature (T_d) is typically observed in protein biopolymers which is often very similar to the representative melt temperature observed in semi-crystalline polymers.

Above the T_d protein polymers will either unfold to a less ordered conformation associated with the partial dissociation of these ordered structures (α-helices and β-sheets) or the increased temperature can induce the production of these ordered regions, mainly β-sheets, known as aggregation. For protein polymers these transitions can be poorly defined, occurring over a broad temperature range, and depend upon the degree of hydrogen bonding, electrostatic forces and hydrophobic interactions which exist within the protein. Therefore, these transitions are protein dependent.

It is important to emphasize the production of a protein thermoplastic melt does not necessarily indicate the complete disruption of α-helices and β-sheet conformations to an amorphous phase (unlike traditional thermoplastics) and in some cases may promote the formation of β-sheets [15]. However, mechanical shear may result in a reduction or dispersion of these crystalline regions, enabling a thermoplastic melt. The crystallinity of a protein-based thermoplastic affects the melt, the corresponding rheological properties and its behaviour with respect to solubility and diffusivity of gases. The stabilisation of foams will also be effected by the ability of these materials to recreate these ordered regions on cooling to freeze the resulting foam structure.

Solubility and Diffusivity

Foaming relies on the introduction and retention of a gaseous phase within a polymer matrix. The solubility and diffusivity of gases strongly influences the foaming ability of these materials. Gases can be incorporated through the use of either physical blowing agents (PBA) or chemical blowing agents (CBA). Common PBAs used to include chlorofluorocarbons (CFC), hydrochloro-fluorocarbons (HCFC), and hydrocarbons such as pentane and butane. However, there has been a recent migration towards more environmentally benign agents such as CO_2, O_2 and N_2. Azodicarbonamide, sodium bicarbonate or sodium bicarbonate/citric acid systems are examples of CBAs where gases are evolved by heating to decompose the CBA.

Factors which affect solubility and diffusivity within polymer systems have been reported as the phase of the polymer (molten or glassy), the structure and degree of crosslinking, crystallinity of the polymer and type, concentration and behaviour of plasticisers and fillers. An excellent review [54] defines the relationship between solubility and diffusivity of gases within polymers for foaming. The production of a fine cell homogenous foam is produced by the complete dissolution of the gaseous phase while partial dissolution results in a heterogeneous foam structure.

Normally, the solubility of a gas can be improved by increasing pressure and decreasing temperature. However, CO_2 or N_2 in a polymeric systems does not necessarily behave as expected. For example, the solubility of CO_2 has been observed to decrease with an increase in temperature while N_2 displays an increase in solubility with temperature in polystyrene. Consequently they are often used as a blend during batch foaming, for example 75% vol. N_2 and 25% vol. CO_2 for zein proteins [6].

The crystallinity of the polymer dictates the gas its solubility and diffusivity as the crystalline regions oppose the movement of absorbed gas within the amorphous phase. Additional factors, which have been noted as affecting the properties of semi-crystalline polymers, include orientation of the crystalline phase, chain stretching and free volume changes. It is also believed that these crystalline regions can promote nucleation by acting as heterogenous nucleating agents.

Adsorption and absorption of gaseous molecules is greatly influenced by the temperature of the polymer system in relation to its glass transition temperature. When $T>T_g$ the amorphous regions of the polymer or protein system become fluid although motion is still restricted by the crystalline phase (if present) if $T<T_m$. The solubility of the gaseous phase in the polymer matrix is similar to that of gas into a viscous fluid. When $T<T_g$ the polymer remains glassy with limited chain mobility. When gases are taken up they become trapped within polymer chains and physically disrupt the bonding between the polymer chains. In this case the

free volume of the polymer is increased and the absorbed gas has a temporary plasticising effect. During this process the solubility of these gases into the solid phase is governed by surface adsorption rates. High solubility of a gas within a polymer results in more blowing agent which benefits nucleation and cell growth.

The uptake of gases within a solid matrix is a controlling factor in batch foaming. During this process, plasticisation of the solid matrix can occur which provides the system with sufficient mobility to enable foaming. The resulting cellular structure is solidified by the loss of this plasticisation as the gaseous phase moves out of the polymer and into the bubble resulting in an increase in polymer viscosity.

In extrusion systems where a fluid melt is produced, the melt experiences additional plasticisation by the introduction of a gas phase from PBA or CBA. This decreases the viscosity of the polymer melt, reducing the melt pressure with respect to the internal bubble pressure, favouring faster bubble growth. The solidification of the foamed structure once again corresponds to a change in polymer viscosity. Once the melt has exited the extruder and depressurises, the gas migrates into the nucleated bubbles resulting in cell growth, and the material expands. The melt also cools which induces crystallisation and increases the viscosity slowing cell growth, and solidifies the foam structure.

Diffusivity of gas within a polymeric material is dictated by the polymer chain mobility, packing, crystallinity and degree of crosslinking, which are temperature dependent. The diffusion rate is dependent on the gas concentration gradient which decreases as the material approaches saturation. The presence of side chains in a polymer decreases packing resulting in a more amorphous structure, increasing diffusivity. An increase in the crosslinking within a polymer decreases diffusivity and solubility of the gas [54].

High diffusivity of gases within a polymer is ideal for nucleation but has a detrimental effect on bubble growth. High diffusivity results in greater nucleation as the gas leaves the polymer/BA phase quickly, giving a finer cellular structure, but during bubble growth, the gas can escape from the bubble and material easily, reducing bubble size. Diffusivity is particularly important in extrusion processes as the material has a short time before it cools. Gas diffusivity can be manipulated by controlling polymer viscosity and crystallinity by adjusting temperature, enabling bubble size and cell density to be controlled.

The solubility and diffusivity of gases within traditional polymers is well defined and gases with a variety of phases (subcritical, supercritical and gaseous) have been used. Gas solubility and diffusivity in protein-based thermoplastics is not widely reported, but will depend on the blowing agent, denaturants and plasticisers used. For example proteins have been selected for film and coating applications due to their low oxygen permeabilities, but protein films have high water vapour permeabilities [55].

Viscoelastic Behavior and Melt Rheology in Foaming

Polymers and proteins have defined thermal transitions where they change from a solid to a melt. The ability of polymers to foam is closely related to their temperature dependent viscoelastic behaviour which is a controlling factor in both batch and extrusion foaming systems. Foaming behaviour is greatly influenced by viscosity and is related to melt strength, extensional viscosity and the ability to display strain hardening. The mechanisms which control foaming are linked to the amorphous and semi-crystalline nature of these materials and the presence of side chains.

As previously described, bubble growth during foaming is directly linked to the viscosity of the polymer and the pressure difference between the bubble and the matrix. As such, polymer viscosity is a controlling factor for foaming and influences nucleation and stabilisation as well. Ideally viscosity of the polymer is low during the nucleation phase to enable high nucleation rates. It should then increase during bubble growth to restrain bubble growth.

This is commonly achieved through loss of the gaseous phase from the polymer melt or a decrease in temperature to produce high viscosities which stabilise the foam structure in amorphous polymers. If the viscosity of the polymer is too low and the melt strength is poor, the cells are prone to rupture and collapse. In materials with chain branching, *e.g.* polypropylene, strain hardening occurs which can be used to overcome low melt strength [56]. For semi-crystalline polymers the stabilisation effects are also related to crystallization of the material in addition to the factors listed above.

Many attempts have been made to describe the required rheological properties for foaming in relation to melt strength and extensional viscosity. Extensional viscosity is described as being the factor which allows two materials with the same shear viscosity to behave differently in extensional flow. Extensional flows are experienced by polymers in processes such as such as fiber spinning, film blowing and extrusion foaming [57]. The relationship between extensional viscosity and foaming ability is better defined by transient behaviour, as during foaming deformation occurs within a limited timeframe.

Extensional viscosity can be determined from Rheotens curves if there is optical monitoring of the extrudate size, however the results are qualitative. Therefore, the Rheotens method is more commonly used to determine melt strength [57]. The melt strength is defined as the maximum tensile strength required to break the strand extruded through a capillary with a gradual increase in haul off speed. Sufficient melt strength is required for foaming in addition to extensional viscosity.

Strain Hardening

The correlation between melt strength, melt elasticity and strain hardening has been reviewed in depth for polyolefin materials, which are favoured in many applications due to their toughness, flexibility and chemical resistance. For foam extrusion, polyethylene (PE) has typically been favoured over polypropylene due to cost, thermal and chemical stability and processability (PE has a wide foaming window). Polypropylene foams demonstrate high stiffness and can be used static load bearing applications. However, their manufacture is limited as they demonstrate low melt strength and little elasticity. Consequently the cells within polypropylene foams rupture easily as the cell walls are not able to withstand the extensional forces which occur during foaming.

The melt strength of PP can be improved by introducing long-chain branches which restrict the movement of polymer chains under strain and introduce strain hardening effects, stabilising the bubbles during growth and reducing cell coalescence and rupture. An increase in the long chain branching (LCB) content of PP resulted in lower and broader DSC melting peaks, a reduction in crystallinity, and increased zero-shear and elongational viscosities. This resulted in lower foam densities with well-developed and uniform foam structures compared to unbranched PP [58]. Another study looking at blends of linear PP with LCB PP found that increasing LCB PP content past 50% had little effect on the degree of foaming because elongational viscosity of 100% LCB PP was similar to 50% linear PP:50% LCB PP blends [59].

The extensional viscosity of proteins and thermoplastics thereof have been linked to the secondary structure of the proteins for zein and kafirin under film blowing conditions. The film blowing process requires good melt deformability and elongational properties which is similar to the properties required for foaming. Two studies have been conducted to correlate the molecular structure, rheological behaviour and film blowing abilities of these proteins. Zein-based materials containing a high proportion of α-helical structures produced the best blown films, with the highest extensional viscosity, best strain hardening behaviour and haul of force. It was concluded that the change in extensional behaviour could be contributed to the relative content of α-helical to β-sheets [53].

Another study examined film formation from kafirin protein, where it was found that heat treatments promoted β-sheets formation which reduced film forming ability. Freeze-dried material contained the highest proportion of α-helical structures. The higher strain and tensile strength films produced from freeze-dried kafirin is a result of not only the greater ratio of α-helical structures to β-sheets and other protein conformations but also their homogenous distribution and the ability of native-like structures to pack together more efficiently than denatured and aggregated proteins [55].

The correlation between secondary structure and foaming ability has been discussed for the foaming ability of zein proteins. Plasticised zein protein (75% zein: 25% PEG400) has been shown to depend heavily on the concentration of β-sheets within the original zein protein. During this study zein was obtained from the same producer and extracted using the same technique, however FTIR analysis detected differences in the secondary structure of the zein protein. This suggests that variability within protein supplies such as small changes in growth conditions, extraction or protein aging may influence foaming results. The foaming ability of these zein proteins was once again linked to the ratio of α-helical to β-sheet content. The governing conformation was linked to the β-sheet content which influenced how deformable the plasticised material was. Zein with a higher concentration of β-sheets resulted in a collapsed foam under batch foaming conditions which was contributed to the material being unable to withstand deformation during foaming. In comparison the zein a lower β-sheet content resulted in a far more deformable material and a fine cell low density foam [6].

BATCH FOAMING

An effective method for producing cellular foams is the batch foaming technique, also known as the autoclave method. Batch foaming can produce microcellular foams, structures with cell sizes smaller than 10 μm and cell population density greater than 10^9 cells.cm^{-3} [60].

Microcellular foams are attractive because the material's bulk density is reduced while possessing high impact strength, toughness, stiffness-to-weight ratio, low thermal conductivity as well as low weight and cost [61].

There are two approaches to batch foaming. The first is the ***temperature soak*** method where the polymer at room or low temperature is placed in a high-pressure gas environment which forces the gas to diffuse into the polymer matrix. The vessel is depressurised and material transferred to a mould or water bath and heated past the glass transition temperature, resulting in the gas nucleating and forming micro-bubbles. Subsequently, the material is cooled resulting in a stable foam.

The second approach is the ***pressure quench*** method where the material is heated in an autoclave past its glass transition temperature and exposed to gas at high pressure, resulting in the gas diffusing into the material. The pressure is released resulting in the gas forming micro-bubbles in the polymer, and the temperature lowered resulting in a stable foam. Fig. **4-A** shows a typical set-up of a supercritical batch foaming system [62].

The setup consists of an autoclave which is temperature controlled by circulating oil heated/cooled by a cryo-thermostat. The CO_2 is supplied by a CO_2 cylinder connected to a gas booster. Pressure and temperature are monitored and computer controlled. For batch foaming, cell morphology, size and number can be manipulated by controlling pressure, temperature, gas type, saturation time, nucleating agents and depressurisation rate. Polymer crystallinity, viscosity, and melt strength also influences cell size, density and the resulting foam mechanical properties.

Gas Absorption

Gas absorption can be measured directly by change in weight of the material or indirectly by measuring foam density after the material has been foamed. Absorption will depend on gas pressure, absorption time, gas solubility, and diffusion rate.

Gas diffusion rate and solubility will be dependent on gas and material type, decrease with increasing polymer viscosity (which decreases with increasing temperature) and crystallinity (which is dependent on the melting temperature of the crystalline regions and whether or not absorption is carried out above the melting temperature of the crystalline regions).

Doourdiani, *et al.*, were able to manipulate the crystallinity of HDPE, PB, PP and PET by heating them above the melting temperature and cooling them at different rates following compression moulding into a film. Slow cooling rates induced high crystallinity while cold water quenching on the material gave the lowest crystallinity. CO_2 solubility was halved and diffusivity decreased by 5% for PP and up to 100 fold for PET when crystallinity was increased from 45% to 69% for PP and 36-58% for PET [65]. Increasing gas pressure will increase gas absorption and polymer swelling and decrease in foam cell size. PLA (grades 3001D, 8051D and 4060D) showed an increase in swelling ratio of 1.04 to 1.16-1.20 when CO_2 gas pressure was increased from 6 to 22 MPa (at 190°C), while gas solubility increased from 0.04 to 0.14 g/g [66].

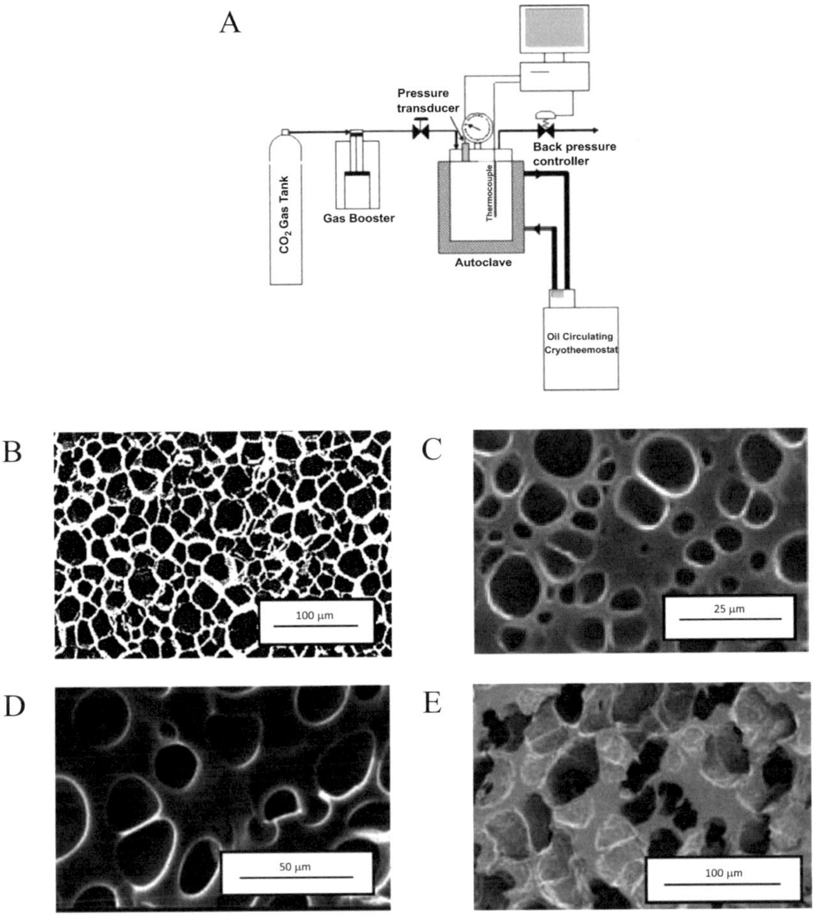

Fig. (4). A: Typical batch foaming system [62] **B:** SEM image of TPZ (Thermoplastic Zein) foamed with N_2-CO_2 80-20 vol%, P_{sat} = 180 bar, T_f =79°C [20] **C:** SEM image of TPZ (Thermoplastic Zein) foamed with N_2-CO_2 80-20 vol%, P_{sat} = 170 bar, T_f =55°C [63] **D:** SEM image of TPG (Thermoplastic Gluten) foamed with CO_2, P_{sat} =85 bar, T_f =100°C [19] **E:** SEM image of NTP/PLA blend foamed with supercritical CO_2, P_{sat} = 200 bar, T_f =55°C [64]. Images reproduced with permission, see note[2].

Aionicesei *et al.* showed similar results for poly(L-lactide), PLLA (42 kDa, semi-crystalline) and poly(D, L-lactide-co-glycolide) PLGA (72 kDa, 50:50 D and L-lactide, amorphous) where CO_2 solubility doubled or tripled when CO_2 pressure was increased from 10-30 MPa. PLLA had almost twice the CO_2 solubility of PLGA (0.75 g/g compared to 0.4 g/g for PLGA at 30 MPa), but both had similar diffusion coefficients, probably due PLLA being semi-crystalline and a lower molecular weight, while PLGA is amorphous, but a larger molecular weight [67]. Similarly, CO_2 solubility in polymethyl methacrylate (PMMA) increased from 0.08 g/g at 15 MPa to 0.24 g/g at 25 MPa [68]. LLDPE showed a foam density of

0.27 g/cm^3 when saturated with N_2 at 12 MPa compared to 0.17 g/cm^3 at 20 MPa [69]. Polystyrene showed a decrease in average cell size with increased CO_2 pressure from 440 μm at 7.5 MPa to 40 μm at 10 MPa when foamed at 80°C [56].

Increasing gas absorption time will increase the amount of gas adsorbed in the matrix up until it reaches saturation, which will decrease foam density. This behaviour is also linked to diffusion rate which is dependent on the polymer, gas and temperature. Higher temperatures will reduce polymer viscosity and increase diffusivity, reducing saturation time. For example, LDPE at 125°C saturated with N_2 took longer to reach saturation (8 min at 12 MPa) than LDPE at 135°C (5 min), while LLDPE at 119°C took 10 min and 8 min at 135°C [69].

Foaming

Extent of foaming will depend on the amount of gas absorbed (see previous section), temperature and pressure at which the material is foamed, depressurisation rate, and polymer melt strength, viscosity and elasticity. These effects on foaming can be measured by foam density, cell density, cell size and cell size distribution.

Kwon and Bae showed higher CO_2 gas solubility (manipulated by ranging CO_2 pressure between 15-25 MPa) in PMMA will result in greater cell density and lower cell mean diameter when the material is foamed [68].

A higher foaming temperature will generally reduce foam density, but this will be polymer dependent. For example LLDPE foamed at 123°C had a foam density of 0.08 g/cm^3 compared to 0.17 g/cm^3 at 119°C, while LDPE showed little difference in foam density when foamed at 125°C and 135°C [69]. The foaming temperature window will depend on the melt strength of the polymer, which is dependent on polymer viscosity and temperature. Higher viscosities will slow bubble growth, while high foaming temperatures will reduce viscosity and increase gas diffusion and bubble growth [56, 70]. As the bubble grows, and the matrix stretches around it, the matrix will undergo strain hardening as previously described, which will slow bubble growth. If bubble growth is too fast for the polymer, it will result in collapsed bubbles. For example LLDPE foamed at temperatures greater than 123°C resulted in collapsed bubbles, whereas LDPE, which has a greater melt strength than LLDPE, had a broad foaming window of 105 to 160°C, and cell size and morphology was largely independent of temperature tested [69]. LLDPE foamed at lower temperatures and showed more irregular cell morphology due to the presence of crystalline regions in the polymer melt.

Polymer viscosity and melt strength can be increased by grafting long chain branches (LCB) onto the polymer back bone, or extending chain length. Grafting

LCB onto polystyrene resulted in a narrower cell size distribution and smaller average cell diameters, and higher cell densities (2 wt% LCB grafted PS gave a 100 fold increase in cell density compared to ungrafted PS foamed with CO_2 at 10 MPa and 110°C). Cell size and distribution increased with increasing foaming temperature (0-19 μm at 70°C to 70-230 μm at 110°C), while the effect of chain branching on cell density was more pronounced for the higher foaming temperatures (80-110°C) [56].

Increased depressurisation rate increases cell nucleation, resulting in higher cell densities and lower cell size [61]. Bimodal foams, *i.e.* foams with two discrete cell size distributions, can be generated by using two step depressurisation after the polymer has been saturated with gas. These have the advantage of having a low thermal conductivity and low weight from the larger average cell size while retaining the high mechanical strength imparted by the microcellular foam. Bimodal foams were produced from PS saturated with CO_2 at 20 MPa and 100°C and depressurised to 15 MPa, held for a period of time, and then depressurised to ambient pressure. Increasing holding time from 0 to 60 min increased the average cell diameter from 4 μm to 22 μm and reduced cell density from $8x10^8$ cells per cm^3 to ~ $1x10^6$ cells per cm3 [61].

Dorourdiani *et al.*, showed that HDPE, PB, PP and PET with increased crystallinities (see previous section) had quite different foam morphologies. Samples with lower crystallinities had a more regular cell structure and smaller cell size, while samples with the higher crystallinities were more difficult to foam only exhibiting foam structures in a few localised regions in the polymer. This was partly due to lower gas solubility and diffusivity at higher crystallinities, but Dorourdiani *et al.* also speculated that CO_2 acted to plasticise the polymer, resulting in easier foaming at lower crystallinities and higher CO_2 concentrations in the matrix [65]

Liao *et al.* manipulated the crystallinities of unbranched and branched PP by heating it to 180°C, exposing it to CO_2 at 14 MPa for 4 hours, cooling it to 130°C at holding at that temperature for different periods of time before depressurising. Longer holding times for unbranched and branched PP resulted in increased crystallinities (observed by measuring the spherulite size and density) and consequently smaller cell sizes and larger cell densities. Reducing holding time resulted in ellipsoid shaped cells, while the longer holding times gave more spherical cells [70]. Unfortunately their foaming windows for unbranched and branched PP were quite different so a direct comparison between using the same conditions for the two types was not possible.

EXAMPLES OF BATCH FOAMED PROTEINS

There are only a few examples of successfully batch foamed proteins using the autoclave method. These used zein, gelatin and bloodmeal (which was blended with PLA).

Salerno *et al.* [20] foamed protein based discs cut from hot-pressed gelatin with 20 wt% glycerol and zein with 25 wt% PEG 400 (Fig. **4-B**). They explored varying the gas composition using a mixture of N_2 and CO_2, saturation pressures between 6-18 MPa at 70°C, foaming temperatures between 50-140°C, and pressure drop rates of 25 and 70 MPa/s. The best foams from zein were obtained using 80% N_2 and 20% CO_2 at 18 MPa at 70°C, with a foam density of 0.1 g/cm^3 and average cell size of 32 μm. At lower foaming temperatures (44°C), the cells were smaller (14 μm) with thicker walls with a foam density of 0.65 g/cm^3, due to the increased viscosity of the material, while higher temperatures (100°C) resulted in apparent cell shrinkage (15 mm) due to gas diffusion through the cell walls. Gelatin foam densities were between 0.5 to 0.7 g/cm^3 and cell sizes ranged between 3 to 25 mm when foamed at 44-120°C, but temperatures above this resulted in browning and loss of glycerol.

Olivierio *et al.* [63] found adding acid or alkali derived lignin to zein thermoplastics was detrimental to foaming (using similar conditions to Salerno *et al.*), and were only able to produce microcellular foams (Fig. **4-C**) with addition of lignin up to 1% by weight. Olivierio *et al.* [19] also foamed hot pressed gelatin (with 20% glycerol)/poly(butylene succinate) discs (Fig. **4-D**) using CO_2 at 6.5-8.5 MPa at 100 and 120°C and foaming temperatures between 75-100°C. Gelatin was immiscible in PBS and *vice versa*, but the presence of PBS reduced the cell size compared to plasticised gelatin foams which was between 15-40 μm.

Trujillo-de Santiago *et al.* [30] produced foams from zein (plasticised with PEG 400), starch (with urea and formaldehyde (UF) and sorbital and glycerol (SG), a blend of starch and zein (with UF and SG), and blue maize (SG)). The materials were saturated with CO_2 at 15 MPa and 70°C, and foamed at temperatures between 75 to 140°C. Foamed zein increased eight fold in volume, followed by starch (UF) which increased in volume by six times. Starch (SG), the zein/starch blend, and blue maize showed only a slight increase in volume and the morphology appeared to be fractures in the material rather than regular cells.

Walallavita *et al.* [64] blended Novatein® thermoplastic (NTP) (a bloodmeal protein based bioplastic) with PLA (Ingeo™ 4060D, fully amorphous) at NTP concentrations between 5-60 wt%. Extruded rods were saturated with CO_2 at 6 MPa, transferred to a hot water bath and heated to 70°C. Rods were also foamed by saturating them with CO_2 at 20 MPa at 50°C, followed by depressurisation. Expansion ratio decreased with increased NTP content (Fig. **4-E**).

FOAM EXTRUSION

Extrusion foaming is highly economical due to the continuous nature of this operation. Foam extrusion can be carried out in any of the common extruder configurations: twin screw extruders (TSE), single screw extruders (SSE) or tandem lines [71], but the equipment must produce a sufficient melt temperature to decompose the blowing agent, the pressure of the melt must be kept high to prevent gas dissolution and the back pressure in the hopper must be controllable to ensure consistent results [71]. Each of the extruder configurations has unique advantages and disadvantages and are typically optimised for a particular polymer/blowing agent system through a trial-and-error approach. Specialised extruders have been designed for foaming applications, although to do so the behaviour of the material and the subsequent interaction of the blowing agent with this material must be well defined. Typically, the extruder contains several zones along the barrel where temperature, pitch, element type can be varied to better control melt properties and pressure (Fig. **5-A**).

The polymeric material is melted and blended with any CBAs within the first zone, the material is then mixed and PBAs can be added. As a general rule of thumb the feed zone within a foaming system must be lower than the initial decomposition temperature of the CBA if it is added with the feed [71]. The material is extruded across the remaining zones, where changes in temperature further alter the rheology of the material to enable foaming such that when the material reaches the die nucleation and bubble growth can occur [72].

Until then the gaseous phase remains dissolved within the polymer melt due to pressure within the barrel. Within these systems foaming occurs under free expansion (atmospheric pressure) unlike foaming conducted *via* injection moulding were the limited volume of the mould dictates the pressure experienced by the expanding melt. Once the material has exited the die, the material must solidify quickly before the gaseous phase causes excessive cell rupture or collapse.

The operating parameters such as screw speed, extrusion temperature, die size and the configuration of the screw elements all influence the properties of the resulting foams such as cell size, cell density, expansion ratio and radial expansion ratio. There are a limited number of examples of extrusion foaming of thermoplastic proteins, as such parameter optimisation has yet to be presented for foaming thermoplastic proteins and these are expected to be highly protein dependent. Therefore, the influence that these parameters have are described in relation to traditional polyolefin foaming and biopolymers such as starch and PLA.

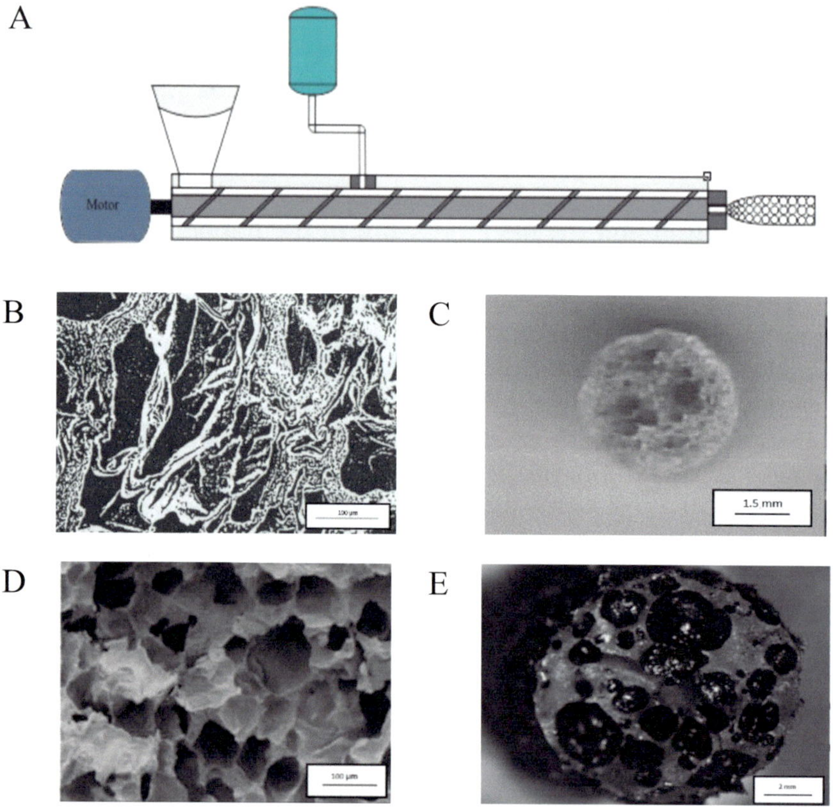

Fig. (5). A: diagram of a foam extruder with PBA capability, **B**: SEM image of Thermoplastic Soy foamed at 150-160°C [73], **C:** Image of 70% Zein 30% Pearl Millet Flour, Screw Speed 80 RPM, Die Size 2 mm [74], **D.** SEM image of PLA/SPC, 2 phr CBA, Screw Speed 60 RPM, Die Size 1 mm [23], **E.** Image of NTP/PE--MAH/LLDPE Foam, Steam Blown, Screw Speed 150 RPM, Die Size 10 mm [75]. Images reproduced with permission see note[3].

During extrusion foaming the behaviour of all thermoplastics are relatively similar therefore it is expected that thermoplastic protein foams will display similar responses to changes in extrusion speed, die size, CBA, PBA and nucleating agent content.

Foam extrusion behaviour of polyethylene and polypropylene is well defined and has been reviewed extensively [7, 51, 76, 77]. One review outlines the behaviour of polyolefins within extrusion foaming using linear low density polyethylene (LDPE) and polypropylene (PP) as examples. Sodium bicarbonate and citric acid was used as a blowing agent. The material was extruded through a single screw extruder, 19 mm in diameter with a length/diameter (L/D) ratio of 25. The extruded foam was then submerged in a water bath to stabilise the foam structure [78]. This particular study discusses a comprehensive set of process variables including melt temperature, cooling water temperature, and blowing agent content on the resulting cell size and distribution. Material characteristics such as molecular weight, structure, melt viscosity and melt tension was also investigated.

The foaming ability of these materials was shown to be directly correlated with the molecular weight of the polymer. Higher molecular weights resulted in smaller cells and lower overall expansion due to the increase in polymer viscosity. The highest foaming ability was demonstrated for polypropylenes with a molecular weight of 231,000 (degree of foaming: 98%) and for LDPE with a similar molecular weight (degree of foaming: 65%). Melt tension was observed to be the governing factor which dictated the differences in foaming ability of the PP when compared to LDPE [78]. PP has a lower melt strength than PE hence lower degree of foaming was observed in PE systems, as an increase in melt tension restrains the bubble growth and results in smaller cells and a corresponding decrease in the foam expansion.

The effect of melt temperature on extrusion foaming was determined between 180-240°C for PP and LDPE. Foaming was observed to increase for both polyolefins with increasing melt temperature from 180-240°C due to a reduction in viscosity. Changes in blowing agent content (sodium bicarbonate/citric acid blend) in polypropylene systems resulted in a decrease in cell size as more gas was generated with greater nucleation rates. An addition of 5 wt% of these CBAs resulted in a degree of foaming of approximately 98%, while 2 wt% achieved 45%.

Stabilisation of the foam structure *via* cooling is also particularly important. In polypropylene systems, the degree of foaming observed was directly linked to the cooling temperature. Highest degrees of foaming were observed when the cooling temperature was lowered to 0-20°C in comparison to a melt temperature of 200°C. While a reduction in foaming degree was evident at higher cooling temperatures >30°C with the lowest foaming degree of 38% observed when the cooling temperature was only 65°C. This emphasizes the stabilising effect which higher cooling rates can have on foamed extrudates. Despite the greater solidification of the extrudate surface with lower temperature no change in cell size was observed.

Extrusion foaming has also been applied to amorphous PLA with azodicarbonamide (azo) as the blowing agent. A single screw extruder with an

L/D of 25 with five temperature zones of 138, 193, 232, 177, 149°C from feeder to die was used. Optimal CBA content for PLA was 0.8% which gave foam densities of 0.55 g/cm^3 compared to LDPE, which was 2% and a foam density of 0.45 g/cm^3. CBA content above and below this gave higher foam densities, which could be explained by phase separation occurring during extrusion resulting in lower amounts of gas within the polymer melt thereby decreasing expansion. A corresponding increase in rod diameter was observed with small quantities of CBA but degassing occurred at the die for blends with increased CBA content [79].

Extrusion foaming has also been used to produce starch based products for loose fill applications. Starch alone is high hydrophilic in nature which is detrimental to the physical and mechanical properties of the foam because it will readily absorb water in a humid environment. Therefore, starch is either modified before extrusion or blended with materials like polystyrene. The water which is incorporated with the starch typically acts as the blowing agent. For a 70% starch 30% polystyrene blend the optimum water content was 18% which gave a radial expansion ratio of 28-29 at processing temperatures of 140 and 160°C respectively, with large cells of up to 1 mm. Increasing water content further plasticised the material so its melt viscosity was too low to sustain foaming and the cell walls lacked the strength to contain the expanding bubbles [80]. Addition of talc, a nucleating agent and filler, reduced the radial expansion ratio and overall expansion ratio of these foams, but increased cell density from 24 cells/cm^2 to 342 cells/cm^2 at 3% talc to a blend with 20% moisture content (MC), narrowed cell size distribution, but increased bulk density from 14.6 to 17.7 kg/m^3 and made the foam more brittle. For foams containing talc, an optimal composition was found to be 18% MC, 1% talc and an extrusion temperature of 160°C [80].

THERMOPLASTIC PROTEIN EXTRUSION FOAMING

Previous reviews of foam extrusion of proteins have typically discussed the necessity of blending proteins with carbohydrates, flours or PLA, and there are very few examples of foamed unblended thermoplastic proteins [1, 76]. In this section four examples of protein foam extrusion will be discussed. These studies have all been conducted with the final product intended for packaging applications. The first is a true thermoplastic protein foam produced from soy protein, the second and third studies examine the foaming ability of blends of zein protein with pearl millet flour and soy protein concentrate - PLA blends. The final study investigated blends of Novatein® Thermoplastic Protein, a thermoplastic derived from bloodmeal with linear low density polyethylene.

Thermoplastic SPI was prepared by extruding SPI, water, glycerol and salts

($CaCl_2$, sodium tripolyphosphate (STP), and $ZnSO_4$) in a co-rotating twin screw extruder and granulated. This was foamed using a CBA in a single screw extruder equipped with a six inch flat sheet die at 150-160°C. The resulting foams had a density of 0.4-0.6 g/cm^3 (Fig. **5-B**). Blends which were heavily plasticised had a lower Young's modulus while those with lower levels of glycerol addition produced a rigid foam structure [73]. Using STP resulted in a more processable and consolidated thermoplastic material, possibly due to disruption of ionic bonding between or within protein chains, but a higher foam density (0.62 g/cm^3). Adding $CaCl_2$, or $ZnSO_4$ increased Young's modulus, possibly due to Ca^{2+} and Zn^{2+} ions crosslinking negatively charged groups on the proteins. Adding $CaCl_2$ gave lower densities (0.53 g/cm^3), higher tensile strength (3.56 MPa) and elongation at break (24.1%) compared to adding $ZnSO_4$ (0.58 g/cm^3, 3.03 MPa and 21.19% respectively). All foams showed similar elongational to break despite the increase in tensile modulus which was due to a similar degree of crosslinking.

Zein protein (70-75%) can be combined with pearl millet flour (25-30%), derived from cereal crops, to produce foams [74]. The composition of pearl millet flour can vary significantly depending upon growing conditions, however, the material typically contains 8.5-19.5% protein, and 63.1-78.5% starch. The foamed blends contained 4.6 g water, 20 g polyethylene glycol (PEG-400) and 3.0 g ammonium bicarbonate per 100 g of zein/flour. Foaming was conducted using a compact extruder with a 1.5-2.0 mm die, screw speeds between 20 and 150 rpm with a temperature profile from hopper to die of 65, 105, 95, 80°C. A foam expansion ratio of 2-3.2 was achieved with a decrease in moisture content from 12% to 6.5%. Zein formed a continuous matrix while the pearl millet remained as discrete particles (Fig. **5-C**). The average cell size was 5 μm although the cell distribution observed was broad and foam density ranged from 0.35-0.5 g/cm^3. The elastic modulus of these foams under compression was 25-50 MPa.

Another study investigated the ability of polylactic acid (PLA) and soy protein concentrate (SPC) to foam (Fig. **5-D**) through twin screw extrusion using poly(2-ethyl-2-oxazoline) as a compatibiliser [23]. The material was foamed in a secondary extrusion step following compounding with a CBA (4-methylbenzen--1-sulfonohydrazide with 5% $CaCO_3$ for nucleation) and co-compatibliser of polymeric methylene diphenyl diiscocynate (pMDI). PLA/SPC blends were prepared using a co-rotating twin screw extruder with a screw diameter of 18 mm and an L/D of 40 with a screw speed of 150 RPM. The temperature profile was 90, 120, 140, 150, 160, 160, 155°C from hopper to die. The blends also contained a compatibiliser (PEOX, 3 phr), and citroflex A-4 (6 parts per hundred parts resin) and glycolube WP2200 (2 phr) as a plasticiser. Extrusion foaming was conducted using a screw speed of 60 RPM, die size of 1 mm and with a variable die temperature of 115-160°C. Cell rupture was a problem during extrusion and

SP/PLA blends displayed severe degradation at higher processing temperatures.

Gas from the CBA lowered melt viscosity enabling lower extrusion temperatures to be used. Adding pMDI as an interfacial modifier promoted cross linking between the hydroxyl groups of PLA and the amine groups of SPC which resulted in a higher viscosity. Cell densities in foams without pMDI decreased with increasing foaming temperature but increased by 1.7-1.8 times when pMDI was added. For foams without pMDI, the cell density increased with increasing CBA content up to 0.5-2 phr, after which cell density decreased. With pMDI, cell density was higher overall, possibly due to a decrease in nucleation activation energy through a heterogenous nucleation mechanism, but decreased with increasing CBA content. CBA content and extrusion temperature affected the pressure drop by altering viscosity. The material without pDMI had its lowest bulk density (0.53 g/cm^3) when extruded through a die at 140°C while blends wih pMDI displayed the lowest bulk densities (0.77 g/cm^3) at 150°C with 2 phr CBA.

In other research, expansion ratios of 1.85 and less were observed for an extrusion foamed blend of a thermoplastic bloodmeal (Novatein®) and linear low density polyethylene (LLDPE) in the presence of polyethylene grafted maleic anhydride (Fig. **5-E**) as a compatibiliser (PE-g-MAH) [75]. Foaming was conducted by adding Novatein® granules to LLDPE and PE-g-MAH in the hopper of a twin screw extruder. Sodium bicarbonate was used as the blowing agent and the material subjected to an extrusion profile of 70,100,100,110,120°C feeder to die. The die size was 10 mm.

As the proportion of NTP was increased the foaming ability reduced until consolidation of the extrudate was lost. A blend of 50% NTP, 10% PE-g-MAH and 40% LLDPE without additional blowing agent resulted in a bulk density of 0.58 g/cm^3. The addition of compatibliser at 6, 8 and 10% had no effect on the foam expansion ratio. Increasing the die temperature from 120 to 140°C resulted in a decrease in radial expansion ratio from 1.37 to 1 and the extrudate surface appeared to be more degraded. Addition of sodium bicarbonate as a blowing agent resulted in a very fine cellular structure while the bulk density remained constant, and the final material was more flexible than foams without the blowing agent.

Foam extrusion cooking is an analogous technique to extrusion foaming and is highly relevant to extrusion foaming processes involving proteins in particular. Extrusion foam cooking is common for starch/carbohydrate based snack foods and whey or soy protein are commonly added to improve the nutritional value of the snack food [81, 82]. Examples of research using foam extrusion cooking are listed in Table **1**, and two examples are discussed below.

Table 1. Extrusion cooking of protein foams. Carb-carbohydrate SPC- soy protein concentrate, SWS- sweet whey solids, WPC- whey protein concentrate, WPI- whey protein isolate. TS- twin screw, SS- single screw. ER – expansion ratio, EI- expansion index.

Blend Properties								Extrusion Conditions		Foam Properties			
Protein	Protein Purity (%)	Protein Added (wt. %)	Max Protein (%)	Carb	Protein in Carb Phase (%)	Carb Added (wt. %)	Total Protein (%)	Screw Type	Screw Speed	ER	EI	Radial ER	Ref
Acid Casein	88.5	50	44.25	Potato Starch	0.02	50	44.35	SS	80	-		3.98-4.96	[83]
Casein Casinate	92	25	23	Corn Meal	9.0	75	29.75	TS	300	-	5.5	-	[84]
Cow Pea	27	100	27	Sorghum Meal	8.5	0	27	TS	200	2.0-5.4		-	[85]
Soy Bean Protein	51.5	50	25.75	Potato Starch	0.02	50	25.85	SS	100	-		1.93-2.70	[83]
SPC	71	50	35.5	Corn starch	-	-	35.5	TS	230-300	3.5-5.2		-	[86]
SPC	-	-	30	Corn Meal	-	-	-	TS	250	-		-	[87]
SPC	64.7	14.00	9.0	Corn Meal	6.5	64.15	13.16	TS	300	-		-	[88]
SPI	-	-	30	Corn Meal	-	-	-	TS	250	-		-	[87]
SPI/SPC	89.3/ 64.7	8.70/ 4.0	10.36	Corn Meal	6.5	66.25	14.66	TS	300	-		-	[88]
SWS	12.9	50	6.45	Potato Flour	8	50	10.45	TS	300		1.0-2.2	-	[89]
SWS	12.9	50	6.45	Rice Flour	5.7	50	9.3	TS	300		0.7-2.8	-	[89]
SWS	12.9	50	6.45	Corn	5.6	50	9.25	TS	300		0.6-2.3	-	[89]
WPC	58.3	25	14.58	Potato Starch	0.02	75	14.59	SS	80	-		3.05-3.45	[83]
WPC	80	-	40	Corn Starch	-	-	40	TS	200	4-26		-	[90]
WPC	80-86	30	25.80	Corn Starch	-	-	25.80	TS	140	-		5.43-10.9	[91]
WPC	34.9	50	17.45	Potato Flour	8	50	21.45	TS	300		1.2-2.2	-	[89]
WPC	34.9	50	17.45	Rice Flour	5.7	50	20.3	TS	300		0.8-2.8	-	[89]
WPC	34.9	50	17.45	Corn	5.6	50	20.25	TS	300		0.5-2.3	-	[89]
WPC	80	25	20	Corn meal	-	-	20	TS	300	-		-	[92]
WPC	78.7	25	19.68	Corn Meal	9.0	75	26.43	TS	300	-	5.2	-	[84]

(Table 1) contd.....

Blend Properties								Extrusion Conditions		Foam Properties			
Protein	Protein Purity (%)	Protein Added (wt. %)	Max Protein (%)	Carb	Protein in Carb Phase (%)	Carb Added (wt. %)	Total Protein (%)	Screw Type	Screw Speed	ER	EI	Radial ER	Ref
WPC	70.2	7.5	5.27	Pearl Millet Flour	10.50	92.50	14.98	TS	350	-	4.34	-	[82]
WPC	34	15	5.1	Corn starch	-	-	5.1	-	-	-		-	[93]
WPI	93.3	35	32.66	Corn Meal	7.2	65	37.34	TS	800	4.5-9		-	[94]
WPI	93.6	25	23.4	Corn Meal	9.0	75	30.15	TS	300	-	4.5	-	[84]
WPI	92	18	16.56	Corn starch	-	-	16.56	TS	300	-		-	[81]

Typical production conditions for expanded products for protein carbohydrate blends use a twin screw extruder operating around 300 RPM. Die sizes are typically limited to 3-3.2 mm in size and typical processing temperatures are between 70-150°C with lower feed zone temperatures of 30°C occurring in some systems (Table **1**). Moisture content, protein type, protein purity and the composition of the carbohydrate phase all affect the mechanical and physical properties of the resulting foams.

When foaming WPI and corn starch blends, an increase in the water content, which acts as a plasticiser and blowing agent, reduced the required specific mechanical energy from 315-359 kJ/kg at 23% moisture content to 274-314 kJ/kg for 27% moisture [81].

At the higher moisture content (27%) expansion ratio and radial expansion ratio was reduced as the melt no longer had the required melt strength to sustain the foamed structure. Increased water content in the blend was observed to reduce cell size from 1.06-2.94 mm at 23% MC to 1.00-2.05 mm at 27% MC. Foam hardness decreased with increasing whey protein up to 12% whey, due to the gelling properties of whey [81].

In another study, addition of soy protein concentrate to starch with high amylose content increased foam mechanical strength. Crushing force measured by a texture analyser increased from 10 N at 0% SPC to 42 N at 20% SPC. Increasing SPC further resulted in a lower crush strength than the control foam, probably due to protein agglomeration due to heat and shear, as well as reducing water available to starch to gelatinise during extrusion [86].

CONCLUSION

Protein based foams present a new and exciting area for research. They provide a means to reduce dependence on fossil fuels, and decrease the environmental impact of non-biodegradable foams. Proteins can be sourced from a wide variety of plant or animal products or wastes, reducing the impact on food supply.

Foaming is a trade-off between manipulating polymer properties for increasing gas solubility to reduce foam density and increase cell nucleation, diffusivity which needs to be high initially for nucleation and low later on to reduce bubble coalescence and prevent gas loss while the polymer is cooling. Solubility will depend on polymer properties such as crystallinity and gas type and gas pressure, while diffusivity is dependent on polymer viscosity which is dependent on crystallinity, polymer type and chain branching, and melt temperature. Bubble formation is also dependent on melt strength, *i.e.* the ability of the material to contain the bubble, which is dependent on the same properties as for polymer viscosity.

The main challenges to be overcome with protein based foams are:

1. Controlling protein-protein interactions so that they can extruded or batch foamed at temperatures lower than the protein degradation temperature using combination of denaturants, reducing agents and plasticisers to reduce hydrophobic interactions, hydrogen bonding, and cysteine-cysteine bonds, as well as increasing free volume between chains to increase chain mobility. Protein-protein interactions are dictated by the amino-acid sequence of the proteins, the secondary structures that they form, and environment they are in, so a potential area of future research is manipulating protein amino acid sequence to produce plastics with desired properties.
2. Dried denatured proteins are rich in β-sheets, and will form more β-sheets when exposed to heat, which results in a highly crystalline, brittle material with low melt strength and extremely high extensional viscosity which is detrimental to producing foams. Furthermore, protein based thermoplastics need mechanical shear as well as heat to form a melt, which provides a challenge for batch foaming, while for extrusion foaming, mechanical shear is provided by the extruder. Proteins such as zein with a high α-helical content have been successfully used to produce foams and films.
3. Proteins have poor gas solubility and high gas diffusion properties, which results in low foam cell densities and high bulk densities compared to conventional foams. A combination of CO_2 and N_2 gas as a blowing agent has been successfully used for zein and gelatin in batch foams, while water or chemical blowing agents serve for extruded zein, soy and bloodmeal protein

foams. Blending proteins with starch or other polymeric materials such as PLA is another way of improving gas solubility and diffusion properties.

The main proteins to be successfully foamed are zein, gelatin, soy and bloodmeal. Whey has been used as an additive in food foams where the main material is starch. As research progresses and new proteins tried, we expect to see more successful protein based foams in the future.

NOTES

[1] Reproduced from J.A.Kiernan 'Formaldehyde, formalin, paraformaldehyde and glutaraldehyde: What they are and what they do', Microscopy Today, v.1: 8-12, Fig. **2**, 2000, copyright Cambridge University Press, with permission.

[2] **A.** Reprinted from The Journal of Supercritical Fluids, 58/1, Y.-M Corre, A. Maazouz, J. Duchet, and J. Reignier, "Batch foaming of chain extended PLA with supercritical CO_2: Influence of the rheological properties and the process parameters on the cellular structure,", Page No. 177-188, Copyright (2011), with permission from Elsevier. **B.** Reproduced with permission from the journal International Polymer Processing, IPP, Vol.22, 2007, pp 480-488, by A. Salerno, M. Oliviero, E. Di Maio and S. Iannace, © Carl Hanser Verlag GmbH & Co.KG, Muenchen. **C.** Reproduced with permission M. Oliviero, L. Verdolotti, I. Nedi, F. Docimo, E. Di Maio, and S. Iannace, "Effect of two kinds of lignins, alkaline lignin and sodium lignosulfonate, on the foamability of thermoplastic zein–based bionanocomposites", 48 (6), pp. 516-525, copyright © 2012 by the authors. Reprinted by permission of SAGE Publications, Ltd. **D.** Reproduced with permission, from B. Liu, L. Jian, and J. Zhang, "Extrusion foaming of Poly (lactic acid)/Soy protein concentrate blends", Macromolecular Materials and Engineering, Wiley. Copyright ©2011 WILEY-VCH Verlag GmbH & Co. KGaA, Weinheim. **E.** Reproduced from A. Walallavita, C. J. R. Verbeek, and M. C. Lay, "Blending Novatein Thermoplastic Protein with PLA for carbon dioxide assisted batch foaming," In: Proceedings of PPS-31: The 31st International Conference of the Polymer Processing Society–Conference Papers, Jeju Island, South Korea, 2015., with the permission of AIP Publishing (http://dx.doi.org/10.1063/1.4942311)

[3] **B.** Image reproduced with permission from the authors: P. Mungara, J. W. Zhang, and J. Jane. **C.** Image reproduced from "Development and characterization of extruded biodegradable foams based on zein and pearl millet flour" with permission from the authors. **D.** Reproduced with permission from Wiley, from M. Olivero, L. Sorrentino, L. Cafiero, B. Galzerano, A. Sorrentino, and S.

Iannace, "Foaming behaviour of bio based blends based on thermoplastic gelatin and poly(butylene succinate)", Journal of Applied Polymer Science, Wiley. © Wiley periodicals, Inc. J. Appl. Polym. Sci. 2015, 132, 42704. **E.** Reproduced from C. Gavin, M. C. Lay, and C. J. R. Verbeek, "Extrusion foaming of protein-based thermoplastic and polyethylene blends," In: Proceedings of PPS-31: The 31st International Conference of the Polymer Processing Society–Conference Papers, Jeju Island, South Korea, 2015., with the permission of AIP Publishing (http://dx.doi.org/10.1063/1.4942308)

CONFLICT OF INTEREST

The author (editor) declares no conflict of interest, financial or otherwise.

ACKNOWLEDGEMENTS

Declared none.

REFERENCES

[1] G. Mensitieri, E. Di Maio, G.G. Buonocore, I. Nedi, M. Oliviero, L. Sansone, and S. Iannace, "Processing and shelf life issues of selected food packaging materials and structures from renewable resources", *Trends Food Sci. Technol.*, vol. 22, no. 2–3, pp. 72-80, March 2011. [http://dx.doi.org/10.1016/j.tifs.2010.10.001]

[2] Smithers Rapra, *Polymer Foams Market Expected to Consume 25.3 Million Tonnes by 2019,* 2014. Available at: http://www.smithersrapra.com/news/ 2014/may/polymer-foam-market-to-consume-25-3-million-tonnes Accessed: 2nd Aug. 2015

[3] Freedonia Group Inc, *US Natural Polymer Demand (2002-2012),* 2008. Available at: http://www.plasticsnews.com/article/20081208/FYI/312089989/us-natural-polymer-demand-2002-12 Accessed: 2nd Aug. 2015

[4] MarketsandMarkets.com, *Bioplastics & Biopolymers Market by Type (Bio PET, Bio PE, PLA, PHA, Bio PBS, Starch Blends, and Regenerated Cellulose), by Application (Packaging, Bottles, Fibers, Agriculture, Automotive, and Others) & by Geography – Trends & Forecasts to 2018,* 2014. Available at: http://www.marketsandmarkets.com/Market-Reports/biopolymers-bioplastics-market-88795240.html Accessed: 2nd Aug. 2015

[5] *The Biopolymers Market is Expected to Generate Global Revenue of 3.67 Billion US Dollars By 2018,* 2014. Available at: http://milksci.unizar.es/bioquimica/temas/azucares/disacaridos.html

[6] C. Marrazzo, E. Di Maio, and S. Iannace, "Foaming of synthetic and natural biodegradable polymers", *J. Cell. Plast.,* vol. 43, no. 2, pp. 123-133, 2007. [http://dx.doi.org/10.1177/0021955X06073214]

[7] S.T. Lee, C.B. Park, and N.S. Ramesh, "Introduction to polymeric foams", In: *Polymeric Foams: Science and Technology.* , S.T. Lee, C.B. Park, N.S. Ramesh, Eds., 1st ed CRC Press: Boca Raton, FL, 2007.

[8] S.T. Lee, "History and trends of polymeric foams: From process/product to performance/regulation", In: *Polymeric Foams: Technology and Developments in Regulation, Process and Products.* , S.T. Lee, D. Scholz, Eds., 1st ed CRC Press: Boca Raton, FL, 2009, pp. 1-40.

[9] S. Park, D. Bae, and K. Rhee, "Soy protein biopolymers cross-linked with glutaraldehyde," ", *J. Am. Oil Chem. Soc.,* vol. 77, no. 8, pp. 879-884, 2000. [http://dx.doi.org/10.1007/s11746-000-0140-3]

[10] J.M. Raquez, M. Deléglise, M.F. Lacrampe, and P. Krawczak, "Thermosetting (bio)materials derived from renewable resources: A critical review", *Prog. Polym. Sci.,* vol. 35, no. 4, pp. 487-509, 2010. [http://dx.doi.org/10.1016/j.progpolymsci.2010.01.001]

[11] J.A. Kiernan, "Formaldehyde, formalin, paraformaldehyde and glutaraldehyde: what they are and what they do", *Micros. Today,* vol. 1, no. 5, pp. 8-12, 2000.

[12] C.J. Verbeek, and L.E. van den Berg, "Extrusion processing and properties of protein-based thermoplastics", *Macromol. Mater. Eng.,* vol. 295, no. 1, pp. 10-21, 2010. [http://dx.doi.org/10.1002/mame.200900167]

[13] C.J. Verbeek, and J.M. Bier, "Synthesis and characterization of thermoplastic agro-polymers", In: *A Handbook of Applied Biopolymer Technology: Synthesis, Degradation and Applications.* , S.K. Sharma, A. Mudhoo, Eds., 1st ed Royal Society of Chemistry, 2011. [http://dx.doi.org/10.1039/9781849733458-00197]

[14] J.M. Bier, *"The Eco-Profile of Thermoplastic Protein Derived from Bloodmeal"*, Masters Thesis, University of Waikato, New Zealand, 2010.

[15] J.M. Bier, C.J. Verbeek, and M.C. Lay, "Thermal transitions and structural relaxations in protein-based thermoplastics", *Macromol. Mater. Eng.,* vol. 299, no. 5, pp. 524-539, 2014. [http://dx.doi.org/10.1002/mame.201300248]

[16] J.M. Bier, C.J. Verbeek, and M.C. Lay, "Thermal and mechanical properties of bloodmeal-based thermoplastics plasticized with tri(ethylene glycol)", *Macromol. Mater. Eng.,* vol. 299, no. 1, pp. 85-95, 2013. [http://dx.doi.org/10.1002/mame.201200460]

[17] J.M. Bier, C.J. Verbeek, and M.C. Lay, "Using synchrotron FTIR spectroscopy to determine secondary structure changes and distribution in thermoplastic protein", *J. Appl. Polym. Sci.,* vol. 130, no. 1, pp. 359-369, 2013. [http://dx.doi.org/10.1002/app.39134]

[18] J.M. Bier, C.J. Verbeek, and M.C. Lay, "Thermally resolved synchrotron FT-IR microscopy of structural changes in bloodmeal-based thermoplastics", *J. Therm. Anal. Calorim.,* vol. 115, no. 1, pp. 433-441, 2014. [http://dx.doi.org/10.1007/s10973-013-3340-8]

[19] M. Oliviero, L. Sorrentino, L. Cafiero, B. Galzerano, A. Sorrentino, and S. Iannace, "Foaming behavior of biobased blends based on thermoplastic gelatin and poly(butylene succinate)", *J. Appl. Polym. Sci.,* vol. 132, no. 48, pp. 1-9, 2015. [http://dx.doi.org/10.1002/app.42704] [PMID: 25866416]

[20] A. Salerno, M. Oliviero, E. Di Maio, and S. Iannace, "Thermoplastic foams from zein and gelatin", *Int. Polym. Process.,* vol. 22, no. 5, pp. 480-488, 2007. [http://dx.doi.org/10.3139/217.2065]

[21] J.R. Barone, W.F. Schmidt, and N. Gregoire, "Extrusion of feather keratin", *J. Appl. Polym. Sci.,* vol. 100, no. 2, pp. 1432-1442, 2006. [http://dx.doi.org/10.1002/app.23501]

[22] J.R. Barone, W.F. Schmidt, and C.F. Liebner, "Thermally processed keratin films", *J. Appl. Polym. Sci.,* vol. 97, no. 4, pp. 1644-1651, 2005. [http://dx.doi.org/10.1002/app.21901]

[23] B. Liu, L. Jiang, and J. Zhang, "Extrusion foaming of poly (lactic acid)/soy protein concentrate blends", *Macromol. Mater. Eng.,* vol. 296, no. 9, pp. 835-842, 2011. [http://dx.doi.org/10.1002/mame.201000449]

[24] P. Guerrero, and K. de la Caba, "Thermal and mechanical properties of soy protein films processed at different pH by compression", *J. Food Eng.,* vol. 100, no. 2, pp. 261-269, 2010. [http://dx.doi.org/10.1016/j.jfoodeng.2010.04.008]

[25] J.L. Jane, and S. Wang, Soy protein-based thermoplastic composition for preparing molded articles, 1996.U.S. Patent 5523293

[26] J. Zhang, P. Mungara, and J. Jane, "Mechanical and thermal properties of extruded soy protein sheets", *Polymer (Guildf.)*, vol. 42, no. 6, pp. 2569-2578, 2001. [http://dx.doi.org/10.1016/S0032-3861(00)00624-8]

[27] E. Di Maio, R. Mali, and S. Iannace, "Investigation of thermoplasticity of zein and kafirin proteins: Mixing process and mechanical properties", *J. Polym. Environ.*, vol. 18, no. 4, pp. 626-633, 2010. [http://dx.doi.org/10.1007/s10924-010-0224-x]

[28] D.J. Sessa, G.W. Selling, J.L. Willett, and D.E. Palmquist, "Viscosity control of zein processing with sodium dodecyl sulfate", *Ind. Crops Prod.*, vol. 23, no. 1, pp. 15-22, 2006. [http://dx.doi.org/10.1016/j.indcrop.2005.02.001]

[29] L. Verdolotti, M. Oliviero, M. Lavorgna, V. Iozzino, D. Larobina, and S. Iannace, "Bio-hybrid foams by silsesquioxanes cross-linked thermoplastic zein films", *J. Cell. Plast.*, vol. 51, no. 1, pp. 75-87, 2015. [http://dx.doi.org/10.1177/0021955X14529138]

[30] G. Trujillo-de Santiago, C. Rojas-de Gante, S. García-Lara, L. Verdolotti, E. Di Maio, and S. Iannace, "Thermoplastic processing of blue maize and white sorghum flours to produce bioplastics", *J. Polym. Environ.*, vol. 23, no. 1, pp. 72-82, 2015. [http://dx.doi.org/10.1007/s10924-014-0708-1]

[31] S.H. Richert, "Physical-chemical properties of whey protein foams", *J. Agric. Food Chem.*, vol. 27, no. 4, pp. 665-668, 1979. [http://dx.doi.org/10.1021/jf60224a036]

[32] K.G. Marinova, E.S. Basheva, B. Nenova, M. Temelska, A.Y. Mirarefi, B. Campbell, and I.B. Ivanov, "Physico-chemical factors controlling the foamability and foam stability of milk proteins: Sodium caseinate and whey protein concentrates", *Food Hydrocoll.*, vol. 23, no. 7, pp. 1864-1876, 2009. [http://dx.doi.org/10.1016/j.foodhyd.2009.03.003]

[33] D. Gammariello, J. Laverse, V. Lampignano, A.L. Incoronato, and M.A. Del Nobile, "Effect of compositional formulation on texture and microstructural of whey protein foam", *Food Res. Int.*, vol. 53, no. 1, pp. 496-501, 2013. [http://dx.doi.org/10.1016/j.foodres.2013.05.023]

[34] J. Leman, T. Dołgań, M. Smoczyński, and Z. Dziuba, "Fractal characteristics of microstructure of beta-lactoglobulin preparations and their emulsifying properties", *Food Sci. Technol. (Campinas.)*, vol. 8, no. 3, p. 29, 2005.

[35] H-B. Chen, Y-Z. Wang, and D.A. Schiraldi, "Foam-like materials based on whey protein isolate", *Eur. Polym. J.*, vol. 49, no. 10, pp. 3387-3391, 2013. [http://dx.doi.org/10.1016/j.eurpolymj.2013.07.019]

[36] N. Kaisangsri, O. Kerdchoechuen, and N. Laohakunjit, "Characterization of cassava starch based foam blended with plant proteins, kraft fiber, and palm oil", *Carbohydr. Polym.*, vol. 110, no. 1, pp. 70-77, 2014. [http://dx.doi.org/10.1016/j.carbpol.2014.03.067] [PMID: 24906730]

[37] P.R. Salgado, V.C. Schmidt, S.E. Ortiz, A.N. Mauri, and J.B. Laurindo, "Biodegradable foams based on cassava starch, sunflower proteins and cellulose fibers obtained by a baking process", *J. Food Eng.*, vol. 85, no. 3, pp. 435-443, 2008. [http://dx.doi.org/10.1016/j.jfoodeng.2007.08.005]

[38] X. Li, A. Pizzi, M. Cangemi, P. Navarrete, C. Segovia, V. Fierro, and A. Celzard, "Insulation rigid and elastic foams based on albumin", *Ind. Crops Prod.*, vol. 37, no. 1, pp. 149-154, 2012. [http://dx.doi.org/10.1016/j.indcrop.2011.11.030]

[39] X. Li, A. Pizzi, M. Cangemi, V. Fierro, and A. Celzard, "Flexible natural tannin-based and protein-

based biosourced foams", *Ind. Crops Prod.,* vol. 37, no. 1, pp. 389-393, 2012. [http://dx.doi.org/10.1016/j.indcrop.2011.12.037]

[40] Y. Lin, and F. Hsieh, "Water-blown flexible polyurethane foam extended with biomass materials", *J. Appl. Polym. Sci.,* vol. 65, no. 4, pp. 695-703, 1997. [http://dx.doi.org/10.1002/(SICI)1097-4628(19970725)65:4<695::AID-APP8>3.0.CO;2-F]

[41] S.K. Park, and N. Hettiarachchy, "Physical and mechanical properties of soy protein-based plastic foams", *J. Am. Oil Chem. Soc.,* vol. 76, no. 10, pp. 1201-1205, 1999. [http://dx.doi.org/10.1007/s11746-999-0094-3]

[42] L.C. Chang, Y. Xue, and F.H. Hsieh, "Comparative study of physical properties of water-blown rigid polyurethane foams extended with commercial soy flours", *J. Appl. Polym. Sci.,* vol. 80, no. 1, pp. 10-19, 2001. [http://dx.doi.org/10.1002/1097-4628(20010404)80:1<10::AID-APP1068>3.0.CO;2-Y]

[43] Y. Mu, X. Wan, Z. Han, Y. Peng, and S. Zhong, "Rigid polyurethane foams based on activated soybean meal", *J. Appl. Polym. Sci.,* vol. 124, no. 5, pp. 4331-4338, 2012. [http://dx.doi.org/10.1002/app.35612]

[44] S. Kumar, E. Hablot, J.L. Moscoso, W. Obeid, P.G. Hatcher, B.M. DuQuette, D. Graiver, R. Narayan, and V. Balan, "Polyurethanes preparation using proteins obtained from microalgae," ", *J. Mater. Sci.,* vol. 49, no. 22, pp. 7824-7833, 2014. [http://dx.doi.org/10.1007/s10853-014-8493-8]

[45] Y-Q. Zhang, "Applications of natural silk protein sericin in biomaterials", *Biotechnol. Adv.,* vol. 20, no. 2, pp. 91-100, 2002. [http://dx.doi.org/10.1016/S0734-9750(02)00003-4] [PMID: 14538058]

[46] R. Elshereef, J. Vlachopoulos, and A. Elkamel, "Comparison and analysis of bubble growth and foam formation models", *Eng. Comput.,* vol. 27, no. 3, pp. 387-408, 2010. [http://dx.doi.org/10.1108/02644401011029943]

[47] N.S. Ramesh, "Fundamentals of bubble nucleation and growth in polymers", In: *Polymeric Materials: Mechanisms and Materials.* , S.T. Lee, N.S. Ramesh, Eds., 1st ed CRC Press Boca Raton: FL, 2004. [http://dx.doi.org/10.1201/9780203506141.ch3]

[48] J. Wang, *"Rheology of Foaming Polymers and its influence on Microcellular Processing"*, PhD Thesis, University of Toronto, Canada, 2009.

[49] V. Kumar, *"Process Synthesis for Manufacturing Microcellular Thermoplastic Parts: A Case Study in Axiomatic Design"*, PhD Thesis, Massachusetts Institute of Technology, US, 1988.

[50] D.R. Uhlmann, and B. Chalmers, "The energetic of nucleation", *Ind. Eng. Chem. Res.,* vol. 57, no. 9, pp. 19-31, 1965. [http://dx.doi.org/10.1021/ie50669a006]

[51] S.T. Lee, C.B. Park, and N.S. Ramesh, "Foaming fundamentals", In: *Polymeric Foams: Science and Technology.* 1st ed CRC Press: Boca Raton, FL, 2007.

[52] N.G. McCrum, C. Buckley, and C.B. Bucknall, *Principles of Polymer Engineering.* Oxford University Press, 1997.

[53] M. Oliviero, E. Di Maio, and S. Iannace, "Effect of molecular structure on film blowing ability of thermoplastic zein," ", *J. Appl. Polym. Sci.,* vol. 115, no. 1, pp. 277-287, 2010. [http://dx.doi.org/10.1002/app.31116]

[54] A.V. Nawaby, and Z. Zhang, "Solubility and diffusivity", In: *Thermoplastic Foam Processing: Principles and Development.* , S.T. Lee, Ed., 1st ed CRC Press: Boca Raton, 2004.

[55] C. Gao, J. Taylor, N. Wellner, Y.B. Byaruhanga, M.L. Parker, E.N. Mills, and P.S. Belton, "Effect of preparation conditions on protein secondary structure and biofilm formation of kafirin", *J. Agric. Food Chem.,* vol. 53, no. 2, pp. 306-312, 2005.

[http://dx.doi.org/10.1021/jf0492666] [PMID: 15656666]

[56] R. Liao, W. Yu, and C. Zhou, "Rheological control in foaming polymeric materials: I. Amorphous polymers", *Polymer (Guildf.),* vol. 51, no. 2, pp. 568-580, 2010. [http://dx.doi.org/10.1016/j.polymer.2009.11.063]

[57] R. Gendron, "Rheological behavior relevant to extrusion foaming", In: *Thermoplastic Foam Processing: Principles and Development.* , S.T. Lee, Ed., 1st ed CRC Press: Boca Raton, 2005.

[58] G. Nam, J. Yoo, and J. Lee, "Effect of long-chain branches of polypropylene on rheological properties and foam-extrusion performances", *J. Appl. Polym. Sci.,* vol. 96, no. 5, pp. 1793-1800, 2005. [http://dx.doi.org/10.1002/app.21619]

[59] H. Münstedt, and F.R. Schwarzl, *Deformation and Flow of Polymeric Materials.* Springer: Berlin, 2014. [Germany]

[60] D.F. Baldwin, C.B. Park, and N.P. Suh, "A microcellular processing study of poly(ethylene terephthalate) in the amorphous and semicrystalline states. Part I: Microcellular nucleation", *Polym. Eng. Sci.,* vol. 36, no. 11, pp. 1437-1445, 1996. [http://dx.doi.org/10.1002/pen.10538]

[61] J-B. Bao, T. Liu, L. Zhao, and G-H. Hu, "A two-step depressurization batch process for the formation of bi-modal cell structure polystyrene foams using $scCO_2$", *J. Supercrit. Fluids,* vol. 55, no. 3, pp. 1104-1114, 2011. [http://dx.doi.org/10.1016/j.supflu.2010.09.032]

[62] Y-M. Corre, A. Maazouz, J. Duchet, and J. Reignier, "Batch foaming of chain extended PLA with supercritical CO_2: Influence of the rheological properties and the process parameters on the cellular structure", *J. Supercrit. Fluids,* vol. 58, no. 1, pp. 177-188, 2011. [http://dx.doi.org/10.1016/j.supflu.2011.03.006]

[63] M. Oliviero, L. Verdolotti, I. Nedi, F. Docimo, E. Di Maio, and S. Iannace, "Effect of two kinds of lignins, alkaline lignin and sodium lignosulfonate, on the foamability of thermoplastic zein-based bionanocomposites", *J. Cell Plast.,* vol. 48, no. 6, pp. 516-525, 2012. [http://dx.doi.org/10.1177/0021955X12460043]

[64] A. Walallavita, C.J. Verbeek, and M.C. Lay, "Blending novatein thermoplastic protein with pla for carbon dioxide assisted batch foaming", *Proceedings of PPS-31: The 31st International Conference of the Polymer Processing Society–Conference Papers,* 2015 Jeju Island, South Korea

[65] S. Doroudiani, C.B. Park, and M.T. Kortschot, "Effect of the crystallinity and morphology on the microcellular foam structure of semicrystalline polymers", *Polym. Eng. Sci.,* vol. 36, no. 21, pp. 2645-2662, 1996. [http://dx.doi.org/10.1002/pen.10664]

[66] S.H. Mahmood, M. Keshtkar, and C.B. Park, "Determination of carbon dioxide solubility in polylactide acid with accurate PVT properties", *J. Chem. Thermodyn.,* vol. 70, pp. 13-23, 2014. [http://dx.doi.org/10.1016/j.jct.2013.10.019]

[67] E. Aionicesei, M. Škerget, and Ž. Knez, "Measurement of CO_2 solubility and diffusivity in poly(l-lactide) and poly(d,l-lactide-co-glycolide) by magnetic suspension balance", *J. Supercrit. Fluids,* vol. 47, no. 2, pp. 296-301, 2008. [http://dx.doi.org/10.1016/j.supflu.2008.07.011]

[68] Y-K. Kwon, and H-K. Bae, "Production of microcellular foam plastics by supercritical carbon dioxide", *Korean J. Chem. Eng.,* vol. 24, no. 1, pp. 127-132, 2007. [http://dx.doi.org/10.1007/s11814-007-5022-3]

[69] C. Yu, Y. Wang, B. Wu, and W. Li, "A direct method for evaluating polymer foamability", *Polym. Test.,* vol. 30, pp. 118-123, 2011. [http://dx.doi.org/10.1016/j.polymertesting.2010.11.001]

[70] R. Liao, W. Yu, and C. Zhou, "Rheological control in foaming polymeric materials: II. Semi-

crystalline polymers", *Polymer (Guildf.),* vol. 51, pp. 6334-6345, 2010. [http://dx.doi.org/10.1016/j.polymer.2010.11.001]

[71] D. Scholz, "Development of endothermic chemical foaming/nucleation agents and it processes", In: *Polymeric Foams: Technology and Developments in Regulation, Process, and Products.* , S.T. Lee, D. Scholz, Eds., 1st ed CRC Press USA, 2009, pp. 42-65.

[72] W. Ernst, J.r. Oblotzki, and H. Potente, "Description of the foaming process during the extrusion of foams based on renewable resources", *J. Cell. Plast.,* vol. 42, no. 3, pp. 241-253, 2006. [http://dx.doi.org/10.1177/0021955X06063513]

[73] P. Mungara, J.W. Zhang, and J. Jane, "Extrusion processing of soy protein-based foam", *Polymer Prepr.,* vol. 39, no. 2, pp. 148-149, 1998.

[74] M. Filli, M. Sjoqvist, C. Ohgren, M. Stading, and M. Rigdahl, "Development and characterization of extruded biodegradable foams based on zein and pearl millet flour", *Annu. Trans. NRS,* vol. 19, 2011.

[75] C. Gavin, M.C. Lay, and C.J. Verbeek, "Extrusion foaming of protein-based thermoplastic and polyethylene blends", *Proceedings of PPS-31: The 31st International Conference of the Polymer Processing Society–Conference Papers,* 2015 Jeju Island, South Korea

[76] S.T. Lee, and C.B. Park, *Foam Extrusion: Principles and Practice.* 2nd ed Taylor & Francis: Boca Raton, 2014. [http://dx.doi.org/10.1201/b16784]

[77] S.T. Lee, C.B. Park, and N.S. Ramesh, "General foam process technologies", In: *Polymeric Foams: Science and Technology.* , S.T. Lee, C.B. Park, N.S. Ramesh, Eds., 1st ed CRC Press: Boca Raton, FL, 2007.

[78] C.H. Lee, K-J. Lee, H.G. Jeong, and S.W. Kim, "Growth of gas bubbles in the foam extrusion process", *Adv. Polym. Technol.,* vol. 19, no. 2, pp. 97-112, 2000. [http://dx.doi.org/10.1002/(SICI)1098-2329(200022)19:2<97::AID-ADV3>3.0.CO;2-B]

[79] S. Lee, L. Kareko, and J. Jun, "Study of thermoplastic PLA foam extrusion", *J. Cell. Plast.,* vol. 44, no. 4, pp. 293-305, 2008. [http://dx.doi.org/10.1177/0021955X08088859]

[80] H.A. Pushpadass, G.S. Babu, R.W. Weber, and M.A. Hanna, "Extrusion of starch-based loose-fill packaging foams: effects of temperature, moisture and talc on physical properties", *Packag. Technol. Sci.,* vol. 21, no. 3, pp. 171-183, 2008. [http://dx.doi.org/10.1002/pts.809]

[81] E. Cheng, S. Alavi, T. Pearson, and R. Agbisit, "Mechanical–acoustic and sensory evaluations of cornstarch–whey protein isolate extrudates", *J. Texture Stud.,* vol. 38, no. 4, pp. 473-498, 2007. [http://dx.doi.org/10.1111/j.1745-4603.2007.00109.x]

[82] D.N. Yadav, T. Anand, Navnidhi, and A.K. Singh, "Co-extrusion of pearl millet-whey protein concentrate for expanded snacks", *Int. J. Food Sci. Technol.,* vol. 49, no. 3, pp. 840-846, 2014. [http://dx.doi.org/10.1111/ijfs.12373]

[83] M. Włodarczyk-Stasiak, and J. Jamroz, "Analysis of sorption properties of starch–protein extrudates with the use of water vapour", *J. Food Eng.,* vol. 85, no. 4, pp. 580-589, 2008. [http://dx.doi.org/10.1016/j.jfoodeng.2007.08.019]

[84] C. Onwulata, R. Konstance, P. Smith, and V. Holsinger, "Co-extrusion of dietary fiber and milk proteins in expanded corn products", *LWT-Food Sci. Technol. (Campinas.),* vol. 34, no. 7, pp. 424-429, 2001.

[85] L.A. Pelembe, C. Erasmus, and J.R. Taylor, "Development of a protein-rich composite sorghum–cowpea instant porridge by extrusion cooking process", *LWT - Food Sci. Technol. (Campinas.),* vol. 35, no. 2, pp. 120-127, 2002.

[86] L-J. Zhu, R. Shukri, N.J. de Mesa-Stonestreet, S. Alavi, H. Dogan, and Y-C. Shi, "Mechanical and

microstructural properties of soy protein–high amylose corn starch extrudates in relation to physiochemical changes of starch during extrusion", *J. Food Eng.,* vol. 100, no. 2, pp. 232-238, 2010. [http://dx.doi.org/10.1016/j.jfoodeng.2010.04.004]

[87] J. Faller, B. Klein, and J. Faller, "Acceptability of extruded corn snacks as affected by inclusion of soy protein", *J. Food Sci.,* vol. 64, no. 1, pp. 185-188, 1999. [http://dx.doi.org/10.1111/j.1365-2621.1999.tb09888.x]

[88] R.P. Konstance, C.I. Onwulata, P.W. Smith, D. Lu, M.H. Tunick, E.D. Strange, and V.H. Holsinger, "Nutrient-based corn and soy products by twin-screw extrusion", *J. Food Sci.,* vol. 63, no. 5, pp. 864-868, 1998. [http://dx.doi.org/10.1111/j.1365-2621.1998.tb17915.x]

[89] C.I. Onwulata, P.W. Smith, R.P. Konstance, and V.H. Holsinger, "Incorporation of whey products in extruded corn, potato or rice snacks", *Food Res. Int.,* vol. 34, no. 8, pp. 679-687, 2001. [http://dx.doi.org/10.1016/S0963-9969(01)00088-6]

[90] K.E. Allen, C.E. Carpenter, and M.K. Walsh, "Influence of protein level and starch type on an extrusion-expanded whey product", *Int. J. Food Sci. Technol.,* vol. 42, no. 8, pp. 953-960, 2007. [http://dx.doi.org/10.1111/j.1365-2621.2006.01316.x]

[91] F.P. Matthey, and M.A. Hanna, "Physical and functional properties of twin-screw extruded whey protein concentrate–corn starch blends", *LWT - Food Sci. Technol. (Campinas.),* vol. 30, no. 4, pp. 359-366, 1997.

[92] C.I. Onwulata, and R.P. Konstance, "Extruded corn meal and whey protein concentrate: Effect of particle size", *J. Food Process. Preserv.,* vol. 30, no. 4, pp. 475-487, 2006. [http://dx.doi.org/10.1111/j.1745-4549.2005.00082.x]

[93] A.M. Trater, S. Alavi, and S.S. Rizvi, "Use of non-invasive X-ray microtomography for characterizing microstructure of extruded biopolymer foams", *Food Res. Int.,* vol. 38, no. 6, pp. 709-719, 2005. [http://dx.doi.org/10.1016/j.foodres.2005.01.006]

[94] C.I. Onwulata, R.P. Konstance, J.G. Phillips, and P.M. Tomasula, "Temperature profiling: Solution to problems of co-extrusion with whey proteins", *J. Food Process. Preserv.,* vol. 27, no. 5, pp. 337-350, 2003. [http://dx.doi.org/10.1111/j.1745-4549.2003.tb00521.x]

CHAPTER 10

Biopolymer Electrospinning

Athira K. Sunil*

Indian Institute of Science, Bangalore – 560012, India

Abstract: Electrospinning is a useful tool for producing fine fibers of about 10 nanometers to 10 microns in diameter from both natural and synthetic polymers. Historical developments, theory, and the types equipment used to produce aligned and non-aligned fibres in electrospinning are discussed. Collector composition or substrates, dimension of the collector, electrospinning materials and configurations of nozzle in electrospinning are discussed with their respective importance. The electrospinning process is influenced by the following solution parameters like viscosity, polymer molecular weight, polymer concentration, surface tension, polymer solution conductivity or surface charge density, rate of evaporation of solvents, dielectric constant and vapour pressure. Voltage, needle tip to collector distance, feed-rate, collector material, electrospinning setup, diameter of needle orifice, environmental (ambient) parameters like temperature, humidity, type of atmosphere and pressure also determine electrospun fiber size and morphology.

Keywords: Co-electrospinning, Collector, Electrospinning, Parameters, Spinneret, Voltage.

INTRODUCTION

Electrospinning is a versatile and cost-effective technique for producing multi-functional nanofibers from various polymers, polymer blends, composites, sol-gels, ceramics, *etc.* [1]. The fibres have diameters ranging from a few microns to tens of nanometres [2 - 5].

In a typical electrospinning set-up (Fig. **1**), a solution or melt is first fed through a spinneret with an inner diameter in the order of 100 μm. When a sufficiently high electric field is applied to the spinneret, which simultaneously serves as an electrode, the repulsive force caused by the concentration of similar charges in the solution can overcome the opposing surface tension of the droplet at the tip of the spinneret, and a Taylor cone is formed, from which a polymeric jet initiates.

* **Corresponding author Athira K. Sunil**: Indian Institute of Science, Bangalore – 560012, India; Email: athiraiisc@gmail.com

Martin Masuelli & Denis Renard (Eds.)

Although the jet is stable near to the tip of the spinneret, it soon enters a bending instability stage with the formation of a whipping cone under the influence of the coulombic forces. This electrospinning jet stretches and coils, with an increasing diameter, until the solvent has evaporated, yielding fine fibres with diameters as low as a few tens of nanometres. The fibres are collected as a non-woven mesh of continuous fibres on an earthed collector, usually a metal plate or drum. The distance between nozzle and collector is set large enough to enable most of the solvent to evaporate and form dry fibres on the collection plate. The polymer chain entanglements within the solution prevents the electrospinning jet from breaking up and, depending on the stability of the electrospinning process, fibres with lengths of up to many hundreds of metres can be obtained.

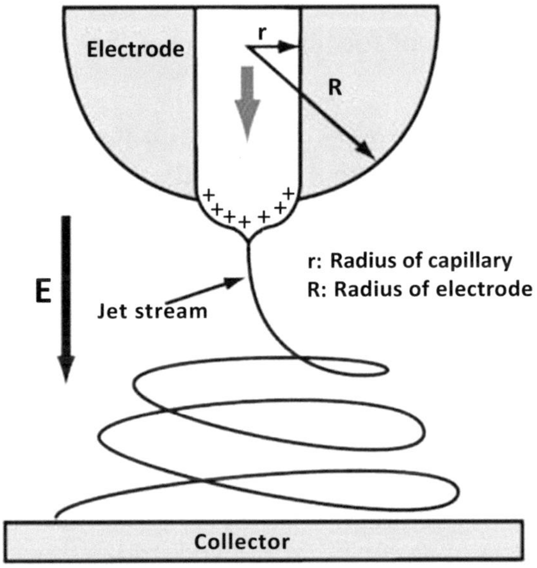

Fig. (1). Schematic diagram of an electrospinning apparatus and Taylor cones.

The versatility of the electrospinning process to produce fibres from a large range of polymeric materials or material combinations, including non-polymeric ones, either through mixing, co-axial or multiphase electrospinning of solutions, or through addition of dopants, provides the scope for this method to be used to fabricate fibres for many different applications. Post-processing of electrospun fibres, *e.g.* heat or solvent treatments for creation of pores in fibres, further increases the field of applications, with more novel functions being realized through the use of different materials.

History of Electrospinning

The discovery of electrical effects on liquids was carried out in the late 1500's when William Gilbert [6] set out to investigate the phenomena of electricity and magnetism. Gilbert observed that a spherical droplet of water was drawn into a conical shape when a charged piece of Amber was placed near it, which caused small droplets of water to be emanated; a phenomenon now called electrospraying.

It was not until 1898 that electrodynamics theory was used by Larmour [7] to explain the excitation of dielectric liquid under the influence of an electric charge. Four years later in 1902, Cooley [8] and Morton [9] patented the first devices to spray liquids by applying an electrical charge. The electrospinning in Coley's patent included a rotating collector for fibres produced from the jet directed by an auxiliary electrode [5].

Electrospinning was further investigated by Antonin Formhals who attained the first patent [10] on electrospinning in 1934 for the fabrication of textile yarns from cellulose acetate using acetone and monomethyl ether of ethylene glycol as solvents [11]. Formhals produced several patents for fibre production and collection, including designs that do not require a spinneret or rotating fibre collection devices [10, 12, 13]. Many of the recent electrospinning setups can in fact be traced back to these inventions and it is clear from these developments that the early researchers had already gained an in-depth knowledge of the electrospinning process [5].

However, it was not until 1964 that Sir Geoffrey Ingram Taylor [14], continuing work started by John Zeleny in 1914 [15], investigated the behaviour of fluid droplets under the influence of an electric field and initiated mathematical studies on the jet forming process [16 - 18]. Taylor described that the onset of the jet in the electrospinning process was initiated by the formation of a conical shape, now known as the Taylor cone. He theoretically derived that a perfect cone formed under the influence of an electric field required a semi-vertical angle of 49.3° and demonstrated that the angle of the cone on a droplet approached this theoretical value just before jet initiation. Two years later in 1966, Simons [19] patented the apparatus for the production of lightweight, ultra-thin, non-woven fabrics using electrospinning. In 1971, Baumgarten [20] reported electrospinning of acrylic fibres from a new set up that produced fibres with diameters in the range of 0.05–1.1 μm. Baumgarten showed how the diameters of the fibres changed with changing electric field. Despite these early discoveries, electrospinning did not gain a lot of interest and was not utilized commercially until the 90's. It is only since the emergence of nanotechnology in the 1990's that electrospinning gained increasing attention as a way of manufacturing fibres at the nanoscale.

Researchers have examined the potential of electrospinning to produce complex and highly functionalized systems, which could be used in a variety of applications and applied on a commercial level. The possibilities presented by electrospinning in the fields of bio-medical research, electronics, catalysis, environmental and energy storage applications has led to an almost exponential increase in the number of scientific publications since the 90's [21, 22]. This research, notably that by Reneker and Rutledge [4, 23], focused on understanding the electrospinning process and system parameters that lead to reduced fibre diameters with different morphologies. Specifically in the last decade, much theoretical work has been carried out to gain an understanding of the electrohydrodynamic mechanism driving the electrospinning of electrically forced jets, such as the growth rate of the Rayleigh and whipping instabilities. Today, there are companies such as eSpin Technologies, NanoTechnics, KATO Tech, and The Electrospinning Company that are using fibres obtained from electrospinning for more commercial applications such as filtration, aerospace, structural composites, healthcare, energy storage, cosmetics, as well as research applications.

Electrospinning Theory

Electrospinning works for liquids that are conducting, such that the theory presented here discusses the behaviour of viscous and ionic solutions under the influence of an applied electric field. When a voltage is applied between the nozzle and the collector, migration of electrons from the cathode to the anode causes a charge build up on the surface of the solution, with the polarity of the charges dependent on the polarity of the nozzle. If the nozzle is negatively charged, cations are attracted to the electrode; if it is positive, anions are attracted (Fig. **2**). The free surface of the liquid maintains an equipotential at all times, with the charges distributed in a way that maintains a zero electric field inside the liquid [24]. When a critical potential is established to enable the coulombic force, resulting from the repulsion of like charges on the liquid surface, to overcome the surface tension that keeps the droplet together, the liquid droplet acquires a conical shape called the Taylor cone [25], having a half angle of 49.3° [24]. Although the size of the coulombic force is correlated to the magnitude of the applied voltage, the density of ions and electrons in the overall neutral solution strongly influences the degree of charge build up.

Once the Taylor cone is established, an electrically charged jet comes out from the tip of the cone and accelerates down or across to the collection plate, depending on the direction of the set up (as shown in Fig. **3** [26]). This jet, if formed from a polymeric solution, undergoes a multi-component jet stretching mode with an initial straight jet portion followed by a radially spiralling conical jet part (see Fig. **4**), which arises as a result of axisymmetric and whipping (or bending) instabilities respectively [27]. These instabilities, which arise as a result of coupling of the coulombic forces in the jet with the electric field, enable stretching of the jet to yield fine fibres.

General Electrospinning Device

An electrospinning machine is shown in Fig. (**5**). Electrospinning devices can be classified into single-fluid or two-fluid electrospinning device based on the capillary arrangement at the upper disk. Single-fluid electrospinning device consists of two aluminium disks spaced apart.

A rigid plastic lab-stand suspends the upper disk above the lower disk. The parallel-disk arrangement provides a uniform external electric field. The disk helps to direct the deposition of the fiber onto the collector. A syringe pump delivers the fluid from

the syringe to the capillary *via* a Teflon tube, which serves as an electrical insulating device. A power supply provides up to 40 kV of direct current to the upper disk with a cut-off feature to prevent electrocution. The lower disk is positioned at a sufficient distance from upper disk, such that the fiber dries by the time it hits the collector. The collector is covered by aluminium foil. A ventilated plastic enclosure houses the entire electrospinning device. The two-fluid electrospinning device has two capillary, one inside the other, forming an annulus. Two syringe pumps deliver two separate fluids to the capillaries simultaneously.

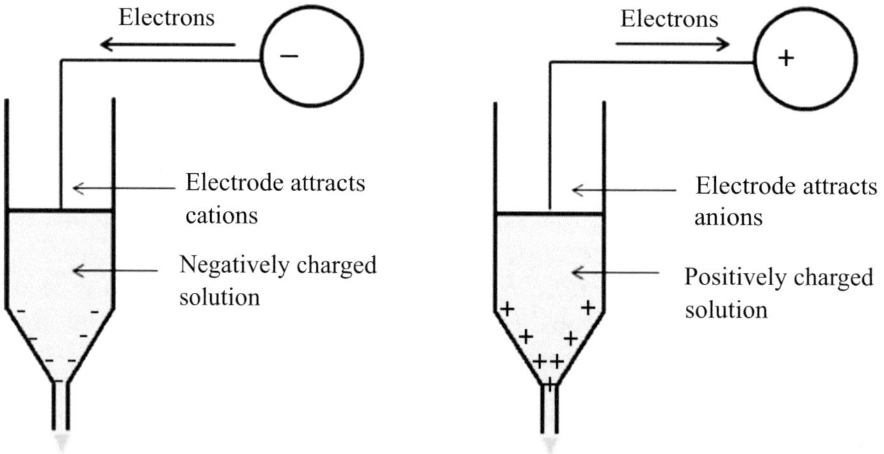

Fig. (2). Schematic diagram of electrospinning nozzle with positive and negatively charged electrode.

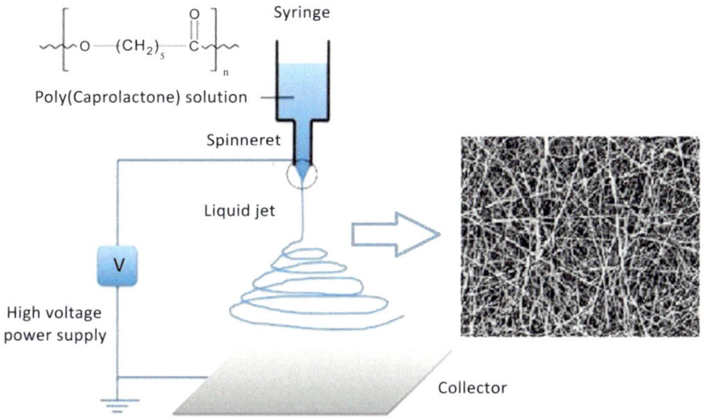

Fig. (3). A schematic diagram showing the process of electrospinning [26].

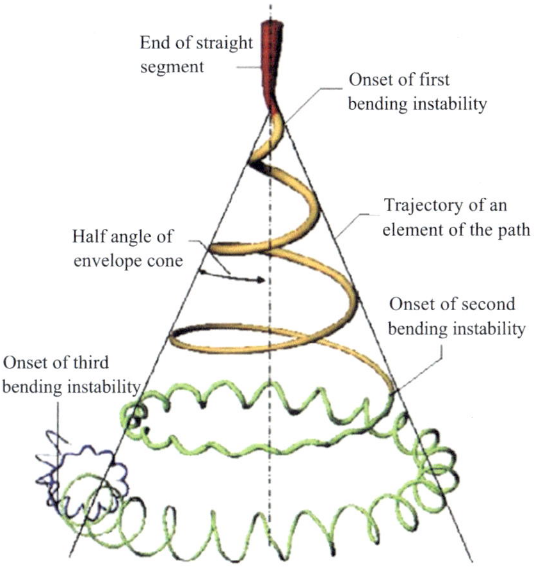

Fig. (4). The electrospinning envelope showing the conical jet path [4].

Collector Types and Properties

Nanofibers can be classified into woven and non-woven nanofibers. The type of collector used plays a big role in differentiating those types or nanofiber alignments. The methods utilized in developing aligned woven nanofibers consist of the use of oriented collector, as well as static double grounded collector [28, 29].

Fig. (5). An electrospinning machine.

Oriented Collector

An oriented collector consist of a rotating drum collector and rotating disk collector. The rotating drum collector is used for collecting the aligned arrays nanofibers, whereas the rotating disk is used to collect uniaxially aligned nanofibers. The alignment of fibers produced *via* rotating drum correspond to the rotational speed applied on the drum [29]. This type of electrospinning method is more complicated since the speed of the rotation needs to be well controlled to produce nanofibers with fine alignment. The rotating disk collector can also be used for collecting continuous nanofibers, since nanofiber is highly attracted to the large electrical field applied at the edge of the disk [28]. According to Baji *et al.* [29], the use of this method can produce aligned nanofibers, even though only a small amount of fiber is produced due to the small area of the disk edge. In addition, Thron *et al.* [30] used the disk as a collector in producing polyethylene oxide nanofibers. They found that the diameter of electrospun nanofibers was 200 nm in average and the distance between the fibers varied from 1000 nm to 2000 nm due to electrostatic repulsion while reaching the collector. This electrostatic repulsion influences the morphology of the formed nanofibers [29].

Double Ground Collector

The double ground collector consists of two parallel plates in the electrospinning system as a collecting device for woven nanofibers. The fibers exiting the syringe tip tend to "straighten" themselves in the region between the two plates, resulting in good alignment [28, 29]. For randomly oriented nanofibers (nonwoven) a stationary collector can be used. The collector can be a vertical flat ground or horizontal flat ground collectors. The electric field/voltage supply will force the jet solution to twirl around before arriving at the top of the collector, which, at this point, the gravitational force is negligible for the polymer solution. Rodoplu and Mutlu [31] claimed that either in vertical or in horizontal direction of the needle, this configuration will still affect the formation of the Taylor cone and the droplet, since at this point the gravity still plays the role. They also claimed that, in vertical direction of the syringe instead of horizontal direction, this configuration will affect the shape of the droplet formation (Taylor cone formation) at the needle tip, which depends on the flat surface of the tip of needle. Kumar [32] conducted an experiment in order to study the effect of the collector type on the electrospun PVA. Four different types of collector were used, such as plane plate collector, drum rotator collector, grid collector, and edge collector. From the experiment, he found that the edge type collector showed a promising result on aligned fibers with good morphological characteristic and no bead was obtained compared with the other types of collector. Thus, this study proved that the collector types give a significant effect on fiber morphology of the electrospun fiber.

Collector Composition/Substrates

As the collector is in metal form, the composition or substrates on the collector for fiber collection can be in various forms such as glass, plastic, metal, or silicon wafer. Collector composition will affect the structure of the electrospun fiber. Liu and Hsieh [33] showed that the structure of electrospun cellulose acetate collected *via* different collector compositions such as copper mesh, aluminium foil, water, and paper differed due to the influence of the conductivity of the collector itself. A conductive collector will help to dissipate the carried charge by the fibers, whereas the nonconductive collectors will not, resulting in more porous fibre structure due to the fibers repelling one another. Tang *et al.* [34] used a small nail as a collector in their study to produce zigzag or helical structured nanofibers. They proposed that by reducing the area of the collector, the electric field can be concentrated at the nail, eventually producing helical nanofibers.

In addition, Pham *et al.* [35] in their review had discussed poly (ethylene-co-vinyl alcohol), which was spun directly onto a human hand. They also studied the effect of the collector composition on the structure of the electrospun fiber. Spinning fiber onto a human hand was also experimented by Kenawy *et al.* [36], in which after 30 minutes, they found that a thick mat of EVOH was formed, coating the hand. Human skin is a dielectric material since it can support an electrostatic field, even though it is a poor electrical conductor [36]. The same effect can be observed on a plastic petri dish.

Dimension of the Collector

The dimension of the collector used in electrospinning process generally depends on the function of the nanofiber itself and the type of collector used. Beachley and Wen [37] performed a study on the effect of the size of two parallel plates on the electrospun fiber diameter and length. In the study, they used three different sizes of collector which were (30.5 × 7.5 × 0.7) cm, (15 × 4 × 0.35) cm, and (7.5 × 2 × 0.15) cm, and they found that, as the plate size increased, the fiber diameter also increased, as well as the fiber length. In reality, the collector dimension can be neglected since the formation of the desirable morphology and diameter of fibers rely more on the type of collector.

Configurations of Nozzle in Electrospinning

In ejection and needle configuration, there are three main types of nozzle configuration such as single [38], side-by-side [39] and coaxial [38], which have become popular in producing different types of nanofibers, such as nanohollow fiber/core-shell fiber, and nanofiber.

Electrospinning Materials

Polymers consist of long chain of molecules (or repeating units called monomers) that are typically connected *via* covalent bonds. The long polymer chains have the tendency to entangle in solutions, with the extent of entanglement dependant on chain length. For this reason polymers and some ceramics are the only materials that enable sustained electrospinning, which is essentially the elongation of polymer chains in a fluid jet. The polymer material to be used is generally the first consideration for the choice of the solution system to be electrospun and is chosen on the basis of desired fibre properties and suitability for the specific application.

Natural polymers are generally preferred over synthetic polymers for biomedical applications, due to their better biocompatibility and immunogenicity. Most natural polymers can be degraded by enzymes and hence electrospun fibres from these polymers are commonly used for biomedical applications such as tissue engineering, biosensors, wound dressing, drug delivery and enzyme immobilization. Most polymers that have been electrospun are proteins and polysaccharides, examples include collagen (a prominent biopolymer used extensively for tissue engineering applications), chitosan, gelatine, cellulose acetate, silk protein, chitin, casein, *etc* [1]. Silk protein obtained from silk worms and spiders provide the best mechanical properties, as well as advantages ofbiocompatibility, biodegradability, good oxygen and water permeability and minimal inflammatory reaction [1, 40]. Synthetic polymers, which are generally more readily available and less costly, are used for a wider range of applications due to their versatility to be tailored to have a wider range of properties, such as high viscoelasticity, strength for improved mechanical properties, and desired degradation rates. Examples of synthetic polymers include polystyrene, poly acrylonitrile, polycarbonate, poly (benzimodozal), pol(vinylidene chloride), poly (vinylidene fluoride), as well as bio-degradable polymers such as poly(e-caprolactone), poly(L-lactic acid), polygylcolide, *etc* [1]. When mechanical properties and control of degradation rates are also important for biological applications, synthetic bio-degradable polymers are instead used.

Electrospinning Parameters

In the electrospinning process, the morphology and uniformity of nanofibers are determined by considering a number of parameters which are investigated under two main parts: a) polymer/solution properties such as molecular weight, molecular weight distribution, viscosity, solution conductivity, dielectric constant of solvent, surface tension; and b) set-up/process parameters such as the applied electric field, the flow rate of solution, diameter of spinneret/needle, type of collector and tip-to-collector distance. Moreover, the environmental parameters (temperature, humidity and air velocity in the chamber) may affect the formation of nanofibers. The polymer/solution properties have the most important influence on the electrospinning process.

Most of these parameters have a range in which they allow for stable electrospinning of polymeric solutions. However, due to the sensitivity of the fibre morphology to small changes in these parameters, optimisation of these parameters is usually necessary for controlling the process to yield the desired fibre morphologies and diameters. However, attempts to control one parameter, especially properties of the solutions, can lead to unwanted changes in another, *e.g.* adding a salt to increase the conductivity can result in a change in the solution viscosity, making mutual optimisation of these parameters difficult. Understanding the effects of each parameter on the process and the resultant fibre morphologies and diameters, followed by the optimisation of each parameter for each system, are important.

Solution Parameters

Viscosity

One of the conditions for electrospinning to yield fibres is that the solution used must have sufficient viscosity, such that polymer entanglement during stretching of the electrically driven jet is sufficient to prevent the jet from breaking up. Viscosity of the polymer-solvent system (solution) is mostly determined by the molecular weight and concentration of the polymer, though the interaction between the polymer and the solvent can also have an effect. Generally the solution viscosity increases with increasing polymer concentration and molecular weight, both of which control the extent of polymer entanglement.

The viscous forces of a solution resist the electrical forces that attempt to stretch the fiber during electrospinning [41]. It has been found that there exists an optimum viscosity range for producing nanofibers [42]. At higher viscosities, the cohesiveness of the solution prohibits the formation of fibers [42, 43]. Also, it can be difficult to pump the solution through the nozzle or the droplet at the tip of the nozzle may dry up too quickly as a result of insufficient amount of solvent. On the contrary, lower viscosities are more dilute and cause the production of droplets instead of fibers [44 - 46]. It is accepted that the viscosity of the polymeric solution influences the fiber diameter. At low viscosity, above the critical point for fibre formation, the fibres formed are beaded (*e.g.* beads attached to a string with gaps) and as the viscosity is increased the shape of the beads change from being spherical to more spindle-like and finally uniform fibres are formed (as shown in Fig. **6**) [47, 48]. Fibre beading is attributed to the effect of surface tension, which acts to reduce the surface area of the jet, and when the surface tension dominates over the repulsive coulombic force on the jet beaded fibres are formed. With increasing viscosity, the resistance to jet breaking increases (*i.e.* chain entanglement increases) while the effect of surface tension exerted by the solvent molecules decreases and, thus, the effect of beading is reduced or removed.

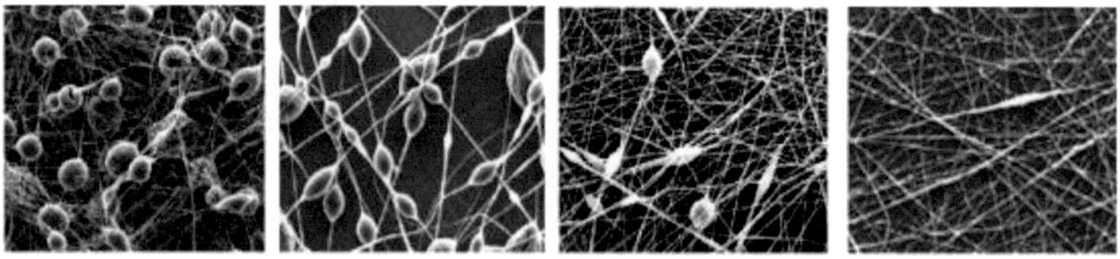

Fig. (6). Change in the beaded structure of fibres electrospun from solutions with increasing viscosity.

As the concentration is proportional to the viscosity, several researchers have sought to establish relationships between the polymer concentration and the final fiber diameter. All data appears to support the idea that higher viscosity solutions produced fibers with larger diameters; however the disagreement occurs in developing a quantitative relationship between these two parameters [44, 49, 50]. Deitzel *et al.* suggested that the there exists a power law relationship between the fiber diameter and the concentration [47]. Other researchers suggest a cubic power relationship between the final fiber diameter and the polymer concentration [44,

51]. Similarly, Mit-uppatham *et al.* proposed that for certain concentration ranges the average fiber diameter has an exponential relationship with concentration [49].

The results from their study showed that at polyurethane (PU) concentration of 6% (w/v), resulted in a bead-on-string morphology with several big beads. At PU concentration of 8%, the shape of the beads became spindle like and the fibers were smooth and homogeneous without beads on thefiber. When PU concentration increased from 8% to 10% (w/v), the electrospun membrane became thicker and exhibited a film-like structural as shown in Fig. (7) [52]. When the viscosity of the solution increased with increasing polymer concentration, the average fiber diameter increased. Therefore, the viscosity of polymer solution was the main factor affecting the average diameter of nanofibers as well as the formation of fiber [52 - 56].

Fig. (7). SEM images of polyurethane fibrous membranes electrospun from (a) 6%, (b) 8%, (c) 10% (w/v) solution in tetrahydrofuran and N,Ndimethyl formamide [52].

Increasing viscosity also typically leads to an increase in the fibre diameters. Thus, to obtain non-beaded low diameter fibres, it is important to control solution viscosity by increasing molecular weight of the polymer (to increase entanglement) while decreasing its concentration.

Surface Tension

In electrospinning, as the solution jet accelerates from the tip of the source to the collector, the solution is stretched while surface tension of the solution may cause the solution to break up into droplets [1]. Also, as the jet travels to the collector, the stretching force by the repelling charges needs to dominate over the surface tension of the jet to prevent beading of the fibres. It is therefore important to reduce the surface tension of the solution to the minimum possible. Lower solution surface tension enables the use of lower solution conductivities or electric fields to conduce jet initiation and production of uniform fibres [48].

Surface tension is mostly determined by the solvent composition and the interaction of the solvent with the polymer and thus varies depending on the solvent used to dissolve a given polymer. Surface tension is determined by the strength of the cohesive force between the surface molecules in a liquid, which bind more closely together than they do to the molecules of the surrounding medium (*e.g.* air). This leads to the formation of a thin film around the fluid, making it harder to break it apart. Since surface tension has dimensions of force per unit length or energy per unit area, and a spherical configuration is favoured to keep the surface energy to minimum. Due to this tendency for solvent molecules to retain a minimum energy configuration, in the presence of a high concentration of free solvent molecules in a solution, there is a greater tendency for solvent molecules to congregate and act against the influence of the charges that push the liquid apart. Conversely, a higher concentration of polymer can reduce this effect by causing solvent molecules to be dispersed more readily over the polymer molecules and hence lowering their tendency to come together [1]. The surface tension of solutions, which is typically 30 mNm^{-1} [40], can be lowered through blending of solvents with low surface tension values or through the addition of surfactants.

Polymer Solution Conductivity/Surface Charge Density

For electrospinning to take place the solution needs to be sufficiently conducting to enable the surface charges to overcome the surface tension of the droplet and initiate jet flow. Thus, non-conducting solutions will not electrospin. Higher stresses on the jet induced by higher conductivities generally result in reduced fiber diameters. The increased amount of charge forces the jet to elongate as a consequence of coulombic repulsion. These forces act against surface tension, hence favouring formation of non-beaded fibers.

The conductivity of the solution is largely determined by the type of solvent and polymer used, although addition of salts and/or surfactants may be needed to increase the charge density (*i.e.* the number of ions). Salt additives in electrospinning solutions create migrating ions in solution that transport charges and on tribute to higher conductivities [57, 59]. While salts and surfactants readily dissolve in most solvents with significant polarity, dissolution of salts in non-polar solvents (such as toluene) may be difficult and hence the ionic conductivity required to initiate electrospinning may be difficult to achieve. In such cases blending of solvents with higher conductivities or those that enable dissolution of salts can be used to increase the conductivity (as we have done to overcome this problem). It should be noted that the size of the ions in the solution may also have an effect on the fibre morphology. Zhong *et al.* has demonstrated the removal of fibre beading on addition of salt, with smallest fibre diameters obtained for solutions containing NaCl and largest diameters obtained for solutions containing KH_2PO_4; while solutions containing NaH_2PO_4 yielded fibres with intermediate diameters [57]. This observation is attributed to the greater elongational force, caused by the greater mobility of the smaller ions, under the influence of the external electrostatic field.

However, even with increased conductivity, there is a limit to the reduction in fibre diameter due to the increasing effect of viscoelastic forces on the stretched solution. As the solution jet is being stretched, the opposing viscoelastic force acts with an increasing magnitude against the coulombic force, halting solution stretching beyond a critical point. Also, in some cases addition of ions can increase the viscosity of the solution, and although conductivity may be increased, the increased viscosity may counteract the effect of increased charge density and hence result in an increase in the fibre diameters [49].

Dielectric Constant

The dielectric constant, also known as static relative permittivity of a material, is a measure of how well the material concentrates electrostatic lines of flux under the influence of an electric field. It is essentially the ratio of amount of stored electrical energy relative to the permittivity of vacuum when a voltage is applied. With an increasing dielectric constant the ability of the material to withstand larger electric fluxes increases and hence the materials capacity to hold larger quantities of charge for longer periods of time also increases. This parameter therefore has a significant influence on electrospinning. Using a solvent with a higher dielectric constant results in enhancement of the bending instability of the electrospinning jet, which enables the reduction of the fibre diameters as facilitated by the increased jet path [60] and helps prevent or reduce bead formation in fibres [61].

Solvent mixtures containing solvents with high dielectric constants, such as dimethylformamide, water or acetonitrile, can be used to increase the dielectric constant of a solution. However, it is important to also consider other parameters of the solvents when forming mixtures, as parameters such as solvent conductivity or vapour pressure could have reverse effects on the fibre morphology or diameters. The dielectric constant of a solvent also has an effect on the solvating ability of the solvent, with higher values enabling easier dissolution of salts or ionic liquids [62]. This is an important point to consider for increasing conductivity of solutions through salt dissolution.

Vapour Pressure

Solvent vapour pressure is another important parameter for success of co-electrospinning and morphology of the as-spun fibres. During the electrospinning process, as the fiber jet is ejected from the tip to the collector, solvent will evaporate, the rate of which is dependent on the volatility of the solvent. A highly volatile solvent will increase the rate of evaporation of the gaseous or liquid phase, leaving only a solid phase (phase separation) which will in turn increase the surface area of the fibers [59]. A very low vapor pressure solvent tend to inhibit solution flow which prevents a fully developed Taylor cone to be maintained due to low charge density or high viscosity. High vapor pressure solvent may bring about irregular multiple jets emerging from the droplet [63]. It will also cause the droplet at the tip of the nozzle to dry, preventing electrospinning.

Lower vapour pressure solvents can enable the electrospinning jet to remain fluid for longer and hence delay the solidification of the jet into fibres. A longer flight time will enable the jet path to increase with further stretching to yield fibres with lower diameters. A too high vapour pressure solvent may also result in the formation of collapsed or elongated fibre cross-sections. This happens as fast evaporation of the solvent from the surface layer of the solution jet leads to formation of a skin layer that buckles under the pressure difference created [64]. It is therefore important to use a sufficiently low vapour pressure solvent to retain a sufficiently liquid droplet at the nozzle tip to enable jet initiation, and for control of the fibre morphologies and diameters. Once again, solvent mixtures can be used to control the vapour pressure.

Process Parameters

Voltage

A crucial element in electrospinning is application of a high voltage to the solution.

The high voltage induces the necessary external electric field to initiate the electrospinning process when the electrostatic force in the solution overcomes the surface tension of the solution. Generally, a voltage potential of more than 6 kV is sufficient to initiate a Taylor cone during electrospinning. Depending on the feed rate of the solution, a higher voltage may be required so that the Taylor cone is stable. The coulombic repulsive force in the jet will then stretch the viscoelastic solution. If the applied voltage is higher, the greater field strength will cause the jet to accelerate faster, and a greater volume of solution will be drawn from the tip of the needle. This may result in a smaller and less stable Taylor cone. When the drawing of the solution to the collection plate is faster than the supply from the source, the Taylor cone may recede into the needle. In most cases, a higher voltage will lead to a greater stretching of the solution due to the greater coulombic forces in the jet as well as the stronger electric field. These have the effect of reducing the diameter of the fibers and also encourage faster solvent evaporation. At a higher voltage, it has been found that there is a greater tendency for bead formation [47]. It was also reported that the shape of the beads changes from spindle-like to spherical-like with increasing voltage [57]. Given the increased stretching of the jet due to higher voltage, there should be less bead formation [65]. The increase in bead density due to increase in voltage may be the result of increased instability of the jet as the Taylor cone recedes into the syringe needle [57]. Increasing voltage will increase the bead density, which at an even higher voltage, beads join together to form a thicker diameter fiber. Fig. (**8**) shows the different regimes that exist as voltage is increased; a saddle point exists in the data [58]. The region before the saddle point (point 1) corresponds to the dripping region when the voltage is too low for the set flow rate (500 μl/hr in this case); in this region the fibre diameters increase gradually as voltage increases up to the saddle point. At the top of the saddle point a stable Taylor cone is obtained at the tip of the nozzle (*i.e.* the rate of draw of the voltage matches the flow rate). With the voltage further increased (point 2) the fibre diameters once again decrease in diameter as the Taylor cone recedes into the nozzle up to the point of the saddle minima. Beyond this point the fibre diameters start increasing again (point 3) as a result of the increased voltage causing more solution to be drawn out of the nozzle than that is available with the set flow rate. In this scenario the electrospinning process is intermittent as the jet breaks with the solution depleting in the nozzle and re-issues as more solution flows through after a short interval. Fig. (**8**) also shows how these effects are much more pronounced for more viscous (*i.e.* higher poly(methyl methacrylate)(PMMA) concentration) solutions.

When a solution of low viscosity is used a high voltage may lead to the formation of multiple jets from the droplet, leading to fibres with lower diameters to be formed, but with a greater fibre diameter distribution. The voltage can also affect the crystallinity of the polymer fibres as the electrostatic field causes stretching of the entangled polymer chains (molecules) [1]. However, this effect is said to be true up to a given voltage, beyond which the jet acceleration is too high (*i.e.* flight-time too low) for polymer molecules to be aligned before they are deposited on the collection plate. Therefore, in order to investigate the effect of voltage on fibre morphology and diameters for a given system, it is important to test a range ofvoltages within the range that enables stable electrospinning. This range is heavily dependent on the solution properties and the feed rate, as well as the collector distance from the tip of the nozzle, which are discussed next.

Tip to Collector Distance

The tip-to-collector distance affects the structure and morphology of electrospun fibers because of the complementing effect of the deposition time, evaporation rate of polymer solution and the electric field strength. For fibers to form, the electrospinning jet must have enough time for the solvent in the polymer solution to evaporate. As the distance increases, thinner fibers can be obtained owing to more stretching time before the collection of nanofibers. Otherwise, closer distances between the needle and collector cause coarse fiber formation or fibre collapse with insufficient solvent evaporation.

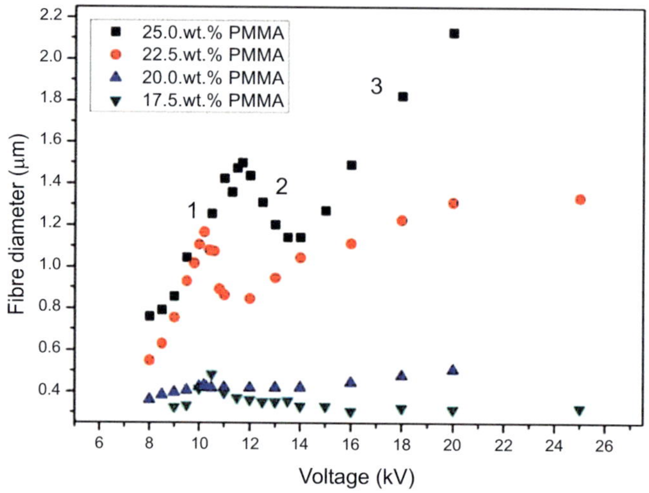

Fig. (8). Fibre diameter *vs.* voltage plot of PMMA fibres electrospun at different concentrations.

Bead formation in electrospun polystyrene fibers decreased with increasing tip to collector distance, while the ribbon shaped morphology was maintained, when tip to collector distance was decreased. Furthermore, Zhang *et al.* [54] found that when chitosan/PVA was electrospun at distance between tip and collector of 10 cm, the average diameter of fibers was 114 ± 35 nm. While with distance between tip and collector of 15 cm, the amount of beads decreased and the image analyzed gave the average diameter of fiber of 99 ± 21 nm. Thereby, increasing the distance in electrospinning process resulted in a decrease of the average fiber diameter. But in extreme cases increasing the distance too far can also lead to increase in fibre diameters, due to reduced stretching of fibres under the lowered electric field strength [1].

It is therefore important, for a given system, to find the optimum tip-to-collector distance that is sufficiently high to enable evaporation of the solvent from the fibres, while enabling formation of a stable jet that stretches out the fibres as much as possible.

Feed-rate

The solution feed-rate (or flow-rate) determines the amount of solution that is available for electrospinning over a given time. For a given voltage, there is a corresponding feed rate if a stable Taylor cone is to be maintained. If the solution extraction rate (as determined by the magnitude of the voltage) is not balanced by the feed rate, the solution at the tip of the nozzle will continuously drip during electrospinning. When the feed rate is increased, there is an increase in the fiber diameter or bead size. This is apparent, as there is a greater volume of solution that is drawn away from the needle tip [57]. A lower feed rate is more desirable, as the solvent will have more time for evaporation [66]. However, there is a limit to the increase in the diameter of the fiber due to higher feed rate. If the feed rate is at the same rate at which the solution is carried away by the jet, there must be a corresponding increase in charges when the feed rate is increased. Thus there is a corresponding increase in the stretching of the solution, which counters the increased diameter due to increased volume. Due to the greater volume of solution drawn from the tip, the jet will take a longer time to dry. As a result, the solvents in the deposited fibers may not have enough time to evaporate given the same flight time. The residual solvent causes the fibers to fuse together where they make contact forming webs. At low feeding rate, the solution is ejected faster and the shift of the mass balance will result in sustained but unstable jet which eventually forms fiber with beads [67]. Thus, once again the fibre diameter is controlled by balancing the effect of separate parameters.

Collector Material

To initiate an electrospinning process, an electric field must be present between the source and the collector. For this reason, most of the collectors in an electrospinning process are made out of a conductive material such as aluminium foil, which is electrically grounded so that there is a stable potential difference between the source and the collector. In the case when a non-conducting material is used as a collector, charges on the electrospinning jet will quickly accumulate on the collector, which will result in fewer fibers deposited. Fibers that are collected on the non-conducting material usually have a lower packing density compared with those collected on a conducting surface. The repulsive forces of the accumulated charges on the collector cause this phenomenon.

For a non-conducting collector, the accumulation of the charges may result in the formation of 3D fiber structures due to the repulsive forces of the like-charges. However, even for conductive collectors, when the deposition rate is high and the fiber mesh becomes thick enough, there will also be a high accumulation of residual charges on the fiber mesh since polymer nanofibers are generally nonconductive. This may result in the formation of dimples on the fiber mesh. The porosity of the collector seems to have an effect on the deposited fibers. Experiments with porous collectors such as paper and metal mesh have shown that the fiber mesh collected had a lower packing density than smooth surfaces such as metal foils [1]. This can be attributed to the diffusion and rate of evaporation of the residual solvents on the fibers collected.

In a porous target, there is faster evaporation of residual fibers due to higher surface area while smooth surfaces may cause an accumulation of solvents around the fibers due to a slower evaporation rate. Due to the wicking and diffusion of the residual solvents on the fibers, the fibers may be pulled together to give a more densely packed structure. Whether or not the collector is static or moving also has an effect on the electrospinning process. While rotating collectors have been used to collect aligned fibers, they were also found to assist in yielding fibers that are dry. This is useful because certain solvents such as di-methyl formamide (DMF), which are good for electrospinning but have a high boiling point, may result in wet fibers during collection. A rotating collector will give the solvent more time to evaporate and also increase the rate of evaporation of the solvents on the fibers. This will improve the morphology of the fiber where distinct fibers are required.

Electrospinning Setup

The arrangement for electrospinning could be horizontal or vertical according to the geometrical arrangements of the spinning needle and collector, where vertical type includes shaft type and converse type. Using different electrospinning system configurations, the obtained fiber properties could be quite different. The vertical setup allows solution flow to be inline or against gravity. Yang *et al.* [68] studied the differences of fibers diameters and fibers mats morphology between the three electrospinning systems. The results showed that in the shaft type system, the electrospinning fibers were the thinnest as the gravitational force strengthen the effect of electric field to maximize fiber extension. While this setup results in the broadest fiber diameter distribution. In the converse type system, the average fiber diameter was the largest and the fiber diameter distribution was the narrowest. The horizontal type system resulted in average fiber diameter and size distribution between shaft and converse types [68].

In another study, Rodoplu and Mutlu [31] indicated that the effect of gravitational force on electrospinning process is negligible with respect to the electric field forces. However, they also found that gravity has an effect on the shape of the polymer droplet and the Taylor cone. This results in a difference in electrospinning parameters used in horizontal and vertical systems. These studies were all based on electrospinning setup used for producing solid nanofibers, the effect of electrospinning configuration on coaxial electrospinning has not been reported.

Diameter of Needle Orifice

The internal diameter of the electrospinning needle will have some effect on the electrospinning process. A smaller internal diameter will reduce the clogging as well as the amount of beads in the electrospun fiber [69]. The reduction in the clogging could be due to the lower degree of exposure of the unspun solution to the atmosphere. Reduction in internal diameter of the needle will reduce the diameter of the electrospun fibers. When the size of the droplet at the tip of the orifice is decreased, such as in the case of a smaller internal diameter of the orifice, the surface tension of the droplet increases. For the same voltage supplied, a greater coulombic force is required to cause jet initiation. As a result, the acceleration of the jet decreases and allows more time for the spun solution to be stretched and elongated before it is collected. However, if the diameter of the orifice is too small, it may not be possible to electrospin a droplet of solution at the tip of the orifice.

Environmental (ambient) Parameters

The effect of ambient conditions on the polymer solution and hence electrospinning process and/or the fibre morphologies is not as thoroughly investigated as the process and system parameters, but increasing amount of research is pointing to the importance of the environmental conditions for controlling fibre morphology and diameters. These parameters include temperature, humidity, atmospheric pressure *etc*. The first two are discussed below.

Temperature

The temperature of the solution has both the effect of increasing its evaporation rate and reducing the viscosity of the polymer solution. Demir *et al.* [51] observed that when polyurethane was electrospun at a higher temperature, the fibers produced had a more uniform diameter. This may be due to the lower viscosity of the solution and greater solubility of the polymer in the solvent that allows even more stretching of the solution. With a lower viscosity, the electrostatic forces are able to exert a greater stretching force on the solution, thus resulting in fibers of smaller diameter. Increased polymer molecule mobility due to increased temperature also allows the electrostatic force to stretch the solution further.

Humidity

Higher humidity (atmospheric water concentration) leads to more water molecules between the tip and collector. These molecules will increase the conductivity of this region, therefore, reducing the electric field strength in the electrospinning process due to polarization of water molecules [70, 71], resulting in a thicker fiber [70, 72]. as shown in Fig. (**9**).

Humidity, above a certain level, results in water condensing on the surface or in the fibre, causing pores on the surface or interior of the fibres. The size of the pores increases with increasing humidity, eventually coalescing to form large, non-uniform structures [73]. The effect of humidity on fibre morphology depends on the solvent-polymer system. A very low humidity may promote rapid evaporation of a highly volatile solvent, as the partial pressure is decreased, causing the polymer jet to dry out too quickly, without sufficient stretching to form fine fibres. In this case the nozzle tip may also become clogged. Therefore, the humidity of the surrounding atmosphere has to be adjusted according to the chosen solvent-solution system to attain smooth (non-porous) fibres.

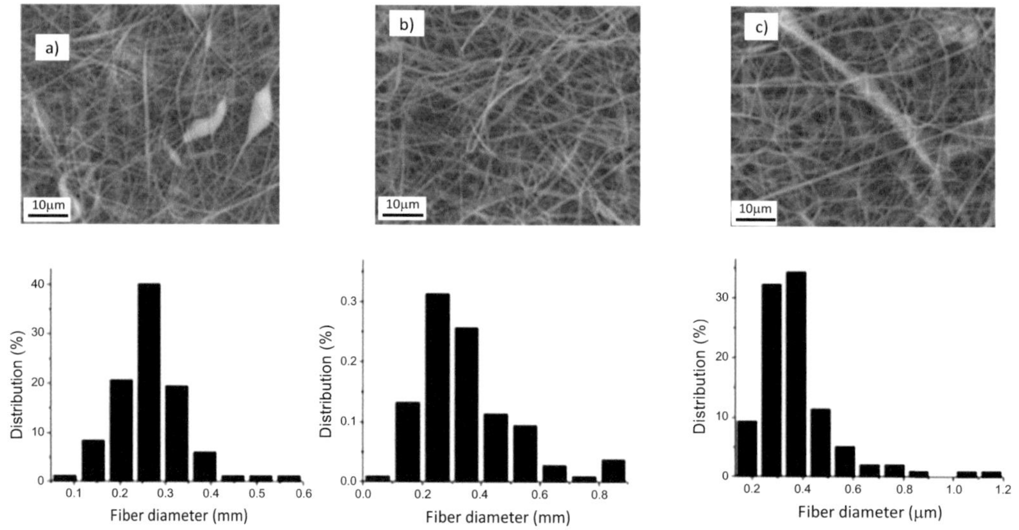

Fig. (9). SEM image and nanofiber diameter distribution of electrospun nanofibers of Polyethersulfone/Dimethylformamide/N-methyl at: a) 45% RH, mean diameter 266 nm, b) 50% RH, mean diameter 349 nm and c) 60% RH, mean diameter 382 nm [70].

Type of Atmosphere

The composition of the gas in the electrospinning chamber will have an effect on the process. Each gas type behaves differently under high electrostatic fields. For example, helium can break down under high electrostatic field, thus making electrospinning impossible [20]. However, Baumgarten [20] found that when a gas with higher breakdown voltage was used (*e.g.* Freon – 12), the fibers obtained were twice the diameter of those electrospun in air given all other conditions equal.

Pressure

It is possible to investigate the effect of pressure on the electrospinning jet. Generally, a reduction in the pressure surrounding the electrospinning jet does not improve the electrospinning process. When the pressure is below atmospheric pressure, the polymer solution in the syringe will have a greater tendency to exhibit unstable flow out of the needle [1]. As the pressure decreases, rapid bubbling of the solution occurs at the needle tip. At a very low pressure, electrospinning is not possible due to direct electrical discharge (sparking) between the needle and collector. The influences of major parameters in electrospinning process on the fiber morphology is shown in Table **1**.

Table 1. Influences of parameters in electrospinning process on the fiber morphology.

Parameter	**Effect on fiber morphology**
Viscosity (polymer solution) ↑	Fiber diameter ↑ (from beads to beaded fibers to smooth fibers)
Surface tension ↑	Number of beaded fibers and beads ↑
Solution conductivity ↑	Fiber diameter ↓
Evaporation of solvents ↑	Fibers exhibit microtexture (pores on fiber surfaces)
Applied voltage ↑	Fiber diameter ↓ initially, then fiber diameter ↑ (not monotonic
Spinneret to collector distance ↑	Fiber diameter ↓ (beaded morphologies occur if the distance between the capillary and collector is too short)
Humidity ↑	Fiber diameter ↓ (pores on fiber surfaces), then fiber diameter ↑
Flow rate ↑	Fiber diameter ↑ (beaded morphologies occur if the flow rate is too high)

Co-electrospinning

Electrospinning has been widely recognised, both at the research and industrial level, as a cost effective and versatile way of producing nanofibres. More recently, as a result of increasing demand for specialised nanostructured composite materials, a more complex variant of this process, called co-electrospinning (or co-axial electrospinning) [74 - 82] has been developed. Co-electrospinning involves the use of two concentric nozzles to deliver two separate solutions to make functionalised bicomponent nanofibres with a core-shell configuration. This process works in a similar way to its single phase variant, such that an electric field is applied to a suspended droplet at the tip of the nozzle to yield co-axial fibres with diameters typically in the sub-micron range. The main difference is that two solutions, instead of one, are delivered independently through the concentric nozzles, with one solution being the precursor for the core and the other for the shell of the fibres. The core material is not required to be polymeric, which enables a wide variety of materials either in melt or solution form to be encapsulated and structured, with properties of the polymeric shell chosen to enhance fibre functionality.

This technique is therefore highly useful for encapsulating and nanostructuring non-polymeric materials for a wide range of uses with the possible benefits of:

- Enabling improved material properties as a result of nanostructuring,
- Isolating the core material that may be unstable in air to minimise its chance of decomposition,
- Enabling controlled release of the core component over time through the use of a shell material that decomposes in a given environment, *e.g.* controlled release of a substance/drug contained in a biodegradable polymer,
- Enabling one material, usually the shell, to act as a scaffold for the other, such that fibres can be functionalised for a specific use, and
- Reinforcing a material to improve its mechanical properties [75].

Co-electrospinning has already gained attention for a variety of applications including biomedical (tissue engineering, controlled drug release systems. Though, co-electrospinning works in a similar way to its single phase variant, such that it involves the control of the same process and system parameters (*e.g.* viscosity, electrical conductivity, voltage, *etc.*), the use of two solutions instead of one, introduces additional complexity to the process. Also, due to the limited amount of work published on co-electrospinning, there is not yet a concrete understanding of the exact mechanisms that enable confinement of the core in the shell solution and hence how the core and the shell parameters have to be balanced to ensure the success of the process.

Process Conditions Required

The degree of miscibility is a necessary but not a complete guide to whether co-axial spinning is possible; it also is required to control the relative scale of a number of other parameters. At the nozzle tip, a stable liquid Taylor cone is needed to enable a consistent and continuous core-shell structure to form. In a normal case for electrospinning compound (core-shell) solutions, the outer solution is chosen to have a higher conductivity than the core; in which case the driving force of the compound jet comes from the electrostatic charge build up on the outer surface of the shell. Stable spinning then depends on the outer shell solution drawing out the inner core solution through shearing and contact forces, viscous dragging and contact friction, which keep the inner solution confined to the core while the fibres are being stretched. For this reason the shell solution should be electrospinnable on its own and also the following conditions must be satisfied:

- The shell solution must have a higher viscosity than the core solution such that the viscous force imparted by the shell on the core is enough to overcome the interfacial tension,
- The shell solution must have a higher flow rate than that of the core, so that the viscous drag applied by the sheath solution is sufficient to confine the core solution within the cone [83],
- The vapour pressure of both solutions must be sufficiently low and comparable so both the core and shell dry at similar rates to stop fibres collapsing,
- The conductivity of the shell solution should ideally be higher than that of the core to enable continuous entrainment of the core by the shell, though higher core conductivity will also allow for co-axial spinning, and
- The two solutions must possess low interfacial tension to prevent large stresses at the core-shell interface reducing fibre stability and to produce a stable Taylor cone [83, 84].

Electrospun Nanofibers and Their Applications

The electrospinning process can be carried with any soluble synthetic or natural polymer type having enough molecular weight. Electrospun nanofibers are easily obtained from synthetic polymer types like; nylon, polyesters, polyurethane *etc.* and also from natural polymers such as; collagen, silk, chitosan, alginate, cellulose *etc* [11]. Nanofibers are not only produced from unique structures. To attain multifunctional properties of composite materials, they can also be blended or combined with additives such as nanoparticles, drugs and different kinds of chemical components [7]. Nowadays, nanofibers has received considerable attention in materials science due to their exclusive properties that can be implemented in specific areas such as catalysis, filtration, nanocomposites, nanofibrous structures, protective textiles, tissue scaffolds, drug delivery systems, *etc* [85]. The unique properties of nanofibers are (1) extremely long length; (2) larger surface area to volume ratio; (3) complex pore size; and (4) alignment on the molecular level (either woven or nonwoven). These characteristics of electrospun nanofiber make it practical in various applications.

Tissue scaffolds and drug delivery are well-known application areas of nanofibers. Nanofibers with high surface area and porosity have enormous potential for applications in engineering mechanically stable and biologically functional tissue scaffolds. The tissue scaffolding material must be selected carefully to ascertain its biocompatibility with the body cells. The biocompatibility depends on the surface chemistry of the scaffolds, which is dictated by the material properties [86]. The high surface to volume ratio of the nanofiber provides more space for the cell attachment than the regular fibers. The dimensions of these engineered scaffolds are in the comparable scale with those of the natural extra cellular matrix. Electrospun biodegradable polymers are considered as suitable scaffolds for tissue engineering, with their big pore diameter and volume.

The high porosity of electrospun nanofiber scaffolds provides enough space for the cell accommodation and an easy passage for the nutrients and metabolic waste excretion. Mechanical properties like elasticity modulus and strain at failure are important for the application of electrospun nanofibers as tissue scaffolds [85, 86]. Moreover, nanofiber mats with their unique functional characteristics find application as drug carriers for the drug delivery system. Controlled delivery of drugs at a defined rate over a span of treatment is possible with biocompatible delivery matrices of either biodegradable or non-biodegradable polymers [87].

As is evident from the aforementioned brief discussion on the applications of nanofibers, the potential of nanostructured materials in advanced applications is highly promising. Researchers are constantly making efforts to exploit the high surface area and porosity properties to develop sophisticated state of art materials.

CONCLUDING REMARKS

This chapter examined the electrospinning process theoretically and discussed the technology for making functional materials by biopolymer electrospinning. The effects of ambient conditions, processing parameters and material properties on the size and structure of electrospun fibers were reviewed. Ways to produce functional electrospun fibers for potential applications was also reviewed along with co-electrospinning and the process condition requirements. It presents the method to electrospin materials into fibers that would otherwise be difficult or impossible to process using conventional extrusion.

Electrospinning uses an electrical charge to draw very fine fibers from a liquid (average diameter ranging from 100 nm to 500 nm). Electrospinning can produce a non-woven mat that has a large ratio of surface area to mass. It also allows the spin solution without compromising the stability during electrospinning. It has very high potential for industrial processing. Electrospinning is cost effective, repeatable and can produce long, continuous nanofibers.

CONFLICT OF INTEREST

The author (editor) declares no conflict of interest, financial or otherwise.

ACKNOWLEDGEMENTS

I am greatly indebted to Dr. Kaushik Chatterjee, Head, Biomaterials and Tissue Engineering Laboratory, Indian Institute of Science, Bangalore for the facility support and supervision and the Department of Science and Technology, Govt. of India for the KVPY fellowship granted to the author for part of the research work included in this manuscript.

REFERENCES

[1] S. Ramakrishna, K. Fujihara, W.E. Teoteik, C. Lim, and Z. Ma, *An Introduction to electrospinning and nanofibers.* World Scientific Publishing: Singapore, 2005. [http://dx.doi.org/10.1142/5894]

[2] I.S. Chronakis, "Novel nanocomposites and nanoceramics based on polymer nanofibers using electrospinning process-A review", *J. Mater. Process. Technol.,* vol. 167, no. 2-3, pp. 283-293, 2005. [http://dx.doi.org/10.1016/j.jmatprotec.2005.06.053]

[3] S. Ramakrishna, K. Fujihara, W.E. Teo, T. Yong, Z. Ma, and R. Ramaseshan, "Electrospun nanofibers: solving global issues", *Mater. Today,* vol. 9, no. 3, pp. 40-50, 2006. [http://dx.doi.org/10.1016/S1369-7021(06)71389-X]

[4] D.H. Reneker, and A.L. Yarin, "Electrospinning jets and polymer nanofibers", *Polymer (Guildf.),* vol. 49, no. 10, pp. 2387-2425, 2008. [http://dx.doi.org/10.1016/j.polymer.2008.02.002]

[5] W.E. Teo, and S. Ramakrishna, "A review on electrospinning design and nanofibre assemblies", *Nanotechnology,* vol. 17, no. 14, pp. R89-R106, 2006. [http://dx.doi.org/10.1088/0957-4484/17/14/R01] [PMID: 19661572]

[6] W. Gilbert, *On the Magnet and Magnetic Bodies, and on That Great Magnet the Earth.* , P. Short, Ed., The Chiswick Press: London, 1628.

[7] J. Larmor, "Note on the complete scheme of electrodynamic equations of a moving material medium, and on electrostriction", *Proc. R. Soc. Lond. A Math. Phys. Sci.,* vol. 63, p. 365, 1898.

[8] J.F. Cooley, Apparatus for Electrically Dispersing Fluids, 1902.U.S. Patent 692,631

[9] W.J. Morton, Method of dispersing fluids, 1902.U.S. Patent 0705691

[10] A. Formhals, Method and apparatus for the production of fibers, 1938.U.S. Patent 2,116,942

[11] N. Bhardwaj, and S.C. Kundu, "Electrospinning: a fascinating fiber fabrication technique", *Biotechnol. Adv.,* vol. 28, no. 3, pp. 325-347, 2010. [http://dx.doi.org/10.1016/j.biotechadv.2010.01.004] [PMID: 20100560]

[12] A. Formhals, A Method and apparatus for the production of artificial fibers, 1939.U.S. Patent 2,158,416

[13] A. Formhals, Method and apparatus for spinning, 1939.U.S. Patent 2,160,962

[14] G. Taylor, "Disintegration of water drops in an electric field", *P. Roy. Soc.Lond. A. Mat.,* vol. 280, pp. 383-397, 1964. [http://dx.doi.org/10.1098/rspa.1964.0151]

[15] J. Zeleny, "Instability of electrified liquid surfaces", *Phys. Rev.,* vol. 10, pp. 69-91, 1917. [http://dx.doi.org/10.1103/PhysRev.10.1]

[16] G. Taylor, "The Force Exerted by an Electric Field on a Long Cylindrical Conductor", *Proc.- Royal Soc., Math. Phys. Eng. Sci.,* vol. 291, pp. 145-158, 1966. [http://dx.doi.org/10.1098/rspa.1966.0085]

[17] G. Taylor, "Studies in electrohydrodynamics: I. The circulation produced in a drop by electrical field", *P. Roy. Soc. Lond. A. Mat.,* vol. 291, pp. 159-166, 1966. [http://dx.doi.org/10.1098/rspa.1966.0086]

[18] G. Taylor, "Electrically driven jets", *P. Roy Soc. Lond. A Mat.,* vol. 313, pp. 453-475, 1969.

[19] H.L. Simons, Process and Apparatus for Producing Patterned Nonwoven Fabrics, 1966. US Patent 3, 280229

[20] P.K. Baumgarten, "Electrostatic spinning of acrylic microfibers", *J. Colloid Interface Sci.,* vol. 36, pp. 71-79, 1971. [http://dx.doi.org/10.1016/0021-9797(71)90241-4]

[21] A. Greiner, and J.H. Wendorff, "Electrospinning: a fascinating method for the preparation of ultrathin fibers", *Angew. Chem. Int. Ed. Engl.,* vol. 46, no. 30, pp. 5670-5703, 2007. [http://dx.doi.org/10.1002/anie.200604646] [PMID: 17585397]

[22] D. Li, and Y. Xia, "Electrospinning of Nanofibers: Reinventing the Wheel?", *Adv. Mater.,* vol. 16, pp. 1151-1170, 2004. [http://dx.doi.org/10.1002/adma.200400719]

[23] D.H. Reneker, A.L. Yarin, H. Fong, and S. Koombhongse, "Bending instability of electrically charged liquid jets of polymer solutions in electrospinning", *J. Appl. Phys.,* vol. 87, no. 9, pp. 4531-4547, 2000. [http://dx.doi.org/10.1063/1.373532]

[24] W.T. Klooster, T.F. Koetzle, T.B. Siegbahn, T.B. Richardson, and R. Crabtree, "Study of the N−H···H−B Dihydrogen Bond Including the Crystal Structure of BH3NH3 by Neutron Diffraction", *J. Am. Chem. Soc.,* vol. 121, no. 27, pp. 6337-6343, 1999. [http://dx.doi.org/10.1021/ja9825332]

[25] A.E. Elcoot, "Nonlinear stability of an axial electric field: Effect of interfacial charge relaxation", *Appl. Math. Model.,* vol. 34, no. 8, pp. 1965-1983, 2010. [http://dx.doi.org/10.1016/j.apm.2009.10.035]

[26] K.S. Athira, P. Sanpui, and K. Chatterjee, "Fabrication of Poly(Caprolactone) Nanofibers by Electrospinning", *J. Polym. Biopolym. Phys. Chem.,* vol. 2, no. 4, pp. 62-66, 2014.

[27] V. Berube, G. Radtke, M. Dresselhaus, and G. Chen, "Size effects on the hydrogen storage properties of nanostructured metal hydrides: A review", *Int. J. Energy Res.,* vol. 31, no. 6-7, pp. 637-663, 2007. [http://dx.doi.org/10.1002/er.1284]

[28] B.P. Sautther, "Continuous Polymer Nanofibers Using Electrospinning", In: *NSF-REU Summer Program,* Chicago, Illinois, USA, 2005.

[29] A. Baji, Y.W. Mai, S.C. Wong, M. Abtahi, and P. Chen, "Electrospinning of polymer nanofibers: effects on oriented morphology, structures and tensile properties", *Compos. Sci. Technol.,* vol. 70, no. 5, pp. 703-718, 2010. [http://dx.doi.org/10.1016/j.compscitech.2010.01.010]

[30] A. Theron, E. Zussman, and A.L. Yarin, "Electrostatic fieldassisted alignment of electrospun nanofibres", *Nanotechnology,* vol. 12, no. 3, pp. 384-390, 2001. [http://dx.doi.org/10.1088/0957-4484/12/3/329]

[31] D. Rodoplu, and M. Mutlu, "Effects of electrospinning setup and process parameters on nanofiber morphology intended for the modification of quartz crystal microbalance surfaces", *J. Eng. Fibers Fabrics,* vol. 7, no. 2, pp. 118-123, 2012.

[32] P. Kumar, *"Effect of Collector on Electrospinning to Fabricate Aligned Nanofiber"*, MTech thesis, National Institute of Technology, Rourkela, India, 2012.

[33] H. Liu, and Y.L. Hsieh, "Ultrafine fibrous cellulose membranes from electrospinning of cellulose acetate", *J. Polym. Sci.Pol. Phys.,* vol. 40, no. 18, pp. 2119-2129, 2002. [http://dx.doi.org/10.1002/polb.10261]

[34] C.C. Tang, J.C. Chen, Y.Z. Long, H.X. Yin, B. Sun, and H.D. Zhang, "Preparation of curled microfibers by electrospinning with tip collector", *Chin. Phys. Lett.,* vol. 28, no. 5, 2011. [http://dx.doi.org/10.1088/0256-307X/28/5/056801]

[35] Q.P. Pham, U. Sharma, and A.G. Mikos, "Electrospinning of polymeric nanofibers for tissue engineering applications: a review", *Tissue Eng.,* vol. 12, no. 5, pp. 1197-1211, 2006. [http://dx.doi.org/10.1089/ten.2006.12.1197] [PMID: 16771634]

[36] R. Kenawy, J.M. Layman, J.R. Watkins, G.L. Bowlin, J.A. Matthews, D.G. Simpson, and G.E. Wnek, "Electrospinning of poly(ethylene-co-vinyl alcohol) fibers", *Biomaterials,* vol. 24, no. 6, pp. 907-913, 2003.
[http://dx.doi.org/10.1016/S0142-9612(02)00422-2] [PMID: 12504511]

[37] V. Beachley, and X. Wen, "Effect of electrospinning parameters on the nanofiber diameter and length", *Mater Sci Eng C Mater Biol Appl,* vol. 29, no. 3, pp. 663-668, 2009. [http://dx.doi.org/10.1016/j.msec.2008.10.037] [PMID: 21461344]

[38] N. Zhu, and X. Chen, *Advances in Biomaterials Science and Biomedical Applications.* In Tech: Hampshire, UK, 2013.

[39] F. Li, Y. Zhao, and Y. Song, *Nano Channel and Capsule by Coaxial Electrospinning Nanofiber.* InTech: Hampshire, UK, 2010.

[40] C.P. Carroll, and Y.L. Joo, "Axisymmetric instabilities in electrospinning of highly conducting, viscoelastic polymer solutions", *Phys. Fluids,* vol. 21, no. 10, pp. 103101-103110, 2009. [http://dx.doi.org/10.1063/1.3246024]

[41] D.H. Reneker, and I. Chun, "Nanometre diameter fibres of polymer, produced by electrospinning", *Nanotechnology,* vol. 7, no. 3, pp. 216-223, 1996. [http://dx.doi.org/10.1088/0957-4484/7/3/009]

[42] Y.Y. Zhao, Q.B. Yang, X.F. Lu, C. Wang, and Y. Wei, "Study on correlation of morphology of electrospun products of polyacrylamide with ultrahigh molecular weight", *J. Polym. Sci. Pol. Phys.,* vol. 43, no. 16, pp. 2190-2195, 2005.
[http://dx.doi.org/10.1002/polb.20506]

[43] Z.Z. Zhao, J. Li, X. Yuan, X. Li, Y. Zhang, and J. Sheng, "Preparation and properties of electrospun poly(vinylidene fluoride) membranes", *J. Appl. Polym. Sci.,* vol. 97, no. 2, pp. 466-474, 2005. [http://dx.doi.org/10.1002/app.21762]

[44] Z.M. Huang, Y. Zhang, M. Kotaki, and S. Ramakrishna, "A review on polymer nanofibers by electrospinning and their applications in nanocomposites", *Compos. Sci. Technol.,* vol. 63, no. 15, pp. 2223-2253, 2003.
[http://dx.doi.org/10.1016/S0266-3538(03)00178-7]

[45] N. Ristolainen, P. Heikkila, A. Harlin, and J. Seppala, "Poly(vinyl alcohol) and polyamide-66 nanocomposites prepared by electrospinning", *Macromol. Mater. Eng.,* vol. 291, no. 2, pp. 114-122, 2006.
[http://dx.doi.org/10.1002/mame.200500213]

[46] P. Gupta, C. Elkins, T.E. Long, and G.L. Wilkes, "Electrospinning of linear homopolymers of poly(methyl methacrylate): exploring relationships between fiber formation, viscosity, molecular weight and concentration in a good solvent", *Polymer (Guildf.),* vol. 46, no. 13, pp. 4799-4810, 2005. [http://dx.doi.org/10.1016/j.polymer.2005.04.021]

[47] J.M. Deitzel, J. Kleinmeyer, D.E. Harris, and N.B. Tan, "The effect of processing variables on the morphology of electrospun nanofibers and textiles", *Polymer (Guildf.),* vol. 42, no. 1, pp. 261-272, 2001.

[http://dx.doi.org/10.1016/S0032-3861(00)00250-0]

[48] A.K. Haghi, and M. Akbari, "Trends in electrospinning of natural nanofibers", *Phys. Status Solidi., A Appl. Mater. Sci.,* vol. 204, no. 6, pp. 1830-1834, 2007. [http://dx.doi.org/10.1002/pssa.200675301]

[49] C. Mit-uppatham, M. Nithitanakul, and P. Supaphol, "Ultratine electrospun polyamide-6 fibers: Effect of solution conditions on morphology and average fiber diameter", *Macromol. Chem. Phys.,* vol. 205, no. 17, pp. 2327-2338, 2004. [http://dx.doi.org/10.1002/macp.200400225]

[50] V. Pornsopone, P. Supaphol, R. Rangkupan, and S. Tantayanon, "Electrospinning of methacrylate-based copolymers: Effects of solution concentration and applied electrical potential on morphological appearance of as-spun fibers", *Polym. Eng. Sci.,* vol. 45, no. 8, pp. 1073-1080, 2005. [http://dx.doi.org/10.1002/pen.20354]

[51] M.M. Demir, I. Yilgor, E. Yilgor, and B. Erman, "Electrospinning of polyurethane fibers", *Int. J. Sci. Tech. Polym.,* vol. 43, no. 11, pp. 3303-3309, 2002.

[52] C. Yao, X. Li, K.G. Neoh, Z. Shi, and E.T. Kang, "Surface modification and antibacterialactivity of electrospun polyurethane", *J. Membr. Sci.,* vol. 320, pp. 259-267, 2008. [http://dx.doi.org/10.1016/j.memsci.2008.04.012]

[53] K. Nakata, S.H. Kim, Y. Ohkoshi, Y. Gotoh, and M. Nagura, "Electrospinning of poly (ether sulfone) and evaluation of the filtration efficiency", *Soc. Fiber. Sci. Technol. Japan,* vol. 63, pp. 307-312, 2007.

[54] M. Kang, and H.J. Jin, "Electrically conducting electrospun silk membranes fabricated by adsorption of carbon nanotube", *Colloid Polym. Sci.,* vol. 285, pp. 1163-1167, 2007. [http://dx.doi.org/10.1007/s00396-007-1668-y]

[55] B. Wang, Y. Sun, and H. Wang, "Preparation and properties of electrospun PAN/Fe_3O_4 magnetic nanofibers", *J. Appl. Polym. Sci.,* vol. 115, pp. 1781-1786, 2010. [http://dx.doi.org/10.1002/app.31288]

[56] H. Fong, I. Chun, and D.H. Reneker, "Beaded nanofibers formed during electrospinning", *Polymer (Guildf.),* vol. 40, pp. 4585-4592, 1999. [http://dx.doi.org/10.1016/S0032-3861(99)00068-3]

[57] K.K. Zhong, D. Fang, S. Ran, B.S. Hsiao, and B. Chu, "Structure and process relationship of electrospun bioabsorbable nanofiber membranes", *Int. J. Sci. Tech. Polym.,* vol. 43, no. 16, pp. 4403-4412, 2002.

[58] Z. Kurban, *"Electrospun nanostructured composite fibres for hydrogen storage applications",* Dissertation, University College London, England, 2011.

[59] S. Megelski, J.S. Stephens, D.B. Chase, and J.F. Rabolt, "Micro- and nano-structured surface morphology on electrospun polymer fibers", *Macromolecules,* vol. 35, pp. 8456-8466, 2002. [http://dx.doi.org/10.1021/ma020444a]

[60] C.M. Hsu, and S. Shivkumar, "N, N-Dimethylformamide Additions to the Solution for the Electrospinning of Poly(ε-caprolactone) Nanofibers", *Macromol. Mater. Eng.,* vol. 289, no. 4, pp. 334-340, 2004. [http://dx.doi.org/10.1002/mame.200300224]

[61] W.K. Son, J.H. Youk, T.S. Lee, and W.H. Park, "The effects of solution properties and polyelectrolyte on electrospinning of ultrafine poly(ethylene oxide) fibers", *Polymer (Guildf.),* vol. 45, no. 9, pp. 2959-2966, 2004. [http://dx.doi.org/10.1016/j.polymer.2004.03.006]

[62] R. Kumar, B. Singh, and S.S. Sekhon, "Effect of dielectric constant of solvent on the conductivity behavior of polymer gel electrolytes", *J. Mater. Sci.,* vol. 40, pp. 1273-1275, 2005. [http://dx.doi.org/10.1007/s10853-005-6950-0]

[63] T. Zhou, *"Bi-component nano-fibrous complexes via coaxial electrospinning"*, MSc thesis, North Carolina State University, Raleigh, North Carolina, 2011.

[64] A. Arinstein, R. Avrahami, and E. Zussman, "Buckling behaviour of electrospun microtubes: a simple theoretical model and experimental observations", *J. Phys. D Appl. Phys.*, vol. 42, no. 1, pp. 015507-1, 015507-7, 2009. [http://dx.doi.org/10.1088/0022-3727/42/1/015507]

[65] T. Jarusuwannapoom, W. Hongrojjanawiwat, S. Jitjaicham, L. Wannatong, M. Nithitanakul, C. Pattamaprom, P. Koombhongse, R. Rangkupan, and P. Supaphol, "Effect of solvents on electro-spinnability of polystyrene solutions and morphological appearance of resulting electrospun polystyrene fibers", *Eur. Polym. J.*, vol. 41, no. 3, pp. 409-421, 2005. [http://dx.doi.org/10.1016/j.eurpolymj.2004.10.010]

[66] X. Yuan, Y. Zhang, C. Dong, and J. Sheng, "Morphology of ultrafine polysulfone fibers prepared by electrospinning", *Polym. Int.*, vol. 53, no. 11, pp. 1704-1710, 2004. [http://dx.doi.org/10.1002/pi.1538]

[67] R. Jalili, S.A. Hosseini, and M. Morshed, "The effects of operating parameters on the morphology of electrospun polyacrilonitrile nanofibres, Iran", *Polym. J.*, vol. 14, no. 12, pp. 1074-1081, 2005.

[68] C. Yang, Z. Jia, Z. Xu, K. Wang, Z. Guan, and L. Wang, "Comparisons of fibers properties between vertical and horizontal type electrospinning systems", *Conference on Electrical Insulation and Dielectric Phenomena,* 2009pp. 204-207 Virginia Beach, Virginia [http://dx.doi.org/10.1109/CEIDP.2009.5377758]

[69] X.M. Mo, C.Y. Xu, M. Kotaki, and S. Ramakrishna, "Electrospun P(LLA-CL) nanofiber: a biomimetic extracellular matrix for smooth muscle cell and endothelial cell proliferation", *Biomaterials,* vol. 25, no. 10, pp. 1883-1890, 2004. [http://dx.doi.org/10.1016/j.biomaterials.2003.08.042] [PMID: 14738852]

[70] Z. Tang, C. Qiu, J.R. McCutcheon, K. Yoon, H. Ma, D. Fang, E. Lee, C. Kopp, B.S. Hsiao, and B. Chu, "Design and fabrication of electrospun poly(ether sulfone) nanofibrous scaffold for high-flux nanofiltration", *J. Polym. Sci. Pol. Phys.*, vol. 47, pp. 2288-2300, 2009. [http://dx.doi.org/10.1002/polb.21831]

[71] E.S. Medeiros, L.H. Mattoso, R.D. Offeman, D.F. Wood, and W.J. Orts, "Effect of Relative Humidity on the Morphology of Electrospun Polymer Fibers", *Can. J. Chem.*, vol. 86, pp. 590-599, 2008. [http://dx.doi.org/10.1139/v08-029]

[72] C. Wang, C.H. Hsu, and L.H. Lin, "Scaling laws in electrospinning of polystyrene solutions", *Macromolecules,* vol. 39, pp. 7662-7672, 2006. [http://dx.doi.org/10.1021/ma060866a]

[73] C.L. Casper, J.S. Stephens, N.G. Tassi, D.B. Chase, and J.F. Rabolt, "Controlling surface morphology of electrospun polystyrene fibers: effect of humidity and molecular weight in the electrospinning process", *Macromolecules,* vol. 37, no. 2, pp. 573-578, 2003. [http://dx.doi.org/10.1021/ma0351975]

[74] Z.M. Huang, C.L. He, A. Yang, Y. Zhang, X.J. Han, J. Yin, and Q. Wu, "Encapsulating drugs in biodegradable ultrafine fibers through co-axial electrospinning", *J. Biomed. Mater. Res. A,* vol. 77, no. 1, pp. 169-179, 2006. [http://dx.doi.org/10.1002/jbm.a.30564] [PMID: 16392131]

[75] A.K. Moghe, and B.S. Gupta, "Co-axial electrospinning for nanofiber structures: preparation and applications", *Polym. Rev. (Phila. Pa.),* vol. 48, no. 2, pp. 353-377, 2008. [http://dx.doi.org/10.1080/15583720802022257]

[76] S.R. Samarasinghe, K. Balasubramanian, and M.J. Edirisinghe, "Encapsulation of silver particles using co-axial jetting", *J. Mater. Sci. Mater. Electron.*, vol. 19, no. 1, pp. 33-38, 2008. [http://dx.doi.org/10.1007/s10854-007-9278-5]

[77] Z. Sun, E. Zussman, A.L. Yarin, J.H. Wendorff, and A. Greiner, "Compound core-shell polymer nanofibers by co-electrospinning", *Adv. Mater.,* vol. 15, no. 22, pp. 1929-1932, 2003. [http://dx.doi.org/10.1002/adma.200305136]

[78] Y. Xin, Z. Huang, W. Li, Z. Jiang, Y. Tong, and C. Wang, "Core-sheath functional polymer nanofibers prepared by coelectrospinning", *Eur. Polym. J.,* vol. 44, no. 4, pp. 1040-1045, 2008. [http://dx.doi.org/10.1016/j.eurpolymj.2008.01.034]

[79] I.G. Loscertales, A. Barrero, I. Guerrero, R. Cortijo, M. Marquez, and A.M. Gañán-Calvo, "Micro/nano encapsulation via electrified coaxial liquid jets", *Science,* vol. 295, no. 5560, pp. 1695-1698, 2002. [http://dx.doi.org/10.1126/science.1067595] [PMID: 11872835]

[80] J.M. López-Herrera, A. Barrero, A. Lopez, I.G. Loscertales, and M. Marquez, "Coaxial jets generated from electrified taylor cones, scaling laws", *J. Aerosol Sci.,* vol. 34, no. 5, pp. 535-552, 2003. [http://dx.doi.org/10.1016/S0021-8502(03)00021-1]

[81] I.G. Loscertales, A. Barrero, M. Márquez, R. Spretz, R. Velarde-Ortiz, and G. Larsen, "Electrically forced coaxial nanojets for one-step hollow nanofiber design", *J. Am. Chem. Soc.,* vol. 126, no. 17, pp. 5376-5377, 2004. [http://dx.doi.org/10.1021/ja049443j] [PMID: 15113206]

[82] M. Lallave, J. Bedia, R. Ruiz Rosas, J. Rodríguez Mirasol, T. Cordero, J.C. Otero, M. Marquez, M. Barrero, and I.G. Loscertales, "Filled and hollow carbon nanofibers by coaxial electrospinning of alcell lignin without binder polymers", *Adv. Mater.,* vol. 19, no. 23, pp. 4292-4296, 2007. [http://dx.doi.org/10.1002/adma.200700963]

[83] J.E. Díaz, A. Barrero, M. Márquez, and I.G. Loscertales, "Controlled Encapsulation of Hydrophobic Liquids in Hydrophilic Polymer Nanofibers by Co-electrospinning", *Adv. Funct. Mater.,* vol. 16, no. 16, pp. 2110-2116, 2006. [http://dx.doi.org/10.1002/adfm.200600204]

[84] J.H. Yu, S.V. Fridrikh, and G.C. Rutledge, "Production of submicrometer diameter fibers by two-fluid electrospinning", *Adv. Mater.,* vol. 16, no. 17, pp. 1562-1566, 2004. [http://dx.doi.org/10.1002/adma.200306644]

[85] T. Subbiah, G.S. Bhat, R.W. Tock, S. Parameswaran, and S.S. Ramkumar, "Electrospinning of nanofibers", *J. Appl. Polym. Sci.,* vol. 96, no. 2, pp. 557-569, 2005. [http://dx.doi.org/10.1002/app.21481]

[86] W.J. Li, C.T. Laurencin, E.J. Caterson, R.S. Tuan, and F.K. Ko, "Electrospun nanofibrous structure: a novel scaffold for tissue engineering", *J. Biomed. Mater. Res.,* vol. 60, no. 4, pp. 613-621, 2002. [http://dx.doi.org/10.1002/jbm.10167] [PMID: 11948520]

[87] R. Kenawy, G.L. Bowlin, K. Mansfield, J. Layman, D.G. Simpson, E.H. Sanders, and G.E. Wnek, "Release of tetracycline hydrochloride from electrospun poly(ethylene-co-vinylacetate), poly(lactic acid), and a blend", *J. Control. Release,* vol. 81, no. 1-2, pp. 57-64, 2002. [http://dx.doi.org/10.1016/S0168-3659(02)00041-X] [PMID: 11992678]

CHAPTER 11

Biopackaging: Tara Gum Films

Martin A. Masuelli* and **María Guadalupe García***

Instituto de Física Aplicada - CONICET, Universidad Nacional de San Luis, Chacabuco 917 (ZP 5700), San Luis, Argentina.

Abstract: In this work, we prepared and characterized water insoluble Tara gum films. Tara gum (TG) is a polysaccharide extracted from the endosperm of *Caesalpinia spinosa* seeds with a performance of 12% wt. TG films were evaluated as barrier material for agricultural and food industries. The films were modified in order to improve water resistance by using glutaraldehyde (Glu) as crosslinker. The crosslinking process consisted of placing the TG films in a bath containing Glu in an acidic medium for a period of 12 and 24 hours (TG-12 and TG-24) at 25ºC. Film properties were structurally studied through XRD, TGA and SEM images and operationally evaluated by measuring water vapor permeability (WVP), gas permeation and mechanical properties. WVP values decreased with increasing crosslinking time. This is due to crosslinking which hinders the diffusion of vapor molecules through the polymer matrix. Gas permeability tests showed permeabilities in the order $CO_2 >> N_2 > O_2$. Mechanical tests indicated an increase in the elastic modulus of the film with increasing crosslinking time; this effect is due to the loss of flexibility of the crosslinked polymeric matrix.

Keywords: Films, Gas permeation, Mechanical properties, Tara gum, Water vapor permeability.

1. INTRODUCTION

Plastic food packaging represents a threat to the environment all around the world due to the fact that they are derived from no renewable source and they are not biodegradable. Global production of packaging materials is estimated in more than 180 million tons per year, with growth and demand increasing annually. Within the plastic packaging market, food packaging is the largest growing sector suggesting the need for creating sustainable packaging alternatives, specifically the design and the development of biodegradable films [1]. The packaging serves several important functions: contains the food, protects from chemical, physical

* **Corresponding author Martin Masuelli and Mariaía Guadalupe García**: Instituto de Física Aplicada - CONICET, Universidad Nacional de San Luis, Chacabuco 917 (ZP 5700), San Luis, Argentina; Email: masuelli@unsl.edu.ar, maggarcia@unsl.edu.ar

Martin Masuelli & Denis Renard (Eds.)

and microbiological deterioration, and provides practical means to inform consumers about the products [2 - 4]. In addition, the packaging preserves the shape and texture of the food, prevents flavor or aroma loss, prolongs storage time and regulates the water content or moisture of the food [5 - 7].

Recent advances in polymer technology promote the use of polymeric films for packaging foodstuffs. In this sense, there have been designed biodegradable materials from agro-industrial polymers obtained from renewable, abundant and low cost resources [8]. For example, cellulose is a biodegradable natural polymer of particular importance due to its application in plastic fibers [9]. From the total biopolymer production 41% is used for packaging industry and 47% of it is particularly applied in food packaging. Due to that, the use of biopolymers emerged as an excellent alternative to reduce existing environmental problems [10].

Certain food products such as coffee show metabolic processes even after packaging. These processes can lead to a faster degradation of food decreasing their useful life. Hence, characteristics of the packaging material play an important role in the food industry to extend the time of consumption. Among the various materials used, Tara gum films have been studied due to thier very good gas barrier and mechanical properties; however they are very permeable to water. In order to solve water permeability, crosslinking agents are often used [7].

Tara gum is a natural polymer obtained by grinding the endosperm of the seeds of *Caesalpinia spinosa* (Fam. Leguminosae); consists chiefly of polysaccharides of high molecular weight composed mainly of galactomannans. Tara gum has been used in food industry as gelling agent in juices, ice cream, sauces, pet food among others. Tara Gum seeds are contained in sheaths of 8 to 10 cm in length and four to seven seeds about 6 to 7 mm in diameter are present in each sheath. The shells of the seeds are about 25 to 27% of endosperm and 11% to 5% of moisture. Tara gum is a white, tasteless powder forming a viscous gel when mixed with water [11]. The viscosity of the solution depends on the chain length containing galactomannan. The hydrolysis with strong acids may reduce or abolish the viscosity of the gum.

Food packaging is concerned with the preservation and protection of all types of foods and their raw materials, particularly from oxidative and microbial spoilage and also to extend their shelf-life characteristics. Polymer cross-linking and graft copolymerization of natural polymers with synthetic monomers are other valuable alternatives in the design of biodegradable packaging. However, chemically modified biopolymers can loss biodegradability so degradation mechanisms of the films before their application need to be studied [12]. During the last decade, the consumer pressure may trigger the use of biobased packaging materials as an alternative to materials produced from non-renewable resources. Biologically based packaging is defined as packaging containing raw materials originating from agricultural sources, *i.e.* produced from renewable biological raw materials such as starch and bioderived monomers. These materials are not necessarily biodegradable. Biobased packaging materials include both edible films and edible coatings along with primary and secondary packaging materials [13].

Galactomannans are preferred hydrocolloids since they are comparatively cheap, non-toxic, eco-friendly and non-polluting during production and application. Galactomannans from seeds of three species of Gleditsia, namely *G. sinensis, G. microphylla and G. melanacantha,* were characterized in terms of structural and thermal properties. Gleditsia polysaccharides consist of D-mannopyranose and D-galactopyranose residues. The mannose/galactose (M/G) ratio of Gleditsia galactomannans was 3.25, 3.31 and 2.30 for *G. sinensis, G. microphylla* and *G. melanacantha*, respectively. These values indicate that Gleditsia gums offer an excellent alternative to locust bean gum [14, 15].

Innovations constantly appear in food packaging, always aiming at creating a more efficient quality preservation system while improving food attractiveness and marketability. The utilization of renewable sources for packaging materials, such as hydrocolloids from biological origin, is one the main trends of the industry. Edible films/coatings have been considered as one of the potential technologies that can be used to increase the storability of foods and to improve the existent packaging technology, helping to ensure the microbial safety and the preservation of food from the influence of external factors. In view of these recent developments, the main objective of this review is to provide information concerning the utilization of galactomannans from Tara gum in the production of films. The most important features of these polysaccharides are discussed, namely: their structure and applications; physical, chemical, thermal and mechanical properties of galactomannan-based films/coatings; transport properties (in particular those related to moisture, oxygen, carbon dioxide exchange through the films/coatings); incorporation of active compounds (*e.g.* natural antimicrobials and/or antioxidants) and applications in food products. It is viewed that in a near future tailored edible packaging based on polysaccharides can be applied to selected foods, partially replacing non-biodegradable/non-edible plastics [16, 17].

Chemical structure of locust bean gum is shown in Fig. (**1A**). It is a galactomannan similar to guar gum consisting of a (1, 4)-linked β-D -mannopyranose backbone with branch points from their 6-positions linked to α-D -galactose (*i.e.* 1,6-linked α-D-galactopyranose). There are about 3.5 (2.8-4.9) mannose residues for every galactose residue. It is less soluble and low viscose than guar gum as it has fewer galactose branch points. Locust bean gum is extracted from the seed (kernels) of the carob tree (*Ceratonia siliqua*). It forms a food reserve for the seeds and helps to retain water under arid conditions. Guar gum (called guaran), shown in Fig. (**1B**), is extracted from the seed of the leguminous shrub *Cyamopsis tetragonoloba*, where it acts as a food and water store. It is made up of non-ionic polydisperse rod-shaped polymers consisting of molecules (longer than found in locust bean gum) made up of about 10,000 residues [18].

Polysaccharides have been finding, in the last decades, very interesting and useful applications in the biomedical and, specifically, in the biopharmaceutical field. Galactomannans are a group of storage polysaccharides from various plant seeds that reserve energy for germination in the endosperm. There are four major sources of seed galactomannans: locust bean (*Ceratonia siliqua*), guar (*Cyamopsis tetragonoloba*), Tara (*Caesalpinia spinosa* Kuntze), and fenugreek (*Trigonella foenum-graecum* L.) [19, 20]. Galactomannans are often used in different forms for human consumption. With their different physicochemical properties, they are versatile materials with many applications. In simple aqueous systems, they are effective viscosifiers and thickeners, and are also excellent stiffeners and stabilizers of emulsions. The absence of toxicity also favors their use in the textile, pharmaceutical, biomedical, cosmetics, and food industries. They are mainly used in pharmaceutical technology and cosmetics, not as pure galactomannan but as crude flour. Commercially sold flours are reported to contain about 85% pure galactomannan on a mass basis and they have an important contribution to the hydrocolloid usage in many products. Galactomannans have sometimes been used in binary mixtures with other polysaccharides, such as with xanthan gum, agar and kappa-carrageenan, to form gels with new properties [21].

Figueiro and co-workers [22] studied dielectric properties of glutaraldehyde crosslinked galactomannan-collagen films using dielectric spectroscopy, and resonance measurement of the piezoelectric strain constant d14 of galactomannan-collagen film (60%) showed 0.081 pC/N.

Figueiro and coworkers [22] performed IR spectrum of galactomannan film (Gal) showing absorption bands at 812 and 871 cm^{-1} indicating the presence of α-linked D-galactopyranose units and β-linked D-mannopyranose units, respectively. Such glycosidic configurations are reported in most seed galactomannan. Other bands are common for polysaccharides, such as the signal at 1150 cm^{-1} assigned to bending vibrational modes δ(C–O) due to the pyranose ring. The spectral range 1150-950 cm^{-1} is characterized by the contribution of bending δ(C–OH) modes. Band at 978 cm^{-1} was assigned to deformation of the axial (C–OH) at C-4 Fig. (**2**).

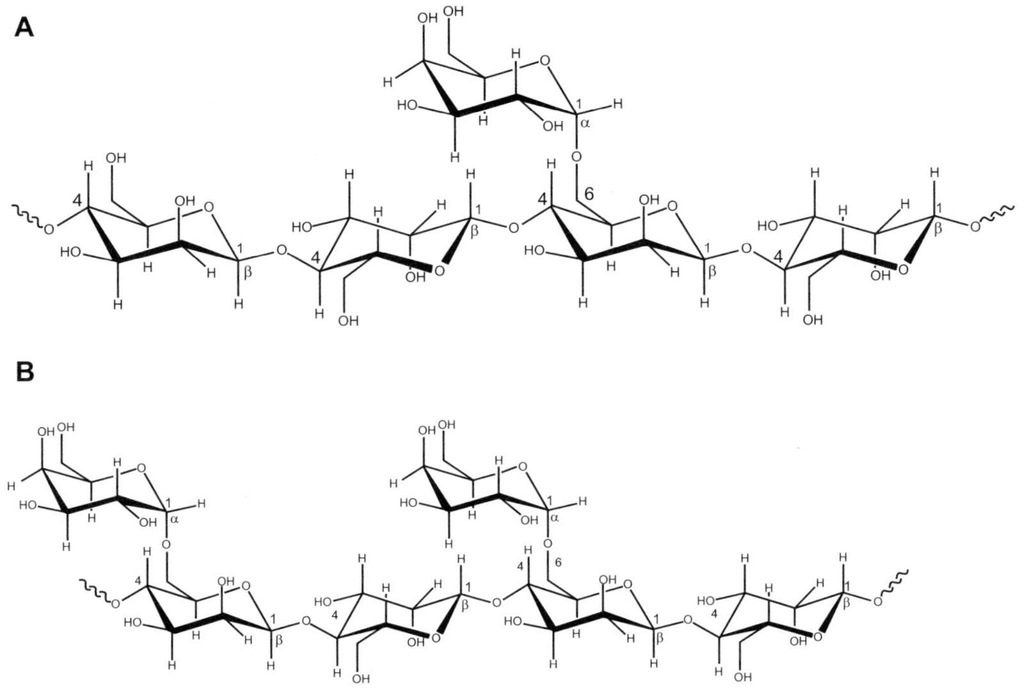

Fig. (1). Chemical structure of (A) locust bean gum and (B) guar gum [18].

Mikkonen and co-workers [23] prepared mixed films with sorbitol, konjac glucomannan (KGM) and poly(vinyl alcohol) (PVOH), and cellulose nanowhiskers (CNW). These films showed permeability properties compared to those films from commercial mannans; the oxygen permeability of films was in the same magnitude as that of commercial polyethylene/ethylene vinyl alcohol/polyethylene laminate film. The aroma permeability of films was low; the films blocked the light transmission in the ultraviolet region of the spectra.

Rao and coworkers [24] prepared films chitosan and guar gum in different ratios through the casting method. Addition of guar gum in varied proportions to chitosan solution led to changes in transparency and opacity of films. The water vapor transmission rate did not change significantly upon addition of guar gum. Films containing 15% (v/v) guar gum showed very low oxygen permeability, good tensile and puncture strength. The antimicrobial activity of films containing 15% (v/v) guar gum was comparable to chitosan films against *Escherichia coli* and *Staphylococcus aureus*. These films may reduce environmental problems associated with synthetic packaging.

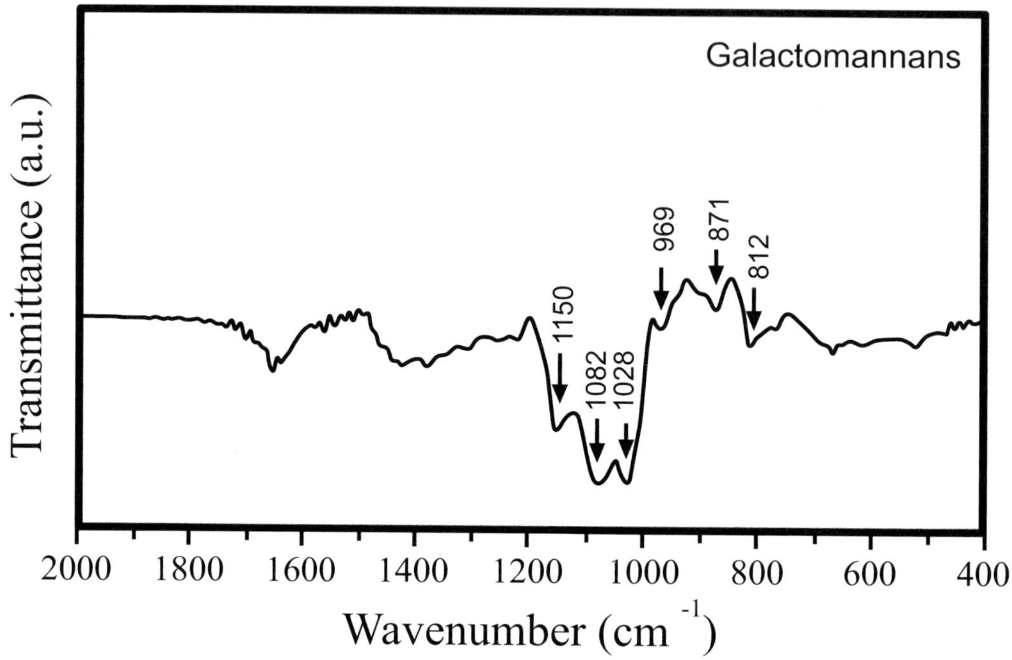

Fig. (2). IR spectrum of galactomannans [22].

Pitkanen and coworkers [25] altered enzymatically (endo-1,4-β-d-mannanase and α-d-galactosidase) the degree of substitution and the degree of polymerization of galactomannan from guar gum. Unmodified guar galactomannan was found to contain aggregates, but partial removal of galactose side units significantly decreased this tendency. The chain length seemed to dictate the solubility of the molecules together with the degree of substitution. The sample with a relatively low degree of polymerization and low degree of substitution formed large assemblies but stayed in solution while the solubility of the samples with longer chain length was reduced.

Saurabha and coworkers [26] investigated biodegradable films prepared from radiation processed guar gum. These films prepared from guar gum irradiated up to 500 Gy demonstrated significantly higher tensile strength as compared to non-irradiated control films. This improvement in tensile strength observed was demonstrated to be due to the ordering of polymer structures as confirmed by small angle X-ray scattering analysis. Exposure to doses higher than 500 Gy, however, resulted in a dose dependent decrease in tensile strength. A dose dependent decrease in puncture strength with no significant differences in the percent elongation was also observed at all the doses studied. Water vapor barrier properties of films improved up to 15% due to radiation processing. Radiation processing at lower doses for improving mechanical and barrier properties of guar based packaging films was demonstrated by Saurabha *et al.* [26] for the first time.

Das and coworkers [27] modified guar gum with benzamide. Benzoylation was carried out by benzoyl chloride reaction in water medium and a propyl amine spacer was used to impart a high degree of hydrophobicity. The new guar gum benzamide was resistant to water and soluble in non aqueous solvent like dimethyl sulfoxide. Cast films of thickness 0.162 mm had a breaking point tensile strength of 21.95 MPa. The water vapor permeability of biomaterial film was 0.28 g mm kPa^{-1} h^{-1} m^{-2} and water contact angle on evaporative surface was 90.35 degree. Qualitative and quantitative biocide activity of film was established against *Salmonella enterica*, *Escherichia coli*, *Staphylococcus aureus* and *Bacillus subtilis*. The new guar gum benzamide absorbed strongly in UV region.

Das & Mukherjee [28], prepared a biomaterial for widespread applications in food and water technology. They reported a new hydrophobic biomaterial film for soluble pollutant sorption and reduction of microbial load in water environment. Guar gum was hydrophobically modified following benzoyl chloride reaction in aqueous alkali. The new biopolymer derivative is insoluble in water and forms castable films with a high degree of surface hydrophobicity. The film water vapor permeability was 0.233 ± 0.029 g mm kPa^{-1} h^{-1} m^{-2} and the tensile property was 38.65 ± 3.66 MPa. The film surface contact angle against water was 77.36 ± 1.09^{o}. The new biomaterial film demonstrated strong antimicrobial contact killing against both gram positive and gram negative bacteria. It also acted as an efficient sorbent for a range of water soluble organic pollutants.

Mikkonen and coworkers [29] synthesized films from a renewable and biodegradable forest biorefinery product, spruce O-acetyl-galactoglucomannans (GGMs), and crosslinked with glyoxal. For the first time, cohesive and self-standing films were obtained from GGM without the addition of polyol plasticizer. In addition, glyoxal-crosslinked films were prepared using sorbitol at 10, 20, 30, and 40% (wt. % of GGM). Glyoxal clearly strengthened the GGM matrix, as detected by tensile testing and dynamic mechanical analysis. The elongation at break of films slightly increased, and Young's modulus decreased with increasing sorbitol content. Interestingly, the tensile strength of films was constant with the increased plasticizer content. The glyoxal-crosslinked GGM films containing 20% sorbitol exhibited the lowest oxygen permeability and WVP of the studied films and showed satisfactory mechanical performance.

Mikkonen and coworkers [30] modified enzymatically guar gum to obtain structurally functionalized guar gum. Cohesive and flexible films were formed from galactomannans plasticized with 20-60% (w/w of polymer) glycerol or sorbitol. Galactomannans with lower galactose content (locust bean gum, modified guar gum) produced films with higher elongation at break and tensile strength. The mechanical properties of films were improved statistically significantly by decreasing the degree of polymerization of guar gum with mannanase treatments and produced films with low elongation at break and tensile strength.

Cerqueira and coworkers [31] evaluated the influence of glycerol and corn oil on physicochemical properties of polysaccharide-based films. They used polysaccharides of galactomannan from *Gleditsia triacanthos* and chitosan. They established structure-properties relationships with the way they interact with water, other film's constituents (glycerol and oil) and the resulting properties. The presence of glycerol and corn oil originated a more hydrophilic structure and a decreased affinity of the film matrix to water, respectively, in both polysaccharides. However, the two polysaccharides presented different behaviors in terms of glass transition temperature, water vapor permeability and elongation-at-break that have been related with the particularities of their structure: while for the galactomannan the specific sorption sites for water are the -OH groups, for chitosan those are -OH and/or $-NH_2$ groups.

Trinetta and coworkers [32] researched the effects of pullulan (Pul), glycerin (Gly), xanthan gum (Xa) and locust bean gum (Lb) concentrations on pullulan film properties, optimal ingredient combination was determined and antimicrobial activity of films combined with sakacin A was confirmed against *Listeria monocytogenes*. They used predictive models, contour plots and the characteristics of commercial LDPE films as constraints, the following combination within the optimal region was selected: Pul 100 g/l, Gly 10 g/l, Xa 1 g/l and Lb 1 g/l. Statistical analysis demonstrated that Pul and Gly significantly influenced the properties of pullulan films. Strong interactions were observed between Pul-Gly and Xa-Gly. When sakacin A was added to the film mixture, no significant influence on films optical properties was reported, while an increase in flexibility was observed. Results obtained indicate the potential application of pullulan film developed in this study as an effective antimicrobial biopolymer in food-packaging systems.

Valenga and coworkers [33] immobilized glucose/mannose-binding lectin from *Concanavalia ensiformis* seeds (ConA) onto a monolayer made of a galactomannan extracted from *Leucaena leucocephala* seeds (GML). The mean thickness of GML monolayer, which polysaccharide consists of linear 1$\rightarrow$ 4-linked β-d-mannopyranosil units partially substituted at C-6 by α-d-galactopyranosyl units, amounted to (1.5±0.2) nm. ConA molecules adsorbed onto GML surfaces forming

(2.0±0.5) nm thick layers. The GML film was also used as support for the adsorption of three serotypes of dengue virus particles. Interactions between polysaccharides thin films, proteins, and viruses are of great relevance since they can provide basis for the development of biotechnological devices. These results indicate that GML is a potential polysaccharide for biomaterials development, as those could involve interactions between ConA in immune system and viruses.

Carneiro-da-Cunha and coworkers [34] modified and functionalized polysaccharide multinanolayers of sodium alginate, carragennan, chitosan, and two galactomannans were characterized in terms of zeta potential, as a function of polysaccharides and NaCl concentrations and pH. For the galactomannans solutions, zeta potential values were found to be neutral (ranging from −13.7 to −2.1mV) independently of the combinations of variables used. They established a relationship between important polysaccharides properties and the values of pH, polysaccharide and salt concentrations, thus providing insight on how to control the structure and texture of multinanolayered films in view of their potential application *e.g.*, in the food, pharmaceutical, and cosmetic industries.

Martins and coworkers [35] developed mixed systems, formed by locust bean gum (LBG) and k-carrageenan (k-car) of edible films with particular properties. Through the addition of k-car to LBG, authors improved the barrier properties of the films leading to a decrease in water vapor permeability (WVP). FTIR results suggested that hydrogen bonds interactions between k-car and LBG have a great influence in films properties *e.g.* moisture content, WVP.

Keshavarao and coworkers [36] blended locust bean gum and alginate 4 to 28 wt. % with 1-5 v. % of glycerine and prepared films. These films had the following mechanical properties: tensile strength of 2.41-3.21 MPa/mm^2 and elongation 21.30-33.12%.

The objective of this chapter is to show the preparation and characterization of biopolymeric films from Tara gum (TG). Crosslinking modifications by using glutaraldehyde were performed. Characterization techniques such as XRD, SEM-EDS, water vapor permeation, gas permeation and mechanical properties were applied. The effect of crosslinking time at constant glutaraldehyde concentration on TG films was studied. Mechanical and permeation properties of modified films were evaluated in order to be considered as barrier materials for packaging applications.

2. EXPERIMENTAL

2.1. Film Preparation

Tara gum (TG) Aglumix 01 were provided by INDUNOR Argentina, with molecular weight of 3,770,000 g/mol determinate by intrinsic viscosity (3595.9 cm^3/g) [37]. Film solution was prepared by dissolving 2 g of Tara gum in 50 ml of distilled water, with the addition of 3% v. of glycerol at 60 °C. Then, the solution of Tara gum was poured into petri dishes (diameter of 13 cm) and dried in an oven at 50 °C for 24 hours yielding pure Tara gum films (named TG). In order to improve water resistance, TG films were crosslinked using glutaraldehyde (GLU, grade II, Sigma). Crosslinking process consisted of placing the TG film in a bath containing Glu 5% v. in acetone (Bioscience) and HCl (Sigma) 1% v., for a period of time of 12 and 24 hours (TG-12 and TG-24). Finally, the film was repeatedly washed with ethanol until the solution was colorless. Fig. (**3**) shows a scheme of TG film preparation.

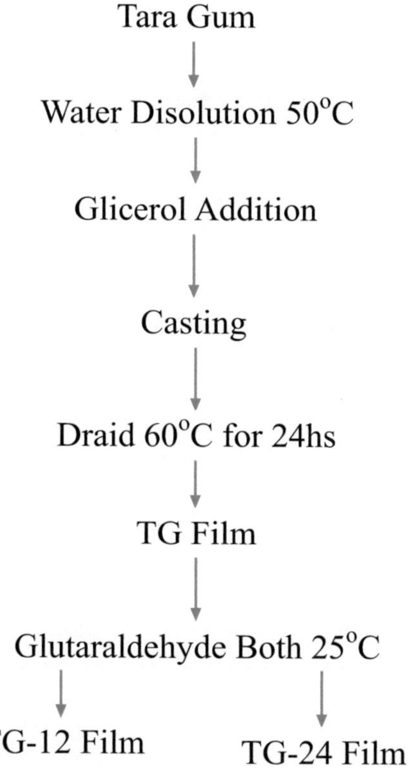

Fig. (3). Procedure of films preparation.

2.2. Solubility Tests

Solubility tests consisted of contacting film samples with different solvents for a period of 48 hours, 25 °C and mechanical stirring. The solvents used were acetone, ethanol, N, N-dimethylformamide (Sigma), NaOH (Tetrahedron) 0.1M, HCl 0.1M and distilled water.

2.3. X-Ray Diffraction

X-ray diffractions (XRD) studies were carried out using an equipment Rigaku model D-Max III C, lamp of Cu K and filter of Nickel. The 2θ operational range was from 0° to 30°. The d-spacing was determined by Bragg's equation,

$$n\lambda = 2d\,sen\theta \quad \textbf{(1)}$$

where d is the average intercatenary distance, n is the integer determined, λ is the wavelength of the X-ray (nm), and θ is the Bragg's angle.

2.4. Thermal Analysis

Thermogravimetric analysis (TGA) was performed to estimate the thermal stability of the TG films. TGA measurements were recorded on TG 2950 analyzers (TA Instruments, Inc., New Castle, USA), respectively. The operating conditions were as follows: (a) heating rate: 10° min^{-1} and, (b) atmosphere: static air or dynamic N_2 (99.99%, flow rate 50 mL/min). The thermogravimetric temperature axes were calibrated with indium (99.99%, melting point of 156.60 °C) and the Curie point of Ni (353 °C), respectively. Empty aluminum pans (40 mL) were used as references. Polysaccharide samples of about 8 mg were employed.

2.5. SEM Images and EDAX Analysis

Morphology of TG films was observed using a scanning electron microscope (SEM) LEO 1450VP, and energy dispersion X-ray analysis was carried out using an EDS Genesis 2000 (EDAX). For the SEM morphological surface analysis, samples were prepared by immersion in liquid nitrogen and afterward coated with gold. They were observed under high vacuum and EDAX spectrums were obtained applying an acceleration voltage of 120 kV.

2.6. Water Vapor Permeation (WVP)

Water vapor permeation tests were carried out according to the ASTM (E95-96) at 37 °C and 85% RH [38]. Gravimetric measurements on an acrylic permeation cell allow knowing the amount of water vapor that pass through the film. The acrylic cell has an upper face in contact with moisture and a lower face in contact with an absorbent material (silica gel, Sigma). Between both compartments is placed the polymeric film acting as barrier. Water Vapor Permeation is calculated as:

$$P_w = \frac{Q\ell}{At\Delta P} \quad \textbf{(2)}$$

where P_w is the water vapor permeation (ng m/m^2 s Pa), Q is permeating mass (ng), ℓ is the thickness (m), A the area of the cell (m^2), t is the measurement time (s), ΔP is Vapor Pressure of pure water (Pa).

2.7. Gas Permeation

Determination of the permeability coefficient "P" of gases was carried out using a "time lag" apparatus at 1 atm and 30 °C [39]. The effective membrane area was 11.34 cm^2 and constant permeate volume was 35.37 cm^3. The gases tested were N_2, O_2 and CO_2 (99.99% purity). The amount of gas transmitted at time "t" through the polymer film was calculated by reading the permeate pressure (p_2) on the low-pressure side of the permeation cell. Plotting low pressure (p_2) *versus* time (t), permeation coefficients "P" for each gas are calculated once reached steady state as:

$$P = \frac{B\ell}{Tp_1}\frac{dp_2}{dt} \quad \textbf{(3)}$$

where B is 11.53 (cm^3(STP) K/cm^2 cmHg) is the cell constant; dp_2/dt (cmHg/s) is the slope of the line at steady state; p_1 (cmHg) is the high pressure side; T_c (K) is cell temperature; and ℓ (cm) is the thickness of the film.

Theoretical separation factors ($\alpha_{i/j}$) were calculated from the ratio between the permeation coefficients of pure gases as:

$$\alpha_{i/j} = \frac{P_i}{P_j} \quad \textbf{(4)}$$

where the subscript "i" indicates a gas different from the gas "j".

2.8. Mechanical Properties

Mechanical properties of TG films were measured using a Comten Industries

Series 94 VC equipment at a constant traction speed of 5 mm/min. Tensile strength (σ) and elongation at break (ε) of the prepared films were obtained from the stress-strain plot. To ensure a complete relaxation of the polymeric structures and standardize experimental procedures, tests were conducted at 25 °C and 40% of relative humidity (RH). The obtained values are the result of averaging at least three measurement for each sample film. The thickness of the films was measured using a micrometer Köfer (accuracy ± 1 micron).

Tensile strength (σ) was calculated by dividing the maximum load by the initial cross-sectional area of the specimen. Percentage elongation at break (% ε) was calculated as the change percentage of the initial gage length of specimen (l_0=40 mm) at the point of sample failure. Finally, elastic module (E) was calculated from the slope of the stress–strain curve when a linear relationship between the stress and strain was observed. The stress–strain relationship is given by the following equation:

$$\sigma = E.\varepsilon \tag{5}$$

where: σ = F/A (F is force and A is area), $\varepsilon = \Delta l/l_0$, and E is the Young's modulus [N m^{-2} = Pa].

3. RESULTS

3.1. Solubility Test

The qualitative analysis of the film samples contacted with different solvents allowed to reach the following conclusions: i) the uncrosslinked films resulted soluble in acid and alkali, ii) all crosslinked films were insoluble in solvents such as acetone, ethanol, dimethylformamide, HCl (0.1M) and NaOH (0.1M) for a contact period of 48 hours, iii) the crosslinked films had varying degrees of insolubility in destilled water, especially when temperature was 50 °C. Depending on the contact time of crosslinking, films were named as TG-12 and TG-24 when crosslinking time was 12 and 24 hours, respectively. Solubility test showed TG-12 was swollen and split in water, while TG-24 swells and expands losing mechanical stability.

3.2. XRD Analysis

Neat and crosslinked TG films were structurally studied through X-ray analysis. Although polymers in general show amorphous X-ray patterns, structural modifications such as crosslinking may be seen through shifting and narrowing of typical broad bands. Bragg's equation is used to estimate an inter-chain distance known as d-spacing. This parameter gives an idea about the spacing between two adjacent polymer chains within the polymer matrix. The d-spacing is obtained from the Bragg's equation as the "d" value taking the 2θ angle from the maximum of the broad band. To analyze structural changes in Tara gum films, X-ray diffraction patterns of the films with different crosslinking times were studied and they are shown in Fig. (**4**).

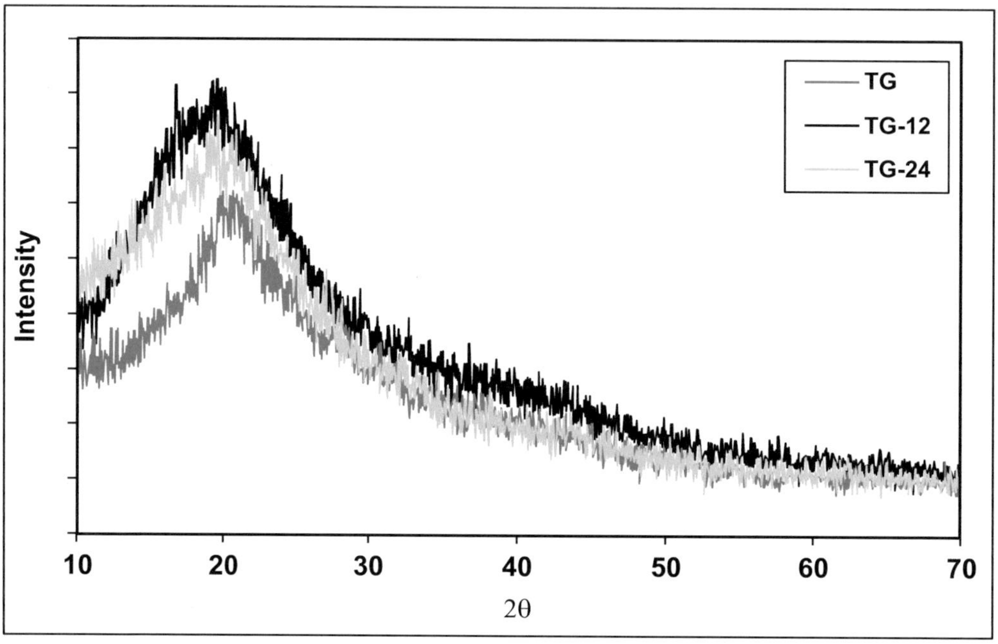

Fig. (4). XRD analysis of films.

TG showed an amorphous XRD pattern and the scattering angle (2θ) was observed at 21.65° which correspond to a d-spacing of 4.10 Å. Modified biopolymers TG-12 and TG-24 also showed amorphous X-ray patterns, besides, they showed broader bands with respect to neat TG. These results indicate that crosslinking modified the original molecular arrange of TG, becoming a less crystalline polymer as the time of crosslinking increases. TG-12 and TG-24 showed 2θ angles at 20.05 and 19.6, respectively, corresponding to d-spacing values of TG-12 = 4.42 Å and TG-24 = 4.52 Å, respectively (see Table **1**). The increase in inter-chain distance as the crosslinking time increases is a way of proving the crosslinking efficiency. Besides, the increase in d-spacing is attributed to the molecular size of crosslinking agent, *i.e.* glutaraldehyde, which is placed between two adjacent polymer chains, spacing them [44, 45]. Crosslinking reactions caused structural changes in the macromolecular arrangement of polysaccharide chains, sometimes called "interruptions". These interruptions can modify the secondary building structure of a polymer or construct a new dimensional arrangement known as tertiary structure. Diffraction patterns of TG-12 and TG-24 depicted the influence of crosslinking by increasing broadness of amorphous bands. This result can be associated with the secondary structure modification. This confirms that glutaraldehyde prevents molecular packing of polysaccharide chains by rigidizing polymeric matrix.

Table 1. Data of θ and d-spacing of TG films.

Films	θ	$d_{spacing}$ (Å)
TG	0.1889	4.10
TG-12	0.1751	4.42
TG-24	0.1711	4.52

Kapoor and coworkers [46] studied gums obtained from the seeds of *Cassia spectabilis* (endosperm). The purified galactomannan shows Mw $1.1x10^6$ g/mol and a mannose to galactose ratio of 2.65. The results of X-ray fiber studies show that the b dimension of the unit cell is very sensitive to relative humidity RH, galactose substitution and orientation of the films. The probable space group symmetry of the unit cell is $P2_1\ 2_1\ 2$. This typical diffraction pattern has much resemblance with *Cassia siamea* [47] Man-Gal=2.55 and Tara gum [c] Man-Gal=3.0 but differs from the patterns obtained for highly substituted galactomannans like *Medicago sativa* and *Trifolium alexandrium* [46].

The diffraction pattern similarities which have been found among mannans and galactomannans indicate a pronounced similarity in both molecular conformation and crystalline packing as already reported [46]. The near constancy of the fiber repeat c. and of the lateral dimensions and suggest a packing model having mannan–mannan interaction and permitting also galactose substitution on mannose residues in such a way that there is little alterations of the main chain organization. The third dimension b. relates to the distance between mannan sheets, although not constant in magnitude and varying with the degree of galactose substitution. The rigid sheets of mannan chains are stabilized by hydrogen bonding and the single galactose units are placed between these sheets. Any gaps created by the absence of galactose substitution are expected to be filled with water at high humidities [48 - 50].

The diffraction diagram of C. spectabilis seed galactomannan has a layer line separation of 10.28 Å in agreement with what is found for other galactomannans. The meridional reflections are missing on the first layer line but appear on the second layer. The tilting of the specimen by 17° from the normal X-ray beam does not cause resolved splitting of the third layer intensity across the meridian. This is consistent with a crystallographic space group which does not have a 2_1 axis parallel to the fiber axis [46].

3.3. Thermogravimetric Analysis

Thermogravimetric (TGA) measurements were carried out in order to better assess the thermal behavior of biopolymeric films. Neat Tara Gum and 12 hs crosslinked Tara Gum films were evaluated. Results are illustrated in Fig. (**5**).

TGA experiments show two events of mass loss for both biopolymer films, being the first step near to 100 °C. This mass loss is attributed to the loss of adsorbed and structural water of biopolymers [40]. TG film shows a second mass loss event at 290 °C resulted in a weight loss of approximately 70% attributed to the polysaccharide decomposition. On the other hand, TG-12 shows a second mass loss at about 209 °C corresponding to 20% of weight loss, following by a higher mass loss at 295 °C attributed to the starting crosslinking break down. The 20% mass loss was associated to the entrapped water molecules within crosslinked TG matrix. Finally, total decomposition of TG-12 was observed at around 470 °C, indicating a higher thermal resistance of the crosslinked polymer regarding to the neat TG film [41 - 43].

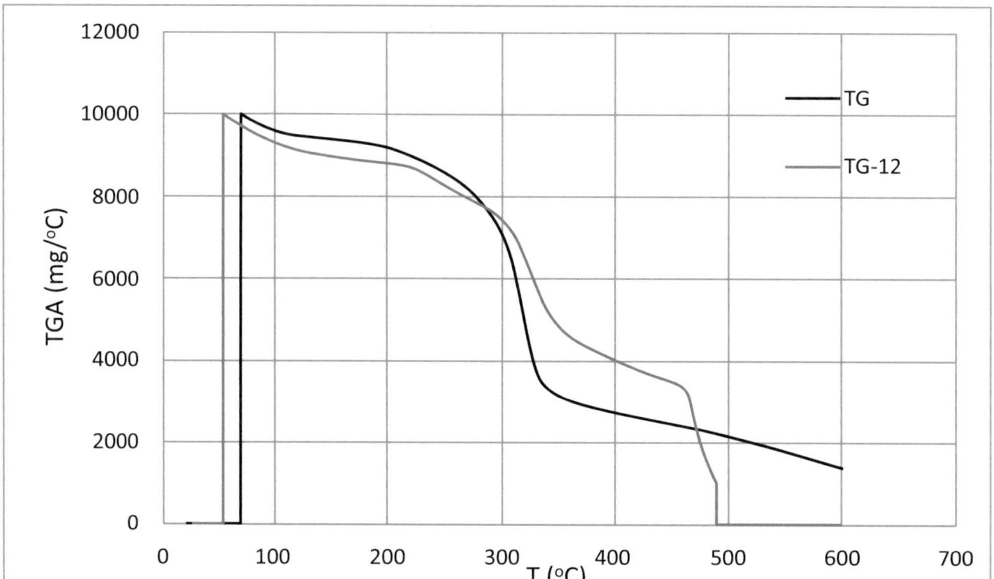

Fig. (5). TGA analysis of films.

3.4. SEM Images

Scanning electron microscopy (SEM) can provide a better understanding of water vapor transmission, mechanical and optical properties with the structural characteristics of the films. Fig. (**6**) shows SEM micrographs of the top surface of TG and TG-12 films. No marked differences were detected between the surface microstructures of the TG, TG-12 and TG-24 films. Surface images revealed that TG films had rough and uneven surfaces without noticeable cracks, breaks, or openings on the surfaces.

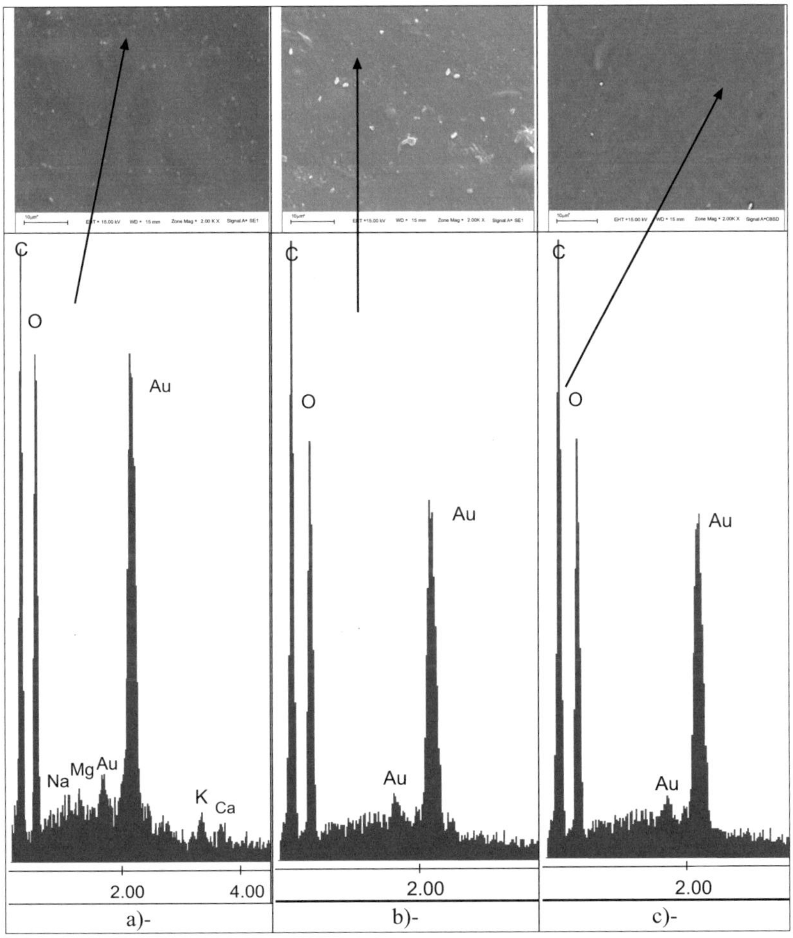

Fig. (6). SEM Images and elemental EDAX analysis of films a)- TG, b)- TG-12 and c)- TG-24.

The presence of glutaraldehyde added to the films was observed as an increase in the roughness of the film surfaces as the crosslinking time increases, however, no marked differences in surface structure was observed when different glutaraldehyde concentrations were proved. Fig. (**6a-6b**) shows a kind of polymer agglomeration in comparison to well dispersed TG-24 surface Fig. (**6c**). The uneven structure and agglomeration of the insoluble fraction of TG can provide an increase in free volume of the polymer matrix leading to higher diffusivity of permeating water molecules in TG films. Higher free volume may cause increments in water vapor transmission rate of TG in comparison to crosslinked TG. White particles on the surface of Tara Gum film are attributed to the remaining salt particles from the extraction and purification process of Tara Gum.

3.5. Water Vapor Permeation (WVP)

Water vapor permeation results for TG, TG-12 and TG-24 are presented in Table **2** and Fig. (**7**). It can be seen that WVP of biopolymers decreases as the crosslinking time increases. Thus, TG-24 showed the lowest WVP value. Permeation phenomenon implies an activated process where gas or vapor molecules have to move through the polymer chains open up spaces to pass through. Considering this, it is coherent that a crosslinked matrix shows lower permeability than an uncrosslinked one. This is due to a crosslinked matrix is more rigid, hence more resistant to the pass of a molecule.

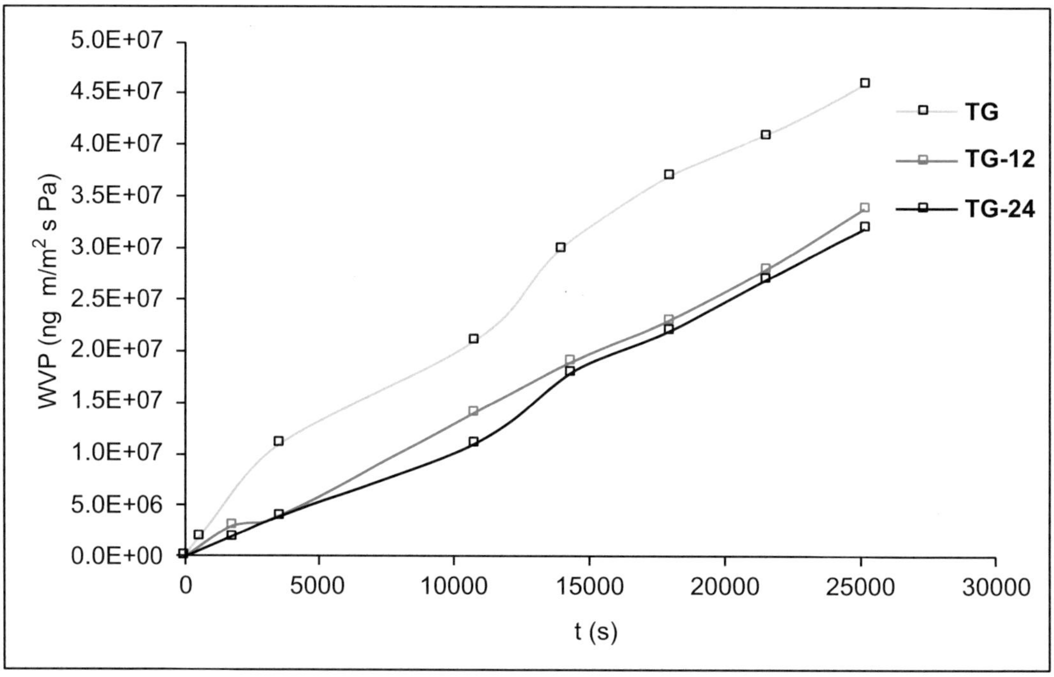

Fig. (7). WVP of films.

Several authors mentioned that polymers which present -OH and -COOH groups have better barrier properties to moisture than those containing sulfonic groups [51]. Besides, other authors [52] have demonstrated that water and glycerin commonly used as plasticizers in biopolymeric films, also are a powerful tool used to controlled barrier properties. They state that plasticization reduces the possible interactions between the polymer and the gas molecule, decreasing gas or vapor permeability. However, on the other hand, it is well known that plasticization increases the mobility of the polymer chains and the free volume of the polymer matrix given as a result an increase in WVP. Similarly, hydrophilic groups favor water permeation.

Table 2. WVP, gas permeability and mechanical properties of TG films.

Films	WVP (ng m m^{-2} s^{-1} Pa^{-1})	P_{N2} (B)	P_{O2} (B)	P_{CO2} (B)	$\alpha_{CO2/O2}$	σ (MPa)	ε (%)	E (MPa)
TG	0.4106	0.088	0.068	0.253	3.72	2.6	89.7	3.2
TG-12	0.2682	0.058	0.033	0.156	4.73	12.0	11.4	11.1
TG-24	0.1783	0.049	0.028	0.129	4.61	33.1	11.0	31.2

1 Barrer (B) = 10^{-10} (cm^3(STP)cm/cm^2 cmHg s)

Karbowiak *et al.* [53] demonstrated that diffusion of small molecules such as water or fluorescein, in edible films depends on the water content which is out of the plasticizer concentration. Higher water content reduces internal hydrogen bridges and intermolecular distance increases, increasing the WVP of the films [54]. TG films as well as biopolymeric films have a structural content of water, which acts as plasticizer. Then, high WVP is measured due to the increase of water content in the films, especially in the pure Tara gum.

3.6. Gas Permeation

Table **2** and Fig. (**8**) show gas permeation results for TG, TG-12 and TG-24. Nitrogen, Oxygen and Carbon dioxide were measured at 1 bar and 30 °C in order to resemble the atmospheric conditions of food packaging expose in shelves. Results showed high permeation of neat TG films for all assayed gases. Crosslinking of TG films reduces the permeability of three gases due to the rigidification of polymeric matrix caused by crosslinking. Thus, TG-24 showed the lowest gas permeability for all gases. Interestingly, the three films showed higher CO_2 permeability than O_2 and N_2 permeability, depicting selectivity to CO_2. This result indicates that TG films would be of great interest to the food preservation in which a fungicide or braking of maturation effect is required [55]. Crosslinked TG films showed higher CO_2/O_2 and CO_2/N_2 selectivity than neat TG, although it was not observed a significant difference between TG-12 and TG-24 selectivity.

3.7. MECHANICAL PROPERTIES

Mechanical properties of prepared films depend on several factors, including: the ratio of water/polymer and the casting temperature. Typical curves of stress-strain for each sample were obtained. These results were used to determine mechanical parameters such as elastic modulus or Young's modulus (E), tensile strength (σ) and elongation at break (ε). Results are showed in Table **2**, indicating that as crosslinking time increases TG films become more rigid due to an increase in the tensile strength and Young module, as well as, a decrease in deformation (ε). These results are consistent with the rigidification observed in crosslinked TG matrix responsible for the permeation results analyzed before. A higher crosslinking produces a higher immobilization of segments polymer chains, consequently higher rigidity [56 - 63].

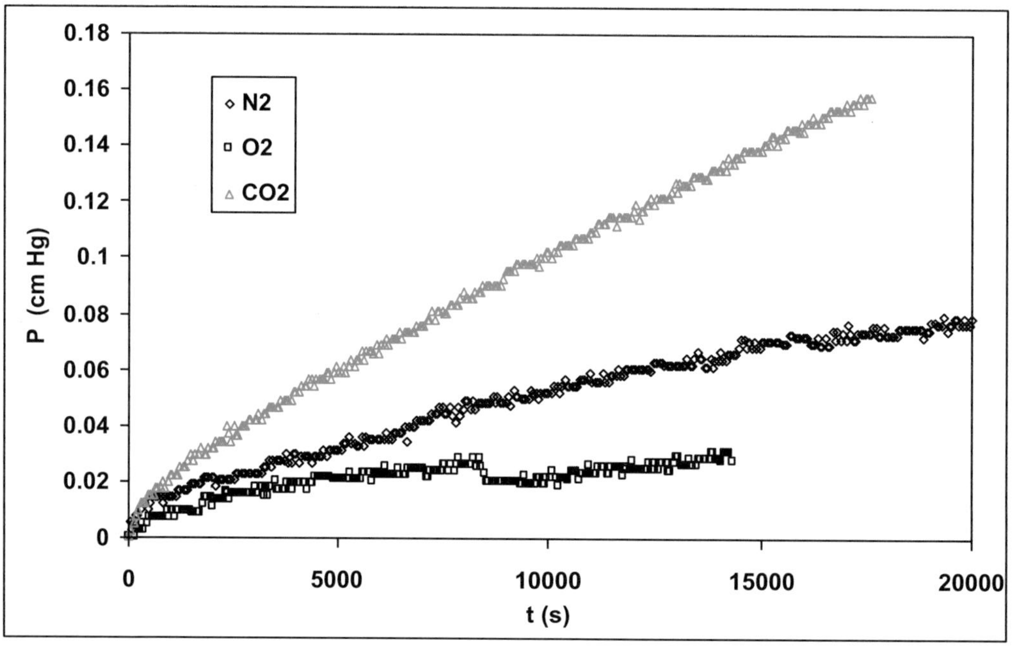

Fig. (8). Gas permeation of TG film.

CONCLUSIONS

In this work we prepared and characterized films of Tara gum. Crosslinking of TG films at different contact times with glutaraldheyde were also performed. Structural and morphological characteristics of TG films were evaluated through XRD, TGA and SEM images. Besides, influence of crosslinking time over WVP, gas permeation and mechanical properties of the films were also studied. Neat TG films were soluble in water, structurally amorphous, less thermal resistant, more gas and water vapor permeable and less rigid than crosslinked TG films. Mechanical properties of the films were affected by the contact time with GLU, longer contact times produced less deformable film such as TG-24 [64]. Besides, crosslinking films showed lower WVP, *i.e.* higher moisture barrier which a desirable characteristic in films for food packaging. To assess the gas barrier properties of TG, TG-12 and TG-24; N_2, O_2 and CO_2 permeation were measured using a "time lag" equipment. Results showed crosslinked TG had lower gas permeability for any N_2, O_2 or CO_2 with regard to neat TG. Besides, as crosslinking time increases gas permeability for all gases decreases. All TG films showed high CO_2 selectivities with respect to O_2 and N_2, being crosslinked TG films most selectives. However, almost no difference in CO_2 selectivity was observed between TG-12 and TG-24. These results are of interest to the food packaging industry due to their applicability to systems of modified atmosphere packaging (MAP) and controlled atmospheres packaging (CAP).

Tara gum polymer is an attractive material to be regional, biodegradable and low cost source. The results obtained from this work indicate that an adequate control of the variables of synthesis and post-synthetic modifications such as crosslinking, allow the production of films with desired physicochemical characteristics for different application, for example food packaging material.

CONFLICT OF INTEREST

The author (editor) declares no conflict of interest, financial or otherwise.

ACKNOWLEDGEMENT

The autors thank Universidad Nacional de San Luis, Instituto de Física Aplicada (INFAP-CONICET). The UNSL projects 2-1712, 2-2116, 2-2414 and 2-1916 for their financial support.

REFERENCES

[1] P. Fairley, F.J. Monahan, B.J. German, and J.M. Krochta, "Mechanical properties and water vapor permeability of edible films from whey protein isolate and sodium dodecyl sulfate", *J. Agric. Food Chem.,* vol. 44, pp. 438-443, 1996. [http://dx.doi.org/10.1021/jf9505234]

[2] S.J. Kim, and Z. Ustunol, "Solubility and moisture sorption isotherms of whey-protein-based edible films as influenced by lipid and plasticizer incorporation", *J. Agric. Food Chem.,* vol. 49, no. 9, pp. 4388-4391, 2001. [http://dx.doi.org/10.1021/jf010122q] [PMID: 11559143]

[3] T.H. McHugh, and J.M. Krochta, "Dispersed phase particle size effects on water vapor permeability of whey protein–beeswax edible emulsion Films", *J. Food Process. Preserv.,* vol. 18, pp. 173-188, 1993. [http://dx.doi.org/10.1111/j.1745-4549.1994.tb00842.x]

[4] T.H. McHugh, and J.M. Krochta, "Sorbitol vs glycerol plasticized whey protein edible films: integrated oxygen permeability and tensile property evaluation", *J. Agric. Food Chem.,* vol. 42, pp. 841-845, 1994. [http://dx.doi.org/10.1021/jf00040a001]

[5] T.H. McHugh, and J.M. Krochta, "Water vapor permeability properties of edible whey protein – lipid emulsion films", *J. Food Process. Preserv.,* vol. 40, pp. 133-138, 1999.

[6] M.B. Pérez, and J.M. Krochta, "Water vapor permeability of whey protein emulsion films as affected by pH", *J. Food Sci.,* vol. 64, pp. 695-698, 1999. [http://dx.doi.org/10.1111/j.1365-2621.1999.tb15112.x]

[7] P. Talens, and J.M. Krochta, "Plasticizing Effects of Beeswax and Carnauba Wax on Tensile and Water Vapor Permeability Properties of Whey Protein Films", *J. Food Sci.,* vol. 70, pp. 638-643, 1999.

[8] Safaa G. Abd Alla, Murat Sen, Abdel Waha, and M. El-Naggar, "Swelling and mechanical properties of superabsorbent hydrogels based on Tara gum/acrylic acid synthesized by gamma radiation", *Carbohyd. Pol.,* vol. 89, pp. 478-485, 2012.

[9] M. S. Lizarraga, D. De Piante Vicin, R. González, A. Rubiolo, and L. G. Santiago, "Rheological behaviour of whey protein concentrate and k-carrageenan aqueous mixtures", *Food Hydr.,* vol. 20, no. 5, pp. 740-748, 2006. [http://dx.doi.org/10.1016/j.foodhyd.2005.07.007]

[10] J.W. Rhim, "Physical and mechanical properties of water resistant sodium alginate films", *LWT,* vol. 37, pp. 323-330, 2004. [http://dx.doi.org/10.1016/j.lwt.2003.09.008]

[11] J. Vartiainen, T. Tammelin, J. Pere, U. Tapper, and A. Harlin, "Biohybrid barrier films from fluidized pectin and nanoclay", *Carbohydr. Polym.,* vol. 82, pp. 989-996, 2010. [http://dx.doi.org/10.1016/j.carbpol.2010.06.031]

[12] R.N. Tharanathan, "Biodegradable films and composite coatings: past, present and future", *Trends Food Sci. Technol.,* vol. 14, pp. 71-78, 2003. [http://dx.doi.org/10.1016/S0924-2244(02)00280-7]

[13] K. Petersen, V. Nielsen, and B. Per, "Grete; Lawther, Mark; Olsen, Mette B.; Nilssonk, Nils H.; Mortensen, Grith. "Potential of biobased materials for food packaging", *Trends Food Sci. Technol.,* vol. 10, pp. 52-68, 1999. [http://dx.doi.org/10.1016/S0924-2244(99)00019-9]

[14] J.X. Jiang, H.L. Jian, C. Cristhian, W.M. Zhang, and R.C. Sun, "Structural and thermal characterization of galactomannans from genus Gleditsia seeds as potential food gum substitutes", *J. Sci. Food Agric.,* vol. 91, no. 4, pp. 732-737, 2011. [http://dx.doi.org/10.1002/jsfa.4243] [PMID: 21213227]

[15] M. Srivastava, and V.P. Kapoor, "Seed galactomannans: an overview", *Chem. Biodivers.,* vol. 2, no. 3, pp. 295-317, 2005. [http://dx.doi.org/10.1002/cbdv.200590013] [PMID: 17191982]

[16] M.A. Cerqueira, A.I. Bourbon, A.C. Pinheiro, J.T. Martins, B.W. Souza, J.A. Teixeira, and A.A. Vicente, "Galactomannans use in the development of edible films/coatings for food applications", *Trends Food Sci. Technol.,* vol. 22, pp. 662-671, 2011. [http://dx.doi.org/10.1016/j.tifs.2011.07.002]

[17] K. Mikkonen, M. Yadav, P. Cooke, S. Wilfor, K. Hicks, and M. Tenkanen, "Films from spruce galactoglucomannan blended with poly(vinylalcohol), corn arabinoxylan and konjak glucomannan", *BioResources,* vol. 3, no. 1, pp. 178-191, 2008.

[18] P. Mormile, L. Petti, M. Rippa, B. Immirzi, M. Malinconico, and G. Santagata, "Monitoring of the degradation dynamics of agricultural films by IR thermography", *Polym. Degrad. Stabil.,* vol. 92, pp.

777-784, 2007.
[http://dx.doi.org/10.1016/j.polymdegradstab.2007.02.015]

[19] V.D. Prajapati, G.K. Jani, N.G. Moradiya, N.P. Randeria, B.J. Nagar, N.N. Naikwadi, and B.C. Variya, "Galactomannan: a versatile biodegradable seed polysaccharide", *Int. J. Biol. Macromol.*, vol. 60, pp. 83-92, 2013.
[http://dx.doi.org/10.1016/j.ijbiomac.2013.05.017] [PMID: 23707734]

[20] D.R. Picout, S.B. Ross-Murphy, K. Jumel, and S.E. Harding, "Pressure cell assisted solution characterization of polysaccharides. 2. Locust bean gum and tara gum", *Biomacromolecules,* vol. 3, no. 4, pp. 761-767, 2002.
[http://dx.doi.org/10.1021/bm025517c] [PMID: 12099820]

[21] J.L. Silveira, and T.M. Bresolin, "Pharmaceutical use of galactomannans", *Quim. Nova,* vol. 34, no. 2, pp. 292-299, 2011.
[http://dx.doi.org/10.1590/S0100-40422011000200023]

[22] S.D. Figueiro, Julio C. Goes, R.A. Moreira, and A.S.B. Sombra, "On the physico-chemical and dielectric properties of glutaraldehyde crosslinked galactomannan–collagen films", *Carbohydr. Polym.,* vol. 56, pp. 313-320, 2004.
[http://dx.doi.org/10.1016/j.carbpol.2004.01.011]

[23] K.S. Mikkonen, M.I. Heikkilä, H. Helén, L. Hyvönen, and M. Tenkanen, "Spruce galactoglucomannan films show promising barrier properties", *Carbohydr. Polym.,* vol. 79, pp. 1107-1112, 2010.
[http://dx.doi.org/10.1016/j.carbpol.2009.10.049]

[24] M.S. Rao, S.R. Kanatt, S.P. Chawla, and A. Sharma, "Chitosan and guar gum composite films: Preparation, physical, mechanical and antimicrobial properties", *Carbohydr. Polym.,* vol. 82, pp. 1243-1247, 2010.
[http://dx.doi.org/10.1016/j.carbpol.2010.06.058]

[25] L. Pitkänen, P. Tuomainen, K.S. Mikkonen, and M. Tenkanen, "The effect of galactose side units and mannan chain length on the macromolecular characteristics of galactomannans", *Carbohydr. Polym.,* vol. 86, pp. 1230-1235, 2011.
[http://dx.doi.org/10.1016/j.carbpol.2011.06.018]

[26] C.K. Saurabh, S. Gupta, J. Bahadur, S. Mazumder, P.S. Variyar, and A. Sharma, "Radiation dose dependent change in physiochemical, mechanical and barrier properties of guar gum based films", *Carbohydr. Polym.,* vol. 98, no. 2, pp. 1610-1617, 2013.
[http://dx.doi.org/10.1016/j.carbpol.2013.07.041] [PMID: 24053847]

[27] D. Das, T. Ara, S. Dutta, and A. Mukherjee, "New water resistant biomaterial biocide film based on guar gum", *Bioresour. Technol.,* vol. 102, no. 10, pp. 5878-5883, 2011.
[http://dx.doi.org/10.1016/j.biortech.2011.02.034] [PMID: 21392974]

[28] D. Das, and A. Mukherjee, "Biomaterial film for soluble organic sorption and anti-microbial activity in water environment", *Bioresour. Technol.,* vol. 110, pp. 412-416, 2012.
[http://dx.doi.org/10.1016/j.biortech.2012.01.147] [PMID: 22357294]

[29] Kirsi S. Mikkonen, Mari I. Heikkila, Stefan M. Willfor, and Maija Tenkanen, "Films from Glyoxal-Crosslinked Spruce Galactoglucomannans Plasticized with Sorbitol", *Int. J. Pol. Sci.,* vol. 2012, pp. 1-8, 2012. Article ID 482810
[http://dx.doi.org/10.1155/2012/482810]

[30] K.S. Mikkonen, H. Rita, H. Helén, R.A. Talja, L. Hyvönen, and M. Tenkanen, "Effect of polysaccharide structure on mechanical and thermal properties of galactomannan-based films", *Biomacromolecules,* vol. 8, no. 10, pp. 3198-3205, 2007.
[http://dx.doi.org/10.1021/bm700538c] [PMID: 17883276]

[31] M.A. Cerqueira, B.W. Souza, J.A. Teixeira, and A.A. Vicente, "Effect of glycerol and corn oil on physicochemical properties of polysaccharide films. A comparative study", *Food Hydrocoll.,* vol. 27, pp. 175-184, 2012.

[http://dx.doi.org/10.1016/j.foodhyd.2011.07.007]

[32] V. Trinetta, C.N. Cutter, and J.D. Floros, "Effects of ingredient composition on optical and mechanical properties of pullulan film for food-packaging applications. LWT - Food", *Sci. Tech. (Paris),* vol. 44, pp. 2296-2301, 2011.

[33] F. Valenga, D.F. Petri, N. Lucyszyn, T.A. Jó, and M.R. Sierakowski, "Galactomannan thin films as supports for the immobilization of Concanavalin A and/or dengue viruses", *Int. J. Biol. Macromol.,* vol. 50, no. 1, pp. 88-94, 2012.
[http://dx.doi.org/10.1016/j.ijbiomac.2011.10.005] [PMID: 22020153]

[34] M.G. Carneiro-da-Cunha, M.A. Cerqueira, B.W. Souza, J.A. Teixeira, and A.A. Vicente, "Influence of concentration, ionic strength and pH on zeta potential and mean hydrodynamic diameter of edible polysaccharide solutions envisaged for multinanolayered films production", *Carbohydr. Polym.,* vol. 85, pp. 522-528, 2011.
[http://dx.doi.org/10.1016/j.carbpol.2011.03.001]

[35] J.T. Martins, M.A. Cerqueira, I. Bourbon, Ana C. Pinheiro, Bartolomeu W.S. Souza, and António A. Vicente, "Synergistic effects between k-carrageenan and locust bean gum on physicochemical properties of edible films made thereof", *Food Hydrocoll.,* vol. 29, pp. 280-289, 2012.
[http://dx.doi.org/10.1016/j.foodhyd.2012.03.004]

[36] K.P. Keshavarao, D. Mudit, L. Krishna, and R. Prudbi, "Development an evaluation of piroxicam loaded biopolymer based transdermal film", *Int. Res. J. Pharmacy,* vol. 2, no. 11, pp. 119-123, 2011.

[37] M.A. Masuelli, "Tara Gum the new biopolymer for different applications. Hydrodynamic Properties", In: *Chapter 2. Handbook of Sustainable Polymers Processing and Applications, ,* Vijay Kumar Thakur, Manju Kumari Thakur, Eds., Pan Stanford Publishing, 2015, pp. 73-79.

[38] ASTM, Standard Test Method for Water Vapor Transmission of Materials, *Annual Books of ASTM Standards, Designation E 96-95,* 1995, pp. 785-792.

[39] J. Marchese, E. Garis, M. Anson, N.A. Ochoa, and C. Pagliero, "ABS copolymer-activated carbon mixed matrix membranes for CO2/CH4 separation", *J. Membr. Sci.,* vol. 43, pp. 19-28, 2004.

[40] C.W. Vendruscolo, C. Ferrero, E.A. Pineda, J.L. Silveira, R.A. Freitas, M.R. Jiménez-Castellanos, and T.M. Bresolin, "Physicochemical and mechanical characterization of galactomannan from Mimosa scabrella: Effect of drying method", *Carbohydr. Polym.,* vol. 76, pp. 86-93, 2009.
[http://dx.doi.org/10.1016/j.carbpol.2008.09.028]

[41] D. Das, T. Ara, S. Dutta, and A. Mukherjee, "New water resistant biomaterial biocide film based on guar gum", *Bioresour. Technol.,* vol. 102, no. 10, pp. 5878-5883, 2011.
[http://dx.doi.org/10.1016/j.biortech.2011.02.034] [PMID: 21392974]

[42] M.A. Cerqueira, B.W. Souza, J. Simões, J.A. Teixeira, M. Domingues, M. Rosário, Manuel A. Coimbra, and António A. Vicente, "Structural and thermal characterization of galactomannans from non-conventional sources", *Carbohydr. Polym.,* vol. 83, pp. 179-185, 2011.
[http://dx.doi.org/10.1016/j.carbpol.2010.07.036]

[43] J. Antoniou, F. Liu, H. Majeed, and F. Zhong, "Characterization of tara gum edible films incorporated with bulk chitosan and chitosan nanoparticles: A comparative study", *Food Hydrocoll.,* vol. 44, pp. 309-319, 2015.
[http://dx.doi.org/10.1016/j.foodhyd.2014.09.023]

[44] S. Naoi, T. Hatakeyama, and H. Hatakeyama, "Phase Transition of Locust Bean Gum, Tara Gum and Guar Gum–Water Systems", *J. Therm. Anal. Calorim.,* vol. 70, pp. 841-852, 2002.
[http://dx.doi.org/10.1023/A:1022260304686]

[45] J. Antoniou, F. Liu, H. Majeed, H.J. Qazi, and F. Zhong, "Physicochemical and thermomechanical characterization of tara gum edible films: effect of polyols as plasticizers", *Carbohydr. Polym.,* vol. 111, pp. 359-365, 2014.
[http://dx.doi.org/10.1016/j.carbpol.2014.04.005] [PMID: 25037362]

[46] V.P. Kapoor, F.R. Taravel, J-P. Joseleau, M. Milas, H. Chanzy, and M. Rinaudo, "Cassia spectabilis DC seed galactomannan: structural, crystallographical and rheological studies", *Carbohydr. Res.*, vol. 306, no. 1-2, pp. 231-241, 1998. [http://dx.doi.org/10.1016/S0008-6215(97)00241-3] [PMID: 9691448]

[47] V.P. Kapoor, H. Chanzy, and F.R. Taravel, "X-ray diffraction studies on some seed galactomannans from India", *Carbohydr. Polym.*, vol. 27, no. 3, pp. 229-233, 1995. [http://dx.doi.org/10.1016/0144-8617(95)00045-9]

[48] Y.Y. Chien, and William T. Winter, "Accurate lattice constants for tara gum", *Macromol.*, vol. 18, pp. 1357-1359, 1985. [http://dx.doi.org/10.1021/ma00148a058]

[49] B.K. Song, W.T. Winter, and F.R. Taravel, "Crystallography of highly substituted galactomannans: fenugreek and lucerne gums", *Macromol.*, vol. 22, pp. 2641-2644, 1989. [http://dx.doi.org/10.1021/ma00196a017]

[50] Y. Cheng, R.K. Prud'homme, J. Chik, and D.C. Rau, "Measurement of Forces between Galactomannan Polymer Chains: Effect of Hydrogen Bonding", *Macromol.*, vol. 35, no. 27, pp. 10155-10161, 2002. [http://dx.doi.org/10.1021/ma020887e]

[51] M.J. Fabra, P. Talens, and A. Chiralt, "Effect of alginate and k-carrageenan on tensile properties and water vapour permeability of sodium caseinate–lipid based films", *Carbohydr. Polym.*, vol. 74, pp. 419-426, 2008. [http://dx.doi.org/10.1016/j.carbpol.2008.03.010]

[52] S. Kokoszka, F. Debeaufort, A. Lenart, and A. Voilley, "Water vapour permeability, thermal and wetting properties of whey protein isolate based edible films", *Int. Dairy J.*, vol. 20, pp. 53-60, 2010. [http://dx.doi.org/10.1016/j.idairyj.2009.07.008]

[53] T. Karbowiak, H. Hervet, L. Léger, D. Champion, F. Debeaufort, and A. Voilley, "Effect of plasticizers (water and glycerol) on the diffusion of a small molecule in iota-carrageenan biopolymer films for edible coating application", *Biomacromolecules,* vol. 7, no. 6, pp. 2011-2019, 2006. [http://dx.doi.org/10.1021/bm060179r] [PMID: 16768427]

[54] F. Artés Calero, "El envasado en atmósfera modificada mejora la calidad de consumo de los productos hortofrutícolas intactos y mínimamente procesados en fresco", *Revista Iberoamericana de Tecnología Postcosecha,* vol. 2, pp. 61-85, 2006. [in spanish].

[55] G.D. Hiatt, and W.J. Rebel, *Cellulose and Cellulose Derivatives, Part V of High Polymers.* , N.M. Bikales, L. Segal, Eds., 1st Wiley-Interscience: New York, 1971, p. 749.

[56] A.L. Brody, "Envasado de alimentos en atmósferas controladas, modificadas y a vacío", Zaragoza (Spain) Ed. Acribia (in spanish)

[57] D. Zagory, and A.A. Kader, "Modified atmosphere packaging of fresh produce", *Food Technol.,* vol. 42, pp. 70-76, 1988.

[58] P.R. Salgado, V.C. Schmidt, S.E. Molina Ortiz, A.N. Mauri, and J.B. Laurindo, "Biodegradable foams based on cassava starch, sunflower proteins and cellulose fibers obtained by a baking process", *J. Food Eng.,* vol. 85, pp. 435-443, 2008. [http://dx.doi.org/10.1016/j.jfoodeng.2007.08.005]

[59] J. Brandrup, E.H. Immergut, Ed., *Polymer Handbook.* 2a J. Wiley and Sons, 1975, pp. V-13 a-V-27.

[60] C.K. Yeom, and K.H. Lee, "Characterization of sodium alginate membrane crosslinked with glutaraldehyde in pervaporation separation", *J. Appl. Polym. Sci.,* vol. 67, pp. 209-219, 1998. [http://dx.doi.org/10.1002/(SICI)1097-4628(19980110)67:2<209::AID-APP3>3.0.CO;2-Y]

[61] E. Khor, "Methods for the treatment of collagenous tissues for bioprostheses", *Biomaterials,* vol. 18, no. 2, pp. 95-105, 1997. [http://dx.doi.org/10.1016/S0142-9612(96)00106-8] [PMID: 9022956]

[62] A. Bigi, G. Cojazzi, S. Panzavolta, K. Rubini, and N. Roveri, "Mechanical and thermal properties of gelatin films at different degrees of glutaraldehyde crosslinking", *Biomaterials,* vol. 22, no. 8, pp. 763-768, 2001.
[http://dx.doi.org/10.1016/S0142-9612(00)00236-2] [PMID: 11246944]

[63] M.A. Masuelli, M.G. Garcia, J. Marchese, and N.A. Ochoa, "Development of dense films from melia azedarach polysaccharides", *J. Appl. Polym. Sci.,* vol. 132, no. 11, p. 41624, 2015.

[64] M.A. Masuelli, "Alcayota films: Effect of crosslinking", *Journal of Advances Chemical Engineering 6, 3,* p. 49, 2016.
[http://dx.doi.org/10.4172/2090-4568.C1.002]

CONCLUSIONS

At the end of this ebook, let us recall the tremendous efforts researchers have put on the development of techniques and methodologies for the molecular characterizations and the determination of the physicochemical properties of biopolymers from different sources in order to diversify as such as possible their end-uses for industrial applications. For that, researchers and engineers have adopted a strategy based on the understanding of the relationship between their structure and their macroscopic properties. Because biopolymers are able to generate macromolecular assemblies such as three-dimensional networks, two-dimensional films or one-dimensional monomolecular layers, in hydrated or solid forms, the structure of biopolymers alone or involved in such macromolecular assemblies have to be characterized at different scales, from the nanoscale to the macroscale including mesoscale, in order to deeply screen the structure-function-property relationships. We hope this first objective is reached in the present ebook.

Authors in this ebook have shown, through the diversity of the countries they belong, that biopolymers, from animal, plant, microbial and seaweed sources, can be judiciously extracted and purified according to as much as possible eco-friendly processes, in order to get reliable macroscopic end-use properties. Polysaccharides from plants, with cellulose being the most abundant on Earth, are the most used biopolymers for applications as their useful properties can be largely exploited for end-use applications. Natural gums from plants such as mesquite, tara and arabic gums are also very good candidates to promote economical development in countries where they originate from as they often are good stabilizers and adhesives for food and non-food products. Plant proteins are shown in this ebook as very promising alternatives to petroleum carbon sources for edible coatings and plastic foams.

The ebook displays also a thoroughly state-of-art in the physicochemical methods of characterization of biopolymers from the molecular level, through the exploration of size, polydispersity, chemical nature of repeating units, type of branching, using separation methods coupled to optical detectors but also mass spectrometry and NMR spectroscopy, to the macroscopic scale, through the exploration of the mechanical properties. A special attention is also given to the methods at the mesoscopic scale with the use of spectroscopic and microscopic methods in order to probe the structure covering the nanometric and micrometric scale. Small angle scattering and positron annihilation lifetime spectroscopy are methods highlighted in the ebook to probe nanostructures of single or complex biopolymeric materials.

The continuing development of new and existing biopolymers for end-uses applications are highly documented and we hope readers will find new opportunities for their teaching and research-development activities with this exhaustive list of biopolymers and related applications presented in the ebook. For instance, the design and engineering of innovative bionanocomposites and bioplastics deserve a special attention because it forms a fascinating interdisciplinary area that brings together biochemistry, materials science and nanotechnology. The understanding of the chemistry and structure at the nanometric scale imply an important knowledge for future developments of novel materials based on biopolymers and nanostructured materials. The extraordinary versatility of these new materials comes from the large choice of biopolymers available. These new materials have been elaborated thanks to the developement of new powerful techniques described in the ebook such as electrospinning.

Revolution in biopolymers applications comes from the use of proteins and polysaccharides in nanostructured food and non-food materials with the elaboration of biocompatible and degradable micro and nanoparticles (with a solid or a liquid core) for active delivery applications and bioplastics for packaging applications. Therefore, the toolbox that micro- and nanotechnologies offer provides new opportunities for product and process innovations in the food and non-food industries. The control of the process and functionality at the nanoscale will lead to more sustainable food and non-food production. This approach will allow in food industry for instance the development of nutrient delivery systems with healthy and/or less caloric value nutrients, sensors, and diagnostic devices that can monitor and ensure the safety of food products throughout the food chain. At least, various enhanced packaging concepts will extend the shelf life of fresh products or indicate quality deterioration of the packaged product. Nevertheless, it is imperative to develop a good communication of the applications of biopolymers in the "nano world" that allows the consumers to make an informed decision whether or not they would like to have the benefits of certain applications of biopolymers in nanotechnologies, or whether they do not accept certain risks.

All these driving forces act as stimuli to develop new materials based on biopolymers, and there are many opportunity areas such as industrial, medical, food, consumer products and pharmaceutical applications for which biopolymers act as stabilizers, thickeners, binders, cross-linkers, dispersants, lubricants, adhesives, drug-delivery agents. We sincerely hope the readers will find stimuli-responsive answers to the huge diversity of bioplymers and their related applications in this ebook.

SUBJECT INDEX

D

E

F

G

H

I

L

M

R

S

T

V

W

X

Made in the USA
Las Vegas, NV
03 November 2026

57424453R00295